The study of complex variables is important for students in engineering and the physical sciences and is a central subject in mathematics. In addition to being mathematically elegant, complex variables provide a powerful tool for solving problems that are either very difficult or virtually impossible to solve in any other way.

Part I of this text provides an introduction to the subject, including analytic functions, integration, series, residue calculus and also includes transform methods, ordinary differential equations in the complex plane, numerical methods, and more. Part II contains conformal mappings, asymptotic expansions, and the study of Riemann–Hilbert problems. The authors also provide an extensive array of applications, illustrative examples, and homework exercises.

This book is ideal for use in introductory undergraduate and graduate level courses in complex variables.

Complex Variables: Introduction and Applications

Cambridge Texts in Applied Mathematics

Complex Variables: Introduction and Applications

MARK J. ABLOWITZ

University of Colorado, Boulder

ATHANASSIOS S. FOKAS

Imperial College of Science and Technology

PUBLISHED BY THE PRESS SYNDICATE OF THE UNIVERSITY OF CAMBRIDGE
The Pitt Building, Trumpington Street, Cambridge CB2 1RP, United Kingdom

CAMBRIDGE UNIVERSITY PRESS
The Edinburgh Building, Cambridge CB2 2RU, United Kingdom
40 West 20th Street, New York, NY 10011-4211, USA
10 Stamford Road, Oakleigh, Melbourne 3166, Australia

© Cambridge University Press 1997

First published 1997

Printed in the United States of America

Typeset in Times Roman

Library of Congress Cataloging-in-Publication Data
Ablowitz, Mark J.
Complex variables : Introduction and applications / Mark J.
Ablowitz, Athanassios S. Fokas.
p.   cm. – (Cambridge texts in applied mathematics)
Includes bibliographical references (p.   –   ) and index.
ISBN 0-521-48058-2 (hb). – ISBN 0-521-48523-1 (pb)
1. Functions of complex variables.   I. Fokas, A. S., 1952–
II. Title.   III. Series.
QA331.7.A25   1997
515′.9 – dc21   96-48902
CIP

A catalogue record for this book is available from
the British Library

ISBN 0-521-48058-2 hardback
ISBN 0-521-48523-1 paperback

Contents

Sections denoted with an asterisk (*) can be either omitted or read independently.

Preface

The study of complex variables is beautiful from a purely mathematical point of view and provides a powerful tool for solving a wide array of problems arising in applications. It is perhaps surprising that to explain real phenomena, mathematicians, scientists, and engineers often resort to the "complex plane." In fact, using complex variables one can solve many problems that are either very difficult or virtually impossible to solve by other means. The text provides a broad treatment of both the fundamentals and the applications of this subject.

This text can be used in an introductory one- or two-semester undergraduate course. Alternatively, it can be used in a beginning graduate level course and as a reference. Indeed, Part I provides an introduction to the study of complex variables. It also contains a number of applications, which include evaluation of integrals, methods of solution to certain ordinary and partial differential equations, and the study of ideal fluid flow. In addition, Part I develops a suitable foundation for the more advanced material in Part II. Part II contains the study of conformal mappings, asymptotic evaluation of integrals, the so-called Riemann–Hilbert and DBAR problems, and many of their applications. In fact, applications are discussed throughout the book. Our point of view is that students are motivated and enjoy learning the material when they can relate it to applications.

To aid the instructor, we have denoted with an asterisk certain sections that are more advanced. These sections can be read independently or can be skipped. We also note that each of the chapters in Part II can be read independently. Every effort has been made to make this book self-contained. Thus advanced students using this text will have the basic material at their disposal without dependence on other references.

We realize that many of the topics presented in this book are not usually covered in complex variables texts. This includes the study of ordinary differential equations in the complex plane, the solution of linear partial differential equations by integral transforms, asymptotic evaluation of integrals, and Riemann–Hilbert problems. Actually some of these topics, when studied at all, are only included in advanced graduate level courses. However, we believe that these topics arise so frequently in applications that early exposure is vital. It is fortunate that it is indeed possible to present this material in such a way that it can be understood with only the foundation presented in the introductory chapters of this book.

We are indebted to our families, who have endured all too many hours of our absence. We are thankful to B. Fast and C. Smith for an outstanding job of word processing the manuscript and to B. Fast, who has so capably used mathematical software to verify many formulae and produce figures.

Several colleagues helped us with the preparation of this book. B. Herbst made many suggestions and was instrumental in the development of the computational section. C. Schober, L. Luo, and L. Glasser worked with us on many of the exercises. J. Meiss and C. Schober taught from early versions of the manuscript and made valuable suggestions.

David Benney encouraged us to write this book and we extend our deep appreciation to him. We would like to take this opportunity to thank those agencies who have, over the years, consistently supported our research efforts. Actually, this research led us to several of the applications presented in this book. We thank the Air Force Office of Scientific Research, the National Science Foundation, and the Office of Naval Research. In particular we thank Arje Nachman, Program Director, Air Force Office of Scientific Research (AFOSR), for his continual support.

Part one

Fundamentals and Techniques of Complex Function Theory

The first portion of this text aims to introduce the reader to the basic notions and methods in complex analysis. The standard properties of real numbers and the calculus of real variables are assumed. When necessary, a rigorous axiomatic development will be sacrificed in place of a logical development based upon suitable assumptions. This will allow us to concentrate more on examples and applications that our experience has demonstrated to be useful for the student first introduced to the subject. However, the important theorems are stated and proved.

1

Complex Numbers and Elementary Functions

This chapter introduces complex numbers, elementary complex functions, and their basic properties. It will be seen that complex numbers have a simple two-dimensional character that submits to a straightforward geometric description. While many results of real variable calculus carry over, some very important novel and useful notions appear in the calculus of complex functions. Applications to differential equations are briefly discussed in this chapter.

1.1 Complex Numbers and Their Properties

In this text we use Euler's notation for the imaginary unit number:

$$i^2 = -1 \qquad (1.1.1)$$

A complex number is an expression of the form

$$z = x + iy \qquad (1.1.2)$$

Here x is the real part of z, $\text{Re}(z)$; and y is the imaginary part of z, $\text{Im}(z)$. If $y = 0$, we say that z is real; and if $x = 0$, we say that z is pure imaginary. We often denote z, an element of the complex numbers as $z \in \mathbb{C}$; where x, an element of the real numbers is denoted by $x \in \mathbb{R}$. Geometrically, we represent Eq. (1.1.2) in a two-dimensional coordinate system called the **complex plane** (see Figure 1.1.1).

The real numbers lie on the horizontal axis and pure imaginary numbers on the vertical axis. The analogy with two-dimensional vectors is immediate. A complex number $z = x + iy$ can be interpreted as a two-dimensional vector (x, y).

It is useful to introduce another representation of complex numbers, namely polar coordinates (r, θ):

$$x = r \cos \theta \qquad y = r \sin \theta \qquad (r \geq 0) \qquad (1.1.3)$$

3

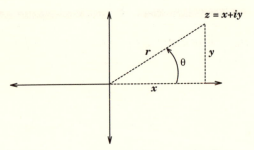

Fig. 1.1.1. The complex plane ("z plane")

Hence the complex number z can be written in the alternative polar form:

$$z = x + iy = r(\cos\theta + i\sin\theta) \qquad (1.1.4)$$

The radius r is denoted by

$$r = \sqrt{x^2 + y^2} \equiv |z| \qquad (1.1.5a)$$

(note: $\equiv$ denotes equivalence) and naturally gives us a notion of the **absolute value** of z, denoted by $|z|$, that is, it is the length of the vector associated with z. The value $|z|$ is often referred to as the **modulus** of z. The angle θ is called the argument of z and is denoted by $\arg z$. When $z \neq 0$, the values of θ can be found from Eq. (1.1.3) via standard trigonometry:

$$\tan\theta = y/x \qquad (1.1.5b)$$

where the quadrant in which x, y lie is understood as given. We note that $\theta \equiv \arg z$ is **multivalued** because $\tan\theta$ is a periodic function of θ with period π. Given $z = x + iy$, $z \neq 0$ we identify θ to have one value in the interval $\theta_0 \leq \theta < \theta_0 + 2\pi$, where θ_0 is an arbitrary number; others differ by integer multiples of 2π. We shall take $\theta_0 = 0$. For example, if $z = -1 + i$, then $|z| = r = \sqrt{2}$ and $\theta = \frac{3\pi}{4} + 2n\pi$, $n = 0, \pm 1, \pm 2, \ldots$. The previous remarks apply equally well if we use the polar representation about a point $z_0 \neq 0$. This just means that we translate the origin from $z = 0$ to $z = z_0$.

At this point it is convenient to introduce a special exponential function. The polar exponential is defined by

$$\cos\theta + i\sin\theta = e^{i\theta} \qquad (1.1.6)$$

Hence Eq. (1.1.4) implies that z can be written in the form

$$z = re^{i\theta} \qquad (1.1.4')$$

This exponential function has all of the standard properties we are familiar with in elementary calculus and is a special case of the complex exponential function to be introduced later in this chapter. For example, using well-known trigonometric identities, Eq. (1.1.6) implies

$$e^{2\pi i} = 1 \qquad e^{\pi i} = -1 \qquad e^{\frac{\pi i}{2}} = i \qquad e^{\frac{3\pi i}{2}} = -i$$
$$e^{i\theta_1} e^{i\theta_2} = e^{i(\theta_1 + \theta_2)} \qquad (e^{i\theta})^m = e^{im\theta} \qquad (e^{i\theta})^{1/n} = e^{i\theta/n}$$

With these properties in hand, it is straightforward to solve an equation of the form

$$z^n = a = |a| e^{i\phi} = |a|(\cos\phi + i\sin\phi)$$

Using the periodicity of $\cos\phi$ and $\sin\phi$, we have

$$z^n = a = |a| e^{i(\phi + 2\pi m)} \qquad m = 0, 1, \ldots, n-1$$

and find the n roots

$$z = |a|^{1/n} e^{i(\phi + 2\pi m)/n} \qquad m = 0, 1, \ldots, n-1.$$

For $m \geq n$ the roots repeat.

If $a = 1$, these are called the n roots of unity: $1, \omega, \omega^2, \ldots, \omega^{n-1}$, where $\omega = e^{2\pi i/n}$. So if $n = 2$, $a = -1$, we see that the solutions of $z^2 = -1 = e^{i\pi}$ are $z = \{e^{i\pi/2}, e^{3i\pi/2}\}$, or $z = \pm i$. In the context of real numbers there are no solutions to $z^2 = -1$, but in the context of complex numbers this equation has two solutions. Later in this book we shall show that an nth-order polynomial equation, $z^n + a_{n-1}z^{n-1} + \cdots + a_0 = 0$, where the coefficients $\{a_j\}_{j=1}^{n-1}$ are complex numbers, has n and only n solutions (roots), counting multiplicities (for example, we say that $(z-1)^2 = 0$ has two solutions, and that $z = 1$ is a solution of multiplicity two).

The **complex conjugate** of z is defined as

$$\bar{z} = x - iy = re^{-i\theta} \tag{1.1.7}$$

Two complex numbers are said to be equal if and only if their real and imaginary parts are respectively equal; namely, calling $z_k = x_k + iy_k$, for $k = 1, 2$, then

$$z_1 = z_2 \quad \Rightarrow \quad x_1 + iy_1 = x_2 + iy_2 \quad \Rightarrow \quad x_1 = x_2, \, y_1 = y_2$$

Thus $z = 0$ implies $x = y = 0$.

Addition, subtraction, multiplication, and division of complex numbers follow from the rules governing real numbers. Thus, noting $i^2 = -1$, we have

$$z_1 \pm z_2 = (x_1 \pm x_2) + i(y_1 \pm y_2) \tag{1.1.8a}$$

and

$$z_1 z_2 = (x_1 + iy_1)(x_2 + iy_2) = (x_1 x_2 - y_1 y_2) + i(x_1 y_2 + x_2 y_1) \quad (1.1.8b)$$

In fact, we note that from Eq. (1.1.5a)

$$z\bar{z} = \bar{z}z = (x + iy)(x - iy) = x^2 + y^2 = |z|^2 \quad (1.1.8c)$$

This fact is useful for division of complex numbers,

$$\begin{aligned}
\frac{z_1}{z_2} &= \frac{x_1 + iy_1}{x_2 + iy_2} = \frac{(x_1 + iy_1)(x_2 - iy_2)}{(x_2 + iy_2)(x_2 - iy_2)} \\
&= \frac{(x_1 x_2 + y_1 y_2) + i(x_2 y_1 - x_1 y_2)}{x_2^2 + y_2^2} \\
&= \frac{x_1 x_2 + y_1 y_2}{x_2^2 + y_2^2} + \frac{i(x_2 y_1 - x_1 y_2)}{x_2^2 + y_2^2} \quad (1.1.8d)
\end{aligned}$$

It is easily shown that the commutative, associative, and distributive laws of addition and multiplication hold.

Geometrically speaking, addition of two complex numbers is equivalent to that of the parallelogram law of vectors (see Figure 1.1.2).

The useful analytical statement

$$||z_1| - |z_2|| \le |z_1 + z_2| \le |z_1| + |z_2| \quad (1.1.9)$$

has the geometrical meaning that no side of a triangle is greater in length than the sum of the other two sides – hence the term for inequality Eq. (1.1.9) is the **triangle inequality**.

Equation (1.1.9) can be proven as follows.

$$\begin{aligned}
|z_1 + z_2|^2 &= (z_1 + z_2)(\bar{z_1} + \bar{z_2}) = z_1\bar{z_1} + z_2\bar{z_2} + z_1\bar{z_2} + \bar{z_1}z_2 \\
&= |z_1|^2 + |z_2|^2 + 2\,\text{Re}(z_1\bar{z_2})
\end{aligned}$$

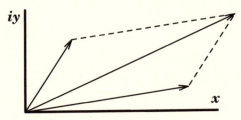

Fig. 1.1.2. Addition of vectors

Hence

$$|z_1 + z_2|^2 - (|z_1| + |z_2|)^2 = 2(\text{Re}(z_1\overline{z_2}) - |z_1||z_2|) \leq 0 \qquad (1.1.10)$$

where the inequality follows from the fact that

$$x = \text{Re}\, z \leq |z| = \sqrt{x^2 + y^2}$$

and $|z_1\overline{z_2}| = |z_1||z_2|$.

Equation (1.1.10) implies the right-hand inequality of Eq. (1.1.9) after taking a square root. The left-hand inequality follows by redefining terms. Let

$$W_1 = z_1 + z_2 \qquad W_2 = -z_2$$

Then the right-hand side of Eq. (1.1.9) (just proven) implies that

$$|W_1| \leq |W_1 + W_2| + |-W_2|$$

or $\qquad |W_1| - |W_2| \leq |W_1 + W_2|$

which then proves the left-hand side of Eq. (1.1.9) if we assume that $|W_1| \geq |W_2|$; otherwise, we can interchange W_1 and W_2 in the above discussion and obtain

$$||W_1| - |W_2|| = -(|W_1| - |W_2|) \leq |W_1 + W_2|$$

Similarly, note the immediate generalization of Eq. (1.1.9)

$$\left| \sum_{j=1}^{n} z_j \right| \leq \sum_{j=1}^{n} |z_j|$$

Problems for Section 1.1

1. Express each of the following complex numbers in polar exponential form:

   (a) 1 (b) $-i$ (c) $1 + i$

   (d) $\dfrac{1}{2} + \dfrac{\sqrt{3}}{2}i$ (e) $\dfrac{1}{2} - \dfrac{\sqrt{3}}{2}i$

2. Express each of the following in the form $a + bi$, where a and b are real:

   (a) $e^{2+i\pi/2}$ (b) $\dfrac{1}{1+i}$ (c) $(1+i)^3$ (d) $|3 + 4i|$

   (e) Define $\cos(z) = (e^{iz} + e^{-iz})/(2)$, and $e^z = e^x e^{iy}$.
      Evaluate $\cos(i\pi/4 + c)$, where c is real

3. Solve for the roots of the following equations:

$$\text{(a) } z^3 = 4 \qquad \text{(b) } z^4 = -1$$

$$\text{(c) } (az + b)^3 = c, \text{ where } a, b, c > 0 \qquad \text{(d) } z^4 + 2z^2 + 2 = 0$$

4. Estabilish the following results:

$$\text{(a) } \overline{z + w} = \bar{z} + \bar{w} \qquad \text{(b) } |z - w| \le |z| + |w| \qquad \text{(c) } z - \bar{z} = 2i \operatorname{Im} z$$

$$\text{(d) } \operatorname{Re} z \le |z| \qquad \text{(e) } |w\bar{z} + \bar{w}z| \le 2|wz| \qquad \text{(f) } |z_1 z_2| = |z_1||z_2|$$

5. There is a partial correspondence between complex numbers and vectors in the plane. Denote a complex number $z = a + bi$ and a vector $\mathbf{v} = a\hat{\mathbf{e}}_1 + b\hat{\mathbf{e}}_2$, where $\hat{\mathbf{e}}_1$ and $\hat{\mathbf{e}}_2$ are unit vectors in the horizontal and vertical directions. Show that the laws of addition $z_1 \pm z_2$ and $\mathbf{v}_1 \pm \mathbf{v}_2$ yield equivalent results as do the magnitudes $|z|^2$, $|\mathbf{v}|^2 = \mathbf{v} \cdot \mathbf{v}$. (Here $\mathbf{v} \cdot \mathbf{v}$ is the usual vector dot product.) Explain why there is no general correspondence for laws of multiplication or division.

1.2 Elementary Functions and Stereographic Projections

1.2.1 Elementary Functions

As a prelude to the notion of a function we present some standard definitions and concepts. A circle with center z_0 and radius r is denoted by $|z - z_0| = r$. A **neighborhood** of a point z_0 is the set of points z for which

$$|z - z_0| < \epsilon \tag{1.2.1}$$

where ϵ is some (small) positive number. Hence a neighborhood of the point z_0 is all the points inside the circle of radius ϵ, not including its boundary. An annulus $r_1 < |z - z_0| < r_2$ has center z_0, with inner radius r_1 and outer radius r_2. A point z_0 of a set of points $\mathcal{S}$ is called an **interior point** of $\mathcal{S}$ if there is a neighborhood of z_0 entirely contained within $\mathcal{S}$. The set $\mathcal{S}$ is said to be an **open set** if all the points of $\mathcal{S}$ are interior points. A point z_0 is said to be a **boundary point** of $\mathcal{S}$ if every neighborhood of $z = z_0$ contains at least one point in $\mathcal{S}$ and at least one point not in $\mathcal{S}$.

A set consisting of all points of an open set and none, some or all of its boundary points is referred to as a **region**. A region is said to be **bounded** if there is a constant $M > 0$ such that all points z of the region satisfy $|z| \le M$, that is, they lie within this circle. A region is said to be **closed** if it contains all of its boundary points. A region that is both closed and bounded is called

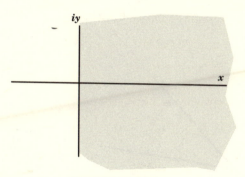

Fig. 1.2.1. Half plane

compact. Thus the region $|z| \leq 1$ is compact because it is both closed and bounded. The region $|z| < 1$ is open and bounded. The half plane $\operatorname{Re} z > 0$ (see Figure 1.2.1) is open and unbounded.

Let $z_1, z_2, \ldots, z_n$ be points in the plane. The $n-1$ line segments $\overline{z_1 z_2}, \overline{z_2 z_3}, \ldots, \overline{z_{n-1} z_n}$ taken in sequence form a broken line. A region is said to be **connected** if any two of its points can be joined by a broken line that is contained in the region. (There are more detailed definitions of connectedness, but this simple one will suffice for our purposes.) For an example of a connected region see Figure 1.2.2.)

A disconnected region is exemplified by all the points interior to $|z| = 1$ and exterior to $|z| = 2$: $S = \{z : |z| < 1, |z| > 2\}$.

A connected open region is called a **domain**. For example the set (see Figure 1.2.3)

$$S = \{z = re^{i\theta} : \theta_0 < \arg z < \theta_0 + \alpha\}$$

is a domain that is unbounded.

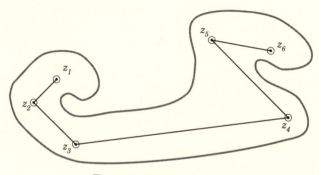

Fig. 1.2.2. Connected region

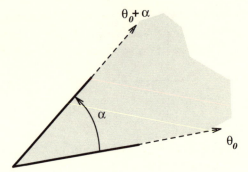

Fig. 1.2.3. Domain – a sector

Because a domain is an open set, we note that no boundary point of the domain can lie in the domain. Notationally, we shall refer to a region as $\mathcal{R}$; the closed region containing $\mathcal{R}$ and all of its boundary points is sometimes referred to as $\overline{\mathcal{R}}$. If $\mathcal{R}$ is closed, then $\mathcal{R} = \overline{\mathcal{R}}$. The notation $z \in \mathcal{R}$ means z is a point contained in $\mathcal{R}$. Usually we denote a domain by $\mathcal{D}$.

If for each $z \in \mathcal{R}$ there is a unique complex number $w(z)$ then we say $w(z)$ is a **function** of the complex variable z, frequently written as

$$w = f(z) \tag{1.2.2}$$

in order to denote the function f. Often we simply write $w = w(z)$, or just w. The totality of values $f(z)$ corresponding to $z \in \mathcal{R}$ constitutes the **range** of $f(z)$. In this context the set $\mathcal{R}$ is often referred to as the **domain of definition** of the function f. While the domain of definition of a function is frequently a domain, as defined earlier for a set of points, it does not need to be so.

By the above definition of a function we disallow multivaluedness; no more than one value of $f(z)$ may correspond to any point $z \in \mathcal{R}$. In Sections 2.2 and 2.3 we will deal explicitly with the notion of multivaluedness and its ramifications.

The simplest function is the **power** function:

$$f(z) = z^n, \qquad n = 0, 1, 2, \ldots \tag{1.2.3}$$

Each successive power is obtained by multiplication $z^{m+1} = z^m z$, $m = 0, 1, 2, \ldots$ A **polynomial** is defined as a linear combination of powers

$$P_n(z) = \sum_{j=0}^{n} a_j z^j = a_0 + a_1 z + a_2 z^2 + \cdots + a_n z^n \tag{1.2.4}$$

where the a_j are complex numbers (i.e.,[1] $a_j \in \mathbb{C}$). Note that the domain of definition of $P_n(z)$ is the entire z plane simply written as $z \in \mathbb{C}$. A **rational** function is a ratio of two polynomials $P_n(z)$ and $Q_m(z)$, where $Q_m(z) = \sum_{j=0}^{m} b_j z^j$

$$R(z) = \frac{P_n(z)}{Q_m(z)} \qquad (1.2.5)$$

and the domain of definition of $R(z)$ is the z plane, *excluding* the points where $Q_m(z) = 0$. For example, the function $w = 1/(1+z^2)$ is defined in the z plane excluding $z = \pm i$. This is written as $z \in \mathbb{C} \setminus \{i, -i\}$.

In general, the function $f(z)$ is complex and when $z = x + iy$, $f(z)$ can be written in the complex form:

$$w = f(z) = u(x, y) + i\, v(x, y) \qquad (1.2.6)$$

The function $f(z)$ is said to have the real part u, $u = \mathrm{Re}\, f$, and the imaginary part v, $v = \mathrm{Im}\, f$. For example,

$$w = z^2 = (x + iy)^2 = x^2 - y^2 + 2ixy$$

which implies

$$u(x, y) = x^2 - y^2 \qquad \text{and} \qquad v = 2xy.$$

As is the case with real variables we have the standard operations on functions. Given two functions $f(z)$ and $g(z)$, we define addition, $f(z) + g(z)$, multiplication $f(z)g(z)$, and composition $f[g(z)]$ of complex functions.

It is convenient to define some of the more common functions of a complex variable – which, as with polynomials and rational functions, will be familiar to the reader.

Motivated by real variables, $e^{a+b} = e^a e^b$, we define the exponential function

$$e^z = e^{x+iy} = e^x e^{iy}$$

Noting the polar exponential definition (used already in section 1.1, Eq. (1.1.6))

$$e^{iy} = \cos y + i \sin y$$

we see that

$$e^z = e^x (\cos y + i \sin y) \qquad (1.2.7)$$

---

[1] Hereafter these abbreviations will frequently be used: i.e. = that is; e.g. = for example.

Equation (1.2.7) and standard trigonometric identities yield the properties

$$e^{z_1+z_2} = e^{z_1} e^{z_2} \quad \text{and} \quad (e^z)^n = e^{nz} \tag{1.2.8}$$

We also note

$$|e^z| = |e^x||\cos y + i \sin y| = e^x \sqrt{\cos^2 y + \sin^2 y} = e^x$$

and

$$\overline{(e^z)} = e^{\bar{z}} = e^{x-iy} = e^x(\cos y - i \sin y)$$

The trigonometric functions $\sin z$ and $\cos z$ are defined as

$$\sin z = \frac{e^{iz} - e^{-iz}}{2i} \tag{1.2.9}$$

$$\cos z = \frac{e^{iz} + e^{-iz}}{2} \tag{1.2.10}$$

and the usual definitions of the other trigonometric functions are taken:

$$\tan z = \frac{\sin z}{\cos z}, \quad \cot z = \frac{\cos z}{\sin z}, \quad \sec z = \frac{1}{\cos z}, \quad \csc z = \frac{1}{\sin z} \tag{1.2.11}$$

All of the usual trigonometric properties such as

$$\sin(z_1 + z_2) = \sin z_1 \cos z_2 + \cos z_1 \sin z_2,$$
$$\sin^2 z + \cos^2 z = 1, \quad \cdots \tag{1.2.12}$$

follow from the above definitions.

The hyperbolic functions are defined analogously

$$\sinh z = \frac{e^z - e^{-z}}{2} \tag{1.2.13}$$

$$\cosh z = \frac{e^z + e^{-z}}{2} \tag{1.2.14}$$

$$\tanh z = \frac{\sinh z}{\cosh z}, \quad \coth z = \frac{\cosh z}{\sinh z}, \quad \operatorname{sech} z = \frac{1}{\cosh z}, \quad \operatorname{csch} z = \frac{1}{\sinh z}$$

Similarly, the usual identities follow, such as

$$\cosh^2 z - \sinh^2 z = 1 \tag{1.2.15}$$

From these definitions we see that as functions of a complex variable, $\sinh z$ and $\sin z$ ($\cosh z$ and $\cos z$) are simply related

$$\sinh iz = i \sin z, \qquad \sin iz = i \sinh z$$
$$\cosh iz = \cos z, \qquad \cos iz = \cosh z \tag{1.2.16}$$

By now it is abundantly clear that the elementary functions defined in this section are natural generalizations of the conventional ones we are familiar with in real variables. Indeed, the analogy is so close that it provides an alternative and systematic way of defining functions, which is entirely consistent with the above and allows the definition of a much wider class of functions. This involves introducing the concept of **power series**. In Chapter 3 we shall look more carefully at series and sequences. However, because power series of real variables are already familiar to the reader, it is useful to introduce the notion here.

A power series of $f(z)$ about the point $z = z_0$ is defined as

$$f(z) = \lim_{n \to \infty} \sum_{j=0}^{n} a_j(z - z_0)^j = \sum_{j=0}^{\infty} a_j(z - z_0)^j \tag{1.2.17}$$

Convergence is of course crucial. For simplicity we shall state (motivated by real variables but without proof at this juncture) that Eq. (1.2.17) converges, via the ratio test, whenever

$$\lim_{n \to \infty} \left| \frac{a_{n+1}}{a_n} \right| |z - z_0| < 1 \tag{1.2.18}$$

that is, it converges inside the circle $|z - z_0| = R$, where

$$R = \lim_{n \to \infty} \left| \frac{a_n}{a_{n+1}} \right|$$

If $R = \infty$, we say the series converges for all finite z; if $R = 0$, we say the series converges only for $z = z_0$. R is referred to as the **radius of convergence**.

The elementary functions discussed above have the following power series representations:

$$e^z = \sum_{j=0}^{\infty} \frac{z^j}{j!}, \qquad \sin z = \sum_{j=0}^{\infty} \frac{(-1)^j z^{2j+1}}{(2j+1)!}, \qquad \cos z = \sum_{j=0}^{\infty} \frac{(-1)^j z^{2j}}{(2j)!}$$
$$\sinh z = \sum_{j=0}^{\infty} \frac{z^{2j+1}}{(2j+1)!}, \qquad \cosh z = \sum_{j=0}^{\infty} \frac{z^{2j}}{(2j)!} \tag{1.2.19}$$

where $j! = j(j-1)(j-2)\cdots 3 \cdot 2 \cdot 1$ for $j \geq 1$, and $0! \equiv 1$. The ratio test shows that these series converge for all finite z.

Complex functions arise frequently in applications. For example, in the investigation of stability of physical systems we derive equations for small deviations from rest or equilibrium states. The solutions of the perturbed equation often have the form e^{zt}, where t is real (e.g. time) and z is a complex number satisfying an algebraic equation (or a more complicated transcendental system). We say that the system is unstable if there are any solutions with $\mathrm{Re}\, z > 0$ because $|e^{zt}| \to \infty$ as $t \to \infty$. We say the system is **marginally stable** if there are no values of z with $\mathrm{Re}\, z > 0$, but some with $\mathrm{Re}\, z = 0$. (The corresponding exponential solution is bounded for all t.) The system is said to be **stable** and **damped** if all values of z satisfy $\mathrm{Re}\, z < 0$ because $|e^{zt}| \to 0$ as $t \to \infty$.

A function $w = f(z)$ can be regarded as a **mapping** or transformation of the points in the z plane ($z = x + iy$) to the points of the w plane ($w = u + iv$). In real variables in one dimension, this notion amounts to understanding the graph $y = f(x)$, that is, the mapping of the points x to $y = f(x)$. In complex variables the situation is more difficult owing to the fact that we really have four dimensions – hence a graphical depiction such as in the real one-dimensional case is not feasible. Rather, one considers the two complex planes, z and w, separately and asks how the region in the z plane transforms or maps to a corresponding region or **image** in the w plane. Some examples follow.

Example 1.2.1 The function $w = z^2$ maps the upper half z-plane including the real axis, $\mathrm{Im}\, z \geq 0$, to the entire w-plane (see Figure 1.2.4). This is particularly clear when we use the polar representation $z = re^{i\theta}$. In the z-plane, θ lies inside $0 \leq \theta < \pi$, whereas in the w-plane, $w = r^2 e^{2i\theta} = Re^{i\phi}$, $R = r^2$, $\phi = 2\theta$ and ϕ lies in $0 \leq \phi < 2\pi$.

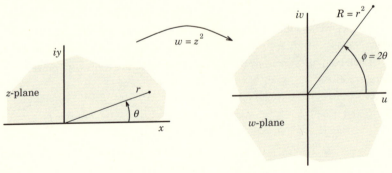

Fig. 1.2.4. Map of $z \to w = z^2$

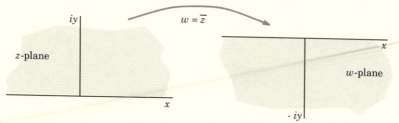

Fig. 1.2.5. Conjugate mapping

Example 1.2.2 The function $w = \bar{z}$ maps the upper half z-plane Im $z > 0$ into the lower half w-plane (see Figure 1.2.5). Namely, $z = x + iy$ and $y > 0$ imply that $w = \bar{z} = x - iy$. Thus $w = u + iv \Rightarrow u = x, v = -y$.

The study and understanding of complex mappings is very important, and we will see that there are many applications. In subsequent sections and chapters we shall more carefully investigate the concept of mappings; we shall not go into any more detail or complication at this juncture.

It is often useful to add the **point at infinity** (usually denoted by ∞ or z_∞) to our, so far open, complex plane. As opposed to a finite point where the neighborhood of z_0, say, is defined by Eq. (1.2.1), here the neighborhood of z_∞ is defined by those points satisfying $|z| > 1/\epsilon$ for all (sufficiently small) $\epsilon > 0$. One convenient way to define the point at infinity is to let $z = 1/t$ and then to say that $t = 0$ corresponds to the point z_∞. An unbounded region $\mathcal{R}$ contains the point z_∞. Similarly, we say a function has values at infinity if it is defined in a neighborhood of z_∞. The complex plane with the point z_∞ included is referred to as the **extended complex plane**.

1.2.2 Stereographic Projection

Consider a unit sphere sitting on top of the complex plane with the south pole of the sphere located at the origin of the z plane (see Figure 1.2.6). In this subsection we show how the extended complex plane can be mapped onto the surface of a sphere whose south pole corresponds to the origin and whose north pole to the point z_∞. All other points of the complex plane can be mapped in a one-to-one fashion to points on the surface of the sphere by using the following construction. Connect the point z in the plane with the north pole using a straight line. This line intersects the sphere at the point P. In this way each point $z(= x + iy)$ on the complex plane corresponds uniquely to a point P on the surface of the sphere. This construction is called the **stereographic projection** and is diagrammatically illustrated in Figure 1.2.6. The extended complex plane

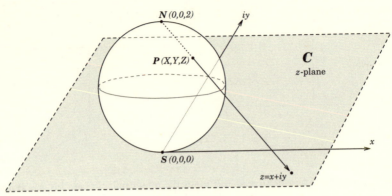

Fig. 1.2.6. Stereographic projection

is sometimes referred to as the **compactified** (closed) complex plane. It is often useful to view the complex plane in this way, and knowledge of the construction of the stereographic projection is valuable in certain advanced treatments.

So, more concretely, the point $P : (X, Y, Z)$ on the sphere is put into correspondence with the point $z = x + iy$ in the complex plane by finding on the surface of the sphere, (X, Y, Z), the point of intersection of the line from the north pole of the sphere, $N : (0, 0, 2)$, to the point $z = x + iy$ on the plane. The construction is as follows. We consider three points in the three-dimensional setup:

$N = (0, 0, 2)$: north pole
$P = (X, Y, Z)$: point on the sphere
$C = (x, y, 0)$: point in the complex plane

These points must lie on a straight line, hence the difference of the points $P - N$ must be a real scalar multiple of the difference $C - N$, namely

$$(X, Y, Z - 2) = s(x, y, -2) \qquad (1.2.20)$$

where s is a real number ($s \neq 0$). The equation of the sphere is given by

$$X^2 + Y^2 + (Z - 1)^2 = 1 \qquad (1.2.21)$$

Equation (1.2.20) implies

$$X = sx, \qquad Y = sy, \qquad Z = 2 - 2s \qquad (1.2.22)$$

Inserting Eq. (1.2.22) into Eq. (1.2.21) yields, after a bit of manipulation

$$s^2(x^2 + y^2 + 4) - 4s = 0 \qquad (1.2.23)$$

This equation has as its only nonvanishing solution

$$s = \frac{4}{|z|^2 + 4} \tag{1.2.24}$$

where $|z|^2 = x^2 + y^2$. Thus given a point $z = x + iy$ in the plane, we have on the sphere the unique correspondence:

$$X = \frac{4x}{|z|^2 + 4}, \qquad Y = \frac{4y}{|z|^2 + 4}, \qquad Z = \frac{2|z|^2}{|z|^2 + 4} \tag{1.2.25}$$

We see that under this mapping, the origin in the complex plane $z = 0$ yields the south pole of the sphere $(0, 0, 0)$, and all points at $|z| = \infty$ yield the north pole $(0, 0, 2)$. (The latter fact is seen via the limit $|z| \to \infty$ with $x = |z| \cos\theta$, $y = |z| \sin\theta$.) On the other hand, given a point $P = (X, Y, Z)$ we can find its unique image in the complex plane. Namely, from Eq. (1.2.22)

$$s = \frac{2 - Z}{2} \tag{1.2.26}$$

and

$$x = \frac{2X}{2 - Z}, \qquad y = \frac{2Y}{2 - Z} \tag{1.2.27}$$

The stereographic projection maps any locus of points in the complex plane onto a corresponding locus of points on the sphere and vice versa. For example, the image of an arbitrary circle in the plane, is a circle on the sphere that does not pass through the north pole. Similarly, a straight line corresponds to a circle passing through the north pole (see Figure 1.2.7). Hence on the sphere the images of straight lines and of circles are not really geometrically different

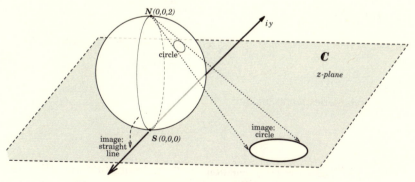

Fig. 1.2.7. Circles and lines in stereographic projection

from one another. Moreover, the images on the sphere of two nonparallel straight lines in the plane intersect at *two* points on the sphere – one of which is the point at infinity. In this framework, *parallel* lines are circles that touch one another at the point at infinity (north pole). We lose Euclidean geometry on a sphere.

Problems for Section 1.2

1. Sketch the regions associated with the following inequalities. Determine if the region is open, closed, bounded, or compact.

$$\text{(a) } |z| \leq 1 \qquad \text{(b) } |2z + 1 + i| < 4 \qquad \text{(c) } \operatorname{Re} z \geq 4$$

$$\text{(d) } |z| \leq |z + 1| \qquad \text{(e) } 0 < |2z - 1| \leq 2$$

2. Sketch the following regions. Determine if they are connected, and what the closure of the region is if they are not closed.

$$\text{(a) } 0 < \arg z \leq \pi \qquad \text{(b) } 0 \leq \arg z < 2\pi$$

$$\text{(c) } \operatorname{Re} z > 0 \quad \text{and} \quad \operatorname{Im} z > 0$$

$$\text{(d) } \operatorname{Re}(z - z_0) > 0 \text{ and } \operatorname{Re}(z - z_1) < 0 \text{ for two complex numbers } z_0, z_1$$

$$\text{(e) } |z| < \tfrac{1}{2} \qquad \text{and} \qquad |2z - 4| \leq 2$$

3. Use Euler's formula for the exponential and the well-known series expansions of the real functions e^x, $\sin y$, and $\cos y$ to show that

$$e^z = \sum_{j=0}^{\infty} \frac{z^j}{j!} .$$

Hint: Use

$$(x + iy)^k = \sum_{j=0}^{k} \frac{k!}{j!(k-j)!} x^j (iy)^{k-j}$$

4. Use the series representation

$$e^z = \sum_{j=0}^{\infty} \frac{z^j}{j!}, \qquad |z| < \infty$$

to determine series representations for the following functions:

$$\text{(a) } \sin z \qquad \text{(b) } \cosh z$$

Use these results to deduce where the power series for $\sin^2 z$ and $\operatorname{sech} z$ would converge. What can be said about $\tan z$?

5. Use any method to determine series expansions for the following functions:

$$\text{(a)} \quad \frac{\sin z}{z} \qquad \text{(b)} \quad \frac{\cosh z - 1}{z^2} \qquad \text{(c)} \quad \frac{e^z - 1 - z}{z}$$

6. Let $z_1 = x_1$ and $z_2 = x_2$, with x_1, x_2 real, and the relationship

$$e^{i(x_1 + x_2)} = e^{ix_1} e^{ix_2}$$

to deduce the known trigonometric formulae

$$\sin(x_1 + x_2) = \sin x_1 \cos x_2 + \cos x_1 \sin x_2$$
$$\cos(x_1 + x_2) = \cos x_1 \cos x_2 - \sin x_1 \sin x_2$$

and therefore show

$$\sin 2x = 2 \sin x \cos x$$
$$\cos 2x = \cos^2 x - \sin^2 x$$

7. Discuss the following transformations (mappings) from the z plane to the w plane; here z is the entire finite complex plane.

$$\text{(a)} \ \ w = z^3 \qquad \text{(b)} \ \ w = 1/z$$

8. Consider the transformation

$$w = z + 1/z \qquad z = x + iy \qquad w = u + iv$$

Show that the image of the points in the upper half z plane $(y > 0)$ that are exterior to the circle $|z| = 1$ corresponds to the entire upper half plane $v > 0$.

9. Consider the following transformation

$$w = \frac{az + b}{cz + d}, \qquad \Delta = ad - bc \neq 1$$

(a) Show that the map can be inverted to find a unique (single-valued) z as a function of w everywhere.

(b) Verify that the mapping can be considered as the result of three successive maps:

$$z' = cz + d, \qquad z'' = 1/z', \qquad w = -\frac{\Delta}{c}z'' + \frac{a}{c}$$

where $c \neq 0$ and is of the form

$$w = \frac{a}{d}z + \frac{b}{d}$$

when $c = 0$.

The following problems relate to the subsection on stereographic projection.

10. To what curves on the sphere do the lines $\operatorname{Re} z = x = 0$ and $\operatorname{Im} z = y = 0$ correspond?

11. Describe the curves on the sphere to which any straight lines on the z plane correspond.

12. Show that a circle in the z plane corresponds to a circle on the sphere.

1.3 Limits, Continuity, and Complex Differentiation

The concepts of limits and continuity are similar to that of real variables. In this sense our discussion can serve as a brief review of many previously understood notions. Consider a function $w = f(z)$ defined at all points in some neighborhood of $z = z_0$, except possibly for z_0 itself. We say $f(z)$ has the limit w_0 if as z approaches z_0, $f(z)$, approaches w_0. Mathematically, we say

$$\lim_{z \to z_0} f(z) = w_0 \tag{1.3.1}$$

if for every (sufficiently small) $\epsilon > 0$ there is a $\delta > 0$ such that

$$|f(z) - w_0| < \epsilon \qquad \text{whenever} \qquad |z - z_0| < \delta \tag{1.3.2}$$

where the absolute value is defined in section 1.1 (see, e.g. Eqs. 1.1.4 and 1.1.5a).

This definition is clear when z_0 is an interior point of a region $\mathcal{R}$ in which $f(z)$ is defined. If z_0 is a boundary point of $\mathcal{R}$, then we require Eq. (1.3.2) to hold only for those $z \in \mathcal{R}$.

Figure 1.3.1 illustrates these ideas. Under the mapping $w = f(z)$, all points interior to the circle $|z - z_0| = \delta$ are mapped to points interior to the circle $|w - w_0| = \epsilon$. The limit will exist only in the case when z approaches z_0 (that is, $z \to z_0$) in an *arbitrary direction*; then this implies that $w \to w_0$.

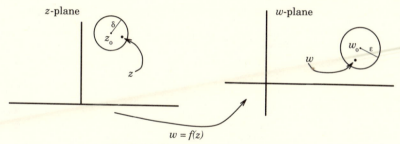

z-plane   w-plane $w = f(z)$

Fig. 1.3.1. Mapping of a neighborhood

This limit definition is standard. Let us consider the following examples.

Example 1.3.1 Show that

$$\lim_{z \to i} 2 \left(\frac{z^2 + iz + 2}{z - i} \right) = 6i. \qquad (1.3.3)$$

We must show that given $\epsilon > 0$, there is a $\delta > 0$ such that

$$\left| 2 \left(\frac{z^2 + iz + 2}{z - i} \right) - 6i \right| = \left| 2 \left(\frac{(z - i)(z + 2i)}{(z - i)} \right) - 6i \right| < \epsilon \qquad (1.3.4)$$

whenever

$$|z - i| < \delta \qquad (1.3.5)$$

Since $z \neq i$, inequality (1.3.4) implies that $2|z - i| < \epsilon$. Thus if $\delta = \epsilon/2$, Eq. (1.3.5) ensures that Eq. (1.3.4) is satisfied. Therefore Eq. (1.3.3) is demonstrated.

This limit definition can also be applied to the point $z = \infty$. We say that

$$\lim_{z \to \infty} f(z) = w_0 \qquad (1.3.6)$$

if for every (sufficiently small) $\epsilon > 0$ there is a $\delta > 0$ such that

$$|f(z) - w_0| < \epsilon \qquad \text{whenever} \qquad |z| > \frac{1}{\delta} \qquad (1.3.7)$$

We assert that the following properties are true. (The proof is an exercise of the limit definition and follows that of real variables.) If for $z \in \mathcal{R}$ we have two functions $w = f(z)$ and $s = g(z)$ such that

$$\lim_{z \to z_0} f(z) = w_0, \qquad \lim_{z \to z_0} g(z) = s_0$$

then

$$\lim_{z \to z_0} (f(z) + g(z)) = w_0 + s_0$$

$$\lim_{z \to z_0} (f(z)g(z)) = w_0 s_0$$

and

$$\lim_{z \to z_0} \frac{f(z)}{g(z)} = \frac{w_0}{s_0} \qquad (s_0 \neq 0)$$

Similar conclusions hold for sums and products of a finite number of functions. As mentioned in Section 1.2, the point $z = z_\infty = \infty$ is often dealt with via the transformation

$$t = \frac{1}{z}$$

The neighborhood of $z = z_\infty$ corresponds to the neighborhood of $t = 0$. So the function $f(z) = 1/z^2$ near $z = z_\infty$ behaves like $f(1/t) = t^2$ near zero; that is, $t^2 \to 0$ as $t \to 0$, or $1/z^2 \to 0$ as $z \to \infty$.

In analogy to real analysis, a function $f(z)$ is said to be **continuous** at $z = z_0$ if

$$\lim_{z \to z_0} f(z) = f(z_0) \tag{1.3.8}$$

Equation (1.3.8) implies that $f(z)$ exists in a neighborhood of $z = z_0$ and that the limit, as z approaches z_0, of $f(z)$ is $f(z_0)$ itself. In terms of ϵ, δ notation, given $\epsilon > 0$, there is a $\delta > 0$ such that $|f(z) - f(z_0)| < \epsilon$ whenever $|z - z_0| < \delta$. The notion of continuity at infinity can be ascertained in a similar fashion. Namely, if $\lim_{z \to \infty} f(z) = w_\infty$, and $f(\infty) = w_\infty$, then the definition for continuity at infinity, $\lim_{z \to \infty} f(z) = f(\infty)$, is the following: Given $\epsilon > 0$ there is a $\delta > 0$ such that $|f(z) - w_\infty| < \epsilon$ whenever $|z| > 1/\delta$.

The theorems on limits of sums and products of functions can be used to establish that sums and products of continuous functions are continuous. It should also be pointed out that since $|f(z) - f(z_0)| = |\overline{f}(z) - \overline{f}(z_0)|$, the continuity of $f(z)$ at z_0 implies the continuity of the complex conjugate $\overline{f}(z)$ at $z = z_0$. (Recall the definition of the complex conjugate, Eq. (1.1.7)). Thus if $f(z)$ is continuous at $z = z_0$, then

$$\operatorname{Re} f(z) = (f(z) + \overline{f}(z))/2$$

$$\operatorname{Im} f(z) = (f(z) - \overline{f}(z))/2i$$

and $\quad |f(z)|^2 = (f(z)\overline{f}(z))$

are all continuous at $z = z_0$.

We shall say a function $f(z)$ is **continuous in a region** if it is continuous at every point of the region. Usually, we simply say that $f(z)$ is continuous when the associated region is understood. Considering continuity in a region $\mathcal{R}$ generally requires that $\delta = \delta(\epsilon, z_0)$; that is, δ depends on both ϵ and the point $z_0 \in \mathcal{R}$. Function $f(z)$ is said to be **uniformly continuous** in a region $\mathcal{R}$ if $\delta = \delta(\epsilon)$; that is, δ is independent of the point $z = z_0$.

As in real analysis, a function that is continuous in a compact (closed and bounded) region $\mathcal{R}$ is uniformly continuous and bounded; that is, there is a $C > 0$ such that $|f(z)| < C$. (The proofs of these statements follow from the analogous statements of real analysis.) Moreover, in a compact region, the modulus $|f(z)|$ actually attains both its maximum and minimum values on $\mathcal{R}$; this follows from the continuity of the real function $|f(z)|$.

Example 1.3.2 Show that the continuity of the real and imaginary parts of a complex function $f(z)$ implies that $f(z)$ is continuous.

$$f(z) = u(x, y) + iv(x, y)$$

We know that

$$\lim_{z \to z_0} f(z) = \lim_{\substack{x \to x_0 \\ y \to y_0}} (u(x, y) + iv(x, y))$$

$$= u(x_0, y_0) + iv(x_0, y_0) = f(z_0)$$

which completes the proof. It also illustrates that we can appeal to real analysis for many of the results in this section.

Conversely, we have

$$|u(x, y) - u(x_0, y_0)| \le |f(z) - f(z_0)|$$

$$|v(x, y) - v(x_0, y_0)| \le |f(z) - f(z_0)|$$

(because $|f|^2 = |u|^2 + |v|^2$) in which case continuity of $f(z)$ implies continuity of the real and imaginary parts of $f(z)$. Namely, this follows from the fact that given $\epsilon > 0$, there is a $\delta > 0$ such that $|f(z) - f(z_0)| < \epsilon$ whenever $|z - z_0| < \delta$ (and note that $|x - x_0| < |z - z_0| < \delta$, $|y - y_0| < |z - z_0| < \delta$).

Let $f(z)$ be defined in some region $\mathcal{R}$ containing the neighborhood of a point z_0. The **derivative** of $f(z)$ at $z = z_0$, denoted by $f'(z_0)$ or $\frac{df}{dz}(z_0)$, is defined by

$$f'(z_0) = \lim_{\Delta z \to 0} \left(\frac{f(z_0 + \Delta z) - f(z_0)}{\Delta z} \right) \tag{1.3.9}$$

provided this limit exists. We sometimes say that f is **differentiable** at z_0.

Alternatively, letting $\Delta z = z - z_0$, Eq. (1.3.9) has another standard form

$$f'(z_0) = \lim_{z \to z_0}\left(\frac{f(z) - f(z_0)}{z - z_0}\right) \tag{1.3.10}$$

If $f'(z_0)$ exists for all points $z_0 \in \mathcal{R}$, then we say $f(z)$ is differentiable in $\mathcal{R}$ – or just differentiable, if $\mathcal{R}$ is understood. If $f'(z_0)$ exists, then $f(z)$ is continuous at $z = z_0$. This follows from

$$\lim_{z \to z_0}(f(z) - f(z_0)) = \lim_{z \to z_0}\left(\frac{f(z) - f(z_0)}{z - z_0}\right)\lim_{z \to z_0}(z - z_0)$$
$$= f'(z_0)\lim_{z \to z_0}(z - z_0) = 0$$

A continuous function is not necessarily differentiable. Indeed it turns out that differentiable functions possess many special properties.

On the other hand, because we are now dealing with complex functions that have a two-dimensional character, there can be new kinds of complications not found in functions of one real variable. A prototypical example follows.

Consider the function

$$f(z) = \bar{z} \tag{1.3.11}$$

Even though this function is continuous, as discussed earlier, we now show that it does not possess a derivative. Consider the difference quotient:

$$\lim_{\Delta z \to 0}\frac{(z_0 + \Delta z) - \bar{z}_0}{\Delta z} = \lim_{\Delta z \to 0}\frac{\overline{\Delta z}}{\Delta z} \equiv q_0 \tag{1.3.12}$$

This limit does not exist because a unique value of q_0 cannot be found; indeed it depends on how Δz approaches zero. Writing $\Delta z = re^{i\theta}$, $q_0 = \lim_{\Delta z \to 0} e^{-2i\theta}$. So if $\Delta z \to 0$ along the positive real axis ($\theta = 0$), then $q_0 = 1$. If $\Delta z \to 0$ along the positive imaginary axis, then $q_0 = -1$ (because $\theta = \pi/2$, $e^{-2i\theta} = -1$), etc. Thus we find the surprising result that the function $f(z) = \bar{z}$ is not differentiable anywhere (i.e., for any $z = z_0$) even though it is continuous everywhere! In fact, this situation will be seen to be the case for general complex functions unless the real and imaginary parts of our complex function satisfy certain compatibility conditions (see Section 2.1). Differentiable complex functions, often called **analytic functions**, are special and important.

Despite the fact that the formula for a derivative is identical in form to that of the derivative of a real-valued function, $f(z)$, a significant point to note is that $f'(z)$ follows from a two-dimensional limit ($z = x + iy$ or $z = re^{i\theta}$). Thus for $f(z)$ to exist, the relevant limit must exist independent of the direction from which z approaches the limit point z_0. For a function of one real variable we only have two directions: $x < x_0$ and $x > x_0$.

If f and g have derivatives, then it follows by similar proofs to those of real variables that

$$(f + g)' = f' + g'$$
$$(fg)' = f'g + fg'$$
$$\left(\frac{f}{g}\right)' = (f'g - fg')/g^2 \qquad (g \neq 0)$$

and if $f'(g(z))$ and $g'(z)$ exist, then

$$[f(g(z))]' = f'(g(z))g'(z)$$

In order to differentiate polynomials, we need the derivative of the elementary function $f(z) = z^n$, n is a positive integer

$$\frac{d}{dz}(z^n) = nz^{n-1} \tag{1.3.13}$$

This follows from

$$\frac{(z + \Delta z)^n - z^n}{\Delta z} = nz^{n-1} + a_1 z^{n-2}\Delta z + a_2 z^{n-3}\Delta z^2 + \ldots + \Delta z^n \rightarrow nz^{n-1}$$

as $\Delta z \rightarrow 0$, where a_1, a_2, ..., are the appropriate binomial coefficients of $(a + b)^n$.

Thus we have as corollaries to this result

$$\frac{d}{dz}(c) = 0, \qquad c = \text{constant} \tag{1.3.15a}$$

$$\frac{d}{dz}(a_0 + a_1 z + a_2 z^2 + \cdots + a_m z^m) = a_1 + 2a_2 z + 3a_3 z^2 + \cdots + ma_m z^{m-1}$$

$$\tag{1.3.15b}$$

Moreover, with regard to the (purely formal at this point) powerseries expansions discussed earlier, we will find that

$$\frac{d}{dz}\left(\sum_{n=0}^{\infty} a_n z^n\right) = \sum_{n=0}^{\infty} na_n z^{n-1} \tag{1.3.15}$$

inside the radius of convergence of the series.

We also note that the derivatives of the usual elementary functions behave in the same way as in real variables. Namely

$$\frac{d}{dz}e^z = e^z, \qquad \frac{d}{dz}\sin z = \cos z, \qquad \frac{d}{dz}\cos z = -\sin z$$

$$\frac{d}{dz}\sinh z = \cosh z, \qquad \frac{d}{dz}\cosh z = \sinh z \tag{1.3.16}$$

etc. The proofs can be obtained from the fundamental definitions. For example,

$$\frac{d}{dz}e^z = \lim_{\Delta z \to 0} \frac{e^{z+\Delta z} - e^z}{\Delta z}$$

$$= e^z \lim_{\Delta z \to 0}\left(\frac{e^{\Delta z} - 1}{\Delta z}\right) = e^z \tag{1.3.17}$$

where we note that

$$\lim_{\Delta z \to 0} \frac{e^{\Delta z} - 1}{\Delta z} = \lim_{\substack{\Delta x \to 0 \\ \Delta y \to 0}} \left(\frac{(e^{\Delta x}\cos \Delta y - 1) + i e^{\Delta x}\sin \Delta y}{(\Delta x + i\Delta y)}\right) = 1 \tag{1.3.18}$$

One can put Eq. (1.3.18) in real/imaginary form and use polar coordinates for Δx, Δy. This calculation is also discussed in the problems given for this section. Later we shall establish the validity of the power series formulae for e^z (see Eq. (1.2.19)), from which Eq. (1.3.18) follows immediately (since $e^z = 1 + z + z^2/2 + \cdots$) without need for the double limit. The other formulae in Eq. (1.3.16) can also be deduced using the relationships (1.2.9), (1.2.10), (1.2.13), (1.2.14).

1.3.1 Elementary Applications to Ordinary Differential Equations

An important topic in the application of complex variables is the study of differential equations. Later in this text we discuss differential equations in the complex plane in some detail, but in fact we are already in a position to see why the ideas already presented can be useful. Many readers will have had a course in differential equations, but it is not really necessary for what we shall discuss. Linear homogeneous differential equations with constant coefficients take the following form:

$$L_n w = \frac{d^n w}{dt^n} + a_{n-1}\frac{d^{n-1}w}{dt^{n-1}} + \cdots a_1\frac{dw}{dt} + a_0 w = 0 \tag{1.3.19}$$

where $\{a_j\}_{j=0}^{n-1}$ are all constant, n is called the order of the equation, and (for our present purposes) t is real. We could (and do, later in section 3.7) allow t

to be complex, in which case the study of such differential equations becomes intimately connected with many of the topics studied later in this text, but for now we keep t real. Solutions to Eq. (1.3.19) can be sought in the form

$$w(t) = ce^{zt} \qquad (1.3.20)$$

where c is a nonzero constant. Substitution of Eq. (1.3.20) into Eq. (1.3.19), and factoring ce^{zt} from each term (note e^{zt} does not vanish), yields the following *algebraic* equation:

$$z^n + a_{n-1}z^{n-1} + \cdots + a_1 z + a_0 = 0 \qquad (1.3.21)$$

There are various subcases to consider, but we shall only discuss the prototypical one where there are n *distinct* solutions of Eq. (1.3.22), which we call $\{z_1, z_2, \ldots, z_n\}$. Each of these values, say z_j, yields a solution to Eq. (1.3.19) $w_j = c_j e^{z_j t}$, where c_j is an arbitrary constant. Because Eq. (1.3.19) is a linear equation, we have the more general solution

$$w(t) = \sum_{j=1}^{n} w_j = \sum_{j=1}^{n} c_j e^{z_j t} \qquad (1.3.22)$$

In differential equation texts it is proven that Eq. (1.3.22) is, in fact, the most general solution. In applications, the differential equations (Eq. (1.3.19)) frequently have real coefficients $\{a_j\}_{j=0}^{n-1}$. The study of algebraic equations of the form (Eq. (1.3.21)), discussed later in this text, shows that there are at most n solutions — precisely n solutions if we count multiplicity of solutions. In fact, when the coefficients are real, then the solutions are either real or come in complex conjugate pairs. Corresponding to complex conjugate pairs, a real solution $w(t)$ is found by taking complex conjugate constants c_j and $\overline{c_j}$ corresponding to each pair of complex conjugate roots z_j and $\overline{z_j}$. For example, consider one such real solution, call it w_p, corresponding to the pair $z, \overline{z}$:

$$w_p(t) = ce^{zt} + \overline{c}e^{\overline{z}t} \qquad (1.3.23)$$

We can rewrite this in terms of trigonometric functions and real exponentials. Let $z = x + iy$:

$$
\begin{aligned}
w_p(t) &= ce^{(x+iy)t} + \overline{c}e^{(x-iy)t} \\
&= e^{xt}[c(\cos yt + i \sin yt) + \overline{c}(\cos yt - i \sin yt)] \\
&= (c + \overline{c})e^{xt} \cos yt + i(c - \overline{c})e^{xt} \sin yt \qquad (1.3.24)
\end{aligned}
$$

Because $c + \bar{c} = A$, $i(c - \bar{c}) = B$ are real, we find that this pair of solutions may be put in the *real* form

$$w_c(t) = Ae^{xt} \cos yt + Be^{xt} \sin yt \tag{1.3.25}$$

Two examples of these ideas are simple harmonic motion (SHM) and vibrations of beams:

$$\frac{d^2w}{dt^2} + \omega_0^2 w = 0 \qquad \text{(SHM)} \tag{1.3.26a}$$

$$\frac{d^4w}{dt^4} + k^4 w = 0 \tag{1.3.26b}$$

where ω_0^2 and k^4 are real nonzero constants, depending on the parameters in the physical model. Looking for solutions of the form of Eq. (1.3.20) leads to the equations

$$z^2 + \omega_0^2 = 0 \tag{1.3.27a}$$

$$z^4 + k^4 = 0 \tag{1.3.27b}$$

which have solutions (see also Section 1.1)

$$z_1 = i\omega_0, \qquad z_2 = -i\omega_0 \tag{1.3.28a}$$

$$z_1 = ke^{i\pi/4} = \frac{k}{\sqrt{2}}(1 + i)$$

$$z_2 = ke^{3i\pi/4} = \frac{k}{\sqrt{2}}(-1 + i)$$

$$z_3 = ke^{5i\pi/4} = \frac{k}{\sqrt{2}}(-1 - i) \tag{1.3.28b}$$

$$z_4 = ke^{7i\pi/4} = \frac{k}{\sqrt{2}}(1 - i)$$

It follows from the above discussion that the corresponding real solutions $w(t)$ have the form

$$w = A \cos \omega_0 t + B \sin \omega_0 t \tag{1.3.29a}$$

$$w = e^{\frac{kt}{\sqrt{2}}}[A_1 \cos kt + B_1 \sin kt] + e^{-\frac{kt}{\sqrt{2}}}[A_2 \cos kt + B_2 \sin kt] \tag{1.3.29b}$$

where A, B, A_1, A_2, B_1, and B_2 are arbitrary constants.

In this chapter we have introduced and summarized the basic properties of complex numbers and elementary functions. We have seen that the theory of functions of a single real variable have so far motivated many of the notions of complex variables; though the two-dimensional character of complex numbers has already led to some significant differences. In subsequent chapters a number of entirely new and surprising results will be obtained, and the departure from real variables will become more apparent.

Problems for Section 1.3

1. Evaluate the following limits:

    (a) $\lim_{z \to i}(z + 1/z)$ (b) $\lim_{z \to z_0} 1/z^m$, m integer

    (c) $\lim_{z \to i} \sinh z$ (d) $\lim_{z \to 0} \dfrac{\sin z}{z}$ (e) $\lim_{z \to \infty} \dfrac{\sin z}{z}$

    (f) $\lim_{z \to \infty} \dfrac{z^2}{(3z + 1)^2}$ (g) $\lim_{z \to \infty} \dfrac{z}{z^2 + 1}$

2. Establish a special case of l'Hopitals rule. Suppose that $f(z)$ and $g(z)$ have formal power series about $z = a$, and

    $$f(a) = f'(a) = f''(a) = \cdots = f^{(k)}(a) = 0$$
    $$g(a) = g'(a) = g''(a) = \cdots = g^{(k)}(a) = 0$$

    If $f^{(k+1)}(a)$ and $g^{(k+1)}(a)$ are not simultaneously zero, show that

    $$\lim_{z \to a} \frac{f(z)}{g(z)} = \frac{f^{(k+1)}(a)}{g^{(k+1)}(a)}$$

    What happens if $g^{(k+1)}(a) = 0$?

3. If $|g(z)| \le M$, $M > 0$ for all z in a neighborhood of $z = z_0$, show that if $\lim_{z \to z_0} f(z) = 0$, then

    $$\lim_{z \to z_0} f(z)g(z) = 0$$

4. Where are the following functions differentiable?

    (a) $\sin z$ (b) $\tan z$ (c) $\dfrac{z - 1}{z^2 + 1}$ (d) $e^{1/z}$ (e) $2\bar{z}$

5. Show that the functions Re z and Imz are nowhere differentiable.

6. Let $f(z)$ be a continuous function for all z. Show that if $f(z_0) \ne 0$, then there must be a neighborhood of z_0 in which $f(z) \ne 0$.

7. Let $f(z)$ be a continuous function where $\lim_{z \to 0} f(z) = 0$. Show that $\lim_{z \to 0}(e^{f(z)} - 1) = 0$. What can be said about $\lim_{z \to 0}((e^{f(z)} - 1)/z)$?

8. Let two polynomials $f(z) = a_0 + a_1 z + \cdots + a_n z^n$ and $g(z) = b_0 + b_1 z + \cdots + b_m z^m$ be equal at all points z in a region R. Use the concept of a limit to show that $m = n$ and that all the coefficients $\{a_j\}_{j=1}^n$ and $\{b_j\}_{j=1}^n$ must be equal. Hint: Consider $\lim_{z \to 0}(f(z) - g(z))$, $\lim_{z \to 0}(f(z) - g(z))/(z)$, etc.

9. (a) Use the real Taylor series formulae

$$e^x = 1 + x + O(x^2), \qquad \cos x = 1 + O(x^2),$$
$$\sin x = x(1 + O(x^2))$$

where $O(x^2)$ means we are omitting terms proportional to power x^2 (i.e, $\lim_{x \to 0}(O(x^2))/(x^2) = C$, where C is a constant), to establish the following:

$$\lim_{z \to 0}(e^z - (1 + z)) = \lim_{r \to 0}(e^{r \cos \theta} e^{ir \sin \theta} - (1 + r(\cos \theta + i \sin \theta))) = 0$$

(b) Use the above Taylor expansions to show that (c.f. Eq. (1.3.18))

$$\lim_{\Delta z \to 0}\left(\frac{e^{\Delta z} - 1}{\Delta z}\right) = \lim_{r \to 0}\left\{\frac{(e^{r \cos \theta} \cos(r \sin \theta) - 1) + i e^{r \cos \theta} \sin(r \sin \theta)}{r(\cos \theta + i \sin \theta)}\right\}$$
$$= 0$$

10. Let $z = x$ be real. Use the relationship $(d/dx)e^{ix} = ie^{ix}$ to find the standard derivative formulae for trigonometric functions:

$$\frac{d}{dx} \sin x = \cos x$$

$$\frac{d}{dx} \cos x = -\sin x$$

11. Suppose we are given the following differential equations:

(a) $\dfrac{d^3 w}{dt^3} - k^3 w = 0$

(b) $\dfrac{d^6 w}{dt^6} - k^6 w = 0$

where t is real and k is a real constant. Find the general real solution of the above equations. Write the solution in terms of *real* functions.

12. Consider the following differential equation:

$$x^2 \frac{d^2 w}{dx^2} + x \frac{dw}{dx} + w = 0$$

where x is real.

(a) Show that the transformation $x = e^t$ implies that

$$x \frac{d}{dx} = \frac{d}{dt},$$

$$x^2 \frac{d^2}{dx^2} = \frac{d^2}{dt^2} - \frac{d}{dt}$$

(b) Use these results to find that w also satisfies the differential equation

$$\frac{d^2 w}{dt^2} + w = 0$$

(c) Use these results to establish that w has the real solution

$$w = C e^{i(\log x)} + \bar{C} e^{-i(\log x)}$$

or

$$w = A \cos(\log x) + B \sin(\log x)$$

13. Use the ideas of Problem 12 to find the real solution of the following equations (x is real and k is a real constant):

(a) $x^2 \dfrac{d^2 w}{dx^2} + k^2 w = 0, \quad 4k^2 > 1$

(b) $x^3 \dfrac{d^3 w}{dx^3} + 3x^2 \dfrac{d^2 w}{dx^2} + x \dfrac{dw}{dx} + k^3 w = 0$

2

Analytic Functions and Integration

In this chapter we study the notion of analytic functions and their properties. It will be shown that a complex function is differentiable if and only if there is an important compatibility relationship between its real and imaginary parts. The concepts of multivalued functions and complex integration are considered in some detail. The technique of integration in the complex plane is discussed and two very important results of complex analysis are derived: Cauchy's theorem and a corollary – Cauchy's integral formula.

2.1 Analytic Functions

2.1.1 The Cauchy–Riemann Equations

In Section 1.3 we defined the notion of complex differentiation. For convenience, we remind the reader of this definition here. The derivative of $f(z)$, denoted by $f'(z)$, is defined by the following limit:

$$f'(z) = \lim_{\Delta z \to 0} \frac{f(z + \Delta z) - f(z)}{\Delta z} \tag{2.1.1}$$

We write the real and imaginary parts of $f(z)$, $f(z) = u(x, y) + iv(x, y)$, and compute Eq. (2.1.1) for (a) $\Delta z = \Delta x$ real and (b) $\Delta z = i\,\Delta y$ pure imaginary (i.e., we take the limit along the real and then along the imaginary axis). Then, for case (a)

$$f'(z) = \lim_{\Delta x \to 0} \left(\frac{u(x + \Delta x, y) - u(x, y)}{\Delta x} + i\frac{v(x + \Delta x, y) - v(x, y)}{\Delta x} \right)$$

$$= u_x(x, y) + iv_x(x, y) \tag{2.1.2}$$

We use the subscript notation for partial derivatives, that is, $u_x = \partial u/\partial x$ and

$v_x = \partial v / \partial x$. For case (b)

$$f'(z) = \lim_{\Delta y \to 0} \frac{u(x, y + \Delta y) - u(x, y)}{i \Delta y} + \frac{i\, (v(x, y + \Delta y) - v(x, y))}{i \Delta y}$$

$$= -i u_y(x, y) + v_y(x, y) \tag{2.1.3}$$

Setting Eqs. (2.1.2) and (2.1.3) equal yields

$$u_x = v_y, \quad v_x = -u_y \tag{2.1.4}$$

Equations (2.1.4) are called the Cauchy–Riemann conditions.

Equations (2.1.4) are a system of partial differential equations that are necessarily satisfied if $f(z)$ has a derivative at the point z. This is in stark contrast to real analysis where differentiability of a function $f(x)$ is only a mild smoothness condition on the function. We also note that if u, v have second derivatives, then we will show that they satisfy the equations $u_{xx} + u_{yy} = 0$ and $v_{xx} + v_{yy} = 0$ (c.f. Eqs. (2.1.11a,b)).

Equation (2.1.4) is a **necessary condition** that must hold if $f(z)$ is differentiable. On the other hand, it turns out that if the partial derivatives of $u(x, y)$, $v(x, y)$ exist, satisfy Eq. (2.1.4), and are continuous, then $f(z) = u(x, y) + iv(x, y)$ must exist and be differentiable at the point $z = x + iy$, that is, Eq. (2.1.4) is a **sufficient condition** as well; namely, if Eq. (2.1.4) holds, then $f'(z)$ exists and is given by Eqs. (2.1.1–2.1.2).

We discuss the latter point next. We use a well-known result of real analysis of two variables, namely, if u_x, u_y and v_x, v_y are continuous at the point (x, y), then

$$\Delta u = u_x \Delta x + u_y \Delta y + \epsilon_1 |\Delta z|$$

$$\Delta v = v_x \Delta x + v_y \Delta y + \epsilon_2 |\Delta z| \tag{2.1.5}$$

where $|\Delta z| = \sqrt{\Delta x^2 + \Delta y^2}$, $\lim_{\Delta z \to 0} \epsilon_1 = \lim_{\Delta z \to 0} \epsilon_2 = 0$, and

$$\Delta u = u(x + \Delta x, y + \Delta y) - u(x, y)$$

$$\Delta v = v(x + \Delta x, y + \Delta y) - v(x, y)$$

Calling $\Delta f = \Delta u + i \Delta v$, we have

$$\frac{\Delta f}{\Delta z} = \frac{\Delta u}{\Delta z} + i \frac{\Delta v}{\Delta z}$$

$$= \left(u_x \frac{\Delta x}{\Delta z} + u_y \frac{\Delta y}{\Delta z} \right) + i \left(v_x \frac{\Delta x}{\Delta z} + v_y \frac{\Delta y}{\Delta z} \right) + \epsilon_1 + i\epsilon_2 \tag{2.1.6}$$

Then, using Eq. (2.1.4), Eq. (2.1.6) yields

$$\frac{\Delta f}{\Delta z} = (u_x + i v_x)\frac{\Delta x + i \Delta y}{\Delta z} + (\epsilon_1 + i\epsilon_2)$$

$$= f'(z) + (\epsilon_1 + i\epsilon_2) \tag{2.1.7}$$

after noting Eq. (2.1.2) and manipulating. Taking the limit of Δz approaching zero yields the desired result.

We state both of the above results as a theorem.

Theorem 2.1.1 The function $f(z) = u(x, y) + i v(x, y)$ is differentiable at a point $z = x + iy$ of a region in the complex plane if and only if the partial derivatives u_x, u_y, v_x, v_y are continuous and satisfy the Cauchy–Riemann conditions (Eq. (2.1.4)) at $z = x + iy$.

A consequence of the Cauchy–Riemann conditions is that the "level" curves of u, that is, the curves $u(x, y) = c_1$ for constant c_1, are orthogonal to the level curves of v, where $v(x, y) = c_2$ for constant c_2, at all points where $f'(z)$ exists and is nonzero. From Eqs. (2.1.2) and (2.1.4) we have

$$|f'(z)|^2 = \left(\frac{\partial u}{\partial x}\right)^2 + \left(\frac{\partial v}{\partial x}\right)^2 = \left(\frac{\partial u}{\partial x}\right)^2 + \left(\frac{\partial u}{\partial y}\right)^2 = \left(\frac{\partial v}{\partial x}\right)^2 + \left(\frac{\partial v}{\partial y}\right)^2$$

hence the two-dimensional vector gradients $\nabla u = \left(\frac{\partial u}{\partial x}, \frac{\partial u}{\partial y}\right)$ and $\nabla v = \left(\frac{\partial v}{\partial x}, \frac{\partial v}{\partial y}\right)$ are nonzero. We know from vector calculus that the gradient is orthogonal to its level curve (i.e., $du = \nabla u \cdot ds = 0$, where ds points in the direction of the tangent to the level curve), and from the Cauchy–Riemann condition (Eq. (2.1.4)) we see that the gradients $\nabla u, \nabla v$ are orthogonal because their vector dot product vanishes:

$$\nabla u \cdot \nabla v = \frac{\partial u}{\partial x}\frac{\partial v}{\partial x} + \frac{\partial u}{\partial y}\frac{\partial v}{\partial y}$$

$$= -\frac{\partial u}{\partial x}\frac{\partial u}{\partial y} + \frac{\partial u}{\partial y}\frac{\partial u}{\partial x} = 0$$

Consequently, the two-dimensional level curves $u(x, y) = c_1$ and $v(x, y) = c_2$ are orthogonal.

The Cauchy–Riemann conditions can be written in other coordinate systems, and it is frequently valuable to do so. Here we quote the result in polar coordinates:

$$\frac{\partial u}{\partial r} = \frac{1}{r}\frac{\partial v}{\partial \theta}$$

$$\frac{\partial v}{\partial r} = -\frac{1}{r}\frac{\partial u}{\partial \theta} \tag{2.1.8}$$

Equation (2.1.8) can be derived in the same manner as Eq. (2.1.4). An alternative derivation uses the differential relationships

$$\frac{\partial}{\partial x} = \cos\theta \frac{\partial}{\partial r} - \frac{\sin\theta}{r}\frac{\partial}{\partial\theta}$$

$$\frac{\partial}{\partial y} = \sin\theta \frac{\partial}{\partial r} + \frac{\cos\theta}{r}\frac{\partial}{\partial\theta}$$

(2.1.9)

which are derived from $x = r\cos\theta$ and $y = r\sin\theta, r^2 = x^2 + y^2, \tan\theta = y/x$.
Employing Eq. (2.1.9) in Eq. (2.1.4) yields

$$\cos\theta \frac{\partial u}{\partial r} - \frac{\sin\theta}{r}\frac{\partial u}{\partial\theta} = \sin\theta \frac{\partial v}{\partial r} + \frac{\cos\theta}{r}\frac{\partial v}{\partial\theta}$$

$$\sin\theta \frac{\partial u}{\partial r} + \frac{\cos\theta}{r}\frac{\partial u}{\partial\theta} = -\cos\theta \frac{\partial v}{\partial r} + \frac{\sin\theta}{r}\frac{\partial v}{\partial\theta}$$

Multiplying the first of these equations by $\cos\theta$, the second by $\sin\theta$, and adding yields the first of Eqs. (2.1.8). Similarly, multiplying the first by $\sin\theta$, the second by $-\cos\theta$, and adding yields the second of Eqs. (2.1.8).

Similarly, using the first relation of Eq. (2.1.9) in $f'(z) = \partial u/\partial x + i\partial v/\partial x$ yields

$$f'(z) = \cos\theta \frac{\partial u}{\partial r} - \frac{\sin\theta}{r}\frac{\partial u}{\partial\theta} + i\cos\theta \frac{\partial v}{\partial r} - i\frac{\sin\theta}{r}\frac{\partial v}{\partial\theta}$$

$$= (\cos\theta - i\sin\theta)\left(\frac{\partial u}{\partial r} + i\frac{\partial v}{\partial r}\right)$$

hence,

$$f'(z) = e^{-i\theta}\left(\frac{\partial u}{\partial r} + i\frac{\partial v}{\partial r}\right)$$

(2.1.10)

Example 2.1.1 Let $f(z) = e^z = e^{x+iy} = e^x e^{iy} = e^x(\cos y + i\sin y)$. Verify Eq. (2.1.4) for all x and y, and then show that $f'(z) = e^z$.

$$u = e^x\cos y, \qquad v = e^x\sin y$$

$$\frac{\partial u}{\partial x} = e^x\cos y = \frac{\partial v}{\partial y}$$

$$\frac{\partial u}{\partial y} = -e^x\sin y = -\frac{\partial v}{\partial x}$$

$$f'(z) = \frac{\partial u}{\partial x} + i\frac{\partial v}{\partial x} = e^x(\cos y + i\sin y)$$

$$= e^x e^{iy} = e^{x+iy} = e^z$$

We have therefore established the fact that $f(z) = e^z$ is differentiable for all finite values of z. Consequently, standard functions like $\sin z$ and $\cos z$, which are linear combinations of the exponential function e^{iz} (see Eqs. (1.2.9–1.2.10)) are also seen to be differentiable functions of z for all finite values of z. It should be noted that these functions do not behave like their real counterparts. For example, the function $\sin x$ oscillates and is bounded between ± 1 for all real x. However, we have

$$\sin z = \sin(x + iy) = \sin x \cos iy + \cos x \sin iy$$
$$= \sin x \cosh y + i \cos x \sinh y$$

Because $|\sinh y|$ and $|\cosh y|$ tend to infinity as y tends to infinity, we see that the real and imaginary parts of $\sin z$ grow without bound.

Example 2.1.2 Let $f(z) = \bar{z} = x - iy$, so that $u(x, y) = x$ and $v(x, y) = -y$. Since $\partial u/\partial x = 1$ while $\partial v/\partial y = -1$, condition (2.1.4) implies $f'(z)$ does not exist anywhere (see also section 1.3).

Example 2.1.3 Let $f(z) = z^n = r^n e^{in\theta} = r^n(\cos n\theta + i \sin n\theta)$, for integer n, so that $u(r, \theta) = r^n \cos n\theta$ and $v(r, \theta) = r^n \sin n\theta$. Verify Eq. (2.1.8) and show that $f'(z) = nz^{n-1}$ ($z \neq 0$ if $n < 0$). By differentiation, we have

$$\frac{\partial u}{\partial r} = nr^{n-1} \cos n\theta = \frac{1}{r}\frac{\partial v}{\partial \theta}$$

$$\frac{\partial v}{\partial r} = nr^{n-1} \sin n\theta = -\frac{1}{r}\frac{\partial u}{\partial \theta}$$

From Eq. (2.1.10),

$$f'(z) = e^{-i\theta}(nr^{n-1})(\cos n\theta + i \sin n\theta)$$
$$= nr^{n-1}e^{-i\theta}e^{in\theta} = nr^{n-1}e^{i(n-1)\theta}$$
$$= nz^{n-1}$$

Example 2.1.4 If a function is differentiable and has constant modulus, show that the function itself is constant. We may write f in terms of real, imaginary, or complex forms where

$$f = u + iv = Re^{i\Theta}$$
$$R^2 = u^2 + v^2, \qquad \tan \Theta = \frac{v}{u}$$
$$R = \text{constant}$$

From Eq. (2.1.8) we have

$$u\frac{\partial u}{\partial r} + v\frac{\partial v}{\partial r} = \frac{1}{r}\left(u\frac{\partial v}{\partial \theta} - v\frac{\partial u}{\partial \theta}\right) = \frac{u^2}{r}\frac{\partial}{\partial \theta}\left(\frac{v}{u}\right)$$

so

$$\frac{\partial}{\partial r}(u^2 + v^2) = \frac{2u^2}{r}\frac{\partial}{\partial \theta}\left(\frac{v}{u}\right)$$

Thus $\partial(v/u)/\partial\theta = 0$ because $R^2 = u^2 + v^2 = $ constant.
  Similarly, using Eq. (2.1.8),

$$u^2\frac{\partial}{\partial r}\left(\frac{v}{u}\right) = \left(u\frac{\partial v}{\partial r} - v\frac{\partial u}{\partial r}\right)$$

$$= -\frac{1}{r}\left(u\frac{\partial u}{\partial \theta} + v\frac{\partial v}{\partial \theta}\right) = -\frac{1}{2r}\frac{\partial}{\partial \theta}(u^2 + v^2) = 0$$

Thus $v/u = $ constant, which implies Θ is constant, and hence so is f.

We have observed that the system of partial differential equations (PDEs), Eq. (2.1.4), that is, the Cauchy–Riemann equations, must hold at every point where $f'(z)$ exists. However, PDEs are really of interest when they hold not only at one point, but rather in a region containing the point. Hence we give the following definition.

Definition 2.1.1 A function $f(z)$ is said to be **analytic** at a point z_0 if $f(z)$ is differentiable in a neighborhood of z_0. The function $f(z)$ is said to be **analytic in a region** if it is analytic at every point in the region.
  Of the previous examples, $f(z) = e^z$ is analytic in the entire finite z plane, whereas $f(z) = \bar{z}$ is analytic *nowhere*. The function $f(z) = 1/z^2$ (Example 2.1.3, $n = -2$) is analytic for all finite $z \neq 0$ (the "punctured" z plane).

Example 2.1.5 Determine where $f(z)$ is analytic when $f(z) = (x + \alpha y)^2 + 2i(x - \alpha y)$ for α real and constant.

$$u(x, y) = (x + \alpha y)^2, \qquad v(x, y) = 2(x - \alpha y)$$

$$\frac{\partial u}{\partial x} = 2(x + \alpha y) \qquad \frac{\partial v}{\partial y} = -2\alpha$$

$$\frac{\partial u}{\partial y} = 2\alpha(x + \alpha y) \qquad \frac{\partial v}{\partial x} = 2$$

The Cauchy–Riemann equations are satisfied only if $\alpha^2 = 1$ and only on the lines $x \pm y = \mp 1$. Because the derivative $f'(z)$ exists only on these lines, $f(z)$ is not analytic *anywhere* since it is not analytic in the neighborhood of these lines.

If we say that that $f(z)$ is analytic in a region, such as $|z| \leq R$, we mean that $f(z)$ is analytic in a domain containing the circle because $f'(z)$ must exist in a neighborhood of every point on $|z| = R$. We also note that some authors use the term **holomorphic** instead of analytic.

An **entire** function is a function that is analytic at each point in the "entire" finite plane. As mentioned above, $f(z) = e^z$ is entire, as is $\sin z$ and $\cos z$. So is $f(z) = z^n$ (integer $n \geq 0$), and therefore, any polynomial.

A **singular point** z_0 is a point where f fails to be analytic. Thus $f(z) = 1/z^2$ has $z = 0$ as a singular point. On the other hand, $f(z) = \bar{z}$ is analytic nowhere and has singular points everywhere in the complex plane. If any region $\mathcal{R}$ exists such that $f(z)$ is analytic in $\mathcal{R}$, we frequently speak of the function as being an **analytic function**. A further and more detailed discussion of singular points appears in Section 3.5.

As we have seen from our examples and from Section 1.3, the standard differentiation formulae of real variables hold for functions of a complex variable. Namely, if two functions are analytic in a domain D, their sum, product, and quotient are analytic in D provided the denominator of the quotient does not vanish at any point in D. Similarly, the composition of two analytic functions is also analytic.

We shall see, in a later section (2.6.1), that an analytic function has derivatives of all orders in the region of analyticity and that the real and imaginary parts have continuous derivatives of all orders as well. From Eq. (2.1.4), because $\partial^2 v/\partial x \partial y = \partial^2 v/\partial y \partial x$, we have

$$\frac{\partial^2 u}{\partial x^2} = \frac{\partial^2 v}{\partial x \partial y} \qquad \frac{\partial^2 v}{\partial y \partial x} = -\frac{\partial^2 u}{\partial y^2}$$

hence

$$\nabla^2 u \equiv \frac{\partial^2 u}{\partial x^2} + \frac{\partial^2 u}{\partial y^2} = 0 \tag{2.1.11a}$$

and similarly

$$\nabla^2 v \equiv \frac{\partial^2 v}{\partial x^2} + \frac{\partial^2 v}{\partial y^2} = 0 \tag{2.1.11b}$$

Equations (2.1.11a,b) demonstrate that u and v satisfy certain uncoupled PDEs. The equation $\nabla^2 w = 0$ is called **Laplace's equation**. It has wide applicability and plays a central role in the study of classical partial differential equations. The function $w(x, y)$ satisfying Laplace's equation in a domain D is called an **harmonic function** in D. The two functions $u(x, y)$ and $v(x, y)$, which are respectively the real and imaginary parts of an analytic function in D, both satisfy Laplace's equation in D. That is, they are **harmonic functions** in D, and v is referred to as the **harmonic conjugate** of u (and vice versa). The function v may be obtained from u via the Cauchy–Riemann conditions. It is clear from the derivation of Eqs. (2.1.11a,b) that $f(z) = u(x, y) + iv(x, y)$ is an analytic function if and only if u and v satisfy Eqs. (2.1.11a,b) and v is the harmonic conjugate of u.

The following example illustrates how, given $u(x, y)$, it is possible to obtain the harmonic conjugate $v(x, y)$ as well as the analytic function $f(z)$.

Example 2.1.6 Suppose we are given $u(x, y) = y^2 - x^2$ in the entire $z = x + iy$ plane. Find its harmonic conjugate as well as $f(z)$.

$$\frac{\partial u}{\partial x} = -2x = \frac{\partial v}{\partial y} \quad \Rightarrow \quad v = -2xy + \phi(x)$$

$$\frac{\partial u}{\partial y} = 2y = -\frac{\partial v}{\partial x} \quad \Rightarrow \quad v = -2xy + \psi(y)$$

where $\phi(x)$, $\psi(x)$ are arbitrary functions of x and y, respectively. Taking the difference of both expressions for v implies $\phi(x) - \psi(y) = 0$, which can only be satisfied by $\phi = \psi = c = $ constant; thus

$$f(z) = y^2 - x^2 - 2ixy + c$$
$$= -(x^2 - y^2 + 2ixy) + c = -z^2 + c$$

It follows from the remark following Theorem 2.1.1, that the two level curves $u = y^2 - x^2 = c_1$ and $v = -2xy = c_2$ are orthogonal to each other at each point (x, y). These are two orthogonal sets of hyperbolae.

Laplace's equation arises frequently in the study of physical phenomena. Applications include the study of two-dimensional ideal fluid flow, steady state heat conduction, electrostatics, and many others. In these applications we are usually interested in solving Laplace's equation $\nabla^2 w = 0$ in a domain D with boundary conditions, typically of the form

$$\alpha w + \beta \frac{\partial w}{\partial n} = \gamma \qquad \text{on } \mathcal{C} \qquad (2.1.12)$$

where $\partial w/\partial n$ denotes the outward normal derivative of w on the boundary of D denoted by $\mathcal{C}$; α, β, and γ are given functions on the boundary. We refer to the solution of Laplace's equation when $\beta = 0$ as the Dirichlet problem, and when $\alpha = 0$ the Neumann problem. The general case is usually called the mixed problem.

2.1.2 Ideal Fluid Flow

Two-dimensional **ideal fluid flow** is one of the prototypical examples of Laplace's equations and complex variable techniques. The corresponding flow configurations are usually easy to conceptualize. **Ideal fluid motion** refers to fluid motion that is steady (time independent), nonviscous (zero friction; usually called inviscid), incompressible (in this case, constant density), and irrotational (no local rotations of fluid "particles"). The two-dimensional equations of motion reduce to a system of two PDEs (see also the discussion in Section 5.4, Example 5.4.1):

(a) incompressibility (divergence of the velocity vanishes)

$$\frac{\partial v_1}{\partial x} + \frac{\partial v_2}{\partial y} = 0 \tag{2.1.13a}$$

where v_1 and v_2 are the horizontal and vertical components of the two-dimensional vector $\mathbf{v}$, that is, $\mathbf{v} = (v_1, v_2)$; and

(b) irrotationality (curl of the velocity vanishes)

$$\frac{\partial v_2}{\partial x} - \frac{\partial v_1}{\partial y} = 0 \tag{2.1.13b}$$

A simplification of these equations is found via the following substitutions:

$$v_1 = \frac{\partial \phi}{\partial x} = \frac{\partial \psi}{\partial y} \qquad v_2 = \frac{\partial \phi}{\partial y} = -\frac{\partial \psi}{\partial x} \tag{2.1.14}$$

We call ϕ the **velocity potential**, and ψ the **stream function**. Equations (2.1.13–2.1.14) show that ϕ and ψ satisfy Laplace's equation. Because the Cauchy–Riemann conditions are satisfied for the functions ϕ and ψ, we have, quite naturally, an associated **complex velocity potential** $\Omega(z)$:

$$\Omega(z) = \phi(x, y) + i\psi(x, y) \tag{2.1.15}$$

The derivative of $\Omega(z)$ is usually called the complex velocity

$$\Omega'(z) = \frac{\partial \phi}{\partial x} + i\frac{\partial \psi}{\partial x} = \frac{\partial \phi}{\partial x} - i\frac{\partial \phi}{\partial y} = v_1 - iv_2 \qquad (2.1.16)$$

The complex conjugate $\overline{\Omega'(z)} = \partial \phi/\partial x + i\partial \phi/\partial y = v_1 + iv_2$ is analogous to the usual velocity vector in two dimensions.

The associated boundary conditions are as follows. The normal derivative of ϕ (i.e., the normal velocity) must vanish on a rigid boundary of an ideal fluid. Because we have shown that the level sets $\phi(x, y) = $ constant and $\psi(x, y) = $ const. are mutually orthogonal at any point (x, y), we conclude that the level sets of the stream function ψ follow the direction of the flow field; namely, they follow the direction of the gradient of ϕ, which are themselves orthogonal to the level sets of ϕ. The level curves $\psi(x, y) = $ const. are called streamlines of the flow. Consequently, boundary conditions in an ideal flow problem at a boundary can be specified by either giving vanishing conditions on the normal derivative of ϕ at a boundary (no flow through the boundary) or by specifying that $\psi(x, y)$ is constant on a boundary, thereby making the boundary a streamline. $\partial \phi/\partial n = \nabla \phi \cdot \hat{n}$, $\hat{n}$ being the unit normal, implies that $\nabla \phi$ points in the direction of the tangent to the boundary. For problems with an infinite domain, some type of boundary condition – usually a boundedness condition – must be given at infinity. We usually specify that the velocity is uniform (constant) at infinity.

Briefly in this section, and in subsequent sections and Chapter 5 (see Section 5.4), we shall discuss examples of fluid flows corresponding to various complex potentials. Upon considering boundary conditions, functions $\Omega(z)$ that are analytic in suitable regions may frequently be associated with two-dimensional fluid flows, though we also need to be concerned with locations of nonanalyticity of $\Omega(z)$. Some examples will clarify the situation.

Example 2.1.7 The simplest example is that of **uniform flow**

$$\Omega(z) = v_0 e^{-i\theta_0} z = v_0(\cos \theta_0 - i \sin \theta_0)(x + iy), \qquad (2.1.17)$$

where v_0 and θ_0 are positive real constants. Using Eqs. (2.1.15, 2.1.16), the corresponding velocity potential and velocity field is given by

$$\phi(x, y) = v_0(\cos \theta_0 x + \sin \theta_0 y) \qquad v_1 = \frac{\partial \phi}{\partial x} = v_0 \cos \theta_0$$

$$v_2 = \frac{\partial \phi}{\partial y} = v_0 \sin \theta_0$$

which is identified with uniform flow making an angle θ_0 with the x axis, as

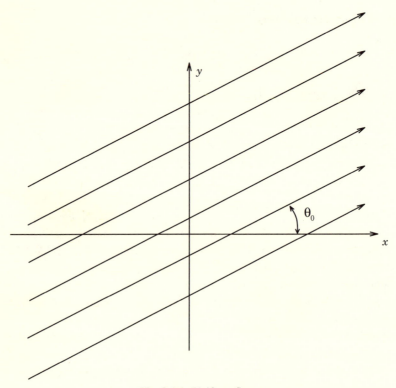

Fig. 2.1.1. Uniform flow

in Figure 2.1.1. Alternatively, the stream function $\psi(x, y) = v_0(\cos \theta_0 y - \sin \theta_0 x) = \text{const.}$ reveals the same flow field.

Example 2.1.8 A somewhat more complicated flow configuration, flow around a cylinder, corresponds to the complex velocity potential

$$\Omega(z) = v_0 \left(z + \frac{a^2}{z} \right) \qquad (2.1.18)$$

where v_0 and a are positive real constants and $|z| > a$. The corresponding velocity potential and stream function are given by

$$\phi = v_0 \left(r + \frac{a^2}{r} \right) \cos \theta \qquad (2.1.19a)$$

$$\psi = v_0 \left(r - \frac{a^2}{r} \right) \sin \theta \qquad (2.1.19b)$$

and for the complex velocity we have

$$\Omega'(z) = v_0 \left(1 - \frac{a^2}{z^2}\right) = v_0 \left(1 - \frac{a^2 e^{-2i\theta}}{r^2}\right) \tag{2.1.20}$$

whereby from Eq. (2.1.16) the horizontal and vertical components of the velocity are given by

$$v_1 = v_0 \left(1 - \frac{a^2 \cos 2\theta}{r^2}\right)$$

$$v_2 = -v_0 \frac{a^2 \sin 2\theta}{r^2} \tag{2.1.21}$$

The circle $r = a$ is a streamline ($\psi = 0$) as is $\theta = 0$ and $\theta = \pi$. As $r \to \infty$, the limiting velocity is uniform in the x direction ($v_1 \to v_0, v_2 \to 0$). The corresponding flow field is that of a uniform stream at large distances modified by a circular barrier, as in Figure 2.1.2, which may be viewed as flow around a cylinder with the same flow field at all points perpendicular to the flow direction.

Note that the velocity vanishes at $r = a, \theta = 0$, and $\theta = \pi$. These points are called **stagnation points** of the flow. On the circle $r = a$, which corresponds to the streamline $\psi = 0$, the normal velocity is zero because the corresponding velocity must be in the tangent direction to the circle. Another way to see this is to compute the normal velocity from ϕ using the gradient in two-dimensional polar coordinates:

$$\mathbf{v} = \nabla\phi = \frac{\partial\phi}{\partial r}\hat{\mathbf{u}}_r + \frac{1}{r}\frac{\partial\phi}{\partial\theta}\hat{\mathbf{u}}_\theta$$

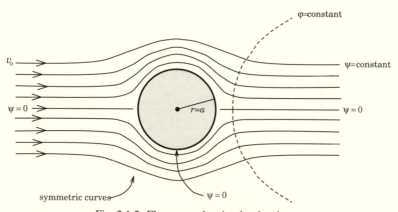

Fig. 2.1.2. Flow around a circular barrier

where $\hat{u}_r$ and $\hat{u}_\theta$ are the unit normal and tangential vectors. So the radial velocity at any point (r, θ) is given by

$$\frac{\partial \phi}{\partial r} = v_0 \left(1 - \frac{a^2}{r^2} \right) \cos \theta$$

which vanishes when $r = a$. As mentioned earlier, as $r \to \infty$ the flow becomes uniform:

$$\phi \longrightarrow v_0 r \cos \theta = v_0 x$$
$$\psi \longrightarrow v_0 r \sin \theta = v_0 y$$

So for large r and correspondingly large y, the curves $y = $ const are streamlines as expected.

Problems for Section 2.1

1. Which of the following satisfy the Cauchy–Riemann (C-R) equations? If they satisfy the C-R equations, give the analytic function of z.

$$\text{(a)} \ \ f(x, y) = x - iy + 1$$
$$\text{(b)} \ \ f(x, y) = y^3 - 3x^2 y + i(x^3 - 3xy^2 + 2)$$
$$\text{(c)} \ \ f(x, y) = e^y (\cos x + i \sin y)$$

2. In the following we are given the real part of an analytic function of z. Find the imaginary part and the function of z.

$$\text{(a)} \ \ 3x^2 y - y^3 \qquad \text{(b)} \ \ 2x(c - y), \qquad c = \text{constant}$$
$$\text{(c)} \ \ \frac{y}{x^2 + y^2} \qquad \text{(d)} \ \ \cos x \cosh y$$

3. Determine whether the following functions are analytic. Discuss whether they have any singular points or if they are entire.

$$\text{(a)} \ \ \tan z \qquad \text{(b)} \ \ e^{\sin z} \qquad \text{(c)} \ \ e^{1/(z-1)} \qquad \text{(d)} \ \ e^{\bar{z}}$$
$$\text{(e)} \ \ \frac{z}{z^4 + 1} \qquad \text{(f)} \ \ \cos x \cosh y - i \sin x \sinh y$$

4. Show that the real and imaginary parts of a twice-differentiable function $f(\bar{z})$ satisfy Laplace's equation. Show that $f(\bar{z})$ is nowhere analytic unless it is constant.

5. Let $f(z)$ be analytic in some domain. Show that $f(z)$ is necessarily a constant if either the function $\overline{f(z)}$ is analytic or $f(z)$ assumes only pure imaginary values in the domain.

6. Consider the following complex potential

$$\Omega(z) = -\frac{k}{2\pi}\frac{1}{z}, \qquad k \text{ real,}$$

referred to as a "doublet." Calculate the corresponding velocity potential, stream function, and velocity field. Sketch the stream function. The value of k is usually called the strength of the doublet. See also Problem 4 of Section 2.3, in which we obtain this complex potential via a limiting procedure of two elementary flows, referred to as a "source" and a "sink."

7. Consider the complex analytic function, $\Omega(z) = \phi(x, y) + i\psi(x, y)$, in a domain D. Let us transform from z to w using $w = f(z)$, $w = u + iv$, where $f(z)$ is analytic in D, with the corresponding domain in the w plane, D'. Establish the following:

$$\frac{\partial\phi}{\partial x} = \frac{\partial u}{\partial x}\frac{\partial\phi}{\partial u} + \frac{\partial v}{\partial x}\frac{\partial\phi}{\partial v}$$

$$\frac{\partial^2\phi}{\partial x^2} = \frac{\partial^2 u}{\partial x^2}\frac{\partial\phi}{\partial u} - \frac{\partial^2 u}{\partial x\partial y}\frac{\partial\phi}{\partial v} + \left(\frac{\partial u}{\partial x}\right)^2\frac{\partial^2\phi}{\partial u^2} - 2\frac{\partial u}{\partial x}\frac{\partial u}{\partial y}\frac{\partial^2\phi}{\partial u\partial v}$$

$$+ \left(\frac{\partial u}{\partial y}\right)^2\frac{\partial^2\phi}{\partial v^2}$$

Also find the corresponding formulae for $\partial\phi/\partial y$ and $\partial^2\phi/\partial y^2$. Recall that $f'(z) = \frac{\partial u}{\partial x} - i\frac{\partial u}{\partial y}$, and $u(x, y)$ satisfies Laplace's equation in the domain D. Show that

$$\nabla^2_{x,y}\phi = \frac{\partial^2\phi}{\partial x^2} + \frac{\partial^2\phi}{\partial y^2} = \left(u_x^2 + u_y^2\right)\left(\frac{\partial^2\phi}{\partial u^2} + \frac{\partial^2\phi}{\partial v^2}\right)$$

$$= |f'(z)|^2\nabla^2_{u,v}\phi$$

Consequently, we find that if ϕ satisfies Laplace's equation $\nabla^2_{x,y}\phi = 0$ in the domain D, then it also satisfies Laplace's equation $\nabla^2_{u,v}\phi = 0$ in domain D'.

8. Given the complex analytic function $\Omega(z) = z^2$, show that the real part
    of Ω, $\phi(x, y) = \text{Re}\,\Omega(z)$, satisfies Laplace's equation, $\nabla^2_{x,y}\phi = 0$. Let
    $z = (1 - w)/(1 + w)$, where $w = u + iv$. Show that $\phi(u, v) = \text{Re}\,\Omega(w)$
    satisfies Laplace's equation $\nabla^2_{u,v}\phi = 0$.

2.2 Multivalued Functions

A single-valued function $w = f(z)$ yields one value w for a given complex
number z. A multivalued function admits more than one value w for a given z.
Such a function is more complicated and frequently requires a great deal of care.
Multivalued functions are naturally introduced as the inverse of single-valued
functions.

The simplest such function is the square root function. If we consider $z = w^2$,
the inverse is written as

$$w = z^{\frac{1}{2}} \tag{2.2.1}$$

From real variables we already know that $x^{1/2}$ has two values, often written as
$\pm\sqrt{x}$ where $\sqrt{x} \geq 0$. For the complex function (Eq. (2.2.1)) we can ascertain
the multivaluedness by letting $z = re^{i\theta}$, and $\theta = \theta_p + 2\pi n$, where, say,
$0 \leq \theta_p < 2\pi$

$$w = r^{1/2}e^{i\theta_p/2}e^{n\pi i} \tag{2.2.2}$$

where $r^{1/2} \equiv \sqrt{r} \geq 0$ and n is an integer. For a given value z, the function
$w(z)$ takes two possible values corresponding to n even and n odd, namely

$$\sqrt{r}e^{i\theta_p/2} \quad \text{and} \quad \sqrt{r}e^{i\theta_p/2}e^{i\pi} = -\sqrt{r}e^{i\theta_p/2}$$

An important consequence of the multivaluedness of w is that as z traverses
a small circuit around $z = 0$, w does not return to its original value. Indeed,
suppose we start at $z = \epsilon$ for real $\epsilon > 0$. Let us see what happens to w as
we return to this point after going around a circle with radius ϵ. Let $n = 0$.
When we start, $\theta_p = 0$ and $w = \sqrt{\epsilon}$; when we return to $z = \epsilon$, $\theta_p = 2\pi$ and
$w = \sqrt{\epsilon}e^{\frac{2i\pi}{2}} = -\sqrt{\epsilon}$. We note that the value $-\sqrt{\epsilon}$ can also be obtained from
$\theta_p = 0$ provided we take $n = 1$. In other words, we started with a value w
corresponding to $n = 0$ and ended up with a value w corresponding to $n = 1$!
(Any even/odd values of n suffice for this argument.) The point $z = 0$ is called
a **branch point**. A point is a branch point if the multivalued function $w(z)$ is
discontinuous upon traversing a small circuit around this point. It should be
noted that the point $z = \infty$ is also a branch point. This is seen by using the
transformation $z = \frac{1}{t}$, which maps $z = \infty$ to $t = 0$. Using arguments such

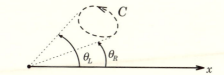

Fig. 2.2.1. Closed circuit away from branch cut

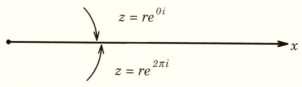

Fig. 2.2.2. Cut plane, $z^{1/2}$

as that above, it follows that $t = 0$ is a branch point of the function $t^{-1/2}$, and hence $z = \infty$ is a branch point of the function $z^{1/2}$. The points $z = 0$ and $z = \infty$ are the only branch points of the function $z^{1/2}$. Indeed, if we take a closed circuit C (see Figure 2.2.1) that does *not* enclose $z = 0$ or $z = \infty$, then $z^{1/2}$ returns to its original value as z traverses C. Along C the phase θ will vary continuously between $\theta = \theta_R$ and $\theta = \theta_L$. So if we begin at $z_R = r_R e^{i\theta_R}$ and follow the curve C, the value z will return to exactly its previous value with no phase change. Hence $z^{1/2}$ will not have a jump as the curve C is traversed.

The analytic study of multivalued functions usually is best effected by expressing the multivalued function in terms of a single-valued function. One method of doing this is to consider the multivalued function in a restricted region of the plane and choose a value at every point such that the resulting function is single-valued and continuous. A continuous function obtained from a multivalued function in this way is called a **branch** of the multivalued function. For the function $w = z^{1/2}$ we can carry out this procedure by taking $n = 0$ and restricting the region of z to be the open or cut plane in Figure 2.2.2. For this purpose the real positive axis in the z plane is cut out. The values of $z = 0$ and $z = \infty$ are also deleted. The function $w = z^{1/2}$ is now continuous in the cut plane that is an open region. The semiaxis $\operatorname{Re} z > 0$ is referred to as a **branch cut**.

It should be noted that the location of the branch cut is arbitrary save that it ends at branch points. If we restrict θ_p to $-\pi \le \theta_p < \pi$, $n = 0$ in the polar representation of $z = re^{i\theta}$, $\theta = \theta_p + 2n\pi$, then the branch cut would naturally be on the negative real axis. More complicated curves (e.g. spirals) could equally well be chosen as branch cuts but rarely do we do so because a cut is chosen for convenience. The simplest choice (sometimes motivated by physical application) is generally satisfactory. We reiterate that the main purpose of a branch cut is to artificially create a region in which the function is single-valued and continuous.

On the other hand, if we took a closed circuit that didn't enclose the branch point $z = 0$, then the function $z^{1/2}$ would return to its same value. We depict, in Figure 2.2.1, a typical closed circuit C not enclosing the origin, with the choice of branch cut ($z = re^{i\theta}, 0 \le \theta < 2\pi$) on the positive real axis.

Note that if we had chosen $w = (z-z_0)^{1/2}$ as our prototype example, a (finite) branch point would have been at $z = z_0$. Similarly, if we had investigated $w = (az + b)^{1/2}$, then a (finite) branch point would have been at $-b/a$. (In either case, $z = \infty$ would be another branch point.) We could deduce these facts by translating to a new origin in our coordinate system and investigating the change upon a circuit around the branch point, namely, letting $z = z_0 + re^{i\theta}$, $0 \le \theta < 2\pi$. We shall see that multivalued functions can be considerably more exotic than the ones described above.

A somewhat more complicated situation is illustrated by the inverse of the exponential function, that is, the logarithm (see Figure 2.2.3). Consider

$$z = e^w \tag{2.2.3}$$

Let $w = u + iv$. We have, using the properties of the exponential function

$$z = e^{u+iv} = e^u e^{iv} = e^u(\cos v + i \sin v) \tag{2.2.4a}$$

in polar coordinates $z = re^{i\theta_p}$ for $0 \le \theta_p < 2\pi$, so

$$r = e^u$$
$$v = \theta_p + 2\pi n, \qquad n \text{ integer} \tag{2.2.4b}$$

From the properties of real variables

$$u = \log r$$

Thus, in analogy with real variables, we write $w = \log z$, which is

$$w = \log z = \log r + i\theta_p + 2n\pi i \tag{2.2.4c}$$

where $n = 0, \pm 1, \pm 2, \ldots$ and where θ_p takes on values in a particular range of 2π. Here we take

$$0 \le \theta_p < 2\pi$$

When $n = 0$, Eq. (2.2.4) is frequently referred to as the principal branch of the logarithm; the corresponding value of the function is referred to as the **principal value**. From (2.2.4) we see that, as opposed to the square root example, the function is infinitely valued; that is, n takes on an infinite number of integer values. For example, if $z = i$, then $|z| = r = 1, \theta_p = \pi/2$; hence

$$\log i = \log 1 + i \left(\frac{\pi}{2} + 2n\pi \right) \qquad n = 0, \pm 1, \pm 2, \dots \qquad (2.2.5)$$

Similarly, if $z = x$, a real quantity, $|z| = r = |x|$, then

$$\log z = \log |x| + 2n\pi i \qquad n = 0, \pm 1, \pm 2, \dots \qquad (2.2.6)$$

The complex logarithm function differs from the real logarithm by additive multiples of $2\pi i$. If z is real and positive, we normally take $n = 0$ so that the principal branch of the complex logarithm function agrees with the usual one for real variables.

Suppose we consider a given point $z = x_0$, x_0 real and positive, and fix a branch of $\log z$, $n = 0$. So $\log z = \log |x_0|$. Let us now allow z to vary on a circle about $z = 0$: $z = |x_0|e^{i\theta}$. As θ varies from $\theta = 0$ to $\theta = 2\pi$ the value of $\log z$ varies from $\log |x_0|$ to $\log |x_0| + 2\pi i$. Thus we see that $z = 0$ is a branch point: A small circuit (x_0 can be as small as we wish) about the origin results in a change in $\log z$. Indeed, we see that after one circuit we come to the $n = 1$ branch of $\log z$. The next circuit would put us on the $n = 2$ branch of $\log z$ and so on. The function $\log z$ is thus seen to be infinitely branched, and the line $\operatorname{Re} z > 0$ is a branch cut (see Figure 2.2.3).

We reiterate that the branch cut $\operatorname{Re} z > 0$ is arbitrarily chosen, although in a physical problem a particular choice might be indicated. Had we defined $\log z$ as

$$\log z = \log |x_0| + i(\theta_p + 2n\pi), \qquad -\pi \leq \theta_p < \pi \qquad (2.2.7)$$

this would be naturally related to values of $\log z$ that have a jump on the negative real axis. So $n = 0$, $\theta_p = -\pi$ corresponds to $\log z = \log |x_0| - i\pi$. A full

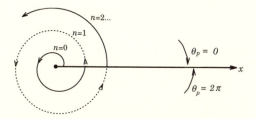

Fig. 2.2.3. Logarithm function and branch cut

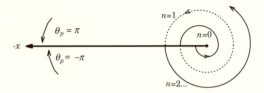

Fig. 2.2.4. Logarithm function and alternative branch cut

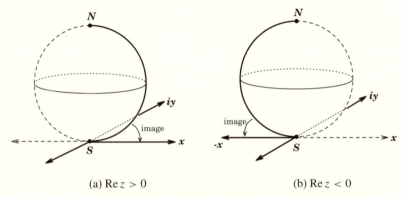

(a) Re $z > 0$ (b) Re $z < 0$

Fig. 2.2.5. Branch cuts, stereographic projection

circuit in the counterclockwise direction puts us on the first branch $\log z = \log |x_0| + i\pi$ (see Figure 2.2.4).

It should be noted that the point $z = \infty$ is also a branch point for $\log z$. As we have seen, the point at infinity is easily understood via the transformation $z = 1/t$, so that t near zero corresponds to z near ∞. The above arguments, which are used to establish whether a point is in fact a branch point, apply at $t = 0$. The use of this transformation and the properties of $\log |z|$ yields $\log z = \log 1/t = -\log t$. We establish $t = 0$ as a branch point by letting $t = re^{i\theta}$, varying θ by 2π, and noting that this function does not return to its original values.

It is convenient to visualize the branch cut as joining the two branch points $z = 0$ and $z = \infty$. For those who studied the stereographic projection (Section 1.2.2), this branch cut is a (great circle) curve joining the south ($z = 0$) and the north ($z = \infty$) poles (see Figure 2.2.5).

The analyticity of $\log z$ in the cut plane can be established using the Cauchy–Riemann conditions. We shall also show the important relationship $d/dz(\log z) = 1/z$. Using (2.2.3) and (2.2.4a,b,c), we see that for $z = x + iy$, $w = \log z$, $w = u + iv$

$$e^{2u} = x^2 + y^2, \qquad \tan v = \frac{y}{x} \qquad (2.2.8)$$

Note that in deriving Eq. (2.2.8) we use $w = \log[|z|e^{i \arg z}]$, $|z| = (x^2 + y^2)^{1/2}$, and $\theta = \arg z = \tan^{-1} y/x$. A branch is fixed by assigning suitable values for the *real* functions u and v. The function u is given by

$$u = \frac{1}{2} \log(x^2 + y^2) \qquad (2.2.9)$$

To fix the branch of v corresponding to the inverse tangent of y/x takes is more subtle. Suppose we fix $\tan^{-1}(y/x)$ to be the standard real-valued function taking values between $-\pi/2$ and $\pi/2$; that is

$$\frac{-\pi}{2} \le \tan^{-1}(y/x) < \frac{\pi}{2}$$

Thus the value of v will have a jump whenever x passes through zero (e.g. a jump of π when we pass from the first to the second quadrant).

Alternatively we could have written

$$v = \tan^{-1}\left(\frac{y}{x}\right) + C_i \qquad (2.2.10)$$

with $C_1 = 0$, $C_2 = C_3 = \pi$, $C_4 = 2\pi$, where the values of the constant C_i correspond to suitable values in each of the four quadrants.

It can be verified that v is continuous in the z plane apart from $\operatorname{Re} z > 0$ where there is a jump of 2π across the $\operatorname{Re} z > 0$ axis. Figure 2.2.6 depicts the choice of $v = \tan^{-1}(y/x)$ that will make $\log z$ continuous off the real axis, $\operatorname{Re} z \ne 0$.

From real variables we know that

$$\frac{d}{ds} \tan^{-1} s = \frac{1}{1 + s^2} \qquad (2.2.11)$$

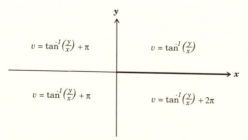

Fig. 2.2.6. A branch choice for inverse tangent

and with this from Eq. (2.2.10) we can verify that the Cauchy–Riemann conditions are satisfied for Eq. (2.2.8). The partial derivatives of u and v are given by

$$u_x = \frac{x}{x^2 + y^2}, \qquad u_y = \frac{y}{x^2 + y^2} \qquad (2.2.12a)$$

$$x = \frac{-y}{x^2 + y^2}, \qquad v_y = \frac{x}{x^2 + y^2} \qquad (2.2.12b)$$

hence the Cauchy–Riemann conditions $u_x = v_y$ and $u_y = -v_x$ are satisfied and the function $\log z$ is analytic in the cut plane $\operatorname{Re} z > 0$ (as implied by the properties of the inverse tangent function). Alternatively we could have used $u = \log r$, $v = \theta$ and Eq. (2.1.8).

Because $\log z$ is analytic in the cut plane, its derivative can be easily calculated. We need only to calculate the derivative along the x direction

$$\frac{d}{dz} \log z = \frac{\partial u}{\partial x} + i \frac{\partial v}{\partial x} = \frac{x - iy}{x^2 + y^2} = \frac{1}{x + iy} = \frac{1}{z} \qquad (2.2.13)$$

Hence the expected result is obtained for the derivative of $\log z$ in a cut plane. Indeed, this development can be carried out for any of the branches (suitable cut planes) of $\log z$. Alternatively from (2.1.10): $f'(z) = e^{-i\theta} \frac{\partial}{\partial r}(\log r) = \frac{1}{re^{i\theta}} = \frac{1}{z}$.

The generalized power function is defined in terms of the logarithm

$$z^a = e^{a \log z} \qquad (2.2.14)$$

for any complex constant a. When $a = m =$ integer, the power function is simply z^m. Using Eq. (2.2.4) and $e^{2k\pi i} = \cos 2k\pi + i \sin 2k\pi = 1$, where k is an integer, we have

$$z^m = e^{m[\log r + i(\theta_p + 2\pi n)]} = e^{m \log r} e^{mi\theta_p} = (re^{i\theta_p})^m$$

whereupon we have the usual integer power function with no branching and no branch points. If, however, a is a rational number

$$a = \frac{m}{l}$$

m and l are integers with no common factor then we have

$$z^{m/l} = \exp\left[\frac{m}{l}(\log r + i(\theta_p + 2\pi n))\right]$$

$$= \exp\left[\frac{m}{l}(\log r + i\theta_p)\right] \exp\left[2\pi i \left(\frac{mn}{l}\right)\right] \qquad (2.2.15)$$

It is evident that when $n = 0, 1, \ldots, (l-1)$, the expression (2.2.15) takes on different values corresponding to the term $e^{2\pi i (mn/l)}$. Thus $z^{(m/l)}$ takes on l different values. If n increases beyond $n = l - 1$, say $n = l, (l+1), \ldots, (2l - 1)$, the above values are correspondingly repeated, and so on. The formula (2.2.15) yields l branches for the function $z^{m/l}$. The function $z^{m/l}$ has branch points at $z = 0$ and $z = \infty$. Similar considerations apply to the function $w = (z - z_0)^{m/l}$ with a (finite) branch point now being located at $z = z_0$. A cut plane can be fixed by choosing θ_p appropriately. Hence a branch cut on $\mathrm{Re}\, z > 0$ is fixed by requiring $0 \le \theta_p < 2\pi$. Similarly, a cut for $\mathrm{Re}\, z < 0$ is fixed by assigning $-\pi \le \theta_p < \pi$. Thus if $m = 1, l = 4$, the formula (2.2.15) yields four branches of the function $z^{1/4}$.

Values of a that are neither integer nor rational result in functions that are infinitely branched with branch points at $z = 0, z = \infty$. Branch cuts can be defined via choices of θ_p as above. For any suitable branch, standard differentiation formulae give

$$\frac{d}{dz} z^a = \frac{d}{dz} e^{a \log z} = z^a \left(\frac{a}{z}\right) = a z^{a-1} \qquad (2.2.16)$$

From Eq. (2.2.4) we also have

$$\begin{aligned}
\log(z_1 z_2) &= \log(r_1 e^{i\theta_1} r_2 e^{i\theta_2}) \\
&= \log r_1 r_2 + i(\theta_{1p} + \theta_{2p}) + 2n\pi i \\
&= \log r_1 + i(\theta_{1p} + 2n_1\pi) + \log r_2 + i(\theta_{2p} + 2n_2\pi) \\
&= \log z_1 + \log z_2
\end{aligned}$$

where $n_1 + n_2 = n$. The other standard algebraic properties of the complex logarithm, which are analogous to the real logarithm, follow in a similar manner.

The inverse of trigonometric and hyperbolic functions can be computed via logarithms. It is another step in complication regarding multivalued functions. For example

$$w = \cos^{-1} z \qquad (2.2.17)$$

satisfies

$$\cos w = z = \frac{e^{iw} + e^{-iw}}{2}$$

Thus

$$e^{2iw} - 2z e^{iw} + 1 = 0 \qquad (2.2.18)$$

Hence solving this quadratic equation for e^{iw} yields

$$e^{iw} = z + (z^2 - 1)^{\frac{1}{2}} = z + i(1 - z^2)^{\frac{1}{2}}$$

and then

$$w(z) = -i \log(z + i(1 - z^2)^{\frac{1}{2}}) \qquad (2.2.19)$$

This function $w(z)$ has two sources of multivaluedness; one due to the logarithm, the other due to $f(z) = (1 - z^2)^{\frac{1}{2}}$. The function $f(z)$ has two branches and two branch points, at $z = \pm 1$. We can deduce that $z = \pm 1$ are branch points of $f(z)$ by investigating the local behavior of $f(z)$ near the points $z = \pm 1$. Namely, use $z = 1 + r_1 e^{i\theta_1}$ and $z = -1 + r_2 e^{i\theta_2}$ for small values of r_1 and r_2. For, say, $z = -1$, we have $f(z) \approx (2r_2)^{1/2} e^{i\theta_2/2}$ (dropping r_2^2 terms as much smaller than r_2), which certainly has a discontinuity as θ_2 changes by 2π. The function $f(z)$ has two branches. The log function has an infinite number of branches, hence so does w; sometimes we say that $w(z)$ is doubly infinite because for each of the infinity of branches of the log we also have two branches of $f(z)$. In the finite plane the only branch points of $w(z)$ are at $z = \pm 1$ because the function $g(z) = z + i(1 - z^2)^{1/2}$ has no solutions of $g(z) = 0$. (Equating both sides, $z = -i(1 - z^2)^{1/2}$ leads to a contradiction.) The branch structure of $w(z)$ in Eq. (2.2.19) is discussed further in Section 2.3 (c.f. Eq. (2.3.8)).

Because the log function is determined up to additive multiples of $2\pi i$, it follows that for a *fixed* value of $(1 - z^2)^{1/2}$, and a particular branch of the log function, $w = \cos^{-1} z$ is determined only to within multiples of 2π. Namely, if we write $w_1 = -i \log(z + i(1 - z^2)^{1/2})$ for a particular branch, then the general form for w satisfies

$$w = -i \log(z + i(1 - z^2)^{1/2}) + 2n\pi$$

or $w = w_1 + 2n\pi$, which expresses the periodicity of the cosine function. Similarly, from the quadratic equation (2.2.18) we find that the product of the two roots e^{iw_1} and e^{iw_2} satisfies

$$e^{iw_1} e^{iw_2} = 1 \qquad (2.2.20)$$

or by taking the logarithm of Eq. (2.2.20) with $1 = e^{i0}$ or $1 = e^{2\pi i}$ we see that the two solutions of Eq. (2.2.18) are simply related:

$$w_1 + w_2 = 0 \quad \text{or} \quad w_1 + w_2 = 2\pi, \text{ etc.} \qquad (2.2.21)$$

Equation (2.2.21) reflects the fact that the cosine of an angle, say α, equals the cosine of $-\alpha$ or the cosine of $2\pi - \alpha$, etc.

Differentiation establishes the relationship

$$\frac{d}{dz}\cos^{-1}z = \frac{-i}{z + i(1 - z^2)^{1/2}}\left(1 - \frac{iz}{(1 - z^2)^{1/2}}\right)$$

$$= \frac{-i}{z + i(1 - z^2)^{1/2}}(-i)\frac{(z + i(1 - z^2)^{1/2})}{(1 - z^2)^{1/2}}$$

$$= \frac{-1}{(1 - z^2)^{1/2}} \tag{2.2.22}$$

Formulae for the other inverse trigonometric and hyperbolic functions can be established in a similar manner. For reference we list some of them below.

$$\sin^{-1}z = -i\log(iz + (1 - z^2)^{1/2}) \tag{2.2.23a}$$

$$\tan^{-1}z = \frac{1}{2i}\log\frac{i - z}{i + z} \tag{2.2.23b}$$

$$\sinh^{-1}z = \log(z + (1 + z^2)^{1/2}) \tag{2.2.23c}$$

$$\cosh^{-1}z = \log(z + (z^2 - 1)^{1/2}) \tag{2.2.23d}$$

$$\tanh^{-1}z = \frac{1}{2}\log\frac{1 + z}{1 - z} \tag{2.2.23e}$$

In the following section we shall discuss the branch structure of more complicated functions such as $\sqrt{(z - a)(z - b)}$ and $\cos^{-1}z$.

In Section 2.1 we mentioned that the real and imaginary parts of an analytic function in a domain D satisfy Laplace's equation in D. In fact, some simple complex functions yield fundamental and physically important solutions to Laplace's equation.

For example, consider the function

$$\Omega(z) = A\log z + iB \tag{2.2.24}$$

where A and B are real and we take the branch cut of the logarithm along the real axis with $z = re^{i\theta}$, $0 \le \theta < 2\pi$. The imaginary part of $\Omega(z)$: $\Omega(z) = \phi(x, y) + i\psi(x, y)$, satisfies Laplace's equation

$$\nabla^2\psi = \frac{\partial\psi}{\partial x^2} + \frac{\partial\psi}{\partial y^2} = 0 \tag{2.2.25}$$

in the upper half plane: $-\infty < x < \infty$, $y > 0$. (Note that the function is analytic for $y > 0$, i.e., there is no branch cut for $y > 0$.) From Eq. (2.2.24) a solution of Laplace's equation is

$$\psi(x, y) = A\theta + B$$

$$= A \tan^{-1}\left(\frac{y}{x}\right) + B, \tag{2.2.26}$$

where $\tan^{-1}(y/x)$ stands for the identifications in Eq. (2.2.10) (see Figure 2.2.6). Thus, for $y > 0$, $0 < \tan^{-1}(y/x) < \pi$.

Note that as $y \to 0^+$, then $\theta = \tan^{-1}(y/x) \to 0$ for $x > 0$, and $\to \pi$ for $x < 0$. Taking $B = 1$ and $A = -1/\pi$, we find that

$$\psi(x, y) = 1 - \frac{1}{\pi}\tan^{-1}\left(\frac{y}{x}\right) \tag{2.2.27}$$

is the solution of Laplace's equation in the upper half plane bounded at infinity, corresponding to the boundary conditions

$$\psi(x, 0) = \begin{cases} 1 & \text{for } x > 0 \\ 0 & \text{for } x < 0 \end{cases} \tag{2.2.28}$$

Physically speaking, Eq. (2.2.27) corresponds to the steady state heat distribution of a plate with the prescribed temperature distribution (Eq. (2.2.28)) on the bottom of the plate (steady state heat flow satisfies Laplace's equation).

We also mention briefly that in many applications it is useful to employ suitable transformations that have the effect of transforming Laplace's equation in a complicated domain to a "simple" one, that is, one for which Laplace's equation can be easily solved such as in a half plane or inside a circle. In terms of two-dimensional ideal fluid flow, this means that a flow in a complicated domain would be converted to one in a simpler domain under the appropriate transformation of variables. (A number of physical applications are discussed in Chapter 5.)

The essential idea is the following. Suppose we are given a complex analytic function in a domain D:

$$\Omega(z) = \phi(x, y) + i\psi(x, y)$$

where ϕ and ψ satisfy Laplace's equation in D. Let us transform to a new independent complex variable w, where $w = u + iv$, via the transformation

$$z = F(w) \tag{2.2.29}$$

where $F(w)$ is analytic in the corresponding domain D' in the u, v-plane. Then $\Omega(F(w))$, which we shall call $\Omega(w)$

$$\Omega(w) = \phi(u, v) + i\psi(u, v)$$

is also analytic in D'. Hence the function ϕ and ψ will satisfy Laplace's equation in D'. (A direct verification of this statement is included in the problem section.) For this transformation to be useful, D' must be a simplified domain in which Laplace's equation is easily solved.

The complication inherent in this procedure is that of returning back from the w plane to the z plane in order to obtain the required solution $\Omega(z)$, or $\phi(x, y)$ and $\psi(x, y)$. We must invert Eq. (2.2.29) to find w as a function of z. In general, this introduces multivaluedness, which we shall discuss in Section 2.3. From a general point of view we can deduce where the "difficulties" in the transformation occur by examining the derivative of the function $\Omega(w)$. We denote the inverse of the transformation (2.2.29) by

$$w = f(z) \tag{2.2.30}$$

where $f(z)$ is assumed to be analytic in D. By the chain rule, we find that

$$\frac{d\Omega}{dw} = \frac{d\Omega}{dz}\frac{dz}{dw} = \frac{d\Omega}{dz}\Big/\frac{dw}{dz} = \frac{d\Omega}{dz}\Big/\frac{df(z)}{dz} \tag{2.2.31}$$

Consequently, $\Omega(w)$ will be an analytic function of w in D' so long as there are no points in the w plane that correspond to points in the z plane via Eq. (2.2.30) where $df/dz = 0$.

In Chapter 5 we shall discuss in considerable detail transformations or mappings of the form of Eqs. (2.2.29)–(2.2.30). There it will be shown that if two curves intersect at a point z_0, then their angle of intersection is preserved by the mapping (i.e., the angle of intersection in the z plane equals the angle between the corresponding images of the intersecting curves in the w plane) so long as $f'(z_0) \neq 0$. Such mappings are referred to as **conformal mappings**, and as mentioned above they are important for applications.

A simple example of an ideal fluid flow problem (see Section 2.1) is one in which the complex flow potential is given by

$$\Omega(z) = z^2 \tag{2.2.32}$$

As discussed in Section 2.1, the streamlines correspond to the imaginary part of $\Omega(z) = \phi + i\psi$, hence

$$\psi = r^2 \sin 2\theta = 2xy \tag{2.2.33}$$

Clearly, the streamline $\psi = 0$ corresponds to the edges of the quarter plane, $\theta = 0$ and $\theta = \pi/2$ (see Figure 2.2.7) and the streamlines of the flow inside the quarter plane are the hyperbolae $xy = $ const.

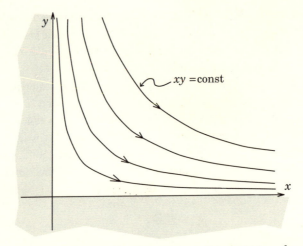

Fig. 2.2.7. Flow configuration corresponding to $\Omega(z) = z^2$

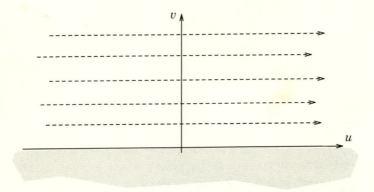

Fig. 2.2.8. Uniform flow

On the other hand, we can introduce the transformation

$$z = w^{1/2} \tag{2.2.34}$$

which converts the flow configuration $\Omega = z^2$ to the "standard" problem

$$\Omega(z(w)) = w \tag{2.2.35}$$

discussed in Section 2.1. This equation corresponds to uniform straight line flow (see Eq. (2.1.17) with $v_0 = 1$ and $\theta_0 = 0$). Equation (2.2.35) may be viewed as uniform flow over a flat plate with $w = u + iv$, with the boundary streamline $v = 0$ (see Figure 2.2.8). The speed of the flow is $|\Omega'(z)| = 2|z| = 2r$, which can also be obtained from Eq. (2.2.35) via $\dfrac{d\Omega}{dz} = \dfrac{d\Omega}{dw}\dfrac{dw}{dz}$.

The transformation (2.2.34) is an elementary example of a conformal mapping.

Problems for Section 2.2

1. Find the location of the branch points and discuss possible branch cuts for the following functions:

    (a) $\dfrac{1}{(z-1)^{1/2}}$ (b) $(z+1-2i)^{1/4}$ (c) $2\log z^2$ (d) $z^{\sqrt{2}}$

2. Determine all possible values and give the principal value of the following numbers (put in the form $x+iy$):

    (a) $i^{1/2}$ (b) $\dfrac{1}{(1+i)^{1/2}}$ (c) $\log(1+\sqrt{3}i)$

    (d) $\log i^3$ (e) $i^{\sqrt{3}}$ (f) $\sin^{-1}\frac{1}{\sqrt{2}}$

3. Solve for z:

    (a) $z^5 = 1$ (b) $3 + 2\log(z-i) = 1$ (c) $\tan^{-1} z = 1$

4. Let α be a real number. Show that the set of all values of the multivalued function $\log(z^\alpha)$ is not necessarily the same as that of $\alpha \log(z)$.

5. Derive the following formulae:

    (a) $\coth^{-1} z = \dfrac{1}{2}\log\dfrac{z+1}{z-1}$ (b) $\operatorname{sech}^{-1} z = \log\left(\dfrac{1+(1-z^2)^{1/2}}{z}\right)$

6. Deduce the following derivative formulae:

    (a) $\dfrac{d}{dz}\tan^{-1} z = \dfrac{1}{1+z^2}$ (b) $\dfrac{d}{dz}\sin^{-1} z = \dfrac{1}{(1-z^2)^{1/2}}$

    (c) $\dfrac{d}{dz}\sinh^{-1} z = \dfrac{1}{(1+z^2)^{1/2}}$

7. Consider the complex velocity potential

    $$\Omega(z) = k\log(z - z_0)$$

    where k is real and z_0 is a complex constant. Find the corresponding velocity potential and stream function. Show that the velocity is purely radial relative to the point $z = z_0$, and sketch the flow configuration. Such a flow is called a "source" if $k > 0$ and a "sink" if $k < 0$. The strength M is defined as the outward rate of flow of fluid, with unit density, across a circle enclosing $z = z_0$: $M = \oint_C V_r \, ds$, where V_r is the radial velocity and ds is the increment of arc length in the direction tangent to the circle C. Show that $M = 2\pi k$.

8. Consider the complex velocity potential $\Omega(z) = -ik \log(z - z_0)$, where k is real. Find the corresponding velocity potential and stream function. Show that the velocity is purely circumferential relative to the point $z = z_0$, being counterclockwise if $k > 0$. Sketch the flow configuration. The strength of this flow, called a **point vortex**, is defined to be $M = \oint_C V_\theta \, ds$, where V_θ is the velocity in the circumferential direction and ds is the increment of arc length in the direction tangent to the circle C. Show that $M = 2\pi k$.

9. (a) Show that the solution to Laplace's equation $\nabla^2 T = \partial^2 T / \partial u^2 + \partial^2 T / \partial v^2 = 0$ in the region $-\infty < u < \infty$, $v > 0$, with the boundary conditions $T(u, 0) = T_0$ if $u > 0$ and $T(u, 0) = -T_0$ if $u < 0$, is given by

$$T(u, v) = T_0 \left(1 - \frac{2}{\pi} \tan^{-1} \frac{v}{u} \right)$$

   (b) We shall use the result of part (a) to solve Laplace's equation inside a circle of radius $r = 1$ with the boundary conditions

$$T(r, \theta) = \begin{cases} T_0 & \text{on } r = 1, \quad 0 < \theta < \pi \\ -T_0 & \text{on } r = 1, \quad \pi < \theta < 2\pi \end{cases}$$

   Show that the transformation

$$w = i \left(\frac{1 - z}{1 + z} \right) \qquad \text{or} \qquad z = \frac{i - w}{i + w}$$

   where $w = u + iv$, maps the interior of the circle $|z| = 1$ onto the upper half of the w plane ($-\infty < u < \infty$, $v > 0$) and maps the boundary conditions $r = 1$, $0 < \theta < \pi$ onto $0 < u < \infty$, $v = 0$, and $r = 1$, $\pi < \theta < 2\pi$ onto $-\infty < u < 0$, $v = 0$. Show that consequently we must solve

$$\frac{\partial^2 T}{\partial u^2} + \frac{\partial^2 T}{\partial v^2} = 0 \qquad -\infty < u < \infty, \qquad v > 0$$
$$T = \{T_0 \quad \text{if } u > 0, -T_0 \quad \text{if } u < 0\}$$

   to find the solution obtained in part (a),

$$T(u, v) = T_0 \left(1 - \frac{2}{\pi} \tan^{-1} \frac{v}{u} \right)$$

(c) Use the result of part (b) and the mapping function to show that the solution of the boundary value problem in the circle is given by

$$T(x, y) = T_0\left(1 - \frac{2}{\pi}\cot^{-1}\left(\frac{2y}{1 - (x^2 + y^2)}\right)\right)$$

$$= T_0\left(1 - \frac{2}{\pi}\tan^{-1}\left(\frac{1 - (x^2 + y^2)}{2y}\right)\right),$$

or, in polar coordinates,

$$T(r, \theta) = T_0\left(1 - \frac{2}{\pi}\cot^{-1}\left(\frac{2r\sin\theta}{1 - r^2}\right)\right)$$

$$= T_0\left(1 - \frac{2}{\pi}\tan^{-1}\left(\frac{1 - r^2}{2r\sin\theta}\right)\right)$$

*2.3 More Complicated Multivalued Functions and Riemann Surfaces

We begin this section by discussing the branch structure associated with the function

$$w = (z - a)^{1/2}(z - b)^{1/2} \tag{2.3.1}$$

Functions such as Eq. (2.3.1) arise very frequently in applications. The function (2.3.1) is obviously the solution of the equation $w^2 = (z - a)(z - b)$, for real values a and b, $a < b$. Hence we expect square root type branch points at $z = a, b$. Indeed, $z = a, b$ are branch points as can be verified by letting z be near, say, a, $z = a + \epsilon_1 e^{i\theta_1}$. Formula (2.3.1) implies that $w \approx q^{1/2}e^{i\theta_1/2}$ with $q = \epsilon_1(a - b)$ and, as θ_1 varies between $\theta_1 = 0$ and $\theta_1 = 2\pi$, w jumps from $q^{1/2}$ to $-q^{1/2}$ (similarly near $z = b$). Perhaps surprising is the fact that $z = \infty$ is not a branch point. Letting $z = 1/t$, formula (2.3.1) yields

$$w = \frac{(1 - ta)^{1/2}(1 - tb)^{1/2}}{t} \tag{2.3.2}$$

and hence there is no jump near $t = 0$, because near $t = 0$, $w \approx 1/t$, which is single valued.

We can fix a branch cut for Eq. (2.3.1) as follows. We define the local polar coordinates

$$z - b = r_1 e^{i\theta_1}$$

$$z - a = r_2 e^{i\theta_2} \qquad 0 \le \theta_1, \theta_2 < 2\pi \tag{2.3.3}$$

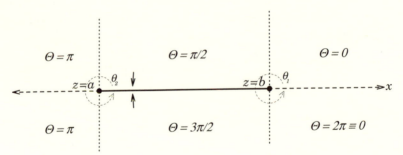

Fig. 2.3.1. A branch cut for $w = (z - a)^{1/2}(z - b)^{1/2}$

Note that the magnitudes r_1 and r_2 are fixed uniquely by the location of the point z: $r_1 = |z - b|$, $r_2 = |z - a|$. However, there is freedom in the choice of angles. In Eq. (2.3.3) we have taken $0 \le \theta_1, \theta_2 < 2\pi$, but another branch could be specified by choosing θ_1 and θ_2 differently; as we discuss below.

Then Eq. (2.3.1) yields

$$w = (r_1 r_2)^{1/2} e^{i(\theta_1 + \theta_2)/2} \tag{2.3.4}$$

In Figure 2.3.1 we denote values of the function w and the respective phases θ_1, θ_2 in those regions where a jump could be expected, that is, on the Re $z = x$ axis. (A heavy solid line denotes a branch cut.) We denote $\Theta = (\theta_1 + \theta_2)/2$.

For the above choice of angles θ_1, θ_2, the only jump of w (which depends on $(\theta_1 + \theta_2)/2$) occurs on the real axis between a and b, $a \le$ Re $z \le b$. Hence the branch cut is located on the Re $z = x$ axis between (a, b). The points $z = a, b$ are square root branch points. Increasing θ_1, θ_2 to 4π, 6π, etc., would only put us on either side of the two branches of Eq. (2.3.1). Sometimes the branch depicted in Figure 2.3.1 is referred to as the one for which $w(z)$ is real and positive for $z = x$, $x > a, b$.

Other branches can be obtained by taking different choices of the angles θ_1, θ_2. For example, if we choose θ_1, θ_2 as follows, $0 \le \theta_1 < 2\pi$, $-\pi \le \theta_2 < \pi$, we would have a branch cut in the region $(-\infty, a) \cup (b, \infty)$ whereas the function is continuous in the region (a, b). In Figure 2.3.2 we give the phase angles in the respective regions that indicate why the branch cut is in the above-mentioned location.

The branch cut in the latter case is best thought of as passing from $z = a$ to $z = b$ through the point at infinity. As mentioned earlier (see Eq. (2.3.2)), infinity is not a branch point. An alternative and useful view follows from the stereographic projection. The stereographic projection of the plane to a Riemann sphere corresponding to the branch cuts of Figures 2.3.1 and 2.3.2 is depicted in Figure 2.3.3.

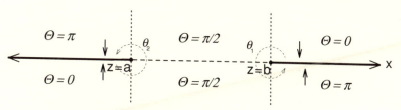

Fig. 2.3.2. Another branch cut for $w = (z - a)^{1/2}(z - b)^{1/2}$

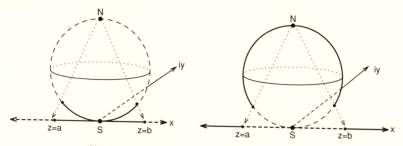

Fig. 2.3.3. Projection of w onto Riemann sphere

More complicated functions are handled in similar ways. For example, consider the function

$$w = ((z - x_1)(z - x_2)(z - x_3))^{1/2} \quad \text{with } x_k \text{ real}, x_1 < x_2 < x_3 \quad (2.3.5)$$

If we let

$$z - x_k = r_k e^{i\theta_k}, \qquad 0 \le \theta_k < 2\pi \quad (2.3.6)$$

then

$$w = \sqrt{r_1 r_2 r_3}\, e^{i(\theta_1 + \theta_2 + \theta_3)/2} \quad (2.3.7)$$

Defining $\Theta = (\theta_1 + \theta_2 + \theta_3)/2$, the phase diagram is given in Figure 2.3.4. From the choices of phase (see Figure 2.3.4) it is clear that the branch cuts lie in the region $\{x_1 < \operatorname{Re} z < x_2\} \cup \{\operatorname{Re} z > x_3\}$.

A somewhat more complicated example is given by Eq. (2.2.19)

$$w = \operatorname{ccs}^{-1} z = -i \log(z + i(1 - z^2)^{1/2})$$
$$= -i \log(z + (z^2 - 1)^{1/2}) \quad (2.3.8)$$

It is clear from the previous discussion that the points $z = \pm 1$ are square root branch points. However, $z = \infty$ is a logarithmic branch point. Letting

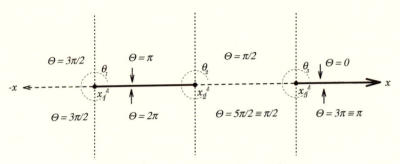

Fig. 2.3.4. Triple choice of phase angles

$z = 1/t$, we have

$$w = -i \log\left(\frac{1 + i(t^2 - 1)^{1/2}}{t}\right) = -i[\log(1 + i(t^2 - 1)^{1/2}) - \log t]$$

which demonstrates the logarithmic branch point behavior near $t = 0$. (We assume that the square root is such that the first logarithm does not have a vanishing modulus, with the other sign of the square root more work is required.) There are no other branch points because $z + i(1 - z^2)^{1/2}$ never vanishes in the finite z plane. It should also be noted that owing to the fact that $(1 - z^2)^{1/2}$ has two branches, and the logarithm has an infinite number of branches, the function $\cos^{-1}$ can be thought of as having a "double infinity" of branches.

A particular branch of this function can be obtained by first taking

$$z + 1 = r_1 e^{i\theta_1}, \qquad z - 1 = r_2 e^{i\theta_2}, \qquad 0 \le \theta_i < 2\pi, \qquad i = 1, 2$$

Then, by adding the above relations,

$$z = (r_1 e^{i\theta_1} + r_2 e^{i\theta_2})/2$$

and the function $q(z) = z + (z^2 - 1)^{1/2}$ is given by

$$q(z) = (r_1 e^{i\theta_1} + r_2 e^{i\theta_2})/2 + \sqrt{r_1 r_2} e^{i(\theta_1 + \theta_2)/2} \qquad (2.3.9)$$

whereupon

$$q(z) = \frac{r_1 e^{i\theta_1}}{2}\left(1 + \frac{r_2}{r_1} e^{i(\theta_2 - \theta_1)} + 2\sqrt{\frac{r_2}{r_1}} e^{i(\theta_2 - \theta_1)/2}\right) \qquad (2.3.10)$$

We further make the choice

$$1 + \frac{r_2}{r_1} e^{i(\theta_2 - \theta_1)} + 2\sqrt{\frac{r_2}{r_1}} e^{i(\theta_2 - \theta_1)/2} = R e^{i\Theta}, \qquad 0 \le \Theta < 2\pi$$

Θ can be chosen to be any interval of length 2π, which determines the particular branch of the logarithm. Here we made a convenient choice: $0 \leq \Theta < 2\pi$.

With these choices of phase angle it is immediately clear that the function (2.3.8), $\log q(z)$, has a branch cut for $\operatorname{Re} z > -1$. In this regard we note that $\log(R\, e^{i\Theta})$ has no jump for $\operatorname{Re} z < -1$, nor does $\log(r_1 e^{i\theta_1})$, but for $\operatorname{Re} z > -1$, $\log(r_1 e^{i\theta_1})$ does have a jump.

In what follows we give a brief description of the concept of a Riemann surface. Actually, for the applications in this book, the preceding discussion of branch cuts and branch points is sufficient. Nevertheless, the notion of a Riemann surface for a multivalued function is helpful and arises sometimes in application. By a Riemann surface we mean an extension of the ordinary complex plane to a surface that has more than one "sheet." The multivalued function will have only one value corresponding to each point on the Riemann surface. In this way the function is single valued, and standard theory applies.

For example, consider again the square root function

$$w = z^{1/2} \tag{2.3.11}$$

Rather than considering the normal complex plane for z, it is useful to consider the two-sheeted surface depicted in Figure 2.3.5. This is the Riemann surface for Eq. (2.3.11).

Referring to Figure 2.3.5 we have double copies **I** and **II** of the z plane with a cut along the positive x axis. Each copy of the z plane has identical coordinates z placed one on top of the other. Along the cut plane we have the planes joined in the following way. The cut along **Ib** is joined with the cut on **IIc**, while **Ia** is joined with the cut on **IId**. In this way, we produce a continuous one-to-one map from the Riemann surface for the function $z^{1/2}$ onto the w plane, that is, the set of values $w = u + iv = z^{1/2}$. If we follow the curve C in Figure 2.3.5, we begin on sheet **Ia**, wind around the origin (the branch point) to **Ib**; we then *go through the cut and come out on* **IIc**. We again wind around the origin to

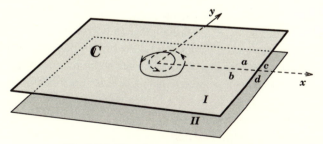

Fig. 2.3.5. Two-sheeted Riemann surface

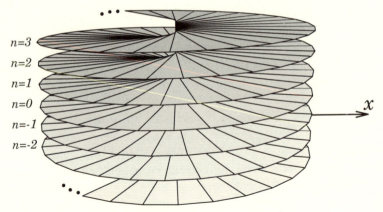

Fig. 2.3.6. Infinitely sheeted Riemann surface

IId, *go through the cut and come out on* **Ia**. The process obviously repeats after this.

In a similar manner we can construct an n-sheeted Riemann surface for the function $w = z^{m/n}$, where m and n are integers with no common factors. This would contain n identical sheets stacked one on top of the other with a cut on the positive x axis and each successive sheet is connected in the same way that **Ia** is connected to **IIc** in Figure 2.3.5 and the nth sheet would be connected to the first in the same manner as **IId** is connected to **Ia** in Figure 2.3.5.

The logarithmic function is infinitely multivalued, as discussed in Section 2.2. The corresponding Riemann surface is infinitely sheeted. For example, Figure 2.3.6 depicts an infinitely sheeted Riemann surface with the cut along the positive x axis.

Each sheet is labeled $n = 0, n = 1, n = 2, \ldots$, corresponding to the branch of the log function (2.3.12):

$$w = \log z = \log |z| + i(\theta_p + 2n\pi), \qquad 0 \le \theta_p < 2\pi \qquad (2.3.12)$$

The branch $n = 0$ is connected to $n = 1$, the branch $n = 1$ to $n = 2$, the branch $n = 2$ to $n = 3$, etc., in the same fashion that **Ib** is connected to **IIc** in Figure 2.3.7. A continuous closed circuit around the branch point $z = 0$ continuing on all the sheets $n = 0$ to $n = 1$ to $n = 2$ and so on resembles an "infinite" spiral staircase. The main point here is that because the logarithmic function is infinitely branched (we say it has a branch point of infinite order) it has an infinitely sheeted Riemann surface.

This beautiful geometric description, while useful, will be of far less importance for our purposes than the analytical understanding of how to specify

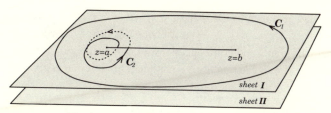

Fig. 2.3.7. Riemann surface of two sheets

particular branches and how to work with these multivalued functions in examples and concrete applications.

Finally we remark that more complicated multivalued functions can have very complicated Riemann surfaces. For example, the function given by formula (2.3.1) with local coordinates given by Eq. (2.3.3) has a two-sheeted Riemann surface depicted in Figure 2.3.7.

A closed circuit, for example, C_1 in Figure 2.3.7, enclosing both branch points $z = a$ and $z = b$, stays on the same sheet. However, a circuit enclosing either branch point, for example, the $z = a$ circuit C_2 in Figure 2.3.7, would start on sheet I; then after encircling the branch point would go through the cut onto sheet II and encircling the branch point again would end up on sheet I, and so on.

As described in Section 2.2, elementary analytic functions may yield physically interesting solutions of Laplace's equation. For example, we shall find the solution to Laplace's equation

$$\frac{\partial^2 \psi}{\partial x^2} + \frac{\partial^2 \psi}{\partial y^2} = 0 \qquad (2.3.13)$$

for $-\infty < x < \infty$, $y > 0$, with the boundary conditions

$$\psi(x, y = 0) = \begin{cases} 0 & \text{for } x < -\ell \\ 1 & \text{for } -\ell < x < \ell \\ 0 & \text{for } x > \ell \end{cases} \qquad (2.3.14)$$

which are bounded at infinity.

A typical physical application is the following: the steady state temperature distribution of a two-dimensional plate with an imposed nonzero temperature (unity) on a portion of the bottom of the plate.

Consider the function

$$\Omega(z) = A \log(z + \ell) + B \log(z - \ell) + iC, \qquad A, B, C \text{ real constants}$$
$$(2.3.15)$$

with branch cuts taken by choosing $z + l = r_1 e^{i\theta_1}$ and $z - l = r_2 e^{i\theta_2}$, where $0 \leq \theta_i < 2\pi$ for $i = 1, 2$. The function (2.3.15) is therefore analytic in the upper half plane, and consequently, we know that the imaginary part ψ of $\Omega(z) = \phi + i\psi$ satisfies Laplace's equation. This solution is given by

$$\psi(x, y) = A\theta_1 + B\theta_2 + C$$

$$= A \tan^{-1}\left(\frac{y}{x + \ell}\right) + B \tan^{-1}\left(\frac{y}{x - \ell}\right) + C \quad (2.3.16)$$

where we are taking $0 < \tan^{-1}\alpha < \pi$ (see Eq. (2.2.10)).

It remains to fix the boundary conditions on $y = 0$ given by Eq. (2.3.14). For $x > \ell$ and $y = 0$, we have $\theta_1 = \theta_2 = 0$; hence we take $C = 0$. For $-\ell < x < \ell$ and $y = 0$ we have $\theta_1 = 0$ and $\theta_2 = \pi$; hence $B = 1/\pi$. For $x < -\ell$ and $y = 0$ we have $\theta_1 = \theta_2 = \pi$; hence $A + 1/\pi = 0$. The boundary value solution is therefore given by

$$\psi(x, y) = \frac{1}{\pi}\left[\tan^{-1}\left(\frac{y}{x - \ell}\right) - \tan^{-1}\left(\frac{y}{x + \ell}\right)\right] \quad (2.3.17)$$

Problems for Section 2.3

1. Find the location of the branch points and discuss the branch cut structure of the following functions:

$$\text{(a)} \ (z^2 + 1)^{1/2} \quad \text{(b)} \ ((z + 1)(z - 2))^{1/3}$$

2. Find the location of the branch points and discuss the branch cuts associated with the following functions:

$$\text{(a)} \ \log((z - 1)(z - 2)) \quad \text{(b)} \ \coth^{-1}\frac{z}{a} = \frac{1}{2}\log\frac{z + a}{z - a}, \quad a > 0$$

Related to the second function, show that, when n is an integer

$$\text{(c)} \ \coth^{-1}\frac{z}{a} = \frac{1}{4}\log\frac{(x + a)^2 + y^2}{(x - a)^2 + y^2}$$

$$+ \frac{i}{2}\tan^{-1}\left(\left(\frac{2ay}{a^2 - x^2 - y^2}\right) + 2n\pi\right)$$

3. Given the function

$$\log(z - (z^2 + 1)^{1/2})$$

discuss the branch point/branch cut structure and where this function is analytic.

4. Consider the complex velocity potential

$$\Omega(z, z_0) = \frac{M}{2\pi}[\log(z - z_0) - \log z]$$

for $M > 0$, which corresponds to a source at $z = z_0$ and a sink at $z = 0$. (See also Exercise 6 in Section 2.1, and Exercises 7 and 8 of Section 2.2.) Find the corresponding velocity potential and stream function. Let $M = k/|z_0|$, $z_0 = |z_0|e^{i\theta_0}$, and show that

$$\Omega(z, z_0) = -\frac{k}{2\pi}\left(\frac{\log z - \log(z - z_0)}{z_0}\right)\frac{z_0}{|z_0|}$$

Take the limit as $z_0 \to 0$ to obtain

$$\Omega(z) = \lim_{z \to z_0} \Omega(z, z_0) = -\frac{ke^{i\theta_0}}{2\pi}\frac{1}{z}$$

This is called a "doublet" with strength k. The angle θ_0 specifies the direction along which the source/sink coalesces. Find the velocity potential and the stream function of the doublet, and sketch the flow.

5. Consider the complex velocity potential

$$\Omega(w) = -\frac{i\Gamma}{2\pi}\log w, \qquad \Gamma \text{ real}$$

(a) Show that the transformation $z = \frac{1}{2}(w + \frac{1}{w})$ transforms the complex velocity potential to

$$\Omega(z) = -\frac{i\Gamma}{2\pi}\log(z + (z^2 - 1)^{1/2})$$

(b) Choose a branch of $(z^2 - 1)^{1/2}$ as follows:

$$(z^2 - 1)^{1/2} = (r_1 r_2)^{1/2}e^{i(\theta_1 + \theta_2)/2}$$

where $0 \le \theta_i < 2\pi$, $i = 1, 2$, so that there is a branch cut on the x axis, $-1 < x < 1$, for $(z^2 - 1)^{1/2}$. Show that a positive circuit around a closed curve enclosing $z = -1$ and $z = +1$ increases Ω by Γ (we say the circulation increases by Γ).

(c) Establish that the velocity field $v = (v_1, v_2)$ satisfies

$$v_1 = -\frac{\Gamma}{2\pi\sqrt{1-x^2}} \quad \text{on } y = 0^+ \quad \text{for } -1 < x < 1, \quad \text{and}$$

$$v_2 = \begin{cases} \dfrac{\Gamma}{2\pi\sqrt{x^2-1}} & \text{for } x > 1, \, y = 0 \\[3mm] -\dfrac{\Gamma}{2\pi\sqrt{x^2-1}} & \text{for } x < -1, \, y = 0 \end{cases}$$

6. Consider the transformation (see also Problem 5 above) $z = \frac{1}{2}(w + \frac{1}{w})$. Show that $T(x, y) = -\text{Im}\,\Omega(z)$, where $\Omega = 1/w$ satisfies Laplace's equation and satisfies the following conditions:

$$T(x, y = 0^+) = \sqrt{1 - x^2} \quad \text{for} \quad |x| \le 1$$
$$T(x, y = 0^-) = -\sqrt{1 - x^2} \quad \text{for} \quad |x| \le 1$$
$$T(x, y = 0) = 0 \quad \text{for} \quad |x| \ge 1$$

and

$$T(x = 0, y) = \begin{cases} \dfrac{1}{y + \sqrt{y^2 - 1}} & \text{for } y > 0 \\[3mm] -\dfrac{1}{-y + \sqrt{y^2 - 1}} & \text{for } y < 0 \end{cases}$$

2.4 Complex Integration

In this section we consider the evaluation of integrals of complex variable functions along appropriate curves in the complex plane. We shall see that some of the analysis bears a similarity to that of functions of real variables. However, for analytic functions, very important new results can be derived, namely Cauchy's Theorem (sometimes called the Cauchy–Goursat Theorem). Complex integration has wide applicability, and we shall describe some of the applications in this book.

We begin by considering a complex-valued function f of a real variable t on a fixed interval, $a \le t \le b$:

$$f(t) = u(t) + iv(t) \tag{2.4.1}$$

where $u(t)$ and $v(t)$ are real valued. The function $f(t)$ is said to be **integrable** on the interval $[a, b]$ if the functions u and v are integrable. Then

$$\int_a^b f(t)\, dt = \int_a^b u(t)\, dt + i \int_a^b v(t)\, dt \qquad (2.4.2)$$

The usual rules of integration for real functions apply; in particular, from the fundamental theorems of calculus, we have for continuous functions $f(t)$

$$\frac{d}{dt} \int_a^t f(\tau)\, d\tau = f(t) \qquad (2.4.3a)$$

and for $f'(t)$ continuous

$$\int_a^b f'(t)\, dt = f(b) - f(a) \qquad (2.4.3b)$$

Next we extend the notion of complex integration to integration on a curve in the complex plane. A curve in the complex plane can be described via the parameterization

$$z(t) = x(t) + iy(t), \qquad a \le t \le b \qquad (2.4.4)$$

For each given t in $[a, b]$ there is a set of points $(x(t), y(t))$ that are the image points of the interval. The image points $z(t)$ are ordered according to increasing t. The curve is said to be continuous if $x(t)$ and $y(t)$ are continuous functions of t. Similarly, it is said to be differentiable if $x(t)$ and $y(t)$ are differentiable.

A curve or arc C is **simple** (sometimes called a **Jordan arc**) if it does not intersect itself, that is, $z(t_1) \ne z(t_2)$ if $t_1 \ne t_2$ for $t \in [a, b]$, except that $z(b) = z(a)$ is allowed; in the latter case we say that C is a **simple closed curve** (or **Jordan curve**). Examples are seen in Figure 2.4.1.

(a) Simple, not closed
(Jordan Arc)

(b) Not simple, not closed

(c) Simple, closed
(Jordan Curve)

Fig. 2.4.1. Examples of curves

Next we shall discuss evaluation of integrals along curves. When the curve is closed, our convention shall be to take the positive direction to be the one in which the interior remains to the left of C. Integrals along a closed curve will be taken along the positive direction unless otherwise specified. The function $f(z)$ is said to be **continuous** on C if $f(z(t))$ is continuous for $a \leq t \leq b$, and f is said to be **piecewise continuous** on $[a, b]$ if $[a, b]$ can be broken up into a finite number of subintervals in which $f(z)$ is continuous. A **smooth arc** C is one for which $z'(t)$ is continuous. A **contour** is an arc consisting of a finite number of connected smooth arcs; that is, a contour is a piecewise smooth arc. Hereafter we shall only consider integrals along such contours unless otherwise specified. Frequently, a simple closed contour is referred to as a **Jordan contour**.

The contour integral of a piecewise continuous function is defined to be

$$\int_C f(z)\,dz = \int_a^b f(z(t))z'(t)\,dt \qquad (2.4.5)$$

where the right-hand side of Eq. (2.4.5) is obtained via the formal substitution $dz = z'(t)\,dt$. In general, Eq. (2.4.5) depends on $f(z)$ and the contour C. Thus the integral (2.4.5) is really a line integral in the (x, y) plane and is naturally related to the study of vector calculus in the plane. As mentioned earlier, the complex variable $z = x + iy$ can be thought of as a two-dimensional vector.

We remark that values of the above integrals are invariant if we redefine the parameter t appropriately. Namely, if we make the change of variables $t \to s$ by $t = T(s)$ where $T(s)$ maps the interval $A \leq s \leq B$ to interval $a \leq t \leq b$, $T(s)$ is continuously differentiable, and $T'(s) > 0$ (needed to ensure that t increases with s), then only the form the integrals take on is modified, but its value is invariant. The importance of this remark is that one can evaluate integrals by the most convenient choice of parameterization. Examples discussed later in this section will serve to illustrate this point.

The usual properties of integration apply. We have

$$\int_C [\alpha f(z) + \beta g(z)]\,dz = \alpha \int_C f(z)\,dz + \beta \int_C g(z)\,dz \qquad (2.4.6)$$

for constants α and β and piecewise continuous functions f and g. The arc C traversed the opposite direction, that is, from $t = b$ to $t = a$, is denoted by $-C$. We then have

$$\int_{-C} f(z)\,dz = -\int_C f(z)\,dz \qquad (2.4.7)$$

because the left-hand side of Eq. (2.4.7) is equivalent to $\int_b^a f(z(t))z'(t)dt$. Similarly, if C consists of n connected contours with endpoints from z_1 to z_2 for C_1, from z_2 to z_3 for C_2, ..., from z_n to z_{n+1} for C_n, then we have

$$\int_C f = \sum_{j=1}^n \int_{C_j} f$$

The fundamental theorem of calculus yields the following result.

Theorem 2.4.1 Suppose $F(z)$ is an analytic function and that $f(z) = F'(z)$ is continuous in a domain D. Then for a contour C lying in D with endpoints z_1 and z_2

$$\int_C f(z)\,dz = F(z_2) - F(z_1) \tag{2.4.8}$$

Proof Using the definition of the contour integral (2.4.5) and the chain rule, we have

$$\int_C f(z)\,dz = \int_C F'(z)\,dz = \int_a^b F'(z(t))z'(t)\,dt$$

$$= \int_a^b \frac{d}{dt}[F(z(t))]\,dt$$

$$= F(z(b)) - F(z(a))$$

$$= F(z_2) - F(z_1) \qquad \blacksquare$$

As a consequence of Theorem 2.4.1, for closed curves we have

$$\oint_C f(z)\,dz = \oint_C F'(z)\,dz = 0 \tag{2.4.9}$$

where $\oint_C$ denotes a closed contour C (that is, the endpoints are equal).

If the function $f(z)$ satisfies the hypothesis of Theorem 2.4.1, then for *all* contours C lying in D beginning at z_1 and ending at z_2 we have Eq. (2.4.8). Hence the result demonstrates that the integral is independent of path. Indeed, Figure 2.4.2 illustrates this fact.

Referring to Figure 2.4.2, we have $\int_{C_1} f\,dz = \int_{C_2} f\,dz$ because

$$\oint_C f\,dz = \int_{C_1} f\,dz - \int_{C_2} f\,dz = 0 \tag{2.4.10}$$

where the closed curve $C = C_1 - C_2$.

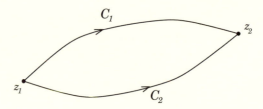

Fig. 2.4.2. Independent paths forming closed curve

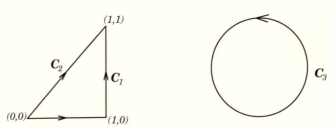

Fig. 2.4.3. Contours C_1, C_2, and C_3

The hypothesis in Theorem 2.4.1 requires the existence of $F(z)$ such that $f(z) = F'(z)$. Later in this chapter we shall show this for a large class of functions $f(z)$.

Sometimes it is convenient to evaluate the complex integral by reducing it to two real-line integrals in the x, y plane. In the definition (2.4.5) we use $f(z) = u(x, y) + iv(x, y)$ and $dz = dx + idy$ to obtain

$$u = x \quad v = y$$

$$\int_C f(z)\, dz = \int_C [(u\, dx - v\, dy) + i(v\, dx + u\, dy)] \qquad (2.4.11)$$

$$\int x\,dx - y\,dy + i(y\,dx + x\,dy)$$

This can be shown, via parameterization, to be equivalent to

$$\frac{x^2}{2} - \frac{y^2}{2} + i(yx + xy) = \frac{1}{2} - \frac{1}{2} + i(1 + 1)$$

$$\int_a^b f(z(t))z'(t)\, dt \qquad = 2i$$

Later in this chapter we shall use Eq. (2.4.11) in order to derive one form of Cauchy's Theorem.

In the following examples we illustrate how line integrals may be calculated in prototypical cases.

Example 2.4.1 Evaluate $\int_C \bar{z}\, dz$ for (a) $C = C_1$, a contour from $z = 0$ to $z = 1$ to $z = 1 + i$; (b) $C = C_2$, the line from $z = 0$ to $z = 1 + i$; and (c) $C = C_3$, the unit circle (see Figure. 2.4.3).

(a)
$$\int_{C_1} \bar{z}\,dz = \int_{C_1}(x - iy)(dx + i\,dy)$$

$$= \int_{x=0}^{1} x\,dx + \int_{y=0}^{1}(1 - iy)(i\,dy)$$

$$= \frac{1}{2} + i[y - iy^2/2]_0^1$$

$$= 1 + i$$

Note in the integral from $z = 0$ to $z = 1$, $y = 0$, hence $dy = 0$. In the integral from $z = 1$ to $z = 1 + i$, $x = 1$, hence $dx = 0$.

(b)
$$\int_{C_2} \bar{z}\,dz = \int_{x=0}^{1}(x - ix)(dx + i\,dx)$$

$$= (1 - i)(1 + i)\int_0^1 x\,dx$$

$$= 1$$

Note that C_2 is the line $y = x$, hence $dy = dx$. Since $\bar{z}$ is *not* analytic we see that $\int_{C_2} \bar{z}\,dz$ and $\int_{C_1} \bar{z}\,dz$ need not be equal.

(c)
$$\int_{C_3} \bar{z}\,dz = \int_{\theta=0}^{2\pi} e^{-i\theta} i e^{i\theta}\,d\theta = 2\pi i$$

Note that $z = e^{i\theta}$, $\bar{z} = e^{-i\theta}$, and $dz = i e^{i\theta}\,d\theta$, on the unit circle, $r = 1$.

Example 2.4.2 Evaluate $\int_C z\,dz$ along the three contours described above and as illustrated in Figure 2.4.3. Because z is analytic in the region containing z, and $z = (d/dz)(z^2/2)$, we immediately have, from Theorem 2.4.1

$$\int_{C_1} z\,dz = \int_{C_2} z\,dz = \frac{(1 + i)^2}{2} = i$$

$$\int_{C_3} z\,dz = 0$$

These results can be calculated directly via the line integral methods described above – which we will leave for the reader to verify.

Example 2.4.3 Evaluate $\int_C (1/z)\,dz$ for (a) any closed contour C not enclosing the origin, and; (b) any closed contour C enclosing the origin.

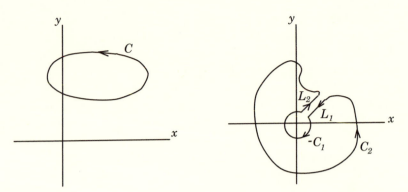

(a) *C* not enclosing origin   (b) "deforming" C_2, which encloses origin

Fig. 2.4.4. Integration contours in Example 2.4.3

(a) Because $1/z$ is analytic for all $z \neq 0$, we immediately have, from Theorem 2.4.1 and from $1/z = (d/dz)(\log z)$

$$\int_C \frac{1}{z}\, dz = 0$$

because $[\log z]_C = 0$ so long as C does not enclose the branch point of $\log z$ at $z = 0$ (see Figure 2.4.4a)

(b) Any closed contour, call it C_2, around the origin can be *deformed* into a small, but finite circle of radius r as follows. Introduce a "crosscut" (L_1, L_2) as in Figure 2.4.4b. Then in the limit of the crosscut width tending to zero we have a closed contour: $C = C_2 + L_1 + L_2 - C_1$. (Note that for C_1 we take the positive counterclockwise orientation.) In Figure 2.4.4b we have taken care to distinguish the positive and negative directions of C_1 and C_2, respectively. From part (a) of this problem

$$\int_C \frac{1}{z}\, dz = 0$$

then, because $\int_{L_1} + \int_{L_2} = 0$, we have (using $z = re^{i\theta}$ and $dz = rie^{i\theta}\,d\theta$)

$$\int_{C_2} \frac{1}{z}\, dz = \int_{C_1} \frac{1}{z}\, dz = \int_0^{2\pi} r^{-1} e^{-i\theta} i e^{i\theta} r\, d\theta = 2\pi i$$

Thus the integral of $1/z$ around any closed curve enclosing the origin is $2\pi i$. We also note that if we formally use the antiderivative of $1/z$ (namely,

$1/z = d/dz(\log z))$, we can also find $\int_{C_2} (1/z)\, dz = 2\pi i$. In this case, even though we enclose the branch point of $\log z$, the argument θ_p of $\log z = \log r + i(\theta_p + 2n\pi)$ increases by 2π as we enclose the origin. In this case, we need only select a convenient branch of $\log z$.

Example 2.4.4 Evaluate $\int_C z^n\, dz$ for integer n and some closed contour C that encloses the origin.

Using the crosscut segment as indicated in Figure 2.4.4, the integral in question is equal to that on C_1, a small, but finite circle of radius r. Thus

$$\int_{C_1} z^n\, dz = \int_0^{2\pi} r^{n+1} e^{in\theta} i e^{i\theta}\, d\theta$$

$$= i \int_0^{2\pi} r^{n+1} e^{i(n+1)\theta}\, d\theta$$

$$= \left\{ \begin{matrix} 0 & n \neq -1 \\ 2\pi i & n = -1 \end{matrix} \right\}$$

Hence even though z^n is nonanalytic at $z = 0$ for $n < 0$, only the value $n = -1$ gives a nontrivial contribution. We remark that use of the antiderivative

$$z^n = \frac{d}{dz}\left(\frac{z^{n+1}}{n+1} \right), \quad n \neq -1$$

yields the same results.

As mentioned earlier, complex line integrals arise in many physical applications. For example, in ideal fluid flow problems (in Section 2.1 we briefly discussed ideal fluid flows), the real-line integrals

$$\Gamma = \int_C (\phi_x\, dx + \phi_y\, dy) = \int_C \mathbf{v} \cdot \frac{d\hat{\mathbf{l}}}{ds}\, ds \qquad (2.4.12)$$

$$\mathcal{F} = \int_C (\phi_x\, dy - \phi_y\, dx) = \int_C \mathbf{v} \cdot \frac{d\hat{\mathbf{n}}}{ds}\, ds \qquad (2.4.13)$$

where $\mathbf{v} = (\phi_x, \phi_y)$ is the velocity vector, $\mathbf{v} \cdot \frac{d\hat{\mathbf{l}}}{ds} = (\frac{dx}{ds}, \frac{dy}{ds})$ is the unit tangent vector to C, and $\mathbf{v} \cdot \frac{d\hat{\mathbf{n}}}{ds} = (\frac{dy}{ds}, -\frac{dx}{ds})$ is the unit normal vector to C, represent (Γ) the circulation around the curve C (when C is closed), and ($\mathcal{F}$) the flux across the curve C. We note that in terms of analytic complex functions we have the simple equation

$$\Gamma + i\mathcal{F} = \int_C (\phi_x - i\phi_y)(dx + i\, dy) = \int_C \Omega'(z)\, dz \qquad (2.4.14)$$

Recall from (2.1.16) that the complex velocity is given by $\Omega'(z) = \phi_x + i\psi_x = \phi_x - i\phi_y$ (the latter follows from the Cauchy–Riemann conditions). Using complex function theory to evaluate Eq. (2.4.14) often provides an easy way to calculate the real-line integrals (2.4.12–2.4.13), which are the real and imaginary parts of the integral in Eq. (2.4.14). An example is discussed in the problem section.

Next we derive an important inequality that we shall use frequently.

Theorem 2.4.2 Let $f(z)$ be continuous on a contour C. Then

$$\left| \int_C f(z)\, dz \right| \leq ML \qquad (2.4.15)$$

where L is the length of C and M is an upper bound for $|f|$ on C.

Proof

$$I = \left| \int_C f(z)\, dz \right| = \left| \int_a^b f(z(t)) z'(t)\, dt \right| \qquad (2.4.16)$$

From real variables we know that, for $a \leq t \leq b$,

$$\left| \int_a^b G(t)\, dt \right| \leq \int_a^b |G(t)|\, dt$$

hence

$$I \leq \int_a^b |f(z(t))|\, |z'(t)|\, dt$$

Then if $|f|$ is bounded on C, that is, if $|f(t)| \leq M$ on C, where M is a constant, then

$$I \leq M \int_a^b |z'(t)|\, dt$$

However, because

$$|z'(t)|\, dt = |x'(t) + iy'(t)|\, dt$$
$$= \sqrt{(x'(t))^2 + (y'(t))^2}\, dt = ds \qquad (2.4.17)$$

where s represents arc length along C, we have Eq. (2.4.15). ∎

We also remark that the preceding developments of contour integration could also have been derived using limits of appropriate sums. This would be in

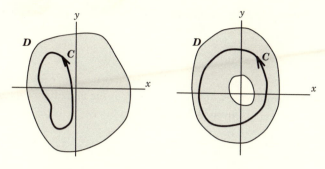

(a) *D* simply connected   (b) *D* multiply connected

Fig. 2.4.5. Connected regions of domain *D*

analogy to the one-dimensional evaluation of integrals by Riemann sums. More specifically, given a contour C in the z plane beginning at z_a and terminating at z_b, choose any ordered sequence $\{z_j\}$ of $n+1$ points on C such that $z_0 = z_a$ and $z_n = z_b$. Define $\Delta z_j = z_{j+1} - z_j$ and form the sum

$$S_n = \sum_{j=1}^{n} f(\xi_j) \Delta z_j \tag{2.4.18}$$

where ξ_j is any point on C between z_{j-1} and z_j. If $f(x)$ is piecewise continuous on C, then the limit of S_n as $n \to \infty$ and $|\Delta z_j| \to 0$ converges to the integral of $f(z)$, namely

$$\int_C f(z)\,dz = \lim_{\substack{n \to \infty \\ |\Delta z_j| \to 0}} \sum_{j=1}^{n} f(\xi_j) \Delta z_j \tag{2.4.19}$$

Finally, we define a **simply connected** domain D to be one in which every simple closed contour within it encloses only points of D. The points within a circle, square, and polygon are examples of a simply connected domain. An annulus (doughnut) is not simply connected. A domain that is not simply connected is called multiply connected. An annulus is multiply connected, because a contour encircling the inner hole encloses points within and outside D (see Figure 2.4.5).

Problems for Section 2.4

1. From the basic definition of complex integration, evaluate the integral $\oint_C f(z)\,dz$, where C is the parametrized unit circle enclosing the origin,

$C : x(t) = \cos t, \ y(t) = \sin t$ or $z = e^{it}$, and where $f(z)$ is given by

$$\text{(a) } z^2 \qquad \text{(b) } \bar{z}^2 \qquad \text{(c) } \frac{z+1}{z^2}$$

2. Evaluate the integral $\oint_C f(z) \, dz$, where C is the unit circle enclosing the origin, and $f(z)$ is given as follows:

$$\text{(a) } 1 + 2z + z^2 \quad \text{(b) } 1/(z - 1/2)^2 \quad \text{(c) } 1/\bar{z} \quad \text{(d) } z\bar{z} \quad \text{(e) } e^{\bar{z}}$$

3. Let C be the unit square with diagonal corners at $-1 - i$ and $1 + i$. Evaluate $\oint_C f(z) \, dz$, where $f(z)$ is given by the following:

$$\text{(a) } \sin z \qquad \text{(b) } \frac{1}{2z+1} \qquad \text{(c) } \bar{z} \qquad \text{(d) } \operatorname{Re} z$$

4. Use the principal branch of $\log z$ and $z^{1/2}$ to evaluate

$$\text{(a) } \int_{-1}^{1} \log z \, dz \qquad \text{(b) } \int_{-1}^{1} z^{1/2} \, dz$$

5. Show that the integral $\int_C (1/z^2) \, dz$, where C is a path beginning at $z = -a$ and ending at $z = b, a, b > 0$, is independent of path so long as C doesn't go through the origin. Explain why the real-valued integral $\int_{-a}^{b} (1/x^2) \, dx$ doesn't exist, but the value obtained by formal substitution of limits agrees with the complex integral above.

6. Consider the integral $\int_0^b (1/z^{1/2}) dz, \ b > 0$. Let $z^{1/2}$ have a branch cut along the positive real axis. Show that the value of the integral obtained by integrating along the top half of the cut is exactly minus that obtained by integrating along the bottom half of the cut. What is the difference between taking the principal versus the second branch of $z^{1/2}$?

7. Let C be an open (upper) semicircle of radius R with its center at the origin, and consider $\int_C f(z) \, dz$. Let $f(z) = 1/(z^2 + a^2)$ for real $a > 0$. Show that $|f(z)| \leq 1/(R^2 - a^2), \ R > a$, and

$$\left| \int_C f(z) \, dz \right| \leq \frac{\pi R}{R^2 - a^2}, \qquad R > a$$

8. Let C be an arc of a circle of radius R $(R > 1)$ of angle $\pi/3$. Show that

$$\left| \int_C \frac{dz}{z^3 + 1} \right| \leq \frac{\pi}{3} \left(\frac{R}{R^3 - 1} \right)$$

and deduce $\lim_{R \to \infty} \int_C \frac{dz}{z^3 + 1} = 0$

9. Consider $I_R = \int_{C_R} \frac{e^{iz}}{z^2} dz$, where C_R is the semicircle with radius R in the upper half plane with endpoints $(-R, 0)$ and $(R, 0)$ (C_R is open, it does not include the x axis). Show that $\lim_{R \to \infty} I_R = 0$.

10. Consider

$$I_\epsilon = \oint_{C_\epsilon} z^\alpha f(z)\, dz, \qquad \alpha > -1, \quad \alpha \text{ real}$$

    where C_ϵ is a circle of radius ϵ centered at the origin and $f(z)$ is analytic inside the circle. Show that $\lim_{\epsilon \to 0} I_\epsilon = 0$.

11. (a) Suppose we are given the complex flow field $\Omega(z) = -ik \log(z - z_0)$, where k is a real constant and z_0 a complex constant. Show that the circulation around a closed curve C_0 encircling $z = z_0$ is given by $\Gamma = 2\pi k$. (Hint: from Section 2.4, $\Gamma + i\mathcal{F} = \oint_{C_0} \Omega'(z)dz$.)

    (b) Suppose $\Omega(z) = k \log(z - z_0)$. Find the circulation around C_0 and the flux through C_0.

2.5 Cauchy's Theorem

In this section we study Cauchy's Theorem, which is one of the most important theorems in complex analysis. In order to prove Cauchy's Theorem in the most convenient manner, we will use a well-known result from vector analysis in real variables, known as Green's Theorem in the plane, which can be found in advanced calculus texts; see, for example, Buck (1956).

Theorem 2.5.1 (Green) Let the real functions $u(x, y)$ and $v(x, y)$ along with their partial derivatives $\partial u/\partial x$, $\partial u/\partial y$, $\partial v/\partial x$, $\partial v/\partial y$, be continuous throughout a simply connected region $\mathcal{R}$ consisting of points interior to and on a simple

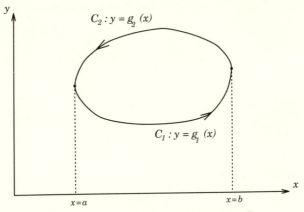

Fig. 2.5.1. Deriving Eq. (2.5.1) for region $\mathcal{R}$

closed contour C in the x-y plane. Let C be described in the positive (counter-clockwise) direction, then

$$\oint_C (u\,dx + v\,dy) = \iint_R \left(\frac{\partial v}{\partial x} - \frac{\partial u}{\partial y} \right) dx\,dy \qquad (2.5.1)$$

We remark for those readers who may not recall or have not seen this formula, Eq. (2.5.1) is a two-dimensional version of the divergence theorem of vector calculus (taking the divergence of a vector $\vec{v} = (v, -u)$).

An elementary derivation of Eq. (2.5.1) can be given if we restrict the region $\mathcal{R}$ to be such that every vertical and horizontal line intersects the boundary of $\mathcal{R}$ in at most two points. Then if we call the "top" and "bottom" curves defining C, $y = g_2(x)$ and $y = g_1(x)$, respectively (see Figure 2.5.1)

$$-\iint_{\mathcal{R}} \frac{\partial u}{\partial y}\,dx\,dy = -\int_a^b \int_{g_1(x)}^{g_2(x)} \frac{\partial u}{\partial y}\,dy\,dx$$

$$= -\int_a^b [u(x, g_2(x)) - u(x, g_1(x))]\,dx$$

$$= +\int_{C_2} u(x, y)\,dx + \int_{C_1} u(x, y)\,dx$$

$$= \oint_C u(x, y)\,dx$$

Following the same line of thought, we also find

$$\iint_R \frac{\partial v}{\partial x}\, dx\, dy = \oint_C v\, dy$$

From these relationships we obtain Eq. (2.5.1).

With Green's Theorem we can give a simple proof of Cauchy's Theorem as long as we make a certain extra assumption to be explained shortly.

Theorem 2.5.2 (Cauchy) If a function f is analytic in a simply connected domain D, then along a simple closed contour C in D

$$\oint_C f(z)\, dz = 0 \qquad (2.5.2)$$

We remark that in the proof given here, we shall also *require* that $f'(z)$ be continuous in D. In fact, a more general proof owing to Goursat enables one to establish Eq. (2.5.2) without this assumption. We discuss Goursat's proof in the optional Section 2.7 of this chapter, which shows that even when $f(z)$ is only assumed analytic, we still have Eq. (2.5.2). From Eq. (2.5.2) one could then derive as a consequence that $f'(z)$ is indeed continuous in D (note so far in our development, analytic only means that $f'(z)$ exists, not that it is necessarily continuous). In a subsequent theorem (Theorem 2.6.5: Morera's Theorem) we show that if $f(z)$ is continuous and Eq. (2.5.2) is satisfied, then in fact $f(z)$ is analytic.

Proof (Theorem 2.5.2) From the definition of $\oint_C f(z)\, dz$, using $f(z) = u + iv$, $dz = dx + i\, dy$, we have

$$\oint_C f(z)\, dz = \oint_C (u\, dx - v\, dy) + i \oint_C (u\, dy + v\, dx) \qquad (2.5.3)$$

Then, using $f'(z)$ continuous, we find that u and v have continuous partial derivatives, hence Theorem 2.5.1 holds, and each of the above line integrals can be converted to the following double integrals for points of D enclosed by C:

$$\oint_C f(z)\, dz = -\iint_D \left(\frac{\partial v}{\partial x} + \frac{\partial u}{\partial y} \right) dx\, dy + i \iint_D \left(\frac{\partial u}{\partial x} - \frac{\partial v}{\partial y} \right) dx\, dy$$

$$(2.5.4)$$

Because $f(z)$ is analytic we find that the Cauchy–Riemann conditions (Eqs. (2.1.4)) hold:

$$\frac{\partial u}{\partial y} = -\frac{\partial v}{\partial x} \quad \text{and} \quad \frac{\partial u}{\partial x} = \frac{\partial v}{\partial y}$$

hence we have $\oint_C f(z)\,dz = 0$.   ■

Knowing that $\oint_C f(z)\,dz = 0$ yields numerous results of interest. In particular, we will see that this condition and continuous $f(z)$ yield an analytic antiderivative for f.

Theorem 2.5.3 If $f(z)$ is continuous in a simply connected domain D and if $\oint_C f(z)\,dz = 0$ for every simple closed contour C lying in D, then there exists a function $F(z)$, analytic in D, such that $F'(z) = f(z)$.

Proof Consider three points within D: z_0, z, and $z + h$. Define F by

$$F(z) = \int_{z_0}^{z} f(z')\,dz' \tag{2.5.5}$$

where the contour from z_0 to z lies within D (see Figure 2.5.2). Then from $\oint_C f(z)\,dz = 0$ we have

$$\int_{z_0}^{z+h} f(z')\,dz' + \int_{z+h}^{z} f(z')\,dz' + \int_{z}^{z_0} f(z')\,dz' = 0 \tag{2.5.6}$$

where again all paths must lie within D. Although it may seem that choosing a contour in this way is special, shortly we will show that when $f(z)$ is analytic in D, the integral over $f(z)$ enclosing the domain D is equivalent to any closed integral along a simple contour inside D.

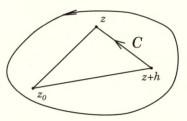

Fig. 2.5.2. Three points lying in D

Fig. 2.5.3. Non-simple contour

Then, using Eq. (2.5.6) and reversing the order of integration of the last two terms,

$$F(z+h) - F(z) = \left(\int_{z_0}^{z+h} - \int_{z_0}^{z} \right) f(z')\, dz' = \int_{z}^{z+h} f(z')\, dz'$$

hence

$$\frac{F(z+h) - F(z)}{h} = \frac{\int_{z}^{z+h} f(z')\, dz'}{h} \tag{2.5.7}$$

Because $f(z)$ is continuous, we find, from the definition of the derivative and the properties of real integration, that as $h \to 0$

$$F'(z) = f(z) \tag{2.5.8}$$

∎

We remark that any (nonsimple) contour that has self-intersections can be decomposed into a sequence of contours that are simple. This fact is illustrated in Figure 2.5.3, where the complete nonsimple contour ("figure eight" contour) can be decomposed into two simple closed contours corresponding to each "loop" of the nonsimple contour. A consequence of this observation is that Cauchy's Theorem can be applied to a nonsimple contour with a finite number of intersections.

In a multiply connected domain with a function $f(z)$ analytic in this domain, we can also apply Cauchy's Theorem. The best way to see this is to introduce crosscuts, as mentioned earlier, such that Cauchy's Theorem can be applied to a simple contour. Consider the multiply connected region depicted in Figure 2.5.4(b) with outer boundary C_0 and n holes with boundaries $C_1, C_2, \ldots, C_n$, and introduce n crosscuts $L_1^1 L_2^1, L_1^2 L_2^2, \ldots, L_1^n L_2^n$, as in Figure 2.5.4(a).

Then Cauchy's Theorem applies to an analytic function in a domain D with the simple contour

$$\tilde{C} = C_0 - \sum_{j=1}^{n} C_j + \sum_{j=1}^{n} \left(L_1^j - L_2^j \right)$$

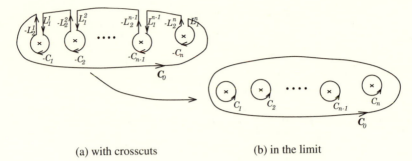

(a) with crosscuts   (b) in the limit

Fig. 2.5.4. Multiply connected domain

where we have used the convention that each closed contour is taken in the positive counterclockwise direction, and we take L_1^j, L_2^j in the same direction.

Because the integrals along the crosscuts vanish as the width between the crosscuts vanishes (i.e., $\int_{L_1^j - L_2^j} f(z)\,dz \to 0$), we have

$$\oint_C f(z)\,dz = 0$$

where $C = C_0 - \sum_{j=1}^n C_j = C_0 + \sum_{j=1}^n (-C_j)$. It is often best to interpret the integral

$$\oint_C = \oint_{C_0} + \sum_{j=1}^n \oint_{-C_j}$$

as one contour with the enclosed region bounded by C as that lying to the *left* of C_0 and to the *right* of C_j (or to the *left* of $-C_j$). From $\oint_C f(z)\,dz = 0$ we have

$$\oint_{C_0} f(z)\,dz = \sum_{j=1}^n \oint_{C_j} f(z)\,dz \qquad (2.5.9)$$

with all the contours taken in the counterclockwise direction as depicted in Figure 2.5.4(b). We often say that the contour C_0 has been **deformed** into the contours C_j, $j = 1, \ldots, n$. A simple case is depicted in Figure 2.5.5.

This is an example of a **deformation** of the contour, deforming C_0 into C_1. By introducing crosscuts it is seen that

$$\oint_{C_0} f(z)\,dz = \oint_{C_1} f(z)\,dz \qquad (2.5.10)$$

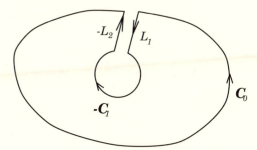

Fig. 2.5.5. Nonintersecting closed curves C_0 and C_1

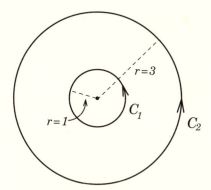

Fig. 2.5.6. Annulus

where C_0 and C_1 are two nonintersecting closed curves in which $f(z)$ is analytic on and in the region between C_0 and C_1. With respect to Eq. (2.5.10) we say that C_0 can be **deformed** into C_1, and for the purpose of this integration they are equivalent contours.

The process of introducing crosscuts, and deformation of the contour, effectively allows us to deal with multiply connected regions and closed contours that are not simple. That is, one can think of integrals along such contours as a sum of integrals along simple contours, as long as $f(z)$ is analytic in the relevant region.

Example 2.5.1 Evaluate

$$\mathcal{I} = \oint_C \frac{e^z}{z(z^2 - 16)}\, dz$$

where C is the boundary of the annulus between the circles of radius 1 and radius 3 (see Figure 2.5.6).

We note that $C = C_2 + (-C_1)$, and in the region between C_1 and C_2 the function $f(z) = e^z/z(z^2 - 16)$ is analytic because its derivative $f'(z)$ exists and is continuous. The only nonanalytic points are at $z = 0$, $z = \pm 4$; hence, $\mathcal{I} = 0$.

Example 2.5.2 Evaluate

$$\mathcal{I} = \frac{1}{2\pi i} \oint_C \frac{dz}{(z-a)^m}, \qquad m = 1, 2, \ldots, M$$

where C is a simple closed contour.

The function $f(z) = 1/(z-a)^m$ is analytic for all $z \neq a$. Hence if C does not enclose $z = a$, then we have $\mathcal{I} = 0$. If C incloses $z = a$, we use Cauchy's Theorem to *deform* the contour to C_a, a small, but finite circle of radius r centered at $z = a$ (see Figure 2.5.7). Namely

$$\int_C f(z)\, dz - \int_{C_a} f(z)\, dz = 0, \qquad f(z) = 1/(z-a)^m$$

We evaluate $\int_{C_a} f(z)\, dz$ by letting

$$z - a = re^{i\theta}, \qquad dz = ie^{i\theta} r\, d\theta$$

in which case

$$\mathcal{I} = \frac{1}{2\pi i} \oint_{C_a} \frac{1}{z-a}\, dz = \frac{1}{2\pi i} \int_0^{2\pi} \frac{1}{r^m e^{im\theta}} ie^{i\theta} r\, d\theta$$

$$= \frac{1}{2\pi i} \int_0^{2\pi} ie^{-i(m-1)\theta} r^{-m+1}\, d\theta = \delta_{m,1} = \begin{cases} 1 & \text{if } m = 1 \\ 0 & \text{otherwise} \end{cases}$$

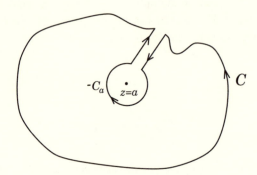

Fig. 2.5.7. Deformed contour around $z = a$

Thus

$$\mathcal{I} = \begin{cases} 0 & z = a \text{ outside } C, \\ 0 & z = a \text{ inside } C, \quad m \neq 1 \\ 1 & z = a \text{ inside } C, \quad m = 1 \end{cases}$$

By considering contour integrals over functions $f(z)$ that enclose many points in which $f(z)$ have the local behavior

$$\frac{g_j(z)}{(z - a_j)^m}, \qquad j = 1, 2, \ldots, N, \quad m = 1, 2, \ldots, M$$

where $g_j(z)$ is analytic, numerous important results can be obtained. In Chapter 3 we discuss functions with this type of local behavior (we say $f(z)$ has a pole of order m at $z = a_j$). In Chapter 4 we discuss extensions of the crosscut concept and the methods described in Example 2.5.2 will be used to derive the well-known Cauchy Residue Theorem (Theorem 4.1.1). The following example is an application of these kinds of ideas.

Example 2.5.3 Let $P(z)$ be a polynomial of degree n, with n simple roots, none of which lie on a simple closed contour C. Evaluate

$$\mathcal{I} = \frac{1}{2\pi i} \oint_C \frac{P'(z)}{P(z)} \, dz$$

Because $P(z)$ is a polynomial with distinct roots, we can factor it as

$$P(z) = A(z - a_1)(z - a_2) \cdots (z - a_n)$$

where A is the coefficient of the term of highest degree. Because

$$\frac{P'(z)}{P(z)} = \frac{d}{dz} (\log P(z))$$

$$= \frac{d}{dz} \log (A(z - a_1)(z - a_2) \cdots (z - a_n))$$

it follows that

$$\frac{P'(z)}{P(z)} = \frac{1}{z - a_1} + \frac{1}{z - a_2} + \cdots \frac{1}{z - a_n}.$$

Hence, using the result from Example 2.5.2 above, we have

$$\mathcal{I} = \frac{1}{2\pi i} \oint_C \frac{P'(z)}{P(z)} \, dz = \text{number of roots lying within } C$$

Problems for Section 2.5

1. Evaluate $\oint_C f(z)\,dz$, where C is the unit circle centered at the origin, and $f(z)$ is given by the following:

$$\text{(a) } e^{iz} \qquad \text{(b) } e^{z^2} \qquad \text{(c) } \frac{1}{z - 1/2} \qquad \text{(d) } \frac{1}{z^2 - 4}$$

$$\text{(e) } \frac{1}{2z^2 + 1} \qquad \text{(f) } \sqrt{z - 4}$$

2. Use partial fractions to evaluate the following integrals $\oint_C f(z)\,dz$, where C is the unit circle centered at the origin, and $f(z)$ is given by the follwing:

$$\text{(a) } \frac{1}{z(z - 2)} \qquad \text{(b) } \frac{z}{z^2 - 1/9} \qquad \text{(c) } \frac{1}{z\left(z + \frac{1}{2}\right)(z - 2)}$$

3. Evaluate the following integral

$$\oint_C \frac{e^{iz}}{z(z - \pi)}\,dz$$

for each of the following four cases:

(a) C is the boundary of the annulus between circles of radius 1 and radius 3.
(b) C is the boundary of the annulus between circles of radius 1 and radius 4.
(c) C is a circle of radius R, where $R > \pi$.
(d) C is a circle of radius R, where $R < \pi$.

4. Discuss how to evaluate

$$\oint_C \frac{e^{z^2}}{z^2}\,dz$$

where C is a simple closed curve enclosing the origin.

5. We wish to evaluate the integral $I = \int_0^\infty e^{ix^2}\,dx$. Consider the contour $I_R = \oint_{C_{(R)}} e^{iz^2}\,dz$, where $C_{(R)}$ is the closed circular sector in the upper half plane with boundary points 0, R, and $Re^{i\pi/4}$. Show that $I_R = 0$ and that $\lim_{R \to \infty} \int_{C_{1(R)}} e^{iz^2}\,dz = 0$, where $C_{1(R)}$ is the line integral along the

circular sector from R to $Re^{i\pi/4}$. (Hint: use $\sin x \geq \frac{2x}{\pi}$ on $0 \leq x \leq \frac{\pi}{2}$.)
Then, breaking up the contour $C_{(R)}$ into three component parts, deduce

$$\lim_{R \to \infty} \left(\int_0^R e^{ix^2}\, dx - \int_0^R e^{-r^2}\, dr \right) = 0$$

and from the well-known result of real integration, $\int_0^\infty e^{-x^2}\, dx = \sqrt{\pi}/2$,
deduce that $I = \sqrt{\pi}/2$.

6. Consider the integral $I = \int_{-\infty}^{\infty} \frac{dx}{x^2 + 1}$. Show how to evaluate this integral
   by considering $\oint_{C_{(R)}} \frac{dz}{z^2 + 1}$, where $C_{(R)}$ is the closed semicircle in the upper
   half plane with endpoints at $(-R, 0)$ and $(R, 0)$ plus the x axis. Hint: use
   $1/(z^2 + 1) = -\frac{1}{2i}\left(\frac{1}{z+i} - \frac{1}{z-i} \right)$, and show that the integral along the
   open semicircle in the upper half plane vanishes as $R \to \infty$. Verify your
   answer by usual integration in real variables.

2.6 Cauchy's Integral Formula, Its $\bar{\partial}$ Generalization and Consequences

In this section we discuss a number of fundamental consequences and extensions of the ideas presented in earlier sections, especially Cauchy's Theorem.
Subsections 2.6.2 and 2.6.3 are more difficult and can be skipped entirely or
returned to when desired.

2.6.1 Cauchy's Integral Formula and Its Derivatives

An important result owing to Cauchy shows that the values of an analytic
function f on the boundary of a closed contour C determine the values of f
interior to C.

Theorem 2.6.1 Let $f(z)$ be analytic interior to and on a simple closed contour
C. Then at any interior point z

$$f(z) = \frac{1}{2\pi i} \oint_C \frac{f(\zeta)}{\zeta - z}\, d\zeta \qquad (2.6.1)$$

Equation (2.6.1) is referred to as Cauchy's Integral Formula.

Proof Inside the contour C, inscribe a small circle C_δ, radius δ with center at
point z (see Figure 2.6.1).

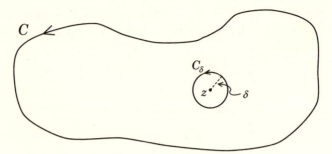

Fig. 2.6.1. Circle C_δ inscribed in contour C

From Cauchy's Theorem we can deform the contour C into C_δ:

$$\oint_C \frac{f(\zeta)}{\zeta - z}\, d\zeta = \oint_{C_\delta} \frac{f(\zeta)}{\zeta - z}\, d\zeta \qquad (2.6.2)$$

We rewrite the second integral as

$$\oint_{C_\delta} \frac{f(\zeta)}{\zeta - z}\, d\zeta = f(z) \oint_{C_\delta} \frac{d\zeta}{\zeta - z} + \oint_{C_\delta} \frac{f(\zeta) - f(z)}{\zeta - z}\, d\zeta \qquad (2.6.3)$$

Using polar coordinates, $\zeta = z + \delta e^{i\theta}$, the first integral on the right in Eq. (2.6.3) is computed to be

$$\oint_{C_\delta} \frac{d\zeta}{\zeta - z} = \int_0^{2\pi} \frac{i\delta e^{i\theta}}{\delta e^{i\theta}}\, d\theta = 2\pi i \qquad (2.6.4)$$

Because $f(z)$ is continuous

$$|f(\zeta) - f(z)| < \epsilon$$

for small enough $|z - \zeta| = \delta$. Then (see also the inequality (2.4.15)

$$\left| \oint_{C_\delta} \frac{f(\zeta) - f(z)}{\zeta - z}\, d\zeta \right| \leq \oint_{C_\delta} \frac{|f(\zeta) - f(z)|}{|\zeta - z|}\, |d\zeta|$$

$$< \frac{\epsilon}{\delta} \int_{C_\delta} |d\zeta|$$

$$= 2\pi\epsilon$$

Thus as $\epsilon \to 0$, the second integral in Eq. (2.6.3) vanishes. Hence Eqs. (2.6.3) and (2.6.4) yield Cauchy's Integral Formula, Eq. (2.6.1). ∎

A particularly simple example of Cauchy's Integral Formula is the following. If on the unit circle $|\zeta| = 1$ we are given $f(\zeta) = \zeta$, then by Eq. (2.6.1)

$$\frac{1}{2\pi i} \oint_C \frac{\zeta}{\zeta - z} d\zeta = z$$

An alternative way to obtain this answer is as follows:

$$\frac{1}{2\pi i} \oint_C \frac{\zeta}{\zeta - z} d\zeta = \frac{1}{2\pi i} \oint_C \left(1 + \frac{z}{\zeta - z} \right) d\zeta$$

$$= \frac{1}{2\pi i} [\zeta + z \log(\zeta - z)]_C$$

$$= z$$

where we use the notation $[\cdot]_C$ to denote the change around the unit circle, and we have selected some branch of the logarithm.

A corollary of Cauchy's Theorem demonstrates that the derivatives of $f(z)$: $f'(z)$, $f''(z)$, ..., $f^{(n)}(z)$ all exist and there is a simple formula for them. Thus the analyticity of $f(z)$ implies the analyticity of all the derivatives.

Theorem 2.6.2 If $f(z)$ is analytic interior to and on a closed contour C, then all the derivatives $f^{(k)}(z)$, $k = 1, 2, \ldots$ all exist in the domain D interior to C, and

$$f^{(k)}(z) = \frac{k!}{2\pi i} \int_C \frac{f(\zeta)}{(\zeta - z)^{k+1}} d\zeta \qquad (2.6.5)$$

Proof Let z be any point in D. It will be shown that all the derivatives of $f(z)$ exist at z. Because z is arbitrary, this establishes the existence of all derivatives in D.

We begin by establishing Eq. (2.6.5) for $k = 1$. Consider the usual difference quotient:

$$\frac{f(z + h) - f(z)}{h} = \frac{1}{2\pi i} \frac{1}{h} \oint_C f(\zeta) \left(\frac{1}{\zeta - (z + h)} - \frac{1}{\zeta - z} \right) d\zeta$$

$$= \frac{1}{2\pi i} \oint_C \frac{f(\zeta)}{(\zeta - (z + h))(\zeta - z)} d\zeta$$

$$= \frac{1}{2\pi i} \oint_C \frac{f(\zeta)}{(\zeta - z)^2} d\zeta + R \qquad (2.6.6)$$

where

$$R = \frac{h}{2\pi i} \oint_C \frac{f(\zeta)}{(\zeta - z)^2(\zeta - z - h)} d\zeta \qquad (2.6.7)$$

We shall call min $|\zeta - z| = 2\delta > 0$. Then, if $|h| < \delta$ for ζ on C, we have

$$|\zeta - (z + h)| \geq |\zeta - z| - |h| > 2\delta - \delta = \delta$$

Because $|f(\zeta)| < M$ on C, then

$$|R| \leq \frac{|h|}{2\pi} \frac{M}{(2\delta)^2\delta} L \qquad (2.6.8)$$

where L is the length of the contour C. Because $|R| \to 0$ as $h \to 0$, we have established Eq. (2.6.5) for $k = 1$:

$$f'(z) = \frac{1}{2\pi i} \oint_C \frac{f(\zeta)}{(\zeta - z)^2} d\zeta \qquad (2.6.9)$$

We may repeat the above argument beginning with Eq. (2.6.9) and thereby prove the existence of $f''(z)$, that is, Eq. (2.6.5) for $k = 2$. This shows that f' has a derivative f'', and so is itself analytic. Consequently we find that if $f(z)$ is analytic, so is $f'(z)$. Applying this argument to f' instead of f proves that f'' is analytic, and, more generally, the analyticity of $f^{(k)}$ implies the analyticity of $f^{(k+1)}$. By induction, we find that all the derivatives exist and hence are analytic. Because $f^{(k)}(z)$ is analytic, Eq. (2.6.1) gives

$$f^{(k)}(z) = \frac{1}{2\pi i} \oint_C \frac{f^{(k)}(\zeta)}{\zeta - z} d\zeta \qquad (2.6.10)$$

Integration by parts (k) times (the boundary terms vanish) yields Eq. (2.6.5).

■

An immediate consequence of this result is the following.

Theorem 2.6.3 All partial derivatives of u and v are continuous at any point where $f = u + iv$ is analytic.

For example, the first derivative of $f(z)$, using the Cauchy–Riemann equations, are

$$f'(z) = u_x + iv_x = v_y - iu_y \qquad (2.6.11)$$

Because $f'(z)$ is analytic, it is certainly continuous. The continuity of $f'(z)$ ensures that u_x, v_y, v_x, and u_y are all continuous. Similar arguments are employed for the higher-order derivatives, $u_{xx}, u_{yy}, u_{xy}, \ldots$.

2.6.2 Liouville, Morera, and Maximum-Modulus Theorems

First we establish a useful inequality. From

$$f^{(n)}(z) = \frac{n!}{2\pi i} \oint_C \frac{f(\zeta)}{(\zeta - z)^{n+1}} \, d\zeta \tag{2.6.12}$$

where C is a circle, $|\zeta - z| = R$, and $|f(z)| < M$, we have

$$|f^{(n)}(z)| \leq \frac{n!}{2\pi} \oint_C \frac{|f(\zeta)|}{|\zeta - z|^{n+1}} \, |d\zeta|$$

$$\leq \frac{n!M}{2\pi R^{n+1}} \oint_C |d\zeta|$$

$$\leq \frac{n!M}{R^n} \tag{2.6.13}$$

With Eq. (2.6.13) we can derive a result about functions that are everywhere analytic in the finite complex plane. Such functions are called **entire**.

Theorem 2.6.4 (Liouville) If $f(z)$ is entire and bounded in the z plane (including infinity), then $f(z)$ is a constant.

Proof Using the inequality (2.6.13) with $n = 1$ we have

$$|f'(z)| \leq \frac{M}{R}$$

Because this is true for any point z in the plane, we can make R arbitrarily large; hence $f'(z) = 0$ for any point z in the plane. Because

$$f(z) - f(0) = \int_0^z f'(\zeta) \, d\zeta = 0$$

we have $f(z) = f(0) = $ constant, and the theorem is proven. ∎

Cauchy's Theorem tells us that if $f(z)$ is analytic inside C, then $\oint_C f(z) \, dz = 0$. Now we prove that the converse is also true.

Theorem 2.6.5 (Morera) If $f(z)$ is continuous in a domain D and if

$$\oint_C f(z) \, dz = 0$$

for every simple closed contour C lying in D, then $f(z)$ is analytic in D.

Proof From Theorem 2.5.3 it follows that if the contour integral always vanishes, then there exists an analytic function $F(z)$ in D such that $F'(z) = f(z)$. Theorem 2.6.2 implies that $F'(z)$ is analytic if $F(z)$ is analytic, hence so is $f(z)$.   ∎

A corollary to Liouville's Theorem is the so-called Fundamental Theorem of Algebra, namely, any polynomial

$$P(z) = a_0 + a_1 z + \cdots + a_m z^m, \qquad (a_m \neq 0) \tag{2.6.14}$$

$m \geq 1$, integer, has at least one point $z = \alpha$ such that $P(\alpha) = 0$; that is, $P(z)$ has at least one root.

We establish this statement by contradiction. If $P(z)$ does not vanish, then the function $Q(z) = 1/P(z)$ is analytic (has a derivative) in the finite z plane. For $|z| \to \infty$, $P(z) \to \infty$; hence $Q(z)$ is bounded in the entire complex plane, including infinity. Liouville's Theorem then implies that $Q(z)$ and hence $P(z)$ is a constant, which violates $m \geq 1$ in Eq. (2.6.14) and thus contradicts the assumption that $P(z)$ does not vanish. In Section 4.4 it is shown that $P(z)$ has m and only m roots, including multiplicities.

There are a number of valuable statements that can be made about the maximum (minimum) modulus an analytic function can achieve, and certain mean value formulae can be ascertained.

For example, using Cauchy's integral formula (Eq. (2.6.1)) with C being a circle centered at z and radius r, we have $\zeta - z = re^{i\theta}$, and $d\zeta = ire^{i\theta}d\theta$; hence Eq. (2.6.1) becomes

$$f(z) = \frac{1}{2\pi} \int_0^{2\pi} f\left(z + re^{i\theta}\right) d\theta \tag{2.6.15}$$

Equation (2.6.15) is a "mean-value" formula; that is, the value of an analytic function at any interior point is the "mean" of the function integrated over the circle centered at z. Similarly, multiplying Eq. (2.6.15) by $r\,dr$, and integrating over a circle of radius R yields

$$f(z) \int_0^R r\,dr = \frac{1}{2\pi} \int_0^R \int_0^{2\pi} f\left(z + re^{i\theta}\right) r\,dr\,d\theta$$

hence

$$f(z) = \frac{1}{\pi R^2} \iint_{D_0} f\left(z + re^{i\theta}\right) dA \tag{2.6.16}$$

where D_0 is the region inside the circle C, radius R, center z.

Thus the value of $f(z)$ also equals its mean value over the area of a circle centered at z.

This result can be used to establish the following maximum-modulus theorem.

Theorem 2.6.6 (Maximum Principles) (i) If $f(z)$ is analytic in a domain D, then $|f(z)|$ cannot have a maximum in D unless $f(z)$ is a constant. (ii) If $f(z)$ is analytic in a bounded region D and $|f(z)|$ is continuous in the closed region $\bar{D}$, then $|f(z)|$ assumes its maximum on the boundary of the region.

Proof Equation (2.6.16) is a useful device to establish this result. Suppose z is an interior point in the region such that $|f(\zeta)| \leq |f(z)|$ for all points ζ in the region. Choose any circle center z radius R such that the circle lies entirely in the region. Calling $\zeta = z + re^{i\theta}$ for any point in the circle, we have (from Eq. (2.6.16))

$$|f(z)| \leq \frac{1}{\pi R^2} \iint_{D_0} |f(\zeta)| \, dA \tag{2.6.17}$$

Actually, the assumed inequality $|f(\zeta)| \leq |f(z)|$ substituted into Eq. (2.6.17) implies that in fact $|f(\zeta)| = |f(z)|$ because if in any subregion, equality did not hold, Eq. (2.6.17) would imply $|f(z)| < |f(z)|$. Thus the modulus of $f(z)$ is constant. Use of the Cauchy–Riemann equations then shows that if $|f(z)|$ is constant, then $f(z)$ is also constant (see Example 2.1.4). This establishes the maximum principle (i) inside C.

Because $f(z)$ is analytic within and on the circle C, then $|f(z)|$ is continuous. A result of real variables states that a continuous function in a bounded region must assume a maximum somewhere in the closed bounded region, including the boundary. Hence the maximum for $|f(z)|$ must be achieved on the boundary of the circle C, and the maximum principle (ii) is established for the circle.

In order to extend these results to more general regions, we may construct appropriate new circles centered at interior points of D and overlapping with the old ones. In this way, by using a sequence of such circles, the region can be filled and the above results follow. ∎

We note that if $f(z)$ does not vanish at any point inside the contour, by considering $1/(f(z)) = g(z)$ it can be seen that $|g(z)|$ also attains its maximum value on the boundary and hence $f(z)$ attains its minima on the boundary.

The real and imaginary parts of an analytic function $f(z) = u(x, y) + iv(x, y)$, u and v, attain their maximum values on the boundary. This follows from the fact that $g(z) = \exp(f(z))$ is analytic, and hence it satisfies the maximum principle. Thus the modulus $|g(z)| = \exp u(x, y)$ must achieve its maximum value on the boundary. Similar arguments for a function $g(z) = \exp(-if(z))$ yield analogous results for $v(x, y)$. Now, because $f(z)$ is analytic, we have from Theorem 2.6.2 and Eq. (2.6.11) that u and v are infinitely differentiable. Furthermore, from the Cauchy–Riemann conditions, u and v are harmonic functions, that is, they satisfy Laplace's equation

$$\nabla^2 u = 0, \qquad \nabla^2 v = 0 \qquad (2.6.18)$$

(see Section 2.1, e.g. Eqs. 2.1.11a,b). Hence the maximum principle says that harmonic functions achieve their maxima (and minima by a similar proof) on the boundary of the region.

*2.6.3 Generalized Cauchy Formula and $\overline{\partial}$ Derivatives

In previous sections we concentrated on analytic functions or functions that are analytic everywhere apart from isolated "singular" points where the function blows up or possesses branch points/cuts. On the other hand, as mentioned earlier (see, for example, Section 2.1 worked Example 2.1.2 and the subsequent discussion) there are functions that are nowhere analytic. For example, the Cauchy–Riemann conditions show that the function $f(z) = \overline{z}$ (and hence any function of $\overline{z}$) is nowhere analytic. The reader might mistakenly think that such functions are mathematical artifacts. However, mathematical formulations of physical phenomena are often described via such complicated nonanalytic functions. In fact, the main theorem (Theorem 2.6.7), described in this section, is used in an essential way to study the scattering and inverse scattering theory associated with certain problems arising in nonlinear wave propagation (Ablowitz, Bar Yaakov, and Fokas, 1983). Despite the fact that Cauchy's Integral Formula (Eq. (2.6.1)) requires that $f(z)$ be an analytic function, there is nevertheless an important extension, which we shall develop below, that extends Cauchy's Integral Theorem to certain nonanalytic functions.

From the coordinate representation $z = x + iy$, $\overline{z} = x - iy$, we have $x = (z + \overline{z})/2$ and $y = (z - \overline{z})/2i$. Using the chain rule, that is

$$\frac{\partial}{\partial z} = \frac{\partial x}{\partial z}\frac{\partial}{\partial x} + \frac{\partial y}{\partial z}\frac{\partial}{\partial y}$$

we find

$$\frac{\partial}{\partial z} = \frac{1}{2}\left(\frac{\partial}{\partial x} - i\frac{\partial}{\partial y}\right) \qquad (2.6.19a)$$

$$\frac{\partial}{\partial \overline{z}} = \frac{1}{2}\left(\frac{\partial}{\partial x} + i\frac{\partial}{\partial y}\right) \qquad (2.6.19b)$$

Sometimes it is convenient to consider the function $f(x, y)$ as depending explicitly on both z and $\bar{z}$; that is, $f = f(z, \bar{z})$. For simplicity we still use the notation $f(z)$ to denote $f(z, \bar{z})$. If f is a differentiable function of z and $\bar{z}$, and

$$\frac{\partial f}{\partial \bar{z}} = 0 \qquad (2.6.20)$$

then we say that $f = f(z)$. Moreover, any $f(z)$ satisfying Eq. (2.6.20) is an analytic function, because from Eq. (2.6.19b) and $f(z) = u + iv$, we find, from Eq. (2.6.20), that $\frac{\partial u}{\partial x} - \frac{\partial v}{\partial y} + i(\frac{\partial u}{\partial y} + \frac{\partial v}{\partial x}) = 0$, hence u and v satisfy the Cauchy–Riemann equations.

In what follows we shall use Green's Theorem (Eq. (2.5.1)) in the following form:

$$\oint_C g \, d\zeta = 2i \iint_R \frac{\partial g}{\partial \bar{\zeta}} \, dA(\zeta) \qquad (2.6.21)$$

where $\zeta = \xi + i\eta$, $d\zeta = d\xi + i \, d\eta$, and $dA(\zeta) = d\xi \, d\eta$. Note in Eq. (2.5.1) use $u = g$, $v = ig$

$$\frac{\partial g}{\partial \bar{\zeta}} = \frac{1}{2}\left(\frac{\partial g}{\partial \xi} + i\frac{\partial g}{\partial \eta}\right)$$

and replace x and y by ξ and η.

Next we establish the following:

Theorem 2.6.7 (Generalized Cauchy Formula) If $\partial f/\partial \bar{\zeta}$ exists and is continuous in a region R bounded by a simple closed contour C, then at any interior point z

$$f(z) = \frac{1}{2\pi i} \oint_C \left(\frac{f(\zeta)}{\zeta - z}\right) d\zeta - \frac{1}{\pi} \iint_R \left(\frac{\partial f/\partial \bar{\zeta}}{\zeta - z}\right) dA(\zeta) \qquad (2.6.22)$$

Proof Consider Green's Theorem in the form of Eq. (2.6.21) in the region R_ϵ depicted in Figure 2.6.2, with $g = f(\zeta)/(\zeta - z)$ and the contour composed of two parts C and C_ϵ.

We have, from Eq. (2.6.21), noting that $\frac{1}{\zeta - z}$ is analytic in this region,

$$\oint_C \frac{f(\zeta)}{\zeta - z} \, d\zeta - \int_{C_\epsilon} \frac{f(\zeta)}{\zeta - z} \, d\zeta = 2i \iint_{R_\epsilon} \frac{\partial f/\partial \bar{\zeta}}{\zeta - z} \, dA \qquad (2.6.23)$$

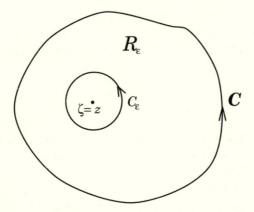

Fig. 2.6.2. Generalized Cauchy formula in region R_ϵ

Note that on C_ϵ: $\zeta = z + \epsilon e^{i\theta}$

$$\oint_{C_\epsilon} \frac{f(\zeta)}{\zeta - z}\, d\zeta \quad = \quad \int_0^{2\pi} \frac{f(z + \epsilon e^{i\theta})}{\epsilon e^{i\theta}} i\epsilon e^{i\theta}\, d\theta$$

$$= \quad \int_0^{2\pi} f\left(z + \epsilon e^{i\theta}\right) i\, d\theta$$

$$\xrightarrow[\epsilon \to 0]{} f(z)(2\pi i) \tag{2.6.24}$$

The limit result is due to the fact that $f(z)$ is assumed to be continuous, and from real variables we find that the limit $\epsilon \to 0$ and the integral of a continuous function over a bounded region can be interchanged. Similarly, because $1/(\zeta - z)$ is integrable over R_ϵ and $\partial f/\partial\bar\zeta$ is continuous, then the double integral over R_ϵ converges to the double integral over the whole region R, the difference tends to zero with ϵ; namely, using polar coordinates $\zeta = z + re^{i\theta}$

$$\left| \iint_{R-R_\epsilon} \left(\frac{\partial f/\partial\bar\zeta}{\zeta - z} \right) i\, dA \right| \le \int_0^\epsilon \int_0^{2\pi} \frac{|\partial f/\partial\bar\zeta|}{r} r\, dr\, d\theta$$

$$\le 2\pi M \epsilon \tag{2.6.25}$$

Using the continuity of $\partial f/\partial\bar\zeta$ in a bounded region implies that

$$\left| \frac{\partial f}{\partial\bar\zeta} \right| \le M$$

Thus Eq. (2.6.23) yields in the limit $\epsilon \to 0$

$$\oint_C \left(\frac{f(\zeta)}{\zeta - z} \right) d\zeta - 2\pi i f(z) = 2i \iint_R \left(\frac{\partial f/\partial \bar{\zeta}}{\zeta - z} \right) i \, dA$$

and hence the generalized Cauchy formula (Eq. (2.6.22)) follows by manipulation. ∎

We note that if $\partial f/\partial \bar{\zeta} = 0$, that is, if $f(z)$ is analytic inside R, then the generalized Cauchy formula reduces to the usual Cauchy Integral Formula (Eq. (2.6.1)).

Problems for Section 2.6

1. Evaluate the integrals $\oint_C f(z) \, dz$, where C is the unit circle centered at the origin and $f(z)$ is given by the following:

   (a) $\dfrac{\sin z}{z}$ (b) $\dfrac{1}{(2z - 1)^2}$ (c) $\dfrac{1}{(2z - 1)^3}$

   (d) $\dfrac{e^z}{z}$ (e) $e^{z^2} \left(\dfrac{1}{z^2} - \dfrac{1}{z^3} \right)$

2. Evaluate the integrals $\oint_C f(z) \, dz$ over a contour C, where C is the boundary of a square with diagonal opposite corners at $z = -(1 + i)R$ and $z = (1+i)R$, where $R > a > 0$, and where $f(z)$ is given by the following:

   (a) $\dfrac{e^z}{z - \frac{\pi i}{4}a}$ (b) $\dfrac{e^z}{\left(z - \frac{\pi i}{2}a \right)^2}$ (c) $\dfrac{z^2}{2z + a}$

   (d) $\dfrac{\sin z}{z^2}$ (e) $\dfrac{\cosh z}{z}$

3. Evaluate the integral

   $$\int_{-\infty}^{\infty} \frac{1}{(x + i)^2} \, dx$$

   by considering $\oint_{C_{(R)}} (1/(z + i)^2) \, dz$, where $C_{(R)}$ is the closed semicircle in the upper half plane with corners at $z = -R$ and $z = R$, plus the x axis. Hint: show that

   $$\lim_{R \to \infty} \oint_{C_{1(R)}} \frac{1}{(z + i)^2} \, dz = 0$$

where $C_{1(R)}$ is the open semicircle in the upper half plane (not including the x axis).

4.   Let $f(z)$ be analytic in a square containing a point w and C be a circle with center ω and radius ρ inside the square. From Cauchy's Theorem show that

$$f(\omega) = \frac{1}{2\pi} \int_0^{2\pi} f\left(\omega + \rho e^{i\theta}\right) d\theta$$

5.   Consider two second-degree polynomials with the same roots and having a ratio equal to unity at infinity. Use Liouville's Theorem to show that they are in fact the same polynomial.

6.   Let $f(z)$ be analytic and nonzero in a region R. Show that $|f(z)|$ has a minimum value in R that occurs on the boundary. (Hint: use the Maximum-Modulus Theorem for the function $1/f(z)$.)

7.   Let $f(z)$ be an entire function, with $|f(z)| \le C|z|$ for all z, where C is a constant. Show that $f(z) = Az$, where A is a constant.

8.   Find the $\bar{\partial}$ (dbar) derivative of the following functions:

$$\text{(a) } e^z \qquad \text{(b) } z\bar{z} \quad (= r^2)$$

Verify the generalized Cauchy formula inside a circle of radius R for both of these functions. Hint: Reduce problem (b) to the verification of the following formula:

$$-\pi \bar{z} = \iint_A \frac{dA}{\zeta - z} = \iint_A \frac{d\xi\, d\eta}{\zeta - z} \equiv I$$

where A is a circle of radius r. To establish this result, transform the integral I to polar coordinates, $\zeta = \xi + i\eta = re^{i\theta}$, and find

$$I = \int_0^{2\pi} \int_0^R \frac{r\, dr\, d\theta}{re^{i\theta} - z}$$

In the θ integral, change variables to $u = e^{i\theta}$, and use $du = ie^{i\theta} d\theta$, $\int_0^{2\pi} d\theta = \frac{1}{i} \oint_{C_0} \frac{du}{u}$, where C_0 is the unit circle. The methods of Section 2.5

can be employed to calculate this integral. Show that we have

$$I = 2\pi \int_0^R r \, dr \left[-\frac{1}{z} + \frac{1}{z} H \left(1 - \frac{|z|}{r} \right) \right]$$

where $H(x) = \{1 \text{ if } x > 0, 0 \text{ if } x < 0\}$. Then show that $I = -\pi |z|^2/z = -\pi \bar{z}$ as is required.

9. Use Morera's Theorem to verify that the following functions are indeed analytic inside a circle of radius R:

$$\text{(a) } z^n, \qquad n \ge 0 \qquad \text{(b) } e^z$$

From Morera's Theorem, what can be said about the following functions?

$$\text{(c) } \frac{\sin z}{z} \qquad \text{(d) } \frac{e^z}{z}$$

10. In Cauchy's Integral Formula (Eq. (2.6.1)), take the contour to be a circle of unit radius centered at the origin. Let $\zeta = e^{i\theta}$ to deduce

$$f(z) = \frac{1}{2\pi} \int_0^{2\pi} \frac{f(\zeta)\zeta}{\zeta - z} \, d\theta$$

where z lies inside the circle. Explain why we have

$$0 = \frac{1}{2\pi} \int_0^{2\pi} \frac{f(\zeta)\zeta}{\zeta - 1/\bar{z}} \, d\theta$$

and use $\zeta = 1/\bar{\zeta}$ to show

$$f(z) = \frac{1}{2\pi} \int_0^{2\pi} f(\zeta) \left(\frac{\zeta}{\zeta - z} \pm \frac{\bar{z}}{\bar{\zeta} - \bar{z}} \right) d\theta$$

whereupon, using the plus sign

$$f(z) = \frac{1}{2\pi} \int_0^{2\pi} f(\zeta) \frac{(1 - |z|^2)}{|\zeta - z|^2} \, d\theta$$

(a) Deduce the "Poisson formula" for the real part of $f(z)$: $u(r, \phi) = \operatorname{Re} f$, $z = re^{i\phi}$

$$u(r, \phi) = \frac{1}{2\pi} \int_0^{2\pi} u(\theta) \frac{1 - r^2}{[1 - 2r \cos(\phi - \theta) + r^2]} d\theta$$

where $u(\theta) = u(1, \theta)$.

(b) If we use the minus sign in the formula for $f(z)$ above, show that

$$f(z) = \frac{1}{2\pi} \int_0^{2\pi} f(\zeta) \left[\frac{1 + r^2 - 2re^{i(\theta - \phi)}}{1 - 2r \cos(\phi - \theta) + r^2} \right] d\theta$$

and by taking the imaginary part

$$v(r, \phi) = C + \frac{1}{\pi} \int_0^{2\pi} u(\theta) \frac{r \sin(\phi - \theta)}{[1 - 2r \cos(\phi - \theta) + r^2]} d\theta$$

where $C = \frac{1}{2\pi} \int_0^{2\pi} v(1, \theta) \, d\theta = v(r = 0)$. (This last relationship follows from the Cauchy Integral formula at $z = 0$ – see the first equation in this exercise.)

(c) Show that

$$\frac{2r \sin(\phi - \theta)}{1 - 2r \cos(\phi - \theta) + r^2} = \operatorname{Im} \left[\frac{1 - r^2 + 2ir \sin(\phi - \theta)}{1 + r^2 - 2r \cos(\phi - \theta)} \right]$$

$$= \operatorname{Im} \left[\frac{\zeta + z}{\zeta - z} \right]$$

and therefore the result for $v(r, \phi)$ from part **(b)** may be expressed as

$$v(r, \phi) = v(0) + \frac{\operatorname{Im}}{2\pi} \int_0^{2\pi} u(\theta) \frac{\zeta + z}{\zeta - z} d\theta$$

This example illustrates that prescribing the real part of $f(z)$ on $|z| = 1$ determines (a) the real part of $f(z)$ everywhere inside the circle and (b) the imaginary part of $f(z)$ inside the circle to within a constant. We *cannot* arbitrarily specify both the real and imaginary parts of an analytic function on $|z| = 1$.

11. The "complex delta function" possesses the following property:

$$\iint_A \delta(z - z_0) F(z) \, dA(z) = F(z_0)$$

or

$$\iint_A \delta(x - x_0)\delta(y - y_0)F(x, y)dA(x, y) = F(x_0, y_0)$$

where $z_0 = x_0 + iy_0$ is contained within the region A.

In formula (2.6.21) of Section 2.6, let $g(z) = F(z)/(z - z_0)$, where $F(z)$ is analytic in A. Show that

$$\oint_C \frac{F(z)}{z - z_0}\, dz = 2i \iint_A F(z)\frac{\partial}{\partial \bar{z}}\left(\frac{1}{z - z_0}\right) dA$$

where C is a simple closed curve enclosing the region A. Use $\oint_C F(z)/(z - z_0)\, dz = 2\pi i F(z_0)$ to establish

$$F(z_0) = \frac{1}{\pi} \iint_A F(z)\frac{\partial}{\partial \bar{z}}\left(\frac{1}{z - z_0}\right) dA$$

and therefore the action of $\partial/\partial \bar{z}\,(1/(z - z_0))$ is that of a complex delta function, that is, $\partial/\partial \bar{z}\,(1/(z - z_0)) = \pi\delta(z - z_0)$.

2.7 Theoretical Developments

In the discussion of Cauchy's Theorem in Section 2.5, we made use of Green's Theorem in the plane that is taken from the vector calculus of real variables. We note, however, that the use of this result and the subsequent derivations of Cauchy's Theorem requires $f(z) = u(x, y) + iv(x, y)$ to be analytic and have a continuous derivative $f'(z)$ in a simply connected region. It turns out that Cauchy's Theorem can be proven without the need for $f'(z)$ to be continuous. This fact was discovered by Goursat many years after the original derivations by Cauchy. Logically, this is especially pleasing because Cauchy's Theorem itself can subsequently be used as a basis to show that if $f(z)$ is analytic in D, then $f'(z)$ is continuous in D.

We next outline the proof of Cauchy's Theorem without making use of the continuity of $f'(z)$; the theorem is frequently referred to as the Cauchy–Goursat Theorem. We refer the reader to Levinson and Redheffer (1970) and appropriate supplementary texts for further details.

Theorem 2.7.1 (*Cauchy–Goursat*) If a function $f(z)$ is analytic at all points interior to and on a simple closed contour, then

$$\oint_C f(z)\, dz = 0 \qquad (2.7.1)$$

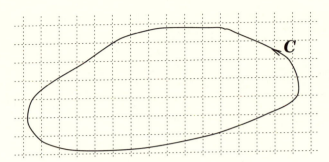

Fig. 2.7.1. Square mesh over region R

Proof Consider a region R consisting of points on and within a simple closed contour C. We form a square mesh over the region R by drawing lines parallel to the x and y axes such that we have a finite number of square subregions in which each point of R lies in at least one subregion. If a particular square contains points not in R, we delete these points. Such partial squares will occur at the boundary (see Figure 2.7.1).

We can refine this mesh by dividing each square in half again and again and redefine partial squares as above. We do this until the length of the diagonal of each square is less than δ. The quantity δ provides a uniform bound on the length of the diagonal or area of any given square.

We note that the integral around the contour C can be replaced by a sum of integrals around the boundary of each square or partial square

$$\oint_C f(z)\,dz = \sum_{j=1}^{n} \oint_{C_j} f(z)\,dz \qquad (2.7.2)$$

where it is noted that all interior contours will mutually cancel because each inner side of a square is covered twice in opposite directions.

Introduce the following equality:

$$f(z) = f(z_j) + (z - z_j)f'(z_j) + (z - z_j)\tilde{f}_j(z) \qquad (2.7.3a)$$

where

$$\tilde{f}(z) = \left(\frac{f(z) - f(z_j)}{z - z_j} \right) - f'(z_j) \qquad (2.7.3b)$$

We remark that

$$\oint_{C_j} dz = 0, \qquad \oint_{C_j} (z - z_j)\,dz = 0 \qquad (2.7.4)$$

which can be established either by direct integration or from the known anti-derivatives:

$$1 = \frac{d}{dz} z, \qquad (z - z_j) = \frac{d^2}{dz} \frac{(z - z_j)^2}{2}, \qquad \ldots$$

then using the results of Theorem 2.4.1 (Eqs. (2.4.8) and (2.4.9)) in Section 2.4. Then, it follows that

$$\left| \oint_C f(z) \, dz \right| \leq \sum_{j=1}^{n} \left| \oint_{C_j} f(z) \, dz \right|$$

$$= \sum_{j=1}^{n} \left| \oint_{C_j} (z - z_j) \tilde{f}_j(z) \, dz \right|$$

$$\leq \sum_{j=1}^{n} \oint_{C_j} |z - z_j| \left| \tilde{f}_j(z) \right| dz \qquad (2.7.5)$$

Note that because for all squares and partial squares we find that $|z - z_j| < \delta$ for z on C_j, and z_j being within C_j, we have a uniform bound on $\tilde{f}_j(z)$, which follows from the analyticity of $f(z)$, that is, the definition of the derivative

$$\left| \tilde{f}_j(z) \right| = \left| \frac{f(z) - f(z_j)}{z - z_j} - f'(z) \right| < \epsilon \qquad (2.7.6)$$

Calling the area of each square A_j, we observe the geometric fact that

$$|z - z_j| \leq \sqrt{2A_j} \qquad (2.7.7)$$

Thus, using Theorem 2.4.2 for all interior squares, we have

$$\oint_{C_j} |z - z_j| \left| \tilde{f}_j(z) \right| dz \leq \left(\sqrt{2A_j} \right) \epsilon \left(4 \sqrt{A_j} \right) = 4 \sqrt{2} \epsilon A_j \qquad (2.7.8)$$

and for all boundary squares, the following upper bound holds:

$$\oint_{C_j} |z - z_j| \left| \tilde{f}_j(z) \right| dz \leq \left(\sqrt{2A_j} \right) \epsilon \left(4 \sqrt{A_j} + L_j \right) \qquad (2.7.9)$$

where L_j is the length of the portion of the contour in the partial square C_j. Then $\oint_C f(z) \, dz$ is obtained by adding over all such contributions Eqs. (2.7.8) and (2.7.9). Calling $A = \sum A_j$, $L = \sum L_j$, quantity A being the area of the square mesh bounded by the contour C and L the length of the contour C, we have

$$\oint_C f(z)\,dz \leq \left(8\sqrt{2}A + \sqrt{2}\sqrt{A}L\right)\epsilon \qquad\qquad (2.7.10)$$

We can refine our mesh indefinitely so as to be able to choose ϵ as small as we wish. Hence the integral $\oint_C f(z)\,dz$ must be zero. ∎

3

Sequences, Series and Singularities of Complex Functions

The representation of complex functions frequently requires the use of infinite series expansions. The best known are Taylor and Laurent series, which represent analytic functions in appropriate domains. Applications often require that we manipulate series by termwise differentiation and integration. These operations may be substantiated by employing the notion of uniform convergence. Series expansions break down at points or curves where the represented function is not analytic. Such locations are termed singular points or singularities of the function. The study of the singularities of analytic functions is vitally important in many applications including contour integration, differential equations in the complex plane, and conformal mapping.

3.1 Definitions of Complex Sequences, Series and Their Basic Properties

Consider the following sequence of complex functions: $f_n(z)$ for $n = 1, 2, 3, \ldots$, defined in a region $\mathcal{R}$ of the complex plane. Usually, we denote the sequence of functions by $\{f_n(z)\}$, where $n = 1, 2, 3, \ldots$. The notion of convergence of a sequence is really the same as that of a limit. We say the sequence $f_n(z)$ **converges** to $f(z)$ on $\mathcal{R}$ or a suitable subset of $\mathcal{R}$, assuming that $f(z)$ exists and is finite, if

$$\lim_{n \to \infty} f_n(z) = f(z) \tag{3.1.1}$$

This means that for each z, given $\epsilon > 0$ there is an N depending on ϵ and z, such that whenever $n > N$ we have

$$|f_n(z) - f(z)| < \epsilon \tag{3.1.2}$$

If the limit does not exist (or is infinite), we say the sequence **diverges** for those values of z.

An infinite series may be viewed as an infinite sequence, $\{s_n(z)\}$, $n = 1, 2, 3, \ldots$, by noting that a sequence of partial sums may be formed by

$$s_n(z) = \sum_{j=1}^{n} b_j(z) \tag{3.1.3}$$

and taking the infinite series as the infinite limit of partial sums:

$$S(z) = \lim_{n \to \infty} s_n(z) = \sum_{j=1}^{\infty} b_j(z) \tag{3.1.4}$$

Conversely, given the sequence of partial sums, we may find the sequence of terms $b_j(z)$ via: $b_1(z) = s_1(z)$, $b_j(z) = s_j(z) - s_{j-1}(z)$, $j \geq 2$. With this correspondence, no real distinction exists between a series or a sequence.

A basic property of a convergent series such as Eq. (3.1.4) is

$$\lim_{j \to \infty} b_j(z) = 0$$

because

$$\lim_{j \to \infty} b_j(z) = \lim_{j \to \infty} s_j(z) - \lim_{j \to \infty} s_{j-1}(z) = S - S = 0 \tag{3.1.5}$$

Thus a necessary condition for convergence is Eq. (3.1.5).

We say that the sequence of functions $s_n(z)$, defined for z in a region $\mathcal{R}$, **converges uniformly** in $\mathcal{R}$ if it is possible to choose N depending on ϵ only (and not z): $N = N(\epsilon)$ in Eq. (3.1.1). In other words, the same estimate for N holds for all z in the domain $\mathcal{R}$; that is, we may establish the validity of the limit process independent of which particular z we choose in $\mathcal{R}$.

For example, consider the sequence of functions

$$f_n(z) = \frac{1}{nz}, \qquad n = 1, 2, \ldots \tag{3.1.6}$$

In the annular region $1 \leq |z| \leq 2$, the sequence of functions $\{f_n\}$ converges uniformly to zero. Namely, given $\epsilon > 0$ for n sufficiently large we have

$$|f_n(z) - f(z)| = \left| \frac{1}{nz} - 0 \right| = \frac{1}{n|z|} < \epsilon$$

Thus, the estimate $1/n|z| < 1/n$ holds in the region $1 \leq |z| \leq 2$ for the first integer n such that $n > N(\epsilon) = 1/\epsilon$. The sequence is therefore uniformly convergent to **zero**.

On the other hand, Eq. (3.1.6) converges to zero, but not uniformly, on the interval $0 < |z| \leq 1$; that is, $|f_n - f| < \epsilon$ only if $n > N(\epsilon, z) = 1/\epsilon|z|$ in the region $0 < |z| \leq 1$. Certainly $\lim_{n \to \infty} f_n(z) = f(z) = 0$, but irrespective of the choice of N there is a value of z (small) such that $|f_n - f| > \epsilon$. Further examples are given in the exercises at the end of the section.

Uniformly convergent sequences possess a number of important properties. In particular, we may employ the notion of uniform convergence to establish the following useful theorem.

Theorem 3.1.1 Let $f_n(z)$ be continuous and let $f_n(z)$ converge to $f(z)$ uniformly in a region $\mathcal{R}$. Then $f(z)$ is continuous, and for any finite contour C inside $\mathcal{R}$

$$\lim_{n \to \infty} \int_C f_n(z)\, dz = \int_C f(z)\, dz \tag{3.1.7}$$

Proof **(a)** First we prove the continuity of $f(z)$. For z and z_0 in $\mathcal{R}$, we write

$$f(z) - f(z_0) = f_n(z) - f_n(z_0) + f_n(z_0) - f(z_0) + f(z) - f_n(z)$$

and hence

$$|f(z) - f(z_0)| \leq |f_n(z) - f_n(z_0)| + |f_n(z_0) - f(z_0)| + |f(z) - f_n(z)|$$

Uniform convergence of $\{f_n(z)\}$ allows us to choose an N independent of z such that for $n > N$

$$|f_n(z_0) - f(z_0)| < \epsilon/3 \qquad \text{and} \qquad |f(z) - f_n(z)| < \epsilon/3$$

Continuity of $f_n(z)$ allows us to choose $\delta > 0$ such that

$$|f_n(z) - f_n(z_0)| < \epsilon/3 \quad \text{for} \quad |z - z_0| < \delta$$

Thus for $n > N$, $|z - z_0| < \delta$

$$|f(z) - f(z_0)| < \epsilon$$

which establishes the continuity of $f(z)$.

(b) Because the function $f(z)$ is continuous, it can be integrated by using the usual definition as described in Chapter 2. Given the continuity of $f(z)$ we shall prove Eq. (3.1.7); namely, for $\epsilon > 0$ we must find N such that when $n > N$

$$\left| \int_C f_n(z)\, dz - \int_C f(z)\, dz \right| < \epsilon \tag{3.1.8}$$

But, for $n > N$

$$\left| \int_C f_n \, dz - \int_C f \, dz \right| \le \int_C |f_n - f| \, |dz| < \epsilon_1 L$$

where the length of C is bounded by L and $|f_n - f| < \epsilon_1$ by uniform convergence of f_n. Taking $\epsilon_1 = \epsilon / L$ establishes Eq. (3.1.8) and hence Eq. (3.1.7).   ∎

An immediate corollary of this theorem applies to series expansions. Namely, if the sequence of continuous partial sums converge uniformly, then we may integrate termwise, that is, for $b_j(z)$ continuous

$$\sum_{j=1}^{\infty} \left(\int_C b_j(z) \, dz \right) = \int_C \left(\sum_{j=1}^{\infty} b_j(z) \right) dz \qquad (3.1.9)$$

Equation (3.1.9) is important and we will use it extensively in our development of power series expansions of analytic functions.

We have already seen that uniformly convergent sequences and series have important and useful properties, for example, they allow the interchange of certain limit processes such as interchanging infinite sums and integrals. In practice it is often unwieldy and frequently difficult to prove that particular series converge uniformly in a given region. Rather, we usually appeal to general theorems that provide conditions under which a series will converge uniformly. In what follows, we shall state one such important theorem, for which the proof is given in Section 3.4. The interested reader can follow the logical development by reading relevant portions of Section 3.4 at this point.

Theorem 3.1.2 (*"Weierstrass M Test"*) Let $|b_j(z)| \le M_j$ in a region $\mathcal{R}$, with M_j constant. If $\sum_{j=1}^{\infty} M_j$ converges, then the series $S(z) = \sum_{j=1}^{\infty} b_j(z)$ converges uniformly in $\mathcal{R}$.

An immediate corollary to this theorem is the so-called **ratio test** for complex series. Namely, suppose $|b_1(z)|$ is bounded, and

$$\left| \frac{b_{j+1}(z)}{b_j(z)} \right| \le M < 1, \qquad j > 1 \qquad (3.1.10)$$

for M constant. Then the series

$$S(z) = \sum_{j=1}^{\infty} b_j(z) \qquad (3.1.11)$$

is uniformly convergent.

In order to prove this statement, we write

$$b_n(z) = b_1(z) \frac{b_2(z)}{b_1(z)} \frac{b_3(z)}{b_2(z)} \cdots \frac{b_n(z)}{b_{n-1}(z)} \tag{3.1.12}$$

Theboundedness of $b_1(z)$ implies

$$|b_1(z)| < B$$

hence

$$|b_n(z)| < BM^{n-1}$$

and therefore

$$\sum_{j=1}^{\infty} |b_j(z)| \le B \sum_{j=1}^{\infty} M^{j-1} = \left(\frac{B}{1-M} \right)$$

We see that the series $\sum_{j=1}^{\infty} |b_j(z)|$ is bounded by a series that converges and is independent of z. Consequently, we see that Theorem 3.1.2, via the assertion Eq. (3.1.10), implies the uniform convergence of Eq. (3.1.11).

We note in the above that if any finite number of terms do not satisfy the hypothesis, they can be added in separately; this will not affect the convergence results.

Problems for Section 3.1

1. In the following we are given sequences. Discuss their limits and whether the convergence is uniform, in the region $\alpha \le |z| \le \beta$, for finite $\alpha, \beta > 0$.

    (a) $\left\{ \dfrac{1}{nz^2} \right\}_{n=1}^{\infty}$ (b) $\left\{ \dfrac{1}{z^n} \right\}_{n=1}^{\infty}$

    (c) $\left\{ \sin \dfrac{z}{n} \right\}_{n=1}^{\infty}$ (d) $\left\{ \dfrac{1}{1+(nz)^2} \right\}_{n=1}^{\infty}$

2. For each sequence in Problem 1, what can be said if

    (a) $\alpha = 0$,   (b) $\alpha > 0$, $\beta = \infty$

3. Compute the integrals

    $$\lim_{n \to \infty} \int_0^1 nz^{n-1} \, dz \quad \text{and} \quad \int_0^1 \lim_{n \to \infty} \left(nz^{n-1} \right) \, dz$$

and show that they are not equal. Explain why this is not a counterexample to Theorem 3.1.1.

4.　In the following, let C denote the unit circle centered at the origin. Let $f(z) = \lim_{n \to \infty} f_n(z)$. Evaluate $\oint_C f(z)\,dz$ and the limit $\lim_{n \to \infty} \oint_C f_n(z)\,dz$, and discuss why they might or might not be equal.

$$\text{(a)} \ \ f_n(z) = \frac{1}{z - n} \qquad \text{(b)} \ \ f_n(z) = \frac{1}{z - (1 - 1/n)}$$

5.　Show that the following series converge uniformly in the given regions:

$$\text{(a)} \ \ \sum_{n=1}^{\infty} z^n, \quad 0 \le |z| < R, \quad R < 1$$

$$\text{(b)} \ \ \sum_{n=1}^{\infty} e^{-nz}, \quad R < |z| \le 1, \quad R > 0 \qquad \text{(c)} \ \ \sum_{n=1}^{\infty} \text{sech } nz, \quad \text{Re } z \ge 1$$

6.　Show that the sequence $\{z^n\}_{n=1}^{\infty}$ converges uniformly inside $0 \le |z| < R$, $R < 1$. (Hint: because $|z| < 1$, we find that $|z| \le R$, $R < 1$. Find $N(\epsilon, R)$ using the definition of uniform convergence.)

3.2 Taylor Series

In a manner similar to a function of a single real variable, as mentioned in Section 1.2, a **power series** about the point $z = z_0$ is defined as

$$f(z) = \sum_{j=0}^{\infty} b_j (z - z_0)^j \tag{3.2.1}$$

or alternatively

$$f(z + z_0) = \sum_{j=0}^{\infty} b_j z^j \tag{3.2.2}$$

Without loss of generality we shall simply work with the series

$$f(z) = \sum_{j=0}^{\infty} b_j z^j \tag{3.2.3}$$

This corresponds to taking $z_0 = 0$. The general case can be obtained by replacing z by $(z - z_0)$.

We begin by establishing the uniform convergence of the above series.

Theorem 3.2.1 If the series Eq. (3.2.3) converges for some $z_* \neq 0$, then it converges for all z in $|z| < |z_*|$. Moreover, it converges uniformly in $|z| \leq R$ for $R < |z_*|$.

Proof For $j \geq J$, $|z| < |z_*|$

$$\left| b_j z^j \right| = \left| b_j z_*^j \right| \left| \frac{z}{z_*} \right|^j < \left| \frac{z}{z_*} \right|^j \leq \left(\frac{R}{|z_*|} \right)^j$$

This follows from the fact that $|b_j z_*^j| < 1$ for sufficiently large j owing to the assumed convergence of the series at $z = z_*$ (i.e., $\lim_{j \to \infty} b_j z_*^j = 0$). We now take

$$M = \frac{R}{|z_*|} < 1 \quad \text{and} \quad M_j \equiv M^j$$

in the Weierstrass M test for $j \geq J$. Thus $\sum_{j=0}^{\infty} b_j z^j$ converges uniformly for $|z| \leq R$, $|R| < |z_*|$, because $\sum_{j=J}^{\infty} |b_j z^j| < \sum_{j=J}^{\infty} M^j = (M^J)/(1 - M)$. ∎

We now establish the Taylor series for an analytic function.

Theorem 3.2.2 (Taylor Series) Let $f(z)$ be analytic for $|z| \leq R$. Then

$$f(z) = \sum_{j=0}^{\infty} b_j z^j \tag{3.2.4}$$

where

$$b_j = \frac{f^{(j)}(0)}{j!}$$

converges uniformly in $|z| \leq R_1 < R$.

We note that this is the Taylor series about $z = 0$. If $z = 0$ is replaced by $z = z_0$, then the result of this theorem would state that $f(z) = \sum b_j (z - z_0)^j$, where $b_j = f^{(j)}(z_0)/j!$ converges uniformly in $|z - z_0| < R$.

Proof The proof is really an application of Cauchy's Integral Formula (Eq. (2.6.1) of Section 2.6). We write

$$f(z) = \frac{1}{2\pi i} \oint_C \frac{f(\zeta)}{\zeta - z} \, d\zeta = \frac{1}{2\pi i} \oint_C \frac{f(\zeta)}{\zeta} \left(1 - \frac{z}{\zeta} \right)^{-1} d\zeta \tag{3.2.5}$$

where C is a circle of radius R. We use the uniformly convergent expansion

$$(1 - z)^{-1} = \sum_{j=0}^{\infty} z^j \tag{3.2.6}$$

Equation (3.2.6) can be established directly. Consider $s_n(z) = \sum_{j=0}^{n} z^n$ for $|z| < 1$. Then

$$s_n(z) - z s_n(z) = 1 - z^{n+1}$$

hence $s_n(z) = (1 - z^{n+1})/(1 - z)$. Because $\lim_{n \to \infty} z^{n+1} = 0$, we have Eq. (3.2.6). Noting that $|z/\zeta| < 1$, we can replace z by z/ζ in Eq. (3.2.6). Using this expansion in Eq. (3.2.5) we deduce

$$f(z) = \frac{1}{2\pi i} \oint_C f(\zeta) \sum_{j=0}^{\infty} \left(\frac{z^j}{\zeta^{j+1}} \right) d\zeta \tag{3.2.7}$$

From Theorem 3.1.1 of Section 3.1 we may interchange $\oint_C$ and $\sum_{j=0}^{\infty}$ to obtain

$$f(z) = \sum_{j=0}^{\infty} b_j z^j \tag{3.2.8}$$

where

$$b_j = \frac{1}{2\pi i} \oint_C \frac{f(\zeta)}{\zeta^{j+1}} d\zeta = \frac{f^{(j)}(0)}{j!}$$

where the right-hand side of Eq. (3.2.8) follows from the corollary of Cauchy's Theorem (Theorem 2.6.2 of Section 2.6). The uniform convergence of the power series follows in the same way as discussed in Theorem 3.2.1 of this section. ∎

Sometimes a Taylor series converges for all finite z. Then $f(z)$ is analytic for $|z| \leq R$, for every R.

We note that **(a)** formula (3.2.4) is the same as that for functions of one real variable and **(b)** the Taylor series about the point $z = z_0$ is given by

$$f(z) = \sum_{j=0}^{\infty} b_j (z - z_0)^j \tag{3.2.9}$$

where

$$b_j = \frac{1}{2\pi i} \oint_C \frac{f(\zeta)}{(\zeta - z_0)^{j+1}} d\zeta = \frac{f^{(j)}(z_0)}{j!}$$

Example 3.2.1

$$e^z = \sum_{j=0}^{\infty} \frac{z^j}{j!} \quad \text{for} \quad |z| < \infty \tag{3.2.10}$$

$$e^{z^2} = \sum_{j=0}^{\infty} \frac{z^{2j}}{j!} \quad \text{for} \quad |z| < \infty \tag{3.2.11}$$

The first of these formulae follows from Eq. (3.2.4) because $f^{(j)}(0) = 1$ for $f(z) = e^z$. Using the "ratio test" discussed in Section 3.1 (Eq. (3.1.10))

$$\lim_{n \to \infty} \left| \frac{\frac{z^{n+1}}{(n+1)!}}{\frac{z^n}{n!}} \right| = \lim_{n \to \infty} \left| \frac{z}{n+1} \right| = 0$$

We see that convergence of Eq. (3.2.10) is obtained for all z. The largest number R for which the power series (3.2.3) converges inside the disk $|z| < R$ is called the **radius of convergence**. The value of R may be *zero* or *infinity* or a finite number. A value for the radius of convergence may be obtained via the usual absolute value tests of calculus such as the ratio test discussed in Section 3.1, Eq. (3.1.10), or via the root test of calculus. In Example 3.2.1 we say the radius of convergence is infinite. The second formula follows by simply using Eq. (3.2.10) and replacing z by z^2. We leave it as an exercise for the reader to verify that formulae (1.2.19) of Section 1.2 are Taylor series representations (about $z = 0$) of the indicated functions.

Taylor series behave just like ordinary polynomials. We may integrate or differentiate Taylor series termwise. Integrating termwise inside its region of convergence,

$$\int f(z)\,dz = \int \left(\sum_{j=0}^{\infty} a_j z^j \right) dz = \sum_{j=0}^{\infty} \frac{a_j z^{j+1}}{j+1} + C \quad C \text{ constant} \tag{3.2.12}$$

is justified by Eq. (3.1.9) of Section 3.1. Similarly, differentiation

$$f'(z) = \sum_{j=0}^{\infty} j a_j z^{j-1} \tag{3.2.13}$$

also follows. We formulate this result as a theorem.

Theorem 3.2.3 Let $f(z)$ be analytic for $|z| \leq R$. Then the series obtained by differentiating the Taylor series termwise converges uniformly to $f'(z)$ in $|z| \leq R_1 < R$.

Proof If $f(z)$ is analytic in $|z| \leq R$, then from our previous results (e.g. Theorem 2.6.2, Section 2.6), $f'(z)$ is analytic in D for $|z| < R$. The Taylor series for $f'(z)$ is given by

$$f'(z) = \sum_{j=0}^{\infty} C_j z^j, \qquad C_j = \frac{f^{(j+1)}(0)}{j!} \qquad (3.2.14)$$

But the Taylor series for $f(z)$ is given by

$$f(z) = \sum_{j=0}^{\infty} \frac{f^{(j)}(0)}{j!} z^j \qquad (3.2.15)$$

hence formal differentiation termwise yields

$$f'(z) = \sum_{j=1}^{\infty} \frac{f^{(j)}(0)}{(j-1)!} z^{j-1} = \sum_{j=0}^{\infty} \frac{f^{(j+1)}(0)}{j!} z^j \qquad (3.2.16)$$

which is equivalent to Eq. (3.2.14). Moreover, the same argument as presented in the proof of Theorem 3.2.2 holds for Eq. (3.2.14), which shows that the differentiated series converges uniformly in $|z| \leq R_1 < R$. ∎

We remark that further differentiation for $f''(z)$, $f'''(z)$, ..., follows in the same manner by reapplying the arguments presented in Theorem 3.2.3.

The Taylor series representing the zero function is also clearly zero. (Because zero is an analytic function, Eq. (3.2.4) applies.) We easily deduce that Taylor series are unique; that is, there cannot be two Taylor series representations of a given function $f(z)$ because if there were two, say, $\sum a_n z^n$ and $\sum b_n z^n$, the difference $\sum c_n z^n$, where $c_n = a_n - b_n$, must represent the zero function, which implies $a_n = b_n$.

Similarly, any convergent power series representation of an analytic function $f(z)$ must be the Taylor series representation of $f(z)$. In order to demonstrate this fact, we first show that any convergent power series can be differentiated termwise.

Theorem 3.2.4 If the power series (3.2.3) converges for $|z| \leq R$, then it can be differentiated termwise to obtain a uniformly convergent series for $|z| \leq R_1 < R$.

Proof From Eq. (2.6.5) of Section 2.6 we have, for any closed contour C, $|z| = R_1 < R$

$$f'(z) = \frac{1}{2\pi i} \oint_C \frac{f(\zeta)}{(\zeta - z)^2} d\zeta$$

$$= \frac{1}{2\pi i} \oint_C \frac{\sum_{j=0}^{\infty} a_j \zeta^j}{(\zeta - z)^2} d\zeta \qquad (3.2.17)$$

Because the series in Eq. (3.2.17) is uniformly convergent, we may interchange the sum and integral (Section 3.1, Eq. (3.1.9)) to find

$$f'(z) = \sum_{j=0}^{\infty} a_j \left(\frac{1}{2\pi i} \oint_C \frac{\zeta^j}{(\zeta - z)^2} d\zeta \right)$$

$$= \sum_{j=0}^{\infty} a_j \frac{d}{dz} (z^j)$$

$$= \sum_{j=0}^{\infty} j a_j z^{j-1} \qquad (3.2.18)$$

where we have employed Eq. (2.6.5) of Section 2.6 for the function $f(z) = z^j$. Uniform convergence follows in the same way as before. ∎

The formula $f(z) = \sum_{j=0}^{\infty} a_j z^j$ clearly may be differentiated over and over again for $|z| \le R_1 < R$. Thus it immediately follows that

$$f(0) = a_0$$

$$f'(0) = a_1$$

$$f^{(j)}(0) = j! a_j \qquad (3.2.19)$$

Hence we have deduced that the power series of $f(z)$ is really the Taylor series of $f(z)$ (about $z = 0$).

The usual properties of series hold; namely, the sum/difference of a series are the sum/difference of the terms. Calling $g(z) = \sum_{j=0}^{\infty} b_j z^j$

$$f(z) \pm g(z) = \sum_{j=0}^{\infty} a_j z^j \pm \sum_{j=0}^{\infty} b_j z^j \qquad (3.2.20)$$

$$= \sum_{j=0}^{\infty} (a_j \pm b_j) z^j \qquad (3.2.21)$$

and it also follows that the product of two convergent series may be written

$$f(z)g(z) = \sum_{j=0}^{\infty} C_j z^j \tag{3.2.22}$$

where

$$C_j = \sum_{k=0}^{j} b_k a_{j-k} \tag{3.2.23}$$

Another result similar to that of real analysis is the **comparison test**.

Theorem 3.2.5 Let the series $\sum_{j=0}^{\infty} a_j z^j$ converge for $|z| < R$. If $|b_j| \le |a_j|$ for $j \ge N$, then the series $\sum_{j=0}^{\infty} b_j z^j$ also converges for $|z| < R$.

Proof For $j \ge N$ and $|z| < |z_*| < R$,

$$\left| b_j z^j \right| \le \left| a_j z^j \right| = \left| a_j z_*^j \right| \left| \frac{z}{z_*} \right|^j < \left| \frac{z}{z_*} \right|^j < 1$$

The latter inequalities follow because $\sum_{j=0}^{\infty} a_j z^j$ converges, and we know that for sufficiently large j, $|a_j R^j| < 1$. Convergence then follows, via the Weierstrass M test. ∎

For example, we know that $\sum_{n=0}^{\infty} (z^n/n!) = e^z$ converges for $|z| < \infty$. Thus by the comparison test the series $\sum_{n=0}^{\infty} [z^n/(n!)^2]$ also converges for $|z| < \infty$ since $(n!)^2 \ge n!$ for all n.

Another example of a Taylor series (about $z = 0$) is given by

$$\frac{1}{1+z} = \sum_{n=0}^{\infty} (-1)^n z^n \quad \text{for} \quad |z| < 1 \tag{3.2.24}$$

(Also see the remark below Eq. (3.2.7).) Equation (3.2.24) is obtained by taking successive derivatives of the function $1/(1 + z)$, evaluating them at $z = 0$, and employing Eq. (3.2.4), or noting formula (3.2.6) and replacing $-z$ with z. The radius of convergence follows from the ratio test. Replacing z with z^2 yields

$$\frac{1}{1+z^2} = \sum_{n=0}^{\infty} (-1)^n z^{2n}, \quad \text{for} \, |z| < 1 \tag{3.2.25}$$

The divergence, for $|z| \ge 1$, of the series given in Eq. (3.2.25) is due to the zeroes of $1 + z^2 = 0$, that is, $z = \pm i$. In the case of real analysis, it was not really clear why the series (3.2.25) with z replaced by x diverges; only when we examine the series in the context of complex analysis do we understand the origins of divergence. Because the function $1/(1 + z^2)$ is nonanalytic only at $z = \pm i$, it is natural to ask whether there is another series representation valid for $|z| > 1$. In fact, there is such a representation that is part of a more general series expansion (Laurent series) to be taken up shortly.

Having the ability to represent a function by a series, such as a Taylor series, allows us to analyze and work with a much wider class of functions than the usual elementary functions (e.g. polynomials, rational functions, exponentials, and logarithms).

Example 3.2.2 Consider the "error function," erf(z):

$$\text{erf}(z) = \frac{2}{\sqrt{\pi}} \int_0^z e^{-t^2} \, dt \qquad (3.2.26)$$

Using Eq. (3.2.11) with z^2 replaced by $-t^2$ and integrating termwise, we have the Taylor series representation

$$\text{erf}(z) = \frac{2}{\sqrt{\pi}} \int_0^z \left(\sum_{n=0}^\infty \frac{(-t^2)^n}{n!} \right) dt$$

$$= \frac{2}{\sqrt{\pi}} \sum_{n=0}^\infty \frac{(-1)^n z^{2n+1}}{(2n+1)n!} = \frac{2}{\sqrt{\pi}} \left[z - \frac{z^3}{3} + \frac{z^5}{5 \cdot 2!} - \frac{z^7}{7 \cdot 3!} + \cdots \right] \qquad (3.2.27)$$

Because we are integrating an exponential function that is entire, it follows that the error function is also entire.

With the results of this section we see that the notion of analyticity of a function $f(z)$ in a region $\mathcal{R}$ may now be broadened. Namely, if $f(z)$ is analytic, then by Definition 2.1.1, $f'(z)$ exists in $\mathcal{R}$. We have seen that this implies that (a) $f(z)$ has derivatives of all orders in $\mathcal{R}$ (an extension of Cauchy's formula, Theorem 2.6.2), and that (b) $f(z)$ has a Taylor series representation in the neighborhood of all points of $\mathcal{R}$ (Theorem 3.2.2).

On the other hand, if $f(z)$ has a convergent power series expansion, that is

$$f(z) = \sum_{n=0}^\infty a_n (z - z_0)^n$$

then integrating $f(z)$ over a simple closed contour C implies $\oint_C f(z)\,dz = 0$ because $\oint_C (z - z_0)^n dz = 0$. Hence by Theorem 2.6.5, $f(z)$ is analytic inside C. This is consistent with point (b) above as we have already shown that the power series is equivalent to the Taylor series representation.

In later chapters we study analytic functions that coincide in a domain or on a curve or that are zero at distinct points. The following theorems will be useful (c.f. Section 3.5 and Chapter 7); we will only sketch the proofs.

Theorem 3.2.6 Let each of two functions $f(z)$ and $g(z)$ be analytic in a common domain D. If $f(z)$ and $g(z)$ coincide in some subportion $D' \subset D$ or on a curve Γ interior to D, then $f(z) = g(z)$ everywhere in D.

Proof Corresponding to any point z_0 in D' (or on Γ) consider the largest circle C contained entirely within D. Both $f(z)$ and $g(z)$ may be represented by a Taylor series inside C, and by the uniqueness of Taylor series, $f(z) = g(z)$ inside C. Next pick a new interior point of C but near its boundary, and repeat the above Taylor series argument to find $f(z) = g(z)$ in an extended domain. In fact, this procedure can be repeated so as to entirely fill up the common domain D. (This statement, while intuitively clear, requires some analysis to substantiate – we shall omit it.)   ∎

Consequently, a function $f(z)$ that vanishes everywhere in a subdomain $D' \subset D$ or on a curve Γ entirely contained within D must vanish everywhere inside D. The discussion in Theorem 3.2.6 provides us with a way of "analytically continuing" a known function in some domain to a larger domain. We remark that one must be careful when continuing a multivalued function; this is discussed further in Section 3.5.

In fact, analytic continuation of a function $f(z)$ to a function $g(z)$, with which it shares a common boundary, is closely related to the above. It can be described via the following theorem, which is proven with the aid of Morera's Theorem.

Theorem 3.2.7 Let D_1 and D_2 be two disjoint domains, whose boundaries share a common contour Γ. Let $f(z)$ be analytic in D_1 and continuous in $D_1 \cup \Gamma$ and $g(z)$ be analytic in D_2 and continuous in $D_2 \cup \Gamma$, and let $f(z) = g(z)$ on Γ. Then the function

$$H(z) = \begin{cases} f(z) & z \in D_1 \\ f(z) = g(z) & z \in \Gamma \\ g(z) & z \in D_2 \end{cases}$$

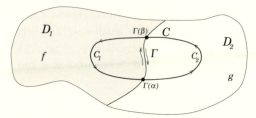

Fig. 3.2.1. Analytic continuation

is analytic in $D = D_1 \cup \Gamma \cup D_2$. We say that $g(z)$ is the **analytic continuation** of $f(z)$.

Proof Consider a closed contour C in D. If C does not intersect Γ, then $\oint_C H(z)\, dz = 0$ because C is entirely contained in D_1 or D_2. On the other hand, if C intersects Γ, then we have (referring to Figure 3.2.1)

$$\oint_C H(z)\, dz = \int_{C_1} f(z)\, dz + \int_{\Gamma(\alpha)}^{\Gamma(\beta)} f(z)\, dz$$

$$+ \int_{\Gamma(\beta)}^{\Gamma(\alpha)} g(z)\, dz + \int_{C_2} g(z)\, dz = 0 \qquad (3.2.28)$$

where we have divided the closed contour C into two other closed contours $C_1 + \Gamma$ and $C_2 - \Gamma$, and the endpoints of the contour Γ inside C are labeled $\Gamma(\alpha)$ and $\Gamma(\beta)$. Note that the two "intermediate" integrals in opposite directions along Γ mutually cancel owing to the fact that $f(z) = g(z)$ on Γ. Then, the fact that $f(z)$ and $g(z)$ are analytic in D_1 and D_2, respectively, ensures that $\oint_C H(z)\, dz = 0$, whereupon from Morera's Theorem (see Section 2.6.2) we find that $H(z)$ is analytic in $D = D_1 \cup \Gamma \cup D_2$. (If f, g are analytic on Γ this statement follows immediately; otherwise some more work is needed). ∎

Theorem 3.2.8 If $f(z)$ is analytic and not identically zero in some domain D containing $z = z_0$, then its zeroes are isolated; that is, there is a neighborhood about $z = z_0$, $f(z_0) = 0$, in which $f(z)$ is nonzero.

Proof Because $f(z)$ is analytic at z_0, it has a Taylor series about $z = z_0$. If it has a zero of order n we write

$$f(z) = (z - z_0)^n g(z)$$

where $g(z)$ has a Taylor series about $z = z_0$, and $g(z_0) \neq 0$. There must exist a maximum integer n, otherwise $f(z)$ would be identically zero in a neighborhood of z_0 and, from Theorem 3.2.6, must vanish everywhere in D. Because $g(z)$ is analytic, it follows that for sufficiently small ϵ, $|g(z) - g(z_0)| < \epsilon$ whenever z is in the neighborhood of z_0; namely, $0 < |z - z_0| < \delta$. Hence $g(z)$ can be made as close to $g(z_0)$ as desired; hence $g(z) \neq 0$ and $f(z) \neq 0$ in this neighborhood.   ∎

Finally, in closing this section we briefly discuss the behavior of functions that are represented by integrals. Such integrals arise frequently in applications, for example, (a) the Fourier transform $F(z)$ of a function $f(t)$

$$F(z) = \int_{-\infty}^{\infty} f(t)e^{-izt}\, dt$$

or (b) the Cauchy type integral $F(z)$ associated with a function $f(t)$ on a closed contour C

$$F(z) = \frac{1}{2\pi i} \oint_C \frac{f(t)}{t - z}\, dt$$

These are two examples we will study in some detail in subsequent chapters, in which a given integral depends on another parameter, in this case z. Frequently one is interested in the question of when the function $F(z)$ is analytic, which we now address in some generality.

Consider integrals of the following form:

$$F(z) = \int_a^b g(z, t)\, dt \qquad a \le t \le b \tag{3.2.29}$$

where (a) for each t, $g(z, t)$ is an analytic function of z in a domain D, and (b) for each z, $g(z, t)$ is a continuous function of t.
With these hypotheses, it follows that $F(z)$ is analytic in D and

$$F'(z) = \int_a^b \frac{\partial g}{\partial z}(z, t)\, dt \tag{3.2.30}$$

Because for each t, $g(z, t)$ is an analytic function of z, we find that

$$g(z, t) = \sum_{j=0}^{\infty} c_j(t)(z - z_0)^j \tag{3.2.31}$$

where z_0 is any point in D and from Eq. (3.2.9) $c_j(t) = \frac{1}{2\pi i} \oint_C \frac{g(\zeta, t)}{(\zeta - z_0)^{j+1}}\, d\zeta$, and C is a circle inside D centered at z_0. The continuity of $g(z, t)$ as a function of t implies the continuity of $c_j(t)$. The function $g(z, t)$ is bounded in D: $|g(z, t)| < M$ because it is analytic there. Hence $|c_j(t)| \leq \frac{M}{2\pi} \oint_C \frac{|d\zeta|}{|\zeta - z_0|^{j+1}} = M/\rho^j$, where $\rho = |\zeta - z_0|$ is the radius of the circle C. Thus the Taylor series given by Eq. (3.2.31) converges uniformly for $|z - z_0| < \rho$ and can therefore be integrated and/or differentiated termwise, from which Eq. (3.2.30) follows, using the series for g with (3.2.29).

We also note that a closed contour in t can be viewed (see Section 2.4) as a special case of a line integral of the form of Eq. (3.2.30) where $a \leq t \leq b$. Hence the above results apply when $F(z) = \oint_{\hat{C}} g(z, t) \, dt$ and $\hat{C}$ is a closed contour (such as a Cauchy type integral).

On the other hand, if the contour becomes infinite (e.g. $a = -\infty$ or $b = \infty$), then, in addition to the hypotheses already stated, it is necessary to specify a uniformity restriction on $g(z, t)$ in order to have analyticity of $F(z)$ and then Eq. (3.2.30). Namely, in this case of infinite limits it is sufficient to add that $g(z, t)$ satisfy $|g(z, t)| \leq G(t)$, where $\int_a^b G(t) \, dt < \infty$. We will not go into further details here.

Problems for Section 3.2

1. Obtain the radius of convergence of the series $\sum_{n=1}^{\infty} s_n(z)$, where $s_n(z)$ is given by

   (a) z^n (b) $\dfrac{z^n}{(n+1)!}$ (c) $n^n z^n$ (d) $\dfrac{z^{2n}}{(2n)!}$ (e) $\dfrac{n!}{n^n} z^n$

2. Find Taylor series expansions around $z = 0$ of the following functions in the given regions:

   (a) $\dfrac{1}{1 - z^2}$, $|z| < 1$ (b) $\dfrac{z}{1 + z^2}$, $|z| < 1$

   (c) $\cosh z$, $|z| < \infty$ (d) $\dfrac{\sin z}{z}$, $0 < |z| < \infty$

   (e) $\dfrac{\cos z - 1}{z^2}$, $0 < |z| < \infty$ (f) $\dfrac{e^{z^2} - 1 - z^2}{z^3}$, $0 < |z| < \infty$

3. Let the Euler numbers E_n be defined by the power series

$$\frac{1}{\cosh z} = \sum_{n=0}^{\infty} \frac{E_n}{n!} z^n$$

   (a) Find the radius of convergence of this series.
   (b) Determine the first six Euler numbers.

4. Show that about any point $z = x_0$

$$e^z = e^{x_0} \sum_{n=0}^{\infty} \frac{(z - x_0)^n}{n!}$$

5. (a)   Use the identity $\frac{1}{z} = 1/((z + 1) - 1)$ to establish

$$\frac{1}{z} = -\sum_{n=0}^{\infty} (z + 1)^n \qquad |z + 1| < 1$$

(b)   Use the above identity to also establish

$$\frac{1}{z^2} = \sum_{n=0}^{\infty} (n + 1)(z + 1)^n \qquad |z + 1| < 1$$

Verify that you get the same result by differentiation of the series in part (a).

6. Evaluate the integrals $\oint_C f(z)\, dz$, where C is the unit circle centered at the origin and $f(z)$ is given by the following:

(a) $\dfrac{\sin z}{z}$ (b) $\dfrac{\sin z}{z^2}$ (c) $\dfrac{\cosh z - 1}{z^4}$

7. Use the Taylor series for $1/(1 + z)$ about $z = 0$ to find the Taylor series expansion of $\log(1 + z)$ about $z = 0$ for $|z| < 1$.

8. Use the Taylor series representation of $1/(1 - z)$ around $z = 0$ for $|z| < 1$ to find a series representation of $1/(1 - z)$ for $|z| > 1$. (Hint: use $1/(1 - z) = -1/(z(1 - 1/z))$)

9. Use the Taylor series representation of $1/(1 - z)$ around $z = 0$, for $|z| < 1$, to deduce the series representation of $1/(1 - z)^2$, $1/(1 - z)^3$, ..., $1/(1 - z)^m$.

10. Use the binomial expansion and Cauchy's Integral Theorem to evaluate

$$\oint_C (z + 1/z)^{2n} \frac{dz}{z}$$

where C is the unit circle centered at the origin. Recall the binomial expansion

$$(a+b)^n = a^n + na^{n-1}b + \cdots = \sum_{k=0}^{n} \binom{n}{k} a^{n-k} b^k,$$

where

$$\binom{n}{k} \equiv \frac{n!}{k!(n-k)!}$$

Use this result to establish the following real integral formula:

$$\frac{1}{2\pi} \int_0^{2\pi} (\cos\theta)^{2n} \, d\theta = \frac{(2n)!}{4^n (n!)^2}$$

3.3 Laurent Series

In many applications we encounter functions that are, in some sense, generalizations of analytic function. Typically, they are not analytic at some point, points, or in some regions of the complex plane, and consequently, Taylor series cannot be employed in the neighborhood of such points. However, another series representation can frequently be found in which both positive and negative powers of $(z - z_0)$ exist. (Recall that Taylor series expansions contain only positive powers of $(z - z_0)$.) Such a series is valid for those functions that are analytic in and on a circular annulus, $R_1 \leq |z - z_0| \leq R_2$ (see Figure 3.3.1).

In the derivation of Laurent series it is convenient to work with the series about an arbitrary point $z = z_0$.

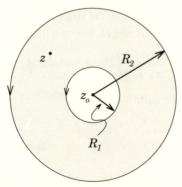

Fig. 3.3.1. Circular annulus $R_1 \leq |z - z_0| \leq R_2$

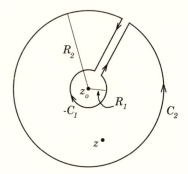

Fig. 3.3.2. Inner contour C_1 and outer contour C_2

Theorem 3.3.1 A function $f(z)$ analytic in an annulus $R_1 \le |z - z_0| \le R_2$ may be represented by the expansion

$$f(z) = \sum_{n=-\infty}^{\infty} C_n (z - z_0)^n \tag{3.3.1}$$

in the region $R_1 < R_a \le |z - z_0| \le R_b < R_2$, where

$$C_n = \frac{1}{2\pi i} \oint_C \frac{f(z)\, dz}{(z - z_0)^{n+1}} \tag{3.3.2}$$

and C is any simple closed contour in the region of analyticity enclosing the inner boundary $|z - z_0| = R_1$.

Proof We introduce the usual cross cut in the annulus (see Figure 3.3.2) where we denote by C_1 and C_2 the inside and outside contours surrounding the point $z = z_0$.

Contour C_1 lies on $|z - z_0| = R_1$, and contour C_2 lies on $|z - z_0| = R_2$. Application of Cauchy's formula to the crosscut region where $f(z)$ is analytic and the crosscut contributions cancel, lets us write $f(z)$ as follows:

$$f(z) = \frac{1}{2\pi i} \oint_{C_2} \frac{f(\zeta)}{\zeta - z}\, d\zeta - \frac{1}{2\pi i} \oint_{C_1} \frac{f(\zeta)}{\zeta - z}\, d\zeta \tag{3.3.3}$$

In the first integral we write

$$\frac{1}{\zeta - z} = \frac{1}{(\zeta - z_0) - (z - z_0)} = \frac{1}{(\zeta - z_0)\left(1 - \left(\frac{z - z_0}{\zeta - z_0}\right)\right)}$$

$$= \left(\frac{1}{(\zeta - z_0)}\right) \sum_{j=0}^{\infty} \frac{(z - z_0)^j}{(\zeta - z_0)^j} \tag{3.3.4}$$

In the above equation we have used Eq. (3.2.6) with z replaced by $\frac{z-z_0}{\zeta-z_0}$, and we note that

$$\left|\frac{z-z_0}{\zeta-z_0}\right| = \frac{|z-z_0|}{R_2} < 1$$

In the second integral we write

$$-\frac{1}{\zeta-z} = \frac{1}{z-z_0-(\zeta-z_0)} = \frac{1}{(z-z_0)}\frac{1}{\left(1-\left(\frac{\zeta-z_0}{z-z_0}\right)\right)}$$

$$= \frac{1}{z-z_0}\sum_{j=0}^{\infty}\left(\frac{\zeta-z_0}{z-z_0}\right)^j \tag{3.3.5}$$

Again in Eq. (3.3.5) we have made use of Eq. (3.2.6), where z is now replaced by $\frac{\zeta-z_0}{z-z_0}$, and we note that

$$\left|\frac{\zeta-z_0}{z-z_0}\right| = \frac{R_1}{|z-z_0|} < 1$$

Using Eq. (3.3.4) in the first integral of Eq. (3.3.3) and Eq. (3.3.5) in the second integral of Eq. (3.3.3) gives us the following representation for $f(z)$:

$$f(z) = \sum_{j=0}^{\infty} A_j(z-z_0)^j + \sum_{j=0}^{\infty} B_j(z-z_0)^{-(j+1)} \tag{3.3.6}$$

where

$$A_j = \frac{1}{2\pi i}\oint_{C_2}\frac{f(\zeta)}{(\zeta-z_0)^{j+1}}\,d\zeta \tag{3.3.7a}$$

$$B_j = \frac{1}{2\pi i}\oint_{C_1} f(\zeta)(\zeta-z_0)^j\,d\zeta \tag{3.3.7b}$$

We make the substitutions $n=j$ in the first sum of Eq. (3.3.6) and $n = -(j+1)$ in the second sum of Eq. (3.3.6) to obtain

$$f(z) = \sum_{n=0}^{\infty} A_n(z-z_0)^n + \sum_{n=-\infty}^{-1} B_{-n-1}(z-z_0)^n \tag{3.3.8}$$

where

$$A_n = \frac{1}{2\pi i}\oint_{C_2}\frac{f(\zeta)}{(\zeta-z_0)^{n+1}}\,d\zeta \tag{3.3.9}$$

and

$$B_{-n-1} = \frac{1}{2\pi i} \oint_{C_1} \frac{f(\zeta)}{(\zeta - z_0)^{n+1}} \, d\zeta \qquad (3.3.10)$$

Because $f(z)$ is analytic in the annulus, it follows from Cauchy's Theorem that each of the integrals $\oint_{C_1}$ and $\oint_{C_2}$ in Eqs. (3.3.9) and (3.3.10) can be deformed into any simple closed contour enclosing C_1. This yields

$$C_n = \frac{1}{2\pi i} \oint_C \frac{f(\zeta)}{(\zeta - z_0)^{n+1}} \, d\zeta \qquad (3.3.11)$$

where $C_n = A_n$ for $n \geq 0$ and $C_n = B_{-n-1}$ for $n \leq -1$. Thus Eq. (3.3.8) becomes

$$f(z) = \sum_{n=-\infty}^{\infty} C_n (z - z_0)^n \qquad (3.3.12)$$

which proves the theorem.   ∎

The coefficient of the term $1/(z - z_0)$, which is C_{-1} in Eq. (3.3.12), turns out to play a very special role in complex analysis. It is given a special name: the **residue** of the function $f(z)$ (see Chapter 4). The negative powers of the Laurent series are referred to as the **principal part** of $f(z)$.

We note two important special cases: **(a)** Suppose $f(z)$ is analytic everywhere *inside* the circle $|z - z_0| = R_1$. Then by Cauchy's Theorem, $C_n = 0$ for $n \leq -1$ because the integrand in Eq. (3.3.11) is analytic; in this case, Eq. (3.3.12) reduces to the Taylor series

$$f(z) = \sum_{n=0}^{\infty} C_n (z - z_0)^n \qquad (3.3.13)$$

where C_n is given by Eq. (3.3.11) for $n \geq 0$. **(b)** Suppose $f(z)$ is analytic everywhere *outside* the circle $|z - z_0| = R_2$. Then the integral (3.3.11) yields $C_n = 0$ for $n \geq 1$. In particular, for $n \geq 1$ we have that the contour C in Eq. (3.3.11) may be deformed to a large circle $|z| = R$. Introducing the transformation (see Section 1.3) $t = 1/z$, we find

$$C_n = \frac{-1}{2\pi i} \oint_{C_\epsilon} \frac{f\left(\frac{1}{t}\right)}{(1 - tz_0)^{n+1}} t^{n-1} \, dt \qquad (3.3.14)$$

where C_ϵ denotes a circle of radius $\epsilon = 1/R$. Because $f(z)$ is analytic at infinity, the function $f(1/t)$ is bounded, that is,

$$\left| f\left(\frac{1}{t}\right) \right| \le M \tag{3.3.15}$$

for sufficiently small t. Hence

$$|C_n| \le \frac{1}{2\pi} \oint_{C_\epsilon} \frac{M|t|^{n-1}}{|1 - tz_0|^{n+1}} |dt| \tag{3.3.16}$$

Using t sufficiently small so that

$$|1 - tz_0|^{n+1} \ge (1 - |t||z_0|)^{n+1} \ge \frac{1}{2}$$

it follows that ($|dt| = \epsilon \, d\theta$)

$$|C_n| \le 2M\epsilon^n \to 0 \qquad \text{as } \epsilon \to 0, n \ge 1.$$

Thus, in this case $f(z)$ has the form:

$$f(z) = \sum_{n=-\infty}^{0} \frac{C_n}{(z - z_0)^n}.$$

In practice, Laurent series frequently may be obtained from the Taylor series of a function by appropriate substitutions. (We consider a number of other examples later in this section.) For example, replacing z in the series expansion for e^z (see Eq. (3.2.10)) by $1/z$ yields a Laurent series for $e^{1/z}$:

$$e^{1/z} = \sum_{j=0}^{\infty} \frac{1}{j! z^j} \tag{3.3.17}$$

which contains an infinite number of negative powers of z. This is an example of case **(b)** above, that is, $e^{1/z}$ is analytic for all $|z| > 0$.

Laurent series have properties very similar to those of Taylor series. For example, the series converges uniformly.

Theorem 3.3.2 The Laurent series, Eqs. (3.3.1) and (3.3.2), of a function $f(z)$ that is analytic in an annulus $R_1 \le |z - z_0| \le R_2$ converges uniformly to $f(z)$ for $\rho_1 \le |z - z_0| \le \rho_2$, where $R_1 < \rho_1$ and $R_2 > \rho_2$.

Proof The derivation of Laurent series shows that $f(z)$ has two representative parts, given by the two sums in Eq. (3.3.6). We write $f(z) = f_1(z) + f_2(z)$.

The first series in Eq. (3.3.6) is the Taylor series part and it converges uniformly to $f_1(z)$ by the proof given in Theorem 3.2.1. For the second sum

$$f_2(z) = \sum_{j=0}^{\infty} B_j (z - z_0)^{-(j+1)} \tag{3.3.18}$$

we can use the M test. For j large enough and for $z = z_1$ on $|z - z_0| = R_1$

$$\left| B_j (z - z_0)^{-(j+1)} \right| = \frac{|B_j|}{|z_1 - z_0|^{j+1}} \left| \frac{z_1 - z_0}{z - z_0} \right|^{j+1}$$

$$< \left| \frac{z_1 - z_0}{z - z_0} \right|^{j+1} < 1 \tag{3.3.19}$$

where $|B_j|/|z_1 - z_0|^{j+1} < 1$ is due to the convergence of the series (3.3.18) (see Theorem 3.3.1), and the fact that $|(z_1 - z_0)/(z - z_0)| < 1$ is due to z being inside the annulus with $\rho_1 > R_1$. ∎

A corollary to this result is the fact that the Laurent series may be integrated termwise, a fact that follows from Theorem 3.1.1. Similarly, the Laurent series may be differentiated termwise; the proof is similar to Theorem 3.2.3 and is therefore omitted. It is also easily shown that the elementary operations such as addition, subtraction, and multiplication for Laurent series behave just like Taylor series.

In fact, we now show that the Laurent expansion given by Eqs. (3.3.11) and (3.3.12) is unique. Namely, if

$$f(z) = \sum_{n=-\infty}^{\infty} b_n (z - z_0)^n \tag{3.3.20}$$

is valid in the annulus $R_1 \leq |z - z_0| \leq R_2$, then $b_n = C_n$, with C_n given by Eq. (3.3.11). This fact will allow us to obtain Laurent expansions by elementary methods. Before turning to some examples, let us prove this result.

Theorem 3.3.3 Suppose $f(z)$ is represented by a uniformly convergent series

$$f(z) = \sum_{n=-\infty}^{\infty} b_n (z - z_0)^n$$

in the annulus $R_1 \leq |z - z_0| \leq R_2$. Then $b_n = C_n$, with C_n given by Eqs. (3.3.11).

Proof Because $f(z)$ is represented by a uniformly convergent series, given $\epsilon \to 0$, there is an $N > 0$ (i.e., $N = N(\epsilon)$) such that when $n \geq N$

$$\left| f(z) - \sum_{n=-N}^{N} b_n(z - z_0)^n \right| < \epsilon \qquad (3.3.21)$$

inside $R_1 \leq |z - z_0| \leq R_2$. Consider

$$I = \left| \frac{1}{2\pi i} \oint_C \left(\sum_{m=-N}^{N} b_m(\zeta - z_0)^m - f(\zeta) \right) \frac{d\zeta}{(\zeta - z_0)^{n+1}} \right| \qquad (3.3.22)$$

where C is the circle $|z - z_0| = R$, and $R_1 \leq R \leq R_2$. Because we know by contour integration (Section 2.4) that

$$\frac{1}{2\pi i} \oint_C (\zeta - z_0)^n \, d\zeta = \begin{cases} 1 & \text{when } n = -1 \\ 0 & \text{when } n \neq -1 \end{cases} \qquad (3.3.23)$$

we find that only the $m = n$ term in the sum in Eq. (3.3.22) is nonzero, hence

$$I = \left| b_n - \frac{1}{2\pi i} \oint_C \frac{f(\zeta)}{(\zeta - z_0)^{n+1}} \right| \qquad (3.3.24)$$

However, from Eq. (3.3.21) we have the following estimate for Eq. (3.3.22):

$$I \leq \frac{\epsilon}{2\pi} \oint_C \frac{|d\zeta|}{|\zeta - z_0|^{n+1}} = \frac{\epsilon}{R^n} \qquad (3.3.25)$$

Then from Eq. (3.3.24), because ϵ may be taken arbitrarily small, we deduce that

$$b_n = C_n = \frac{1}{2\pi i} \oint_C \frac{f(\zeta)}{(\zeta - z_0)^{n+1}} \, d\zeta \qquad (3.3.26)$$

∎

We emphasize that in practice one does not use Eq. (3.3.2) to compute the coefficients of the Laurent expansion of a given function. Instead, one usually appeals to the above uniqueness theorem and uses well-known Taylor expansions and appropriate substitutions.

Example 3.3.1 Find the Laurent expansion of $f(z) = 1/(1 + z)$ for $|z| > 1$.
  The Taylor series expansion (3.2.6) of $(1 - z)^{-1}$ is

$$\frac{1}{1 - z} = \sum_{n=0}^{\infty} z^n \qquad \text{for} \quad |z| < 1 \qquad (3.3.27)$$

We write

$$\frac{1}{1+z} = \frac{1}{z(1+1/z)} \tag{3.3.28}$$

and use Eq. (3.3.27) with z replaced by $-1/z$, noting that if $|z| > 1$ then $|-1/z| < 1$. We find

$$\frac{1}{1+z} = \frac{1}{z}\sum_{n=0}^{\infty}\frac{(-1)^n}{z^n} = \sum_{n=0}^{\infty}\frac{(-1)^n}{z^{n+1}}$$

$$= \frac{1}{z} - \frac{1}{z^2} + \frac{1}{z^3} - \cdots$$

We note that for $|z| < 1$, $f(z) = 1/(1+z) = \sum_{n=0}^{\infty}(-1)^n z^n$. Thus there are different series expansions in different regions of the complex plane. In summary

$$\frac{1}{1+z} = \begin{cases} \sum_{n=0}^{\infty}(-1)^n z^n & |z| < 1 \\ \sum_{n=0}^{\infty}\frac{(-1)^n}{z^{n+1}} & |z| > 1 \end{cases}$$

Example 3.3.2 Find the Laurent expansion of

$$f(z) = \frac{1}{(z-1)(z-2)} \qquad \text{for } 1 < |z| < 2$$

We use partial fraction decomposition to rewrite $f(z)$ as

$$f(z) = -\frac{1}{z-1} + \frac{1}{z-2} \tag{3.3.29}$$

Anticipating the fact that we will use Eq. (3.3.27), we rewrite Eq. (3.3.29) as

$$f(z) = -\frac{1}{z}\left(\frac{1}{1-1/z}\right) - \frac{1}{2}\left(\frac{1}{1-z/2}\right) \tag{3.3.30}$$

Because $1 < |z| < 2$, $|1/z| < 1$, and $|z/2| < 1$, we can use Eq. (3.2.28) to obtain

$$f(z) = -\frac{1}{z}\sum_{n=0}^{\infty}\frac{1}{z^n} - \frac{1}{2}\sum_{n=0}^{\infty}\left(\frac{z}{2}\right)^n$$

$$= -\left(\frac{1}{z} + \frac{1}{z^2} + \frac{1}{z^3} + \cdots\right) - \frac{1}{2}\left(1 + \frac{z}{2} + \left(\frac{z}{2}\right)^2 + \cdots\right) \tag{3.3.31}$$

Thus

$$f(z) = \sum_{n=-\infty}^{\infty} C_n z^n$$

where

$$C_n = \begin{cases} -1 & n \le -1 \\ \dfrac{1}{2^{n+1}} & n \ge 0 \end{cases}$$

As with Example 3.3.1, there exist different Laurent series expansions for $|z| < 1$ and for $|z| > 2$.

A somewhat more complicated example follows.

Example 3.3.3 Find the first two nonzero terms of the Laurent expansion of the function $f(z) = \tan z$ about $z = \pi/2$.

Let us call $z = \pi/2 + u$, so

$$f(z) = \frac{\sin\left(\frac{\pi}{2} + u\right)}{\cos\left(\frac{\pi}{2} + u\right)} = -\frac{\cos u}{\sin u} \qquad . \quad (3.3.32)$$

This can be expanded using the Taylor series for $\sin u$ and $\cos u$:

$$f(z) = -\frac{\left(1 - \frac{u^2}{2!} + \cdots\right)}{\left(u - \frac{u^3}{3!} + \cdots\right)} = -\frac{1}{u}\frac{\left(1 - \frac{u^2}{2!} + \cdots\right)}{\left(1 - \frac{u^2}{3!} + \cdots\right)}$$

The denominator can be expanded via Eq. (3.3.27) to obtain, for the first two nonzero terms

$$f(z) = -\frac{1}{u}\left(1 - \frac{u^2}{2!} + \cdots\right)\left(1 + \frac{u^2}{3!} + \cdots\right)$$

$$= -\frac{1}{u}\left(1 - \frac{u^2}{3} + \cdots\right)$$

$$= -\frac{1}{\left(z - \frac{\pi}{2}\right)} + \frac{\left(z - \frac{\pi}{2}\right)}{3} + \cdots$$

Problems for Section 3.3

1. Expand the function $f(z) = 1/(1 + z^2)$ in

   (a) a Taylor series for $|z| < 1$
   (b) a Laurent series for $|z| > 1$

2. Given the function $f(z) = z/(a^2 - z^2)$, $a > 0$, expand $f(z)$ in a Laurent series in powers of z in the regions

$$\text{(a) } |z| < a \qquad \text{(b) } |z| > a$$

3. Given the function

$$f(z) = \frac{z}{(z - 2)(z + i)}$$

expand $f(z)$ in a Laurent series in powers of z in the regions

$$\text{(a) } |z| < 1 \qquad \text{(b) } 1 < |z| < 2 \qquad \text{(c) } |z| > 2$$

4. Evaluate the integral $\oint_C f(z)\, dz$ where C is the unit circle centered at the origin and $f(z)$ is given as follows:

$$\text{(a) } \frac{e^z}{z^3} \qquad \text{(b) } \frac{1}{z^2 \sin z} \qquad \text{(c) } \tanh z \qquad \text{(d) } \frac{1}{\cos 2z} \qquad \text{(e) } e^{1/z}$$

5. Let

$$e^{\frac{t}{2}(z - 1/z)} = \sum_{n=-\infty}^{\infty} J_n(t) z^n$$

Show from the definition of Laurent series and using properties of integration that

$$J_n(t) = \frac{1}{2\pi} \int_{-\pi}^{\pi} e^{-i(n\theta - t \sin \theta)} \, d\theta$$

$$= \frac{1}{\pi} \int_0^{\pi} \cos(n\theta - t \sin \theta) \, d\theta$$

The functions $J_n(t)$ are called Bessel functions, which are well-known special functions in mathematics and physics.

6. Given the function

$$A(z) = \int_z^{\infty} \frac{e^{-1/t}}{t} \, dt$$

find a Laurent expansion in powers of z for $|z| > R$, $R > 0$. Why will the same procedure fail if we consider

$$E(z) = \int_z^{\infty} \frac{e^{-t}}{t} \, dt$$

(See also Problem 7, below.)

7. Suppose we are given

$$E(z) = \int_z^\infty \frac{e^{-t}}{t} \, dt$$

A formal series may be obtained by repeated integration by parts, that is,

$$E(z) = \frac{e^{-z}}{z} - \int_z^\infty \frac{e^{-t}}{t^2} \, dt$$

$$= \frac{e^{-z}}{z} - \frac{e^{-z}}{z^2} + \int_z^\infty \frac{2e^{-t}}{t^3} \, dt = \cdots$$

If this procedure is continued, show that the series is given by

$$E(z) = \frac{e^{-z}}{z} \left(1 - \frac{1}{z} + \cdots + \frac{(-)^n n!}{z^n} \right) + R_n(z)$$

$$R_n(z) = (-)^{n+1}(n+1)! \int_z^\infty \frac{e^{-t}}{t^{n+2}} \, dt$$

Explain why the series does not converge. (See also Problem **8**, below.)

8. In Problem 7, above, consider $z = x$ real. Show that

$$|R_n(z)| \le (n+1)! \frac{e^{-x}}{x^{n+2}}$$

Explain how to approximate the integral $E(x)$ for large x, given some n. Find suitable values of x for $n = 1, 2, 3$ in order to approximate $E(x)$ to within 0.01, using the above inequality for $|R_n(x)|$. Explain why this approximation holds true for $\mathrm{Re}\, z > 0$. Why does the approximation fail as $n \to \infty$?

*3.4 Theoretical Results for Sequences and Series

In earlier sections of Chapter 3 we introduced the notions of sequences, series, and uniform convergence. Although the Weierstrass M test was stated, a proof was deferred to this section for those interested readers. We begin this section by discussing the notion of a Cauchy sequence.

Definition 3.4.1 A sequence of complex numbers $\{f_n\}$ forms a **Cauchy sequence** if, for every $\epsilon > 0$, there is an $N = N(\epsilon)$, such that whenever $n \ge N$ and $m \ge N$ we have $|f_n - f_m| < \epsilon$.

The same definition as 3.4.1 applies to sequences of complex functions $\{f_n(z)\}$, where it is understood that $f_n(z)$ exists in some region $\mathcal{R}, z \in \mathcal{R}$. Here, in general, $N = N(\epsilon, z)$. Whenever $N = N(\epsilon)$ only, the sequence $\{f_n(z)\}$ is said to be a uniform Cauchy sequence. The following result is immediate.

Theorem 3.4.1 If a sequence converges, then it is a Cauchy sequence.

Proof If $\{f_n(z)\}$ converges to $f(z)$, then for any $\epsilon > 0$ there is an $N = N(\epsilon, z)$ such that whenever $n > N$ and $m > N$

$$|f_n(z) - f(z)| < \frac{\epsilon}{2} \quad \text{and} \quad |f_m(z) - f(z)| < \frac{\epsilon}{2}$$

Hence

$$|f_n(z) - f_m(z)| = |f_n(z) - f(z) - (f_m(z) - f(z))|$$

$$\leq |f_n(z) - f(z)| + |f_m(z) - f(z)|$$

$$< \epsilon$$

and so $\{f_n(z)\}$ is a Cauchy sequence.

We note that if $\{f(z)\}$ converges uniformly to $f(z)$ then $N = N(\epsilon)$ only, and the Cauchy sequence is uniform. ∎

We shall next prove the converse, namely that every Cauchy sequence converges. We shall employ the following result of real analysis, namely *every real Cauchy sequence has a limit.*

Theorem 3.4.2 If $\{f_n(z)\}$ is a Cauchy sequence, then there is a function $f(z)$ such that $\{f_n(z)\}$ converges to $f(z)$.

Proof Let us call $f_n(z) = u_n(x, y) + i v_n(x, y)$. Because

$$|u_n(x, y) - u_m(x, y)| \leq |f_n(z) - f_m(z)|$$

$$|v_n(x, y) - v_m(x, y)| \leq |f_n(z) - f_m(z)|$$

and $\{f_n(z)\}$ is a Cauchy sequence, we find that $\{u_n(x, y)\}$ and $\{v_n(x, y)\}$ are real Cauchy sequences and hence have limits $u(x, y)$ and $v(x, y)$, respectively. Thus the function $f(z) = u(x, y) + i v(x, y)$ exists and is the limit of $\{f_n(z)\}$.

Convergence will be uniform if the number N for the Cauchy sequence $\{f_n(z)\}$ depends only on ϵ ($N = N(\epsilon)$) and not on both ϵ and z ($N = N(\epsilon, z)$). ∎

The above theorem allows us to prove the Weierstrass M test given in Section 3.1. We repeat the statement of this theorem now for the convenience of the reader.

Theorem 3.4.3 Let $|b_j(z)| \leq M_j$ in some region $\mathcal{R}$, with $\{M_j\}$ constant. If $\sum_{j=1}^{\infty} M_j$ converges, then the series $f(z) = \sum_{j=1}^{\infty} b_j(z)$ converges uniformly in $\mathcal{R}$.

Proof Let $n > m$ and $f_n(z) = \sum_{j=1}^{n} b_j(z)$. Then

$$
\begin{aligned}
|f_n(z) - f_m(z)| &= \left| \sum_{j=m+1}^{n} b_j(z) \right| \\
&\leq \sum_{j=m+1}^{n} |b_j(z)| \\
&\leq \sum_{j=m+1}^{n} M_j \\
&\leq \sum_{j=m+1}^{\infty} M_j
\end{aligned}
$$

Because $\sum_{j=1}^{\infty} M_j$ converges, we know that there is an $N = N(\epsilon)$ (the M_j are only constants) such that when $m > N$, $\sum_{j=m+1}^{\infty} M_j < \epsilon$. Thus $\{f_n(z)\}$ is a uniformly convergent Cauchy sequence in $\mathcal{R}$, and Theorem 3.4.2 follows. ∎

Early in this chapter we proved Theorem 3.1.1, which allowed us to interchange the operation of integration with a limit of a uniformly convergent sequence of functions. A corollary of this result is Eq. (3.1.9), which allows the interchange of sum and integral for a uniformly convergent series. A similar theorem holds for the operation of differentiation.

Theorem 3.4.4 Let $f_n(z)$ be analytic in the circle $|z - z_0| < R$, and let $\{f_n(z)\}$ converge uniformly to $f(z)$ in $|z - z_0| < R - \delta$, $\delta > 0$. Then **(a)** $f(z)$ is analytic for $|z - z_0| < R$, and **(b)** $\{f_n'(z)\}$, $\{f_n''(z)\}$, ..., converge uniformly in $|z - z_0| < R - \delta$ to $f'(z)$, $f''(z)$, ...

Proof **(a)** Let C be any simple closed contour lying inside $|z - z_0| \leq R - \delta$ (see Figure 3.4.1) for all $R > \delta > 0$. Because $\{f_n(z)\}$ is uniformly convergent,

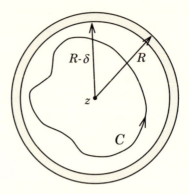

Fig. 3.4.1. Region of analyticity in Theorem 3.4.3(a)

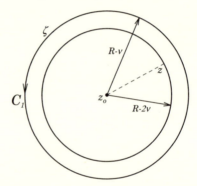

Fig. 3.4.2. For Theorem 3.4.4(b), $|z - \zeta| > \nu$

we have, from Theorem 3.1.1 of Section 3.1

$$\oint_C f(z)\,dz = \lim_{n \to \infty} \oint_C f_n(z)\,dz \tag{3.4.1}$$

Because $f_n(z)$ is analytic, we conclude from Cauchy's Theorem that $\oint_C f_n(z)\,dz = 0$, hence $\oint_c f(z)\,dz = 0$. Now from Theorem 2.6.5 (Morera) of Section 2.6 we find that $f(z)$ is analytic in $|z - z_0| < R$ (because δ may be made arbitrarily small). This proves part **(a)**.

  (b) Let C_1 be the circle $|z - z_0| = R - \nu$ for all $0 < \nu < \frac{R}{2}$ (see Figure 3.4.2). We next use Cauchy's Theorem for $f'(z) - f_n'(z)$ (Theorem 2.6.3, Eq. (2.6.5)), which gives

$$\left(f'(z) - f_n'(z) \right) = \frac{1}{2\pi i} \oint_{C_1} \frac{(f(\zeta) - f_n(\zeta))}{(\zeta - z)^2}\,d\zeta \tag{3.4.2}$$

Because $\{f_n(z)\}$ is a uniformly convergent sequence, we find that for $n > N$ and any z in C_1

$$|f(z) - f_n(z)| < \epsilon_1$$

Thus

$$|f'(z) - f'_n(z)| < \frac{\epsilon_1}{2\pi} \oint_{C_1} \frac{|d\zeta|}{|\zeta - z|^2}$$

If z lies inside C_1, say, $|z - z_0| = R - 2\nu$, then $|\zeta - z| > \nu$, $\oint_{C_1} |d\zeta| < 2\pi R$, hence

$$|f'(z) - f'_n(z)| < \frac{\epsilon_1 R}{\nu^2} \tag{3.4.3}$$

Taking ϵ_1 as small as necessary, that is, $\epsilon_1 = \epsilon \nu^2 / R$, ensures that $|f'(z) - f'_n(z)|$ is arbitrarily small; that is, $|f'(z) - f'_n(z)| < \epsilon$ and hence $\{f'_n(z)\}$ converges uniformly inside $|z - z_0| \le R - 2\nu$. Taking $\delta = 2\nu$ and part (a) above establishes the theorem for the sequence $\{f'_n(z)\}$. (The values of δ and ν can be taken arbitrarily small.) Because the sequence $\{f'_n(z)\}$ converges uniformly inside $|z - z_0| \le R - \delta$, for all $R > \delta > 0$, we can repeat the above procedure in order to establish the theorem for the sequence $\{f''_n(z)\}$; that is, the sequence $\{f''_n(z)\}$ converges uniformly to $f''(z)$ and so on for $\{f'''_n(z)\}$ to $f'''_n(z)$, etc. ■

An immediate consequence of this theorem is the result for series. Namely, call

$$S_n(z) = \sum_{j=1}^{n} f_j(z) \tag{3.4.4}$$

If $S_n(z)$ satisfies the hypothesis of Theorem (3.4.4), then

$$\lim_{n \to \infty} S'_n(z) = \lim_{n \to \infty} \sum_{j=1}^{n} f'_j(z) = \sum_{j=1}^{\infty} f'_j(z) = S(z) \tag{3.4.5}$$

We remark that $\{f_n(z)\}$ being a uniformly convergent sequence of *analytic* functions gives us a much stronger result than we have for uniformly convergent sequences of only real functions. Namely, sequences of derivatives of any order of $f_n(z)$ are uniformly convergent. For example, consider the real sequence $\{u_n(x)\}$, where

$$u_n(x) = \frac{\cos n^2 x}{n}, \qquad |x| < \infty$$

This sequence is uniformly convergent to zero because $|u_n(x)| \leq 1/n$ (independent of x), which converges to *zero*. However, the sequence of functions $\{u'_n(x)\}$

$$u'_n(x) = 2n \sin n^2 x, \qquad |x| < \infty \tag{3.4.6}$$

has no limit whatsoever! The sequence $\{u'_n(x)\}$ is not uniformly convergent. We note also the above sequence $u_n(z)$ for $z = x + iy$ is not uniformly convergent for $|z| < \infty$ because $\cos n^2 z = \cos n^2 x \cosh n^2 y - i \sin n^2 x \sinh n^2 y$; and both $\cosh n^2 y$ and $\sinh n^2 y$ diverge as $n \to \infty$ for $y \neq 0$.

Another corollary of Theorem 3.4.4 is that power series

$$f(z) = \sum_{n=0}^{\infty} a_n (z - z_0)^n$$

may be differentiated termwise inside their radius of convergence. Indeed, we have already shown that any power series is really the Taylor series expansion of the represented function. Hence Theorem 3.4.4 could have alternatively been used to establish the validity of differentiating Taylor series inside their radius of convergence.

We conclude with an example.

Example 3.4.1 We are given

$$\zeta(z) = \sum_{n=1}^{\infty} \frac{1}{n^z} \tag{3.4.7}$$

(The function $\zeta(z)$ is often called the **Riemann zeta function**; it appears in many branches of mathematics and physics.) Show that $\zeta(z)$ is analytic for all $x > 1$, where $z = x + iy$.

By definition, $n^z = e^{z \log n}$, where we take $\log n$ to be the principal branch of the log. Hence

$$n^z = e^{z \log n} = e^{(x+iy) \log n}$$

is analytic for all z because e^{kz} is analytic, and

$$|n^z| = e^{x \log n} = n^x$$

Thus from the Weierstrass M test (Theorem 3.1.2 or 3.4.3, proven in this section), we find that the series representing $\zeta(z)$ converges uniformly because the series $\sum_{n=1}^{\infty} (1/n^x)$ (for $x > 1$) is a convergent series of real numbers. That

is, we may use the integral theorem for a series of real numbers as our upper
bound to establish this. Note that

$$\int_1^\infty \frac{1}{n^x} \, dn = \frac{1}{1-x}$$

Thus from Theorem 3.4.4 we find that because $\{\zeta_m(z)\} = \{\sum_{n=1}^m n^{-z}\}$ is a
uniformly convergent sequence of analytic functions for all $x > 1$, the sum
$\zeta(z)$ is analytic.

Problems for Section 3.4

1. Demonstrate whether or not the following sequences are Cauchy sequences:

    (a) $\{z^n\}_{n=1}^\infty$, $|z| < 1$ (b) $\left\{1 + \dfrac{z}{n}\right\}_{n=1}^\infty$, $|z| < \infty$

    (c) $\{\cos nz\}_{n=1}^\infty$, $|z| < \infty$ (d) $\{e^{-n/z}\}_{n=1}^\infty$, $|z| < 1$

2. Discuss whether the following series converge uniformly in the given do-
    mains:

    (a) $\displaystyle\sum_{j=1}^n z^j$, $|z| < 1$ (b) $\displaystyle\sum_{j=0}^n e^{-jz}$, $\dfrac{1}{2} < |z| < 1$

    (c) $\displaystyle\sum_{j=1}^n j! z^{2j}$, $|z| < a$, $a > 0$

3. Establish that the function $\sum_{n=1}^\infty \frac{1}{e^n n^z}$ is an analytic function of z for all z;
    that is, it is an entire function.

4. Show that the following functions are analytic functions of z for all z; that
    is, they are entire:

    (a) $\displaystyle\sum_{n=1}^\infty \frac{z^n}{(n!)^2}$ (b) $\displaystyle\sum_{n=1}^\infty \frac{\cosh nz}{n!}$ (c) $\displaystyle\sum_{n=1}^\infty \frac{z^{2n+1}}{[(2n+1)!]^{1/2}}$

5. Consider the function $f(z) = \sum_{n=1}^\infty (1/(z^2 + n^2))$. Break the function $f(z)$
    into two parts, $f(z) = f_1(z) + f_2(z)$, where

    $$f_1(z) = \sum_{n=1}^N \left(\frac{1}{z^2 + n^2}\right)$$

and

$$f_2(z) = \sum_{n=N+1}^{\infty} \left(\frac{1}{z^2 + n^2} \right)$$

For $|z| < R$, $N > 2R$, show that in the second sum

$$\left| \frac{1}{z^2 + n^2} \right| \leq \frac{1}{n^2 - R^2} \leq \frac{4}{3n^2}$$

whereupon explain why $f_2(z)$ converges uniformly and consequently, why $f(z)$ is analytic everywhere except at the distinct points $z = \pm in$.

6. Use the method of Problem 5 to investigate the analytic properties of $f(z) = \sum_{n=1}^{\infty} \frac{1}{(z+n)^2}$.

3.5 Singularities of Complex Functions

We begin this section by introducing the notion of an **isolated singular point**. The concept of a singular point was introduced in Section 2.1 as being a point where a given (single-valued) function is not analytic. Namely, $z = z_0$ is a singular point of $f(z)$ if $f'(z_0)$ does not exist. Suppose $f(z)$ (or any single-valued branch of $f(z)$, if $f(z)$ is multivalued) is analytic in the region $0 < |z - z_0| < R$ (i.e., in a neighborhood of $z = z_0$), and *not* at the point z_0. Then the point $z = z_0$ is called an **isolated singular point** of $f(z)$. In the neighborhood of an isolated singular point, the results of Section 3.3 show that $f(z)$ may be represented by a Laurent expansion:

$$f(z) = \sum_{n=-\infty}^{\infty} C_n(z - z_0)^n \qquad (3.5.1)$$

Suppose $f(z)$ has an isolated singular point and in addition it is bounded; that is, $|f(z)| \leq M$ where M is a constant. It is clear that all coefficients $C_n = 0$ for $n < 0$ in order for $f(z)$ to be bounded. Thus such a function $f(z)$ is given by a power series expansion, $f(z) = \sum_{n=0}^{\infty} C_n(z - z_0)^n$, valid everywhere *except possibly at $z = z_0$*. However, because a power series expansion converges at $z = z_0$, it follows that $f(z)$ would be analytic if $C_0 = f(z_0)$ (the $n = 0$ term is the only nonzero contribution), in which case $\sum_{n=0}^{\infty} C_n(z - z_0)^n$ is the Taylor series expansion of $f(z)$. If $C_0 \neq f(z_0)$, we call such a point a **removable singularity**, because by a slight redefinition of $f(z_0)$, the function $f(z)$ is analytic. For example, consider the function $f(z) = (\sin z)/z$, which, strictly speaking, is undefined at $z = 0$. If it were the case that $f(0) \neq 1$, then

$z = 0$ is a removable singularity. Namely, by simply redefining $f(0) = 1$, then $f(z)$ is analytic for all z including $z = 0$ and is represented by the power series

$$f(z) = 1 - \frac{z^2}{3!} + \frac{z^4}{5!} - \frac{z^6}{7!} + \cdots = \sum_{n=0}^{\infty} \frac{(-1)^n z^{2n}}{(2n+1)!}$$

Stated differently, if $f(z)$ is analytic in the region $0 < |z - z_0| < R$, and if $f(z)$ can be made analytic at $z = z_0$ by assigning an appropriate value for $f(z_0)$, then $z = z_0$ is a removable singularity.

An isolated singularity of $f(z)$ is said to be a pole if $f(z)$ has the following representation:

$$f(z) = \frac{\phi(z)}{(z - z_0)^N} \tag{3.5.2}$$

where N is a positive integer, $N \geq 1$, $\phi(z)$ is analytic in a neighborhood of z_0, and $\phi(z_0) \neq 0$. We generally say $f(z)$ has an Nth-**order pole** if $N \geq 2$ and has a **simple pole** if $N = 1$. Equation (3.5.2) implies that the Laurent expansion of $f(z)$ takes the form $f(z) = \sum_{n=-N}^{\infty} C_n (z - z_0)^n$; that is, the first coefficient is $C_{-N} = \phi(z_0)$. Coefficient C_{-N} is often called the **strength of the pole**. Moreover, it is clear that in the neighborhood of $z = z_0$, the function $f(z)$ takes on arbitrarily large values, or $\lim_{z \to z_0} f(z) = \infty$.

Example 3.5.1 Describe the singularities of the function

$$f(z) = \frac{z^2 - 2z + 1}{z(z+1)^3} = \frac{(z-1)^2}{z(z+1)^3}$$

The function $f(z)$ has a simple pole at $z = 0$ and a third order (or **triple**) pole at $z = -1$. The strength of the pole at $z = 0$ is 1 because the expansion of $f(z)$ near $z = 0$ has the form

$$f(z) = \frac{1}{z}(1 - 2z + \cdots)(1 - 3z + \cdots)$$

$$= \frac{1}{z} - 5 + \cdots$$

Similarly, the strength of the third-order pole at $z = -1$ is -4, since the leading term of the Laurent series near $z = -1$ is $f(z) = -4/(z+1)^3$.

Example 3.5.2 Describe the singularities of the function

$$f(z) = \frac{z+1}{z \sin z}$$

Using the Taylor series for $\sin z$

$$f(z) = \frac{z+1}{z\left(z - \frac{z^3}{3!} + \frac{z^5}{5!} - \cdots\right)} = \frac{z+1}{z^2\left(1 - \frac{z^2}{3!} + \frac{z^4}{5!} - \cdots\right)}$$

$$= \frac{z+1}{z^2}\left[1 + \left(\frac{z^2}{3!} - \frac{z^4}{5!} + \cdots\right) + \left(\frac{z^2}{3!} - \frac{z^4}{5!} + \cdots\right)^2 + \cdots\right]$$

$$= \left(\frac{1}{z^2} + \frac{1}{z}\right)\left(1 + \frac{z^2}{3!} + \cdots\right) = \frac{1}{z^2} + \frac{1}{z} + \cdots$$

we find that the function $f(z)$ has a second order (**double**) pole at $z = 0$ with strength 1.

Example 3.5.3 Describe the singularities of the function

$$f(z) = \tan z = \frac{\sin z}{\cos z}$$

Here the function $f(z)$ has simple poles with strength 1 at $z = \pi/2 + m\pi$ for $m = 0, \pm 1, \pm 2, \ldots$. It is sometimes useful to make a transformation of variables to transform the location of the poles to the origin: $z = z_0 + z'$, where $z_0 = \pi/2 + m\pi$, so that

$$f(z) = \frac{\sin(\pi/2 + m\pi + z')}{\cos(\pi/2 + m\pi + z')}$$

$$= \frac{\sin(\pi/2 + m\pi)\cos z' + \cos(\pi/2 + m\pi)\sin z'}{\cos(\pi/2 + m\pi)\cos z' - \sin(\pi/2 + m\pi)\sin z'}$$

$$= \frac{(-1)^m \cos z'}{(-1)^{m+1} \sin z'}$$

$$= -\frac{(1 - (z')^2/2! + \cdots)}{(z' - (z')^3/3! + \cdots)} = -\frac{1}{z'}\frac{(1 - (z')^2/2! + \cdots)}{(1 - (z')^2/3! + \cdots)}$$

$$= -\frac{1}{z'}\left(1 - \left(\frac{1}{2!} - \frac{1}{3!}\right)z'^2 + \cdots\right) = -\frac{1}{z'}\left(1 - \frac{1}{3}z'^2 + \cdots\right)$$

$$= -\frac{1}{z'} + \frac{1}{3}z' + \cdots$$

$$= -\frac{1}{z - (\pi/2 + m\pi)} + \frac{1}{3}(z - (\pi/2 + m\pi)) + \cdots$$

Hence $f(z) = \tan z$ always has a simple pole of strength -1 at $z = \frac{\pi}{2} + m\pi$.

Example 3.5.4 Discuss the pole singularities of the function

$$f(z) = \frac{\log(z+1)}{(z-1)}$$

The function $f(z)$ is multivalued with a branch point at $z = -1$, hence following the procedure in Section 2.2 we make $f(z)$ single valued by introducing a branch cut. We take the cut from the branch point at $z = -1$ to $z = \infty$ along the negative real axis with $z = re^{i\theta}$ for $-\pi \le \theta < \pi$; this branch fixes $\log(1) = 0$. With this choice of branch, $f(z)$ has a simple pole at $z = 1$ with strength $\log 2$. We shall discuss the nature of branch point singularities later in this section.

Sometimes we might have different types of singularities depending on which branch of a multivalued function we select.

Example 3.5.5 Discuss the pole singularities of the function

$$f(z) = \frac{z^{1/2} - 1}{z - 1}$$

We let $z = 1 + t$, so that

$$f(z) = \frac{(1+t)^{1/2} - 1}{t} = \frac{\pm\sqrt{1+t} - 1}{t}$$

where $\pm$ denotes the two branches of the square root function with $\sqrt{x} \ge 0$ for $x \ge 0$. (The point $z = 0$ is a square root branch point).

The Taylor series of $\sqrt{1+t}$ is

$$\sqrt{1+t} = 1 + \frac{1}{2}t - \frac{1}{8}t^2 + \cdots$$

Thus for the "+" branch

$$f(z) = \frac{\frac{t}{2} - \frac{1}{8}t^2 + \cdots}{t} = \frac{1}{2} - \frac{1}{8}t + \cdots$$

whereas for the "−" branch

$$f(z) = \frac{-2 - \frac{t}{2} + \frac{1}{8}t^2 - \cdots}{t} = \frac{-2}{t} - \frac{1}{2} + \frac{1}{8}t - \cdots$$

On the + (principal) branch, $f(z)$ is analytic in the neighborhood of $t = 0$; that is, $t = 0$ is a removable singularity. For the − branch, $t = 0$ is a simple pole with strength -2.

An isolated singular point that is neither removable nor a pole is called an **essential singular point**. An essential singular point has a "full" Laurent series in the sense that given $f(z) = \sum_{n=-\infty}^{\infty} C_n(z - z_0)^n$, then for any $N > 0$ there is an $n < -N$ such that $C_n \neq 0$; that is, the series for negative n does not terminate. If this were not the case, then $f(z)$ would have a pole (if $C_n = 0$ for $n < -N$ and $C_{-N} \neq 0$, then $f(z)$ would have a pole of order N with strength C_{-N}).

The prototypical example of an essential singular point is given by the function

$$f(z) = e^{1/z} \tag{3.5.3}$$

which has the following Laurent series (Eq. (3.3.17)) about the essential singular point at $z = 0$

$$f(z) = \sum_{n=0}^{\infty} \frac{1}{n! z^n} \tag{3.5.4}$$

Because $f'(z) = -e^{1/z}/z^2$ exists for all points $z \neq 0$, it is clear that $f(z)$ is analytic in the neighborhood of $z = 0$; hence it is isolated (as it must be for $z = 0$ to be an essential singular point).

If we use polar coordinates $z = re^{i\theta}$, then Eq. (3.5.3) yields

$$f(z) = e^{\frac{1}{r} e^{-i\theta}} = e^{\frac{1}{r}(\cos\theta - i\sin\theta)}$$

$$= e^{\frac{1}{r}\cos\theta} \left[\cos\left(\frac{\sin\theta}{r}\right) - i\sin\left(\frac{\sin\theta}{r}\right) \right]$$

whereupon the modulus of $f(z)$ is given by

$$|f(z)| = e^{\frac{1}{r}\cos\theta}$$

Clearly for values of θ such that $\cos\theta > 0$, $f(z) \to \infty$ as $r \to 0$, and for $\cos\theta < 0$, $f(z) \to 0$ as $r \to 0$. Indeed, if we let r take values on a suitable curve, namely, $r = (1/R)\cos\theta$ (i.e., the points (r, θ) lie on a circle of diameter $1/R$ tangent to the imaginary axis), then

$$f(z) = e^R [\cos(R\tan\theta) - i\sin(R\tan\theta)] \tag{3.5.5}$$

and

$$|f(z)| = e^R \tag{3.5.6}$$

Thus $|f(z)|$ may take on any positive value other than zero by the appropriate choice of R. As $z \to 0$ on this circle, $\theta \to \pi/2$ (and $\tan\theta \to \infty$) with R fixed, then the coefficient in brackets in Eq. (3.5.5) takes on *all values* on the unit circle *infinitely often*. Hence we see that $f(z)$ takes on *all* nonzero complex values with modulus (3.5.6) infinitely often.

In fact, this example describes a general feature of essential singular points discovered by Picard (Picard's Theorem). He showed that in any neighborhood of an essential singularity of function, $f(z)$ assumes all values, except possibly one of them, an infinite number of times. The following result owing to Weierstrass is similar and more easily shown.

Theorem 3.5.1 If $f(z)$ has an essential singularity at $z = z_0$, then for any complex number w, $f(z)$ becomes arbitrarily close to w in a neighborhood of z_0. That is, given w, and any $\epsilon > 0$, $\delta > 0$, there is a z such that

$$|f(z) - w| < \epsilon \qquad (3.5.7)$$

whenever $|z - z_0| < \delta$.

Proof We prove this by contradiction. Suppose $|f(z) - w| > \epsilon$ whenever $|z - z_0| < \delta$, where δ is small enough such that $f(z)$ is analytic in the region $0 < |z - z_0| < \delta$. Thus in this region

$$h(z) = \frac{1}{f(z) - w}$$

is analytic, and hence bounded; specifically, $|h(z)| < 1/\epsilon$. The function $f(z)$ is not identically constant, otherwise $f(z)$ would be analytic and hence would not possess an essential singular point. Because $h(z)$ is analytic and bounded, it is representable by a power series $h(z) = \sum_{n=0}^{\infty} C_n(z - z_0)^n$, thus its only possible singularity is removable. By choosing $C_0 = h(z_0)$, it follows that $h(z)$ is analytic for $|z - z_0| < \delta$. Consequently

$$f(z) = w + \frac{1}{h(z)}$$

and $f(z)$ is either analytic with $h(z) \neq 0$ or else $f(z)$ has a pole of order N, strength C_N, where C_N is the first nonzero coefficient of the term $(z - z_0)^N$ in the Taylor series representation of $h(z)$. In either case, this contradicts the hypothesis that $f(z)$ has an essential singular point in the neighborhood of $z = z_0$. ∎

Functions that have only isolated singularities, while very special, turn out to be important in applications. An **entire** function is one that is analytic everywhere in the finite z plane. As proved in Chapter 2, the only function analytic everywhere, including the point at infinity, is a constant (Section 2.6, i.e., Liouville's Theorem). Entire functions are either constant functions, or at infinity they have isolated poles or essential singularities. Some of the common entire functions include (a) polynomials, (b) exponential functions, and (c) sine/cosine functions. For example, $f(z) = z$, $f(z) = e^z$, $f(z) = \sin z$ are all entire functions.

As mentioned earlier, one can easily ascertain the nature of the singularity at $z = \infty$ by making the transformation $z = 1/t$ and investigating the behavior of the function near $t = 0$. Polynomials have poles at $z = \infty$, the order of which corresponds to the order of the polynomial. For example, $f(z) = z$ has a simple pole at infinity (strength unity) because $f(t) = 1/t$. Similarly, $f(z) = z^2$ has a double pole at $z = \infty$, etc. The entire functions e^z and $\sin z$ have essential singular points at $z = \infty$. Indeed, the Taylor series for $\sin z$ shows that the Laurent series around $t = 0$ does not terminate in any finite negative power:

$$\sin \frac{1}{t} = \sum_{n=0}^{\infty} \frac{(-1)^n}{t^{2n+1}(2n+1)!}$$

hence it follows that $t = 0$ or $z = \infty$ is an essential singular point.

The next level of complication after an entire function is a function that has only poles in the finite z plane. Such a function is called a **meromorphic function**. As with entire functions, meromorphic functions may have essential singular points at infinity. A meromorphic function is a ratio of entire functions. For example, a **rational function** (i.e., a ratio of polynomials)

$$R(z) = \frac{A_N z^N + A_{N-1} z^{N-1} + \cdots + A_1 z + A_0}{B_M z^M + B_{M-1} z^{M-1} + \cdots + B_1 z + B_0} \tag{3.5.8}$$

is meromorphic. It has only poles as its singular points. The denominator is a polynomial, whose zeroes correspond to the poles of $R(z)$. For example, the function

$$R(z) = \frac{z^2 - 1}{z^5 + 2z^3 + z}$$

$$= \frac{(z+1)(z-1)}{z(z^4 + 2z^2 + 1)} = \frac{(z+1)(z-1)}{z(z^2 + 1)^2}$$

$$= \frac{(z+1)(z-1)}{z(z+i)^2(z-i)^2}$$

has poles at $z = 0$ (simple), at $z = \pm i$ (both double), and zeroes (simple) at $z = \pm 1$.

The function $f(z) = (\sin z)/(1+z)$ is meromorphic. It has a pole at $z = -1$, owing to the vanishing of $(1 + z)$, and an essential singular point at $z = \infty$ due to the behavior of $\sin z$ near infinity (as discussed earlier).

There are other types of singularities of a complex function that are non-isolated. In Chapter 2, Section 2.2–2.3, we discussed at length the various aspects of multivalued functions. Multivalued functions have *branch points*. We recall that their characteristic property is the following. If a circuit is made around a sufficiently small, simple closed contour enclosing the branch point, then the value assumed by the function at the end of the circuit differs from its initial value. A branch point is an example of a nonisolated singular point, because a circuit (no matter how small) around the branch point results in a discontinuity. We also recall that in order to make a multivalued function $f(z)$ single-valued, we must introduce a branch cut. Since $f(z)$ has a discontinuity across the cut, we shall consider the branch cut as a singular curve (it is not simply a point). However, it is important to recognize that a branch cut may be moved, as opposed to a branch point, and therefore the nature of its singularity is somewhat artificial. Nevertheless, once a concrete single-valued branch is defined, we must have an associated branch cut. For example, the function

$$f(z) = \frac{\log z}{z}$$

has branch points at $z = 0$ and $z = \infty$. We may introduce a branch cut along the positive real axis: $z = re^{i\theta}, 0 \le \theta < 2\pi$. We note that $z = 0$ is a branch point and not a pole because $\log z$ has a jump discontinuity as we encircle $z = 0$. It is not analytic in a neighborhood of $z = 0$; hence $z = 0$ is not an isolated singular point. (We note the difference between this example and Example 3.5.4 earlier.)

Another type of singular point is a **cluster point**. A cluster point is one in which a sequence of isolated singular points of a single-valued function $f(z)$ cluster about a point, say, $z = z_0$, in such a way that there are an infinite number of isolated singular points in any arbitrarily small circle about $z = z_0$. The standard example is given by the function $f(z) = \tan(1/z)$. As $z \to 0$ along the real axis, $\tan(1/z)$ has poles at the locations $z_n = 1/(\pi/2 + n\pi)$, which cluster because any small neighborhood of the origin contains an infinite number of them. There is no Laurent series representation valid in the neighborhood of a cluster point.

Another singularity that arises in applications is associated with the case of two analytic functions that are separated by a closed curve or an infinite line. For example, if C is a suitable closed contour and if $f(z)$ is defined as

$$f(z) = \begin{cases} f_i(z) & z \text{ inside } C \\ f_o(z) & z \text{ outside } C \end{cases} \qquad (3.5.9)$$

where $f_i(z)$ and $f_o(z)$ are analytic in their respective regions and not equal on C, then the boundary C is a singular curve across which the function has a jump discontinuity. We shall refer to this as a **boundary jump discontinuity**.

An example of such a situation is given by

$$f(z) = \frac{1}{2\pi i} \oint_C \frac{1}{\zeta - z} \, d\zeta = \begin{cases} f_i(z) = 1 & z \text{ inside } C \\ f_o(z) = 0 & z \text{ outside } C \end{cases} \qquad (3.5.10)$$

The discontinuity depends entirely on the location of C, which is provided in the definition of the function $f(z)$ via the integral representation. We note that the functions $f_i(z) = 1$ and $f_o(z) = 0$ are analytic. Both of these functions can be continued beyond the boundary C in a natural way; just take $f_i(z) = 1$ and $f_o(z) = 0$, respectively. Indeed, functions obtained through integral representations such as Eq. (3.5.10) have a property by which the function $f(z)$ is comprised of functions such as $f_i(z)$ and $f_o(z)$, which are analytic inside and outside the original contour C.

In Chapter 7 we will study questions and applications that deal with equations that are defined in terms of functions that have properties very similar to Eq. (3.5.9). Such equations are called Riemann–Hilbert factorization problems.

3.5.1 Analytic Continuation and Natural Barriers

Frequently, one is given formulae that are valid in a limited region of space, and the goal is to find a representation, either in closed series form, integral representation, or otherwise that is valid in a larger domain. The process of extending the range of validity of a representation or more generally extending the region of definition of an analytic function is called **analytic continuation**. This was briefly discussed at the end of Section 3.2 in Theorems 3.2.6 and 3.2.7. We elaborate further on this important issue in this section.

A typical example is the following. Consider the function defined by the series

$$f(z) = \sum_{n=0}^{\infty} z^n \qquad (3.5.11)$$

when $|z| < 1$. When $|z| \to 1$, the series clearly diverges because z^n does not approach zero as $n \to \infty$. On the other hand, the function defined by

$$g(z) = \frac{1}{1-z} \tag{3.5.12}$$

which is defined *for all z except the point* $z = 1$, is such that $g(z) = f(z)$ for $|z| < 1$ because the Taylor series representation of Eq. (3.5.12) about $z = 0$ is Eq. (3.5.11) inside the unit circle. In fact, we claim that $g(z)$ is the unique analytic continuation of $f(z)$ outside the unit circle. The function $g(z)$ has a pole at $z = 1$. This example is representative of a far more general situation.

The relevant theorem was given earlier as Theorem 3.2.6, which implies the following.

Theorem 3.5.2 A function that is analytic in a domain D is uniquely determined either by values in some interior domain of D or along an arc interior to D.

The fact that a "global" analytic function can be deduced from such a relatively small amount of information illustrates just how powerful the notion of analyticity really is.

The example above (Eqs. (3.5.11), (3.5.12)) shows that the function $f(z)$, which is represented by Eq. (3.5.11) inside the unit circle, uniquely determines the function $g(z)$, which is represented by Eq. (3.5.12) that is valid everywhere.

We remark that the function $1/(1-z)$ is the only analytic function (analytic apart from a pole at $z = 1$) that can assume the values $f(x) = 1/(1-x)$ along the real x axis. This also shows how prescribing values along a curve fixes the analytic extension. Similarly, the function $f(z) = e^{kz}$ (for constant k) is the only analytic function that can be extended from $f(x) = e^{kx}$ on the real x axis.

Chains of analytic continuations are sometimes required, and care may be necessary. For example, consider the regions A, B, and C and the associated analytic functions f, g, and h, respectively (see Figure 3.5.1), and let $A \cap B$ denote the usual intersection of two sets.

Referring to Figure 3.5.1, Theorem 3.5.2 (or Theorem 3.2.6) implies that if $g(z)$ and $f(z)$ are analytic and have a domain $A \cap B$ in common, where $f(z) = g(z)$, then $g(z)$ is the analytic continuation of $f(z)$. Similarly, if $h(z)$ and $g(z)$ are analytic and have a domain $B \cap C$ in common, where $h(z) = g(z)$, then $h(z)$ is the analytic continuation of $g(z)$. However, we cannot conclude that $h(z) = f(z)$ because the intersecting regions A, B, C might enclose a branch point of a multivalued function.

The method of proof (of Theorem 3.2.6 or Theorem 3.5.2) extends the function locally by Taylor series arguments. We note that if we enclose a

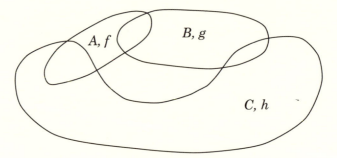

Fig. 3.5.1. Analytic continuation in domains A, B, and C

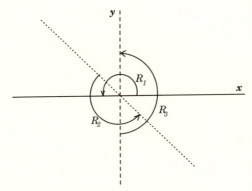

Fig. 3.5.2. Overlapping domains R_1, R_2, R_3

branch point, we move onto the next sheet of the corresponding Riemann surface.

For example, consider the multivalued function

$$f(z) = \log z = \log r + i\theta \qquad (3.5.13)$$

with three regions defined (see Figure 3.5.2) in the sectors

$$R_1: \ 0 \le \theta < \pi$$

$$R_2: \ \frac{3\pi}{4} \le \theta < \frac{7\pi}{4}$$

$$R_3: \ \frac{3\pi}{2} \le \theta < \frac{5\pi}{2}$$

The branches of $\log z$ defined by Eq. (3.5.13) in their respective regions R_1, R_2, and R_3 are related by analytic continuations. Namely, if we call $f_i(z)$ the function Eq. (3.5.13) defined in region R_i, then $f_2(z)$ is the analytic continuation of $f_1(z)$, and $f_3(z)$ is the analytic continuation of $f_2(z)$. Note, however, that $f_3(z) \neq f_1(z)$, that is, the same point $z_0 = Re^{i\pi/4} = Re^{9i\pi/4}$ has

$$f_1(z_0) = \log R + i\pi/4$$
$$f_3(z_0) = \log R + 9i\pi/4$$

This example clearly shows that after analytic continuation the function does not return, upon a complete circuit, to the same value. Indeed, in this example we progress onto the adjacent sheet of the multivalued function because we have enclosed the branch point $z = 0$ of $f(z) = \log z$.

On the other hand, if in a simply connected region, there are no singular points enclosed between any two distinct paths of analytic continuation that together form a closed path, then we could cover the enclosed region with small overlapping subregions and use Taylor series to analytically continue our function and obtain a single valued function. This is frequently called the Monodromy Theorem, which we now state.

Theorem 3.5.3 (Monodromy Theorem) Let D be a simply connected domain and $f(z)$ be analytic in some disk $D_0 \subset D$. If the function can be analytically continued along any two distinct smooth contours C_1 and C_2 to a point in D, and if there are no singular points enclosed within C_1 and C_2, then the result of each analytic continuation is the same and the function is single valued.

In fact, the theorem can be extended to cover the case where the region enclosed by contours C_1 and C_2 contains, at most, isolated singular points, $f(z)$ having a Laurent series of the form (3.5.1) in the neighborhood of any singular point. Thus the enclosed region can have poles or essential singular points.

There are some types of singularities that are, in a sense, so serious that they prevent analytic continuation of the function in question. We shall refer to such a (nonisolated) singularity as a **natural barrier** (often referred to in the literature as a **natural boundary**). A prototypical example of a natural barrier is contained in the function defined by the series

$$f(z) = \sum_{n=0}^{\infty} z^{2^n} \tag{3.5.14}$$

The series (3.5.14) converges for $|z| < 1$, which can be easily seen from the ratio test. We shall sketch an argument that shows that analytic continuation to

$|z| > 1$ is impossible. Because

$$f(z^2) = \sum_{n=0}^{\infty} \left(z^2\right)^{2^n} = \sum_{n=0}^{\infty} z^{2^{n+1}} = \sum_{n=1}^{\infty} z^{2^n} \qquad (3.5.15)$$

it follows that f satisfies the functional equation

$$f(z^2) = f(z) - z \qquad (3.5.16)$$

From Eq. (3.5.14) it is clear that $z_0 = 1$ is a singular point because $f(1) = \infty$. It then follows from Eq. (3.5.16) that $f(z_1) = \infty$, where $z_1^2 = 1$ (i.e., $z_1 = \pm 1$). Similarly, $f(z_2) = \infty$, where $z_2^4 = 1$, because Eq. (3.5.14) implies

$$f(z^4) = f(z^2) - z^2 = f(z) - z - z^2 \qquad (3.5.17)$$

Mathematical induction then yields

$$f(z^{2^m}) = f(z) - \sum_{j=0}^{m-1} z^{2^j} \qquad (3.5.18)$$

Hence the value of the function $f(z)$ at all points z_m on the unit circle satisfying $z^{2^m} = 1$ (i.e., all roots of unity) is infinite: $f(z_m) = \infty$. Therefore all these points are singular points. In order for the function (3.5.14) to be analytically continuable to $|z| \geq 1$, at the very least we need $f(z)$ to be analytic on some small arc of the unit circle $|z| = 1$. However, no matter how small an arc we take on this circle, the above argument shows that there exist points z_m (roots of unity, satisfying $z^{2^m} = 1$) on any such arc such that $f(z_m) = \infty$. Because an analytic function must be bounded, analytic continuation is impossible.

Exotic singularities such as natural barriers are found in solutions of certain nonlinear differential equations arising in physical applications (see, for example, Eqs. (3.7.52) and (3.7.53)). Consequently, their study is not merely a mathematical artifact.

Problems for Section 3.5

1. Discuss the type (removable, pole and order, essential, branch, cluster, natural barrier, etc.); if the type is a pole give the strength of the pole, and give the nature (isolated or not) of all singular points associated with the following functions. Include the point at infinity.

(a) $\dfrac{e^{z^2} - 1}{z^2}$ (b) $\dfrac{e^{2z} - 1}{z^2}$ (c) $e^{\tan z}$ (d) $\dfrac{z^3}{z^2 + z + 1}$

(e) $\dfrac{z^{1/3} - 1}{z - 1}$ (f) $\log(1 + z^{1/2})$ (g) $f(z) = \begin{cases} z^2 & |z| \le 1 \\ 1/z^2 & |z| > 1 \end{cases}$

(h) $f(z) = \displaystyle\sum_{n=1}^{\infty} \dfrac{z^{n!}}{n!}$ (i) $\operatorname{sech} z$ (j) $\coth 1/z$

2. Evaluate the integral $\oint_C f(z)\,dz$, where C is a unit circle centered at the origin and where $f(z)$ is given below.

   (a) $\dfrac{g(z)}{z - w}$, $g(z)$ entire   (b) $\dfrac{z}{z^2 - w^2}$ (c) $z e^{1/z^2}$

   (d) $\cot z$ (e) $\dfrac{1}{8z^3 + 1}$

3. Show that the functions below are meromorphic; that is, the only singularities in the finite z plane are poles. Determine the location, order and strength of the poles.

   (a) $\dfrac{z}{z^4 + 2}$ (b) $\tan z$ (c) $\dfrac{z}{\sin^2 z}$

   (d) $\dfrac{e^z - 1 - z}{z^4}$ (e) $\dfrac{1}{2\pi i} \oint_C \dfrac{w\,dw}{(w^2 - 2)(w - z)}$

   C is the unit circle centered at the origin. First find the function for $|z| < 1$, then analytically continue the function to $|z| \ge 1$.

4. Discuss the analytic continuation of the following functions:

   (a) $\displaystyle\sum_{n=0}^{\infty} z^{2n}$, $|z| < 1$

   (b) $\displaystyle\sum_{n=0}^{\infty} \dfrac{z^{n+1}}{n + 1}$, $|z| < 1$

   Hint: (b) is also represented by the integral

   $$\int_0^z \left(\sum_{n=0}^{\infty} z'^n \right) dz'$$

   (c) $\displaystyle\sum_{n=0}^{\infty} z^{4^n}$

5. Suppose we know a function $f(z)$ is analytic in the finite z plane apart from singularities at $z = i$ and $z = -i$. Moreover, let $f(z)$ be given by the Taylor series

$$f(z) = \sum_{j=0}^{\infty} a_j z^j$$

where a_j is known. Suppose we calculate $f(z)$ and its derivatives at $z = 3/4$ and compute a Taylor series in the form

$$f(z) = \sum_{j=0}^{\infty} b_j \left(z - \frac{3}{4} \right)^j$$

Where would this series converge? How could we use this to compute $f(z)$? Suppose we wish to compute $f(2.5)$; how could we do this by series methods?

*3.6 Infinite Products and Mittag–Leffler Expansions

In previous sections we have considered various kinds of infinite series representations (i.e., Taylor series, Laurent series) of functions that are analytic in suitable domains. Sometimes in applications it is useful to consider infinite products to represent our functions.

If $\{a_k\}$ is a sequence of complex numbers, then an infinite product is denoted by

$$P = \prod_{k=1}^{\infty} (1 + a_k) \qquad (3.6.1)$$

We say that the infinite product (3.6.1) converges if (a) the sequence of partial products P_n

$$P_n = \prod_{k=1}^{n} (1 + a_k) \qquad (3.6.2)$$

converge to a finite limit, and (b) that for N_0 large enough

$$\lim_{N \to \infty} \prod_{k=N_0}^{N} (1 + a_k) \neq 0 \qquad (3.6.3)$$

If Eq. (3.6.3) is violated, that is, $\lim_{N \to \infty} \prod_{k=N_0}^{N} (1 + a_k) = 0$ for all N_0, then we will consider the product to diverge. The reason for this is that the following infinite sum turns out to be intimately connected to the infinite product (see below, Eq. (3.6.4):

$$S = \sum_{k=1}^{\infty} \log(1 + a_k)$$

and it would not make sense if $P = 0$.

Moreover, in analogy with infinite sums, if the infinite product $\prod_{k=1}^{\infty}(1+|a_k|)$ converges, we say P **converges absolutely**. If $\prod_{k=1}^{\infty}(1 + |a_k|)$ diverges but P converges, we say that P **converges conditionally**. Clearly, if one of the $a_k = -1$, then $P_n = P = 0$. For now we shall exclude this trivial case and assume $a_k \neq -1$, for all k.

Equation (3.6.2) implies that $P_n = (1 + a_n)P_{n-1}$, whereupon $a_n = (P_n - P_{n-1})/P_{n-1}$. Thus if $P_n \to P$, we find that $a_n \to 0$, which is a necessary but not sufficient condition for convergence (note also this necessary condition would imply Eq. (3.6.3) for N_0 large enough).

A useful test for convergence of an infinite product is the following.

If the sum

$$S = \sum_{k=1}^{\infty} \log(1 + a_k) \tag{3.6.4}$$

converges, then so does the infinite product (3.6.1). We shall restrict $\log z$ to its principal branch.

Calling $S_n = \sum_{k=1}^{n} \log(1 + a_k)$, the nth partial sum of S, then

$$e^{S_n} = e^{\sum_{k=1}^{n} \log(1+a_k)} = P_n$$

and as $n \to \infty$, $e^{S_n} \to e^S = P$. Note again that if $P = 0$, then $S = -\infty$, which we shall not allow, excluding the case where individual factors vanish.

The above definition applies as well to products of *functions* where, for example, a_k is replaced by $a_k(z)$ for z in a region $\mathcal{R}$. We say that if a product of functions converges for each z in a region $\mathcal{R}$, then it converges in $\mathcal{R}$. The convergence is said to be uniform in $\mathcal{R}$ if the partial sequence of products obey $P_n(z) \to P(z)$ uniformly in $\mathcal{R}$. Uniformity is the same concept as that discussed in Section 3.4; namely, the estimate involved is independent of z. There is a so-called Weierstrass M test for products of functions, which we now give.

Theorem 3.6.1 Let $a_k(z)$ be analytic in a domain D for all k. Suppose for all $z \in D$ and $k \geq N$ either

(a) $|\log(1 + a_k(z))| \leq M_k$

or

(b) $|a_k(z)| \le M_k$

where $\sum_{n=1}^{\infty} M_k < \infty$. Then the product

$$P(z) = \prod_{k=1}^{\infty} (1 + a_k(k))$$

is uniformly convergent to an analytic function $P(z)$ in D. Furthermore $P(z)$ is zero only when a finite number of its factors $1 + a_k(z)$ are zero in D.

Proof For $n \ge N$, define

$$P_n(z) = \prod_{k=N}^{n} (1 + a_k(z))$$

$$S_n(z) = \sum_{k=N}^{n} \log (1 + a_k(z))$$

Using inequality (a) in the hypothesis of Theorem 3.6.1 for any $z \in D$ with $m > N$ yields

$$|S_m(z)| \le \sum_{k=N}^{m} M_k \le \sum_{k=1}^{\infty} M_k = M < \infty$$

Similarly, for any $z \in D$ with $n > m \ge N$ we have

$$|S_n(z) - S_m(z)| \le \sum_{k=m+1}^{n} M_k \le \sum_{k=m+1}^{\infty} M_k \le \epsilon_m$$

where $\epsilon_m \to 0$ as $m \to \infty$, and $S_n(z)$ is a uniformly convergent Cauchy sequence. Because $P_k(z) = \exp S_k(z)$, it follows that

$$(P_n(z) - P_m(z)) = e^{S_m(z)} \left(e^{S_n(z) - S_m(z)} - 1 \right)$$

From the Taylor series, $e^w = \sum_{n=0}^{\infty} w^n/n!$, we have

$$|e^w| \le e^{|w|}, \qquad |e^w - 1| \le e^{|w|} - 1$$

whereupon from the above estimates we have

$$|P_n(z) - P_m(z)| \le e^M (e^{\epsilon_m} - 1)$$

and hence $\{P_n(z)\}$ is a uniform Cauchy sequence. Thus (see Section 3.4) $P_n(z) \to P(z)$ uniformly in D and $P(z)$ is analytic because $P_n(z)$ is a sequence of analytic functions. Moreover, we have

$$|P_m(z)| = |e^{\text{Re}S_m(z)+i\text{Im}S_m(z)}|$$

$$= |e^{\text{Re}S_m(z)}|$$

$$\geq e^{-|\text{Re}S_m(z)|}$$

$$= e^{-|\text{Re}S_m(z)+i\text{Im}S_m(z)|}$$

$$\geq e^{-M}$$

Thus $P_m(z) \geq e^{-M}$. Because M is independent of m, $P(z) \neq 0$ in D.

Because we may write $P(z) = \prod_{k=1}^{N-1}(1 + a_k(z))\tilde{P}(z)$, we see that $P(z) = 0$ only if any of the factors $(1 + a_k(z)) = 0$, for $k = 1, 2, \ldots, N - 1$. (The estimate **(a)** of Theorem 3.6.1 is invalid for such a possibility.) It also follows directly from the analyticity of $a_k(z)$ that

$$P_n(z) = \prod_{k=1}^{N-1}(1 + a_k(z))\tilde{P}_n(z)$$

is a uniformly convergent sequence of analytic functions.

Finally, we note that the first hypothesis, **(a)**, follows from the second hypothesis, **(b)**, as is shown next.

The Taylor series of $\log(1 + w)$ is given by

$$\log(1 + w) = \left(w - \frac{w^2}{2} + \frac{w^3}{3} + \cdots + (-1)^{n-1}\frac{w^n}{n} + \cdots\right)$$

Hence

$$|\log(1 + w)| \leq |w| + \frac{|w|^2}{2} + \frac{|w|^3}{3} + \cdots + \frac{|w|^n}{n} + \cdots$$

and for $|w| \leq 1/2$ we have

$$|\log(1 + w)| \leq |w|\left(1 + \frac{1}{2} + \frac{1}{2^2} + \cdots + \frac{1}{2^n} + \cdots\right)$$

$$\leq |w|\left(\frac{1}{1 - 1/2}\right)$$

$$= 2|w|$$

Thus for $|a_k(z)| < 1/2$

$$|\log(1 + a_k(z))| \leq 2|a_k(z)|$$

If we assume that $|a_k(z)| \leq M_k$, with $\sum_{k=1}^{\infty} M_k < \infty$, it is clear that there is a $k > N$ such that $|a_k(z)| < 1/2$, and we have hypothesis (**a**). The theorem goes through as before simply with M_k replaced by $2M_k$. ■

As an example, consider the product

$$F(z) = \prod_{k=1}^{\infty} \left(1 - \frac{z^2}{k^2} \right) \tag{3.6.5}$$

Theorem 3.6.1 implies that $F(z)$ represents an entire function with simple zeroes as $z = \pm 1, \pm 2, \ldots$. In this case, $a_k(z) = z^2/k^2$. Inside the circle $|z| < R$ we have $|a_k(z)| \leq R^2/k^2$. Because

$$\sum_{k=1}^{\infty} \frac{R^2}{k^2} < \infty$$

Theorem 3.6.1 shows that the function $F(z)$ is analytic for all finite z inside the circle $|z| < R$. Because R can be made arbitrarily large, $F(z)$ is entire, and the only zeroes of $F(z)$ correspond to the vanishing of $(1 - z^2/k^2)$, for $k = 1, 2, \ldots$.

Next we construct a function with simple zeroes at $z = 1, 2, \ldots$, and no other zeroes. We shall show that the function

$$G(z) = \prod_{k=1}^{\infty} \left\{ \left(1 - \frac{z}{k} \right) e^{z/k} \right\} \tag{3.6.6}$$

is one such function. At first it may seem that the "convergence factor" $e^{z/k}$ could be dropped. But we will show that without this term, the product would diverge. In fact, the $e^{z/k}$ term is such that the contributions of the $(1/k)$ term inside the product exactly cancels, that is,

$$\left(1 - \frac{z}{k} \right) e^{z/k} = \left(1 - \frac{z}{k} \right) \left(1 + \frac{z}{k} + \frac{z^2}{2!k^2} + \cdots \right) = 1 - \frac{z^2}{2!k^2} + \cdots$$

We note the Taylor series

$$\log \left((1 - w)e^w \right) = \log(1 - w) + w$$
$$= - \left(\frac{w^2}{2} + \frac{w^3}{3} + \frac{w^4}{4} + \cdots \right)$$
$$= - \left(w^2 \right) \left(\frac{1}{2} + \frac{w}{3} + \frac{w^2}{4} + \cdots \right)$$

hence for $|w| < 1/2$

$$|\log(1 - w)e^w| \leq |w|^2 \left(\frac{1}{2} + \frac{1}{2^2} + \frac{1}{2^3} + \cdots \right)$$

$$= |w|^2 \left(\frac{1}{2} \right) \left(\frac{1}{1 - 1/2} \right)$$

$$= |w|^2$$

Thus for $|z| < R$ and $k > 2R$, for any fixed value R

$$\left| \log \left(1 - \frac{z}{k} \right) e^{z/k} \right| \leq \left| \frac{z}{k} \right|^2 \leq \left(\frac{R}{k} \right)^2$$

hence from Theorem 3.6.1 the product (3.6.6) converges uniformly to an entire function with simple zeroes at $(1 - z/k) = 0$ for $k = 1, 2, \ldots$; that is, for $z = 1, 2, \ldots$.

We now show that $\prod_{k=1}^{\infty} (1 - z/k)$ diverges for $z \neq 0$. We note that for any integer $n \geq 1$

$$H_n = \prod_{k=1}^{n} \left(1 - \frac{z}{k} \right)$$

$$= \prod_{k=1}^{n} \left(1 - \frac{z}{k} \right) e^{z/k} \cdot e^{-z/k}$$

$$= e^{-zS(n)} \prod_{k=1}^{n} \left\{ \left(1 - \frac{z}{k} \right) e^{z/k} \right\}$$

where $S(n) = 1 + 1/2 + 1/3 + \cdots + 1/n$. Thus, using the above result that Eq. (3.6.6) converges and because $S(n) \to \infty$, as $n \to \infty$ we find that for $\operatorname{Re} z < 0$, $H_n \to \infty$; for $\operatorname{Re} z > 0$, $H_n \to 0$; and for $\operatorname{Re} z = 0$ and $\operatorname{Im} z \neq 0$, H_n does not have a limit as $n \to \infty$. By our definition of convergence of an infinite product (Eq. (3.6.1) below) we conclude that H is a divergent product.

Often the following observation is useful. If $F(z)$ and $G(z)$ are two entire functions that have the same zeroes and multiplicities, then there is an entire function $h(z)$ satisfying

$$F(z) = e^{h(z)} G(z) \tag{3.6.7}$$

This follows from the fact that the function $F(z)/G(z)$ is entire with no zeroes; the ratio makes all other zeroes of F and G removable singularities. Because F/G is analytic without zeroes, it has a logarithm that is everywhere analytic: $\log(F/G) = h(z)$.

It is natural to ask whether an entire function can be constructed that has zeroes of specified orders at assigned points with no other zeroes, or similarly, whether a meromorphic function can be constructed that has poles of specified orders at assigned points with no other poles. These questions lead to certain infinite products (the so-called Weierstrass products for entire functions) and infinite series (Mittag–Leffler expansions, for meromorphic functions). These notions extend our ability to represent functions of a certain specified character. Earlier we only had Taylor/Laurent series representations available.

First we shall discuss representations of meromorphic functions. In what follows we shall use certain portions of the Laurent series of a given meromorphic function. Namely, near any pole (of order N_j at $z = z_j$) of a meromorphic function we have the Laurent expansion

$$f(z) = \sum_{n=1}^{N_j} \frac{a_{n,j}}{(z - z_j)^n} + \sum_{n=0}^{\infty} b_{n,j}(z - z_j)^n$$

The first part contains the pole contribution and is called the **principal part** at $z = z_j$, $p_j(z)$:

$$p_j(z) = \sum_{n=1}^{N_j} \frac{a_{n,j}}{(z - z_j)^n} \qquad (3.6.8)$$

We shall order points as follows: $|z_r| \leq |z_s|$ if $r < s$, with $z_0 = 0$ if the origin is one of the points to be included.

If the number of poles of the meromorphic function is finite, then the representation

$$f(z) = \sum_{j=1}^{m} p_j(z) \qquad (3.6.9)$$

is nothing more than the partial fraction decomposition of a rational function, where the right-hand side of Eq. (3.6.9) has poles of specified character at the points $z = z_j$. A more general formula representing a meromorphic function with a finite number of poles is obtained by adding to the right side of Eq. (3.6.9) a function $h(z)$ that is entire. On the other hand, if the number of points z_j is infinite, the sum in Eq. (3.6.9) might or might not converge; for example, the partial sum

$$\sum_{k=-n}^{n} \frac{1}{z - k} = 2z \left(\frac{1}{z^2 - 1^2} + \frac{1}{z^2 - 2^2} + \cdots + \frac{1}{z^2 - n^2} \right)$$

converges uniformly for finite z, whereas the partial sum $\sum_{k=1}^{n} 1/(z - k)$ diverges, as can be verified from the elementary convergence criteria of infinite series. In general, we will need a suitable modification of Eq. (3.6.9) with the addition of an entire function $h(z)$ in order to find a rather general formula for a meromorphic function with prescribed principal parts. In what follows we take the case $\{z_j\}$; $|z_j| \to \infty$ as $j \to \infty$, $z_0 = 0$.

Mittag–Leffler expansions involve the following. One wishes to represent a given meromorphic function $f(z)$ with prescribed principal parts $\{p_j(z)\}_{j=0}^{\infty}$ in terms of suitable functions. The aim is to find polynomials $\{g_j(z)\}_{j=0}^{\infty}$, where $g_0(z) = 0$, such that

$$f(z) = p_0(z) + \sum_{j=1}^{\infty} (p_j(z) - g_j(z)) + h(z) = \tilde{f}(z) + h(z) \qquad (3.6.10)$$

where $h(z)$ is an entire function. The part of Eq. (3.6.10) that is called $\tilde{f}(z)$ has the same principal part (i.e., the same number, strengths and locations of poles) as $f(z)$. The difference $h(z)$, between $f(z)$ and $\tilde{f}(z)$, is necessarily entire. In order to pin down the entire function $h(z)$, more information about the function $f(z)$ is required.

When the function $f(z)$ has only simple poles ($N_j = 1$), the situation is considerably simpler, and we now discuss this situation in detail.

In the case of simple poles,

$$p_j(z) = \frac{a_j}{z - z_j} = -\frac{a_j}{z_j} \left(\frac{1}{1 - z/z_j} \right) \qquad (3.6.11)$$

Then there is an m such that for $|z/z_j| < 1$, the finite series

$$g_j(z) = -\frac{a_j}{z_j} \left(1 + \left(\frac{z}{z_j} \right) + \cdots + \left(\frac{z}{z_j} \right)^{m-1} \right) \qquad (3.6.12)$$

approximates $p_j(z)$ arbitrarily closely; a_j is the residue of the pole $z = z_j$. If we call

$$L(w, m) = \frac{1}{w - 1} + 1 + w + w^2 + \cdots + w^{m-1} \qquad (3.6.13)$$

then, assuming convergence of the infinite series, Eq. (3.6.10) takes the form

$$f(z) = p_0(z) + \sum_{j=1}^{\infty} \left(\frac{a_j}{z_j} \right) L \left(\frac{z}{z_j}, m \right) + h(z) \qquad (3.6.14)$$

where $h(z)$ is an entire function and the following theorem holds.

Theorem 3.6.2 (*Mittag–Leffler – simple poles*) Let $\{z_k\}$ and $\{a_k\}$ be sequences with z_k distinct, $|z_k| \to \infty$ as $k \to \infty$, and m an integer such that

$$\sum_{j=1}^{\infty} \frac{|a_j|}{|z_j|^{m+1}} < \infty \qquad (3.6.15)$$

Then Eq. (3.6.14) represents a meromorphic function whose only singularities are simple poles at z_k with residue a_k for $k = 1, 2, \dots$.

Proof From the fact that

$$1 + w + w^2 + \cdots + w^{m-1} = \frac{1}{1-w} - \frac{w^m}{1-w}$$

we have

$$L(w, m) = -\frac{w^m}{1-w}$$

For $|w| < 1/2$ we have $|1 - w| \geq 1 - |w| \geq 1/2$, and hence

$$|L(w, m)| \leq 2|w|^m \qquad (3.6.16)$$

Let $|z| < R$ and for J large enough take $j > J$, $|z_j| > 2R$, then $|z/z_j| < 1/2$, hence the estimate (3.6.16) holds for $w = z/z_j$, and

$$\left| \frac{a_j}{z_j} L\left(\frac{z}{z_j}, m \right) \right| \leq \left| \frac{a_j}{z_j} \right| 2 \left| \frac{z}{z_j} \right|^m$$
$$\leq \frac{2|R|^m |a_j|}{|z_j|^{m+1}}$$

Thus with Eq. (3.6.15) we find that the series in Eq. (3.6.14) converges uniformly for $|z| < R$ (for arbitrarily large R), and Eq. (3.6.14) therefore represents a meromorphic function with the desired properties. ∎

Using Theorem 3.6.2, we may determine which value of m ensures the convergence of the sum in Eq. (3.6.15), and consequently we may determine the function $L(w, m)$ in Eqs. (3.6.13)–(3.6.14).

For example, let us consider the function

$$f(z) = \pi \cot \pi z$$

This function has simple poles at $z_j = j$, $j = 0, \pm 1, \pm 2, \dots$. The strength of any of these poles is $a_j = 1$, which can be ascertained from the Laurent series of $f(z)$ in the neighborhood of z_j; that is, calling $z' = z - j$

$$f(z') = \pi \frac{\cos \pi z'}{\sin \pi z'} = \frac{\pi \left(1 - \frac{(\pi z')^2}{2!} + \cdots\right)}{\pi \left(z' - \frac{(\pi z')^3}{3!} + \cdots\right)}$$

$$= \frac{1}{z'} \left(1 - \frac{1}{3}(\pi z')^2 + \cdots\right)$$

The principal part at each z_j is therefore given by $p_j(z) = \frac{1}{z-j}$. Then the series (3.6.15) in Theorem 3.6.3

$$\sum_{\substack{j=-\infty \\ j \neq 0}}^{\infty} \frac{1}{j^{m+1}}$$

converges for $m = 1$. Consequently from Theorem 3.6.3 and Eq. (3.6.14) the general form of the function is fixed to be

$$\pi \cot \pi z = \frac{1}{z} + \sum_{j=-\infty}^{\infty}{}' \left(\frac{1}{z-j} + \frac{1}{j}\right) + h(z) \qquad (3.6.17a)$$

$$= \frac{1}{z} + 2\sum_{j=1}^{\infty} \frac{z}{z^2 - j^2} + h(z) = \sum_{j=-\infty}^{\infty} \frac{z}{z^2 - j^2} + h(z) \quad (3.6.17b)$$

where the prime in the sum means that the term $j = 0$ is excluded and where $h(z)$ is an entire function. Note that the $(1/j)$ term in Eq. (3.6.17) is a necessary condition for the series to converge.

In Chapter 4, Section 4.2, we show that by considering the integral

$$\mathcal{I} = \frac{1}{2\pi i} \oint_C \pi \cot \pi \zeta \left(\frac{1}{z - \zeta} + \frac{1}{\zeta}\right) d\zeta \qquad (3.6.18)$$

where C is an appropriate closed contour, that the representation (3.17) holds with $h(z) = 0$.

The general case in which the principal parts contain an arbitrary number of poles – Eq. (3.6.8) with finite N_j – is more complicated. Nevertheless, so long as the locations of the poles are distinct, polynomials $g_j(z)$ can be found that establish the following (see, e.g., Henrici, volume 1, 1977).

Theorem 3.6.3 (Mittag–Leffler – general case) Let $f(z)$ be a meromorphic function in the complex plane with poles $\{z_j\}$ and corresponding principal parts $\{p_j(z)\}$. Then there exist polynomials $\{g_j(z)\}_{j=1}^{\infty}$ such that Eq. (3.6.10) holds and the series $\sum_{j=1}^{\infty} (p_j(z) - g_j(z))$ converges uniformly on every bounded set not containing the points $\{z_j\}_{j=0}^{\infty}$.

Proof We only sketch the essential idea behind the proof; the details are cumbersome. Each of the principal parts $\{p_j(z)\}_{j=1}^{\infty}$ can be expanded in a convergent Taylor series (around $z = 0$) for $|z| < |z_j|$. It can be shown that enough terms can be taken in this Taylor series that the polynomials $g_j(z)$ obtained by truncation of the Taylor series of $p_j(z)$ at order z^{K_j}

$$g_j(z) = \sum_{k=0}^{K_j} B_{k,j} z^k$$

ensure that the difference $|p_j(z) - g_j(z)|$ is suitably small. It can be shown (e.g. Henrici, volume I, 1977) that for any $|z| < R$, the polynomials $g_j(z)$ of order K_j ensure that the series

$$\sum_{j=1}^{\infty} |p_j(z) - g_j(z)|$$

converges uniformly.   ∎

It should also be noted that even when we only have simple poles for the $p_j(z)$, there may be cases where we need to use the more general polynomials described in Theorem 3.6.3; for example, if we have $p_j(z) = 1/(z - z_j)$ where $z_j = \log(1 + j)$, $(a_j = 1)$. Then we see that in this case Eq. (3.6.15) is not true for any integer m.

A similar question to the one we have been asking is how to represent an entire function with specified zeroes at location z_k. We use the same notation as before: $z_0 = 0$, $|z_1| \leq |z_2| \leq \ldots$, and $|z_k| \to \infty$. We specify the order of each zero by a_k. One method to derive such a representation is to use the fact that if $f(z)$ is entire, then $f'(z)/f(z)$ is meromorphic with simple poles. Note near any isolated zero z_k with order a_k of $f(z)$ we have $f(z) \approx b_k(z - z_k)^{a_k}$; hence $f'(z)/f(z) \approx a_k/(z - z_k)$. Thus the order of the zero plays the same role as the residue in the Mittag–Leffler Theorem.

From the proof of Eq. (3.6.10) in the case of simple poles, using Eqs. (3.6.11)–(3.6.15), we have the uniformly convergent series representation

$$\frac{f'(z)}{f(z)} = \frac{a_0}{z} + \sum_{j=1}^{\infty} \left(\frac{a_j}{z - z_j} + \frac{a_j}{z_j} \sum_{k=0}^{m-1} \left(\frac{z}{z_j} \right)^k \right) + h(z) \qquad (3.6.19)$$

where $h(z)$ is an arbitrary entire function. Integrating and taking the exponential yields (care must be taken with regard to the constants of integration, cf. Eq. (3.6.21), below)

$$f(z) = z^{a_0} \prod_{j=1}^{\infty} \left\{ (1 - z/z_j) \exp \left(\sum_{k=0}^{m-1} \frac{\left(\frac{z}{z_j} \right)^{k+1}}{k+1} \right) \right\}^{a_j} g(z) \qquad (3.6.20)$$

where $g(z) = \exp \left(\int h(z) \, dz \right)$ is an entire function without zeroes. Function (3.6.20) is, in fact, the most general entire function with such specified behavior. Equation (3.6.20) could, of course, be proven independently without recourse to the series representations discussed earlier. This result is referred to as the Weierstrass Factor Theorem. When $a_j = 1$, $j = 0, 1, 2, \ldots$, Eq. (3.6.20) with (3.6.15) gives the representation of an entire function with simple zeroes.

Theorem 3.6.4 (Weierstrass) The most general entire function with isolated zeroes at $z_0 = 0$, $\{z_j\}_{j=1}^{\infty}$, $|z_1| \leq |z_2| \leq \ldots$, where $|z_j| \to \infty$ as $j \to \infty$, and with orders a_j, is given by Eq. (3.6.20) with (3.6.15).

We note that z_j cannot have a limit point other than ∞. If z_j has a limit point, say, z_*, then z_j can be taken arbitrarily close to z_*; therefore $f(z)$ would not be entire, resulting in a contradiction. Recall that an analytic function must have its zeroes isolated (Theorem 3.2.8).

In practice, it is usually easiest to employ the Mittag–Leffler expansion for $f'(z)/f(z)$, $f(z)$ entire as we have done above, in order to represent an entire function. Note that the expansion (3.6.17a,b) can be integrated using the principal branch of the logarithm function to find

$$\log \sin \pi z = \log z + A_0 + \sum_{n=1}^{\infty} \left[\log(z^2 - n^2) - A_n \right]$$

where A_0 and A_n are constant. Using

$$\lim_{z \to 0} \frac{\sin \pi z}{z} = \pi$$

the constants can be evaluated: $A_0 = \log \pi$ and $A_n = -\log(-n^2)$. Taking the exponential of both sides yields

$$\frac{\sin \pi z}{\pi} = z \prod_{n=1}^{\infty} \left(1 - \left(\frac{z}{n} \right)^2 \right) \qquad (3.6.21)$$

which provides a concrete example of a Weierstrass expansion.

Problems for Section 3.6

1. Discuss where the following infinite products converge as a function of z:

$$\text{(a)} \prod_{n=0}^{\infty}(1+z^n) \qquad \text{(b)} \prod_{n=0}^{\infty}\left(1+\frac{z^n}{n!}\right)$$

$$\text{(c)} \prod_{n=1}^{\infty}\left(1+\frac{2z}{n}\right) \qquad \text{(d)} \prod_{n=1}^{\infty}\left(1+\left(\frac{2z}{n}\right)^2\right)$$

2. Show that the product

$$\prod_{k=1}^{\infty}\left(1-\frac{z^4}{k^4}\right)$$

represents an entire function with zeroes at $z = \pm k, \pm ik; k = 1, 2, \ldots$..

3. Using the expansion

$$\frac{\sin \pi z}{\pi z} = \prod_{n=1}^{\infty}\left(1-\left(\frac{z}{n}\right)^2\right)$$

show that we also have

$$\frac{\sin \pi z}{\pi z} = \prod_{n=-\infty}^{\infty}{}'\left(1-\frac{z}{n}\right)e^{z/n}$$

where the prime means that the $n = 0$ term is omitted. (Also see Problem **4**, below.)

4. Use the representation

$$\frac{\sin \pi z}{\pi z} = \prod_{n=-\infty}^{\infty}{}'\left(1-\frac{z}{n}\right)e^{z/n}$$

to deduce, by differentiation, that

$$\pi \cot \pi z = \frac{1}{z} + \sum_{n=-\infty}^{\infty}{}'\left(\frac{1}{z-n}+\frac{1}{n}\right)$$

where the prime means that the $n = 0$ term is omitted. Repeat the process to find

$$\pi \csc^2 \pi z = \sum_{n=-\infty}^{\infty}\frac{1}{(z-n)^2}$$

5. Show that if $f(z)$ is meromorphic in the finite z plane, then $f(z)$ must be the ratio of two entire functions.

6. Let $\Gamma(z)$ be given by

$$\frac{1}{\Gamma(z)} = ze^{\gamma z} \prod_{n=1}^{\infty} \left(1 + \frac{z}{n}\right) e^{-z/n}$$

for $z \neq 0, -1, -2, \ldots$ and $\gamma = $ constant.

(a) Show that

$$\frac{\Gamma'(z)}{\Gamma(z)} = -\frac{1}{z} - \gamma - \sum_{n=1}^{\infty} \left(\frac{1}{z+n} - \frac{1}{n}\right)$$

(b) Show that

$$\frac{\Gamma'(z+1)}{\Gamma(z+1)} - \frac{\Gamma'(z)}{\Gamma(z)} - \frac{1}{z} = 0$$

whereupon

$$\Gamma(z+1) = Cz\Gamma(z), \qquad C \text{ constant}$$

(c) Show that $\lim_{z \to 0} z\Gamma(z) = 1$ to find that $C = \Gamma(1)$.

(d) Determine the following representation for the constant γ so that $\Gamma(1) = 1$

$$e^{-\gamma} = \prod_{n=1}^{\infty} \left(1 + \frac{1}{n}\right) e^{-1/n}$$

(e) Show that

$$\prod_{n=1}^{\infty} \left(1 + \frac{1}{n}\right) e^{-1/n} = \lim_{n \to \infty} \frac{2}{1} \frac{3}{2} \frac{4}{3} \cdots \frac{n+1}{n} e^{-S(n)} = \lim_{n \to \infty} (n+1) e^{-S(n)}$$

where $S(n) = 1 + \frac{1}{2} + \frac{1}{3} + \cdots + \frac{1}{n}$. Consequently obtain the limit

$$\gamma = \lim_{n \to \infty} \left(\sum_{k=1}^{n} \frac{1}{k} - \log(n+1)\right)$$

The constant $\gamma = 0.5772157\ldots$ is referred to as Euler's constant.

7. In Section 3.6 we showed that

$$\pi \cot \pi z - \left(\frac{1}{z} + \sum_{j=-\infty}^{\infty}{}' \left(\frac{1}{z-j} + \frac{1}{j} \right) \right) = h(z)$$

where $\sum'$ denotes that the $j = 0$ term is omitted and where $h(z)$ is entire. We now show how to establish that $h(z) = 0$.

(a) Show that $h(z)$ is periodic of period 1 by establishing that the left-hand side of the formula is periodic of period 1. (Show that the second term on the left side doesn't change when z is replaced by $z + 1$.)

(b) Because $h(z)$ is periodic and entire, we need only establish that $h(z)$ is bounded in the strip $0 \le \mathrm{Re}\, z \le 1$ to ensure, by Liouville's Theorem (Section 2.6.2) that it is a constant. For all finite values of $z = x + iy$ in the strip away from the poles, explain why both terms are bounded, and because the pole terms cancel, the difference is in fact bounded. Verify that as $y \to \pm\infty$ the term $\pi \cot \pi z$ is bounded.
To establish the boundedness of the second term on the left, rewrite it as follows:

$$S(z) = \frac{1}{z} + \sum_{n=1}^{\infty} \frac{z}{z^2 - n^2}$$

Use the fact that in the strip $0 < x < 1$, $y > 2$, $|z| \le \sqrt{2}y$, we have $|z^2 - n^2| \ge \frac{1}{\sqrt{2}}(y^2 + n^2)$ (note that these estimates are not sharp), and show that

$$|S(z)| \le \frac{1}{|z|} + 4y \sum_{n=1}^{\infty} \left(\frac{1}{y^2 + n^2} \right)$$

Explain why

$$\sum_{n=1}^{\infty} \frac{y}{y^2 + n^2} = \frac{1}{y} \sum_{n=1}^{\infty} \frac{1}{1 + (n/y)^2} \le \int_0^\infty \frac{1}{1 + u^2}\, du = \frac{\pi}{2}$$

and therefore conclude that $S(z)$ is bounded for $0 < x < 1$ and $y \to \infty$. The same argument works for $y \to -\infty$. Hence $h(z)$ is a constant.

(c) Because both terms on the left are odd in z, that is, $f(z) = -f(-z)$, conclude that $h(z) = 0$.

8. Consider the function $f(z) = (\pi^2)/(\sin^2 \pi z)$.

(a) Establish that near every integer $z = j$ the function $f(z)$ has the singular part $p_j(z) = 1/(z-j)^2$.

(b) Explain why the series

$$S(z) = \sum_{j=-\infty}^{\infty} \frac{1}{(z-j)^2}$$

converges for all $z \neq j$.

(c) Because the series in part **(b)** converges, explain why the representation

$$\frac{\pi^2}{\sin^2 \pi z} = \sum_{j=-\infty}^{\infty} \frac{1}{(z-j)^2} + h(z)$$

where $h(z)$ is entire, is valid.

(d) Show that $h(z)$ is periodic of period 1 by showing that each of the terms $(\pi/\sin \pi z)^2$ and $S(z)$ are periodic of period 1. Explain why $(\pi/\sin \pi z)^2 - S(z)$ is a bounded function, and show that each term vanishes as $|y| \to \infty$. Hence conclude that $h(z) = 0$.

(e) Integrate termwise to find

$$\pi \cot \pi z = \frac{1}{z} + \sum_{n=-\infty}^{\infty}{}' \left(\frac{1}{z-n} + \frac{1}{n} \right)$$

where the prime denotes the fact that the $n = 0$ term is omitted.

9. (a) Let $f(z)$ have simple poles at $z = z_n$, $n = 1, 2, 3, \ldots, N$, with strengths a_n, and be analytic everywhere else. Show by contour integration that (the reader may wish to consult Theorem 4.1.1)

$$\frac{1}{2\pi i} \oint_{C_N} \frac{f(z')}{z'-z} \, dz' = f(z) + \sum_{n=1}^{N} \frac{a_n}{z_n - z} \tag{1}$$

where C_N is a large circle of radius R_N enclosing all the poles. Evaluate (1) at $z = 0$ to obtain

$$\frac{1}{2\pi i} \oint_{C_N} \frac{f(z')}{z'} \, dz' = f(0) + \sum_{n=1}^{N} \frac{a_n}{z_n} \tag{2}$$

(b) Subtract equation (2) from equation (1) of part **(a)** to obtain

$$\frac{1}{2\pi i} \oint_{C_N} \frac{z f(z')}{z'(z'-z)} \, dz' = f(z) - f(0) + \sum_{n=1}^{N} a_n \left(\frac{1}{z_n - z} - \frac{1}{z_n} \right) \tag{3}$$

(c) Assume that $f(z)$ is bounded for large z to establish that the left-hand side of Equation (3) vanishes as $R_N \to \infty$. Conclude that if the sum on the right-hand side of Equation (3) converges as $N \to \infty$, then

$$f(z) = f(0) + \sum_{n=1}^{\infty} a_n \left(\frac{1}{z - z_n} + \frac{1}{z_n} \right) \tag{4}$$

(This is a special case of the Mittag-Leffler Theorems 3.6.2–3.6.3)

(d) Let $f(z) = \pi \cot \pi z - 1/z$, and show that

$$\pi \cot \pi z - \frac{1}{z} = \sum_{n=-\infty}^{\infty}{}' \left(\frac{1}{z - n} + \frac{1}{n} \right) \tag{5}$$

where the prime denotes the fact that the $n = 0$ term is omitted. (Equation (5) is another derivation of the result in this section.) We see that an infinite series of poles can represent the function $\cot \pi z$. Section 3.6 establishes that we have other series besides Taylor series and Laurent series that can be used for representations of functions.

*3.7 Differential Equations in the Complex Plane: Painlevé Equations

In this section we investigate various properties associated with solutions to ordinary differential equations in the complex plane.

In what follows we assume some basic familiarity with ordinary differential equations (ODEs) and their solutions. There are numerous texts on the subject; however, with regard to ODEs in the complex plane, the reader may wish to consult the treatises of Ince (1956) or Hille (1976) for an in-depth discussion, though these books contain much more advanced material. The purpose of this section is to outline some of the fundamental ideas underlying this topic and introduce the reader to concepts which appear frequently in physics and applied mathematics literature.

We shall consider nth-order nonlinear ODEs in the complex plane, with the following structure:

$$\frac{d^n w}{dz^n} = F \left(w, \frac{dw}{dz}, \dots, \frac{d^{n-1} w}{dz^{n-1}}; z \right) \tag{3.7.1}$$

where F is assumed to be a locally analytic function of all its arguments, i.e., F has derivatives with respect to each argument in some domain D; thus F can have isolated singularities, branch points, etc. A system of such ODEs takes the form

$$\frac{dw_i}{dz} = F_i(w_1, \ldots, w_n; z), \qquad i = 1, \ldots, n \qquad (3.7.2)$$

where again F_i is assumed to be a locally analytic function of its arguments. The scalar problem (3.7.1) is a special case of Eq. (3.7.2). To see this, we associate w_1 with w and take

$$\frac{dw_1}{dz} = w_2 \equiv F_1$$

$$\frac{dw_2}{dz} = w_3 \equiv F_2$$

$$\vdots$$

$$\frac{dw_{n-1}}{dz} = w_n \equiv F_{n-1}$$

$$\frac{dw_n}{dz} = F(w_1, \ldots, w_n; z) \qquad (3.7.3)$$

whereupon

$$w_{j+1} = \frac{d^j w_1}{dz^j}, \qquad j = 1, \ldots, n-1 \qquad (3.7.4a)$$

and

$$\frac{d^n w_1}{dz^n} = \left(w_1, \frac{dw_1}{dz}, \ldots, \frac{d^{n-1} w_1}{dz^{n-1}}; z \right) \qquad (3.7.4b)$$

A natural question one asks is the following. Is there an analytic solution to these ODEs? Given bounded initial values, that is, for Eq. (3.7.2), at $z = z_0$

$$w_j(z_0) = w_{j,0} < \infty \qquad j = 1, 2, \ldots, n \qquad (3.7.5)$$

the answer is affirmative in a *small enough* region about $z = z_0$. We state this as a theorem.

Theorem 3.7.1 (Cauchy) The system (3.7.2) with initial values (3.7.5), and with $F_i(w_1, \ldots, w_n; z)$ as an analytic function of each of its arguments in a domain D containing $z = z_0$, has a unique analytic solution in a neighborhood of $z = z_0$.

There are numerous ways to establish this theorem, a common one being the method of **majorants**, that is, finding a convergent series that dominates the true series representation of the solution. The basic ideas are most easily illustrated by the scalar first-order nonlinear equation

$$\frac{dw}{dz} = f(w, z) \tag{3.7.6}$$

subject to the initial conditions $w(0) = 0$. Initial values $w(z_0) = w_0$ could be reduced to this case by translating variables, letting $z' = z - z_0$, $w' = w - w_0$, and writing Eq. (3.7.6) in terms of w' and z'. Function $f(w, z)$ is assumed to be analytic and bounded when w and z lie inside the circles $|z| \leq a$ and $|w| \leq b$, with $|f| \leq M$ for some a, b, and M. The series expansion of the solution to Eq. (3.7.6) may be computed by taking successive derivatives of Eq. (3.7.6), that is,

$$\frac{d^2w}{dz^2} = \frac{\partial f}{\partial z} + \frac{\partial f}{\partial w}\frac{dw}{dz}$$

$$\frac{d^3w}{dz^3} = \frac{\partial^2 f}{\partial z^2} + 2\frac{\partial^2 f}{\partial z \partial w}\frac{dw}{dz} + \frac{\partial^2 f}{\partial w^2}\left(\frac{dw}{dz}\right)^2 + \frac{\partial f}{\partial w}\frac{d^2w}{dz^2}$$

$$\vdots \tag{3.7.7}$$

This allows us to compute

$$w = \left(\frac{dw}{dz}\right)_0 z + \left(\frac{d^2w}{dz^2}\right)_0 \frac{z^2}{2!} + \left(\frac{d^3w}{dz^3}\right)_0 \frac{z^3}{3!} + \cdots \tag{3.7.8}$$

The technique is to consider a comparison equation with the same initial condition

$$\frac{dW}{dz} = F(W, z), \qquad W(0) = 0 \tag{3.7.9}$$

in which each term in the series representation of $F(w, z)$ dominates that of $f(w, z)$. Specifically, the series representation for $f(w, z)$, which is assumed to be analytic in both variables w and z, is

$$f(w, z) = \sum_{j=0}^{\infty}\sum_{k=0}^{\infty} C_{jk} z^j w^k \tag{3.7.10}$$

$$C_{jk} = \frac{1}{j!k!}\left(\frac{\partial^{j+k} f}{\partial z^j \partial w^k}\right)_0 \tag{3.7.11}$$

At $w = b$ and $z = a$, we have assumed that f is bounded, and we take the bound on f to be

$$|f(w, z)| \leq \sum_{j=0}^{\infty} \sum_{k=0}^{\infty} |C_{jk}| a^j b^k = M \tag{3.7.12}$$

Each term of this series is bounded by M; hence

$$|C_{jk}| \leq Ma^{-j}b^{-k} \tag{3.7.13}$$

We take $F(w, z)$ to be

$$F(W, z) = \sum_{j=0}^{\infty} \sum_{k=0}^{\infty} \frac{M}{a^j b^k} z^j W^k \tag{3.7.14}$$

So from Eqs. (3.7.13) and (3.7.10) the function $F(W, z)$ majorizes $f(w, z)$ termwise. Because the solution $W(z)$ is computed exactly the same way as for Eq. (3.7.6), that is, we only replace w and f with W and F in Eq. (3.7.7), clearly the series solution (Eq. (3.7.8) with w replaced by W) for $W(z)$ would dominate that for w. Next we show that $W(z)$ has a solution in a neighborhood of $z = a$. Summing the series (3.7.14) yields

$$F(w, z) = \frac{M}{\left(1 - \frac{z}{a}\right)\left(1 - \frac{W}{b}\right)} \tag{3.7.15}$$

whereupon Eq. (3.7.9) yields

$$\left(1 - \frac{W}{b}\right)\frac{dW}{dz} = \frac{M}{1 - \frac{z}{a}} \tag{3.7.16}$$

hence by integration

$$W(z) - \frac{1}{2b}(W(z))^2 = -Ma \log\left(1 - \frac{z}{a}\right) \tag{3.7.17}$$

and therefore

$$W(z) = b - b\left[1 + \frac{2aM}{b} \log\left(1 - \frac{z}{a}\right)\right]^{1/2} \tag{3.7.18}$$

In Eq. (3.7.18) we take the positive value for the square root and the principal value for the log function so that $W(0) = 0$. The series representation (expanding the log, square root, etc.) of $W(z)$ dominates the series $w(z)$. The series for $W(z)$ converges up and until the nearest singularity: $z = a$ for the log function, or to $z = R$ where $[\cdot]^{1/2} = 0$, whichever is smaller.

Because R is given by

$$1 + \frac{2aM}{b} \log\left(1 - \frac{R}{a}\right) = 0 \qquad (3.7.19a)$$

we have

$$R = a\left(1 - \exp\left(-\frac{b}{2Ma}\right)\right) \qquad (3.7.19b)$$

Because $R < a$, the series representation of Eq. (3.7.9) converges absolutely for $|z| < R$. Hence a solution $w(z)$ satisfying Eq. (3.7.6) must exist for $|z| < R$, by comparison. Moreover, so long as we stay within the class of analytic functions, any series representation obtained this way will be unique because the Taylor series uniquely represents an analytic function.

The method described above can be readily extended to apply to the system of equations (3.7.2). Without loss of generality, taking initial values $w_j = 0$ for $j = 1, 2, \ldots, n$ at $z = 0$ and functions $f_i(w_1, \ldots, w_n, z)$ analytic inside $|z| \leq a$, $|w_j| \leq b$, $j = 1, \ldots, n$, then we can take $|f_i| \leq M$ in this domain. For the majorizing function, similar arguments as before yield

$$\frac{dW_1}{dz} = \frac{dW_2}{dz} = \cdots = \frac{dW_n}{dz}$$
$$= \frac{M}{\left(1 - \frac{z}{a}\right)\left(1 - \frac{W_1}{b}\right)\cdots\left(1 - \frac{W_n}{b}\right)} \qquad (3.7.20)$$

where

$$W_j(z) = 0 \qquad \text{for } 1, \ldots, n$$

Solving

$$\frac{dW_i}{dz} = \frac{dW_{i+1}}{dz}$$

$W_i(0) = W_{i+1}(0) = 0$ for $i = 1, 2, \ldots, n - 1$, implies that

$$W_1 = W_2 = \cdots = W_n \equiv W$$

whereupon Eq. (3.7.20) gives

$$\frac{dW}{dz} = \frac{M}{\left(1 - \frac{z}{a}\right)\left(1 - \frac{W}{b}\right)^n}, \qquad W(0) = 0 \qquad (3.7.21)$$

Solving Eq. (3.7.21) yields

$$W = b - b\left[1 + \frac{(n+1)}{b}Ma \log\left(1 + \frac{z}{a}\right)\right]^{\frac{1}{n+1}} \qquad (3.7.22)$$

with a radius of convergence given by $|z| \le R$ where

$$R = a\left(1 - e^{-\frac{b}{(n+1)Ma}}\right) \qquad (3.7.23)$$

Hence the series solution to the system (3.7.2), $w_j(0) = 0$, converges absolutely and uniformly inside the circle of radius R. ∎

Thus Theorem 3.7.1 establishes the fact that so long as $f_i(w_1, \ldots, w_n; z)$ in Eq. (3.7.2) is an analytic function of its arguments, then there is an analytic solution in a neighborhood (albeit small) of the initial values $z = z_0$. We may analytically continue our solution until we reach a singularity. This is due to the following.

Theorem 3.7.2 (Continuation Principle) The function obtained by analytic continuation of the solution of Eq. (3.7.2), along any path in the complex plane, is a solution of the analytic continuation of the equation.

Proof We note that because $g_i(z) = w_i' - f_i(w_1, \ldots, w_n; z)$, $i = 1, 2, \ldots, n$, is zero inside the domain where we have established the existence of our solution, then any analytic continuation of $g_i(z)$ will necessarily be zero. Because the solution $w_i(z)$ satisfies $g_i(z) = 0$ inside the domain of its existence, and because the operations in $g_i(z)$ maintain analyticity, then analytically extending $w_i(z)$ gives the analytic extension of $g_i(z)$, which is identically zero. ∎

Thus we find that our solution may be analytically continued until we reach a singularity. A natural question to ask is where we can expect a singularity. There are two types: **fixed** and **movable**. A fixed singularity is one that is determined by the explicit singularities of the functions $f_i(., z)$. For example

$$\frac{dw}{dz} = \frac{w}{z^2}$$

has a fixed singular point (**SP**) at $z = 0$. The solution reflects this fact:

$$w = Ae^{-1/z}$$

whereby we have an essential singularity at $z = 0$.

Movable **SP**s, on the other hand, depend on the initial conditions imposed. In a sense they are internal to the equation. For example, consider

$$\frac{dw}{dz} = w^2 \qquad (3.7.24a)$$

There are no fixed singular points, but the solution is given by

$$w = -\frac{1}{z - z_0} \qquad (3.7.24b)$$

where z_0 is arbitrary. The value of z_0 depends on the initial value; that is, if $w(z = 0) = w_0$, then $z_0 = 1/w_0$. Equation (3.7.24b) is an example of a **movable pole** (this is a simple pole). If we consider different equations, we could have different kinds of movable singularities, for example, movable branch points, movable essential singularities, etc. For example

$$\frac{dw}{dz} = w^p, \qquad p \geq 2 \qquad (3.7.25a)$$

has the solution

$$w = ((p - 1)(z_0 - z))^{1/1-p} \qquad (3.7.25b)$$

which has a **movable branch point** for $p \geq 3$.

In what follows we shall, for the most part, quote some well-known results regarding differential equations with fixed and movable singular points. We refer the reader to the monographs of Ince (1956), Hille (1976), for the rigorous development, which would otherwise take us well outside the scope of the present text.

It is natural to ask what happens in the linear case. The linear homogeneous analog of Eq. (3.7.2) is

$$\frac{d\vec{w}}{dz} = A(z)\vec{w}, \qquad \vec{w}(z_0) = \vec{w}_0 \qquad (3.7.26)$$

where $\vec{w}$ is an $(n \times 1)$ column vector and $A(z)$ is an $(n \times n)$ matrix, that is,

$$A = \begin{pmatrix} a_{11} & \cdots & a_{1n} \\ \vdots & \ddots & \vdots \\ a_{n1} & \cdots & a_{nn} \end{pmatrix}, \qquad \vec{w} = \begin{pmatrix} w_1 \\ \vdots \\ w_n \end{pmatrix}$$

The linear homogeneous scalar problem is obtained by specializing Eq. (3.7.26):

$$\frac{d^n w}{dz^n} = p_1(z)\frac{d^{n-1}w}{dz^{n-1}} + p_2(z)\frac{d^{n-2}w}{dz^{n-2}} + \cdots + p_n(z)w \qquad (3.7.27)$$

where we take, in Eq. (3.7.26)

$$A = \begin{pmatrix} 0 & 1 & 0 & \cdots & 0 & 0 \\ 0 & 0 & 1 & \cdots & 0 & 0 \\ \vdots & & & \ddots & \vdots & \vdots \\ 0 & 0 & 0 & \cdots & 0 & 1 \\ p_n(z) & p_{n-1}(z) & p_{n-2}(z) & \cdots & p_2(z) & p_1(z) \end{pmatrix}$$ (3.7.28a)

and

$$w_2 = \frac{dw_1}{dz}, \ldots, w_n = \frac{dw_{n-1}}{dz}, \qquad w_1 \equiv w$$ (3.7.28b)

The relevant result is the following.

Theorem 3.7.3 If $A(z)$ is analytic in a simply connected domain D, then the linear initial value problem (3.7.26) has a unique analytic solution in D.

A consequence of this theorem, insofar as singular points are concerned, is that the general linear equation (3.7.26) has no movable **SP**s; its **SP**s are fixed purely by the singularities of the coefficient matrix $A(z)$, or in the scalar problem (3.7.27), by the singularities in the coefficients $\{p_j(z)\}_{j=1}^n$. One can prove Theorem 3.7.3 by an extension of what was done earlier. Namely, by looking for a series solution about a point of singularity, say, $z = 0$, $w(z) = \sum_{k=0}^\infty c_k z^k$, one can determine the coefficients c_k and show that the series converges until the nearest singularity of $A(z)$. Because this is fixed by the equation, we have the fact that *linear equations have only fixed singularities*.

For example, the scalar first-order equation

$$\frac{dw}{dz} = p(z)w, \qquad w(z_0) = w_0$$ (3.7.29a)

has the explicit solution

$$w(z) = w_0 e^{\int_{z_0}^z p(\zeta)\,d\zeta}$$ (3.7.29b)

Clearly, if $p(z)$ is analytic, then so is $w(z)$.

For linear differential equations there is great interest in a special class of differential equations that arise frequently in physical applications. These are so-called linear differential equations with **regular singular points**. Equation (3.7.26) is said to have a singular point in domain D if $A(z)$ has a singular point in D. We say $z = z_0$ is a regular singular point of Eq. (3.7.26) if the matrix $A(z)$ has a simple pole at $z = z_0$:

$$A(z) = \sum_{k=0}^{\infty} a_k (z - z_0)^{k-1}$$

where a_0 is not the zero matrix. The scalar equation (3.7.27) is said to have a regular singular point at $z = z_0$ if $p_k(z)$ has a kth-order pole, i.e. $(z - z_0)^k p_k(z)$, $k = 1, \ldots, n$ is analytic at $z = z_0$. Otherwise, a singular point of a linear differential equation is said to be an irregular singular point. As mentioned earlier, Eq. (3.7.27) can be written as a matrix equation, Eq. (3.7.26), and the statements made here about scalar and matrix equations are easily seen to be consistent.

We may rewrite Eq. (3.7.27) by calling $Q_j(z) = -(z - z_0)^j p_j(z)$

$$(z - z_0)^n \frac{d^n w}{dz^n} + \sum_{j=1}^{n} Q_j(z)(z - z_0)^{n-j} \frac{d^{n-j} w}{dz^{n-j}} = 0 \qquad (3.7.30)$$

where all the $Q_j(z)$ are analytic at $z = z_0$ for $j = 1, 2, \ldots$.

Fuchs and Frobenius showed that series methods may be applied to solve Eq. (3.7.30) and that, in general, the solution contains branch points at $z = z_0$. Indeed, if we expand $Q_j(z)$ about $z = z_0$ as

$$Q_j(z) = \sum_{k=0}^{\infty} c_{jk}(z - z_0)^k$$

then the solution to Eq. (3.7.30) has the form

$$w(z) = \sum_{k=0}^{\infty} a_k (z - z_0)^{k+r} \qquad (3.7.31)$$

where r satisfies the so-called **indicial equation**

$$r(r-1)(r-2) \cdots (r-n+1)$$
$$+ \sum_{j=1}^{n-1} c_{j0} r(r-1)(r-2) \cdots (r-n+j+1) + c_{n0} = 0 \quad (3.7.32)$$

There is always one solution of (3.7.30) with a root r obtained from (3.7.32). In fact there are n such linearly independent solutions so long as no two roots of this equation differ by an integer or zero (i.e., multiple root). In this special case the solution form (3.7.31) must in general be supplemented by appropriate terms containing powers of $\log(z - z_0)$. Equation (3.7.32) is obtained by inserting the expansion (3.7.31) into Eq. (3.7.30) a recursion relation for the coefficients a_k are obtained by equating powers of $(z - z_0)$. Convergence of the series (3.7.31) is to the nearest singularity of the coefficients $Q_j(z)$, $j = 1, \ldots, n$. If all the functions $Q_j(z)$ were indeed constant, c_{j0}, then Eq. (3.7.32) would lead to the roots associated with the solutions to Euler's equation.

The standard case is the second-order equation $n = 2$, which is covered in most elementary texts on differential equations:

$$(z - z_0)^2 \frac{d^2 w}{dz^2} + (z - z_0)Q_1(z)\frac{dw}{dz} + Q_2(z)w = 0 \tag{3.7.33}$$

where $Q_1(z)$ and $Q_2(z)$ are analytic in a neighborhood of $z = z_0$. The indicial equation (3.7.33) in this case satisfies.

$$r(r - 1) + c_{10}r + c_{20} = 0 \tag{3.7.34}$$

where c_{10} and c_{20} are the first terms in the Taylor expansion of $Q_1(z)$ and $Q_2(z)$ about $z = z_0$, that is, $c_{10} = Q_1(z_0)$ and $c_{20} = Q_2(z_0)$.

Well-known second-order linear equations containing regular singular points include the following:

Bessel's Equation

$$z^2 \frac{d^2 w}{dz^2} + z\frac{dw}{dz} + (z^2 - p^2)w = 0 \tag{3.7.35a}$$

Legendre's Equation

$$\left(1 - z^2\right) \frac{d^2 w}{dz^2} + 2z\frac{dw}{dz} + p(p + 1)w = 0 \tag{3.7.35b}$$

Hypergeometric Equation

$$z(1 - z)\frac{d^2 w}{dz^2} + [c - (a + b + 1)z]\frac{dw}{dz} - abw = 0 \tag{3.7.35c}$$

where a, b, c, p are constant.

We now return to questions involving nonlinear ODEs. In the late 19th and early 20th centuries, there were extensive studies undertaken by mathematicians in order to ennumerate those nonlinear ODEs that had poles as their only movable singularities: We say that ODEs possessing this property are of **Painlevé**

type (named after one of the mathematicians of that time). Mathematically speaking, these equations are among the simplest possible because the solutions apart from their fixed singularities (which are known a priori) are meromorphic; in fact, they can frequently (perhaps always?) be solved exactly. It turns out that equations with this property arise frequently in physical applications, for example, fluid dynamics, quantum spin systems, relativity, etc. (See, for example, Ablowitz and Segur (1981), especially the sections on Painlevé equations.) The historical background and development is reviewed in the monograph of Ince (1956).

The simplest situation occurs with first-order nonlinear differential equations of the following form:

$$\frac{dw}{dz} = F(w, z) = \frac{P(w, z)}{Q(w, z)} \tag{3.7.36}$$

where P and Q are polynomials in w and locally analytic functions of z. Then the *only equation* that is of Painlevé type is

$$\frac{dw(z)}{dz} = A_0(z) + A_1(z)w + A_2(z)w^2 \tag{3.7.37}$$

Equation (3.7.37) is called a **Riccati equation**. Moreover, it can be linearized by the substitution

$$w(z) = \alpha(z) \left(\frac{\frac{d\psi}{dz}}{\psi} \right) \tag{3.7.38a}$$

where

$$\alpha(z) = -1/A_2(z) \tag{3.7.38b}$$

and $\psi(z)$ satisfies the linear equation

$$\frac{d^2\psi}{dz^2} = \left(A_1(z) - A_2'(z)/A_2(z) \right) \frac{d\psi}{dz} - A_0(z)A_2(z)\psi \tag{3.7.38c}$$

Because Eq. (3.7.38c) is linear, it has no movable singularities. But it does have movable zeroes; hence $w(z)$ from Eq. (3.7.38a) has movable poles.

Riccati equations are indeed special equations, and a large literature has been reserved for them. The above conclusions were first realized by Fuchs, but an extensive treatment was provided by the work of Painlevé. For Eqs. (3.7.36), Painlevé proved that the only movable singular points possible were algebraic, that is, no logarithmic or more exotic singular points arise in this case.

Painlevé also considered the question of enumerating those second nonlinear differential equations admitting poles as their only movable singularities. He studied equations of the form

$$\frac{d^2w}{dz^2} = F\left(\frac{dw}{dz}, w, z\right) \tag{3.7.39}$$

where F is rational in w and dw/dz and whose coefficients are locally analytic in z. He found (depending on how one counts) some fifty different types of equations, all of which were either reducible to (a) linear equations, (b) Riccati equations, (c) equations containing so-called elliptic functions, and (d) six "new" equations.

Elliptic functions are single-valued **doubly periodic** functions whose movable singularities are poles. We say $f(z)$ is a doubly periodic function if there are two complex numbers w_1 and w_2 such that

$$f(z + w_1) = f(z)$$
$$f(z + w_2) = f(z) \tag{3.7.40}$$

with a necessarily nonreal ratio: $w_2/w_1 = \gamma$, $\operatorname{Im} \gamma \neq 0$. There are no doubly periodic functions with two real periods and there are no triply periodic functions. The numbers $m_1 w_1 + n w_2$ are periods of $f(z)$ and a lattice formed by the numbers 0, w_1, w_2 with $w_1 + w_2$ as vertices is called the period parallelogram of $f(z)$.

An example of an elliptic function is the function defined by the convergent series:

$$\mathcal{P}(z) = z^{-2} + \sum_{m,n=0}^{\infty}{}' \left[\left(z - \omega_{m,n}\right)^{-2} - \omega_{m,n}^{-2} \right] \tag{3.7.41}$$

($z = 0$ can be translated to $z = z_0$ if we wish) where prime means $(m, n) \neq (0, 0)$, and $\omega_{m,n} = m\omega_1 + n\omega_2$, where ω_1 and ω_2 are the two periods of the elliptic function. The function $\mathcal{P}(z)$ satisfies a simple first-order equation. Calling $w = \mathcal{P}(z)$, we have

$$(w')^2 = 4w^3 - g_2 w - g_3 \tag{3.7.42}$$

where

$$g_2 = 60 \sum_{m,n=0}^{\infty}{}' \omega_{m,n}^{-4}$$

$$g_3 = 140 \sum_{m,n=0}^{\infty}{}' \omega_{m,n}^{-6} \tag{3.7.43}$$

The function $w = \mathcal{P}(z)$ is called the Weierstrass elliptic function.

An alternative representation of elliptic functions is via the so-called Jacobi elliptic functions $w_1(z) = \text{sn}(z, k)$, $w_2(z) = \text{cn}(z, k)$, and $w_3(z) = \text{dn}(z, k)$, the first two of which are often referred to as the Jacobian sine and cosine. These functions satisfy

$$\frac{dw_1}{dz} = w_2 w_3 \qquad w_1(0) = 0 \qquad\qquad (3.7.44a)$$

$$\frac{dw_2}{dz} = -w_1 w_3 \qquad w_2(0) = 1 \qquad\qquad (3.7.44b)$$

$$\frac{dw_3}{dz} = -k^2 w_1 w_2 \qquad w_3(0) = 1 \qquad\qquad (3.7.44c)$$

Multiplying Eq. (3.7.44a) by w_1 and Eq. (3.7.44b) by w_2, and adding, yields (in analogy with the trigonometric sine and cosine)

$$w_1^2(z) + w_2^2(z) = 1$$

Similarly, from Eqs. (3.7.44a) and (3.7.44c)

$$k^2 w_1^2(z) + w_3^2(z) = 1$$

whereupon we see, from these equations, that $\omega_1(z)$ satisfies a scalar first-order nonlinear ordinary differential equation:

$$\left(\frac{dw_1}{dz}\right)^2 = \left(1 - w_1^2\right)\left(1 - k^2 w_1^2\right) \qquad\qquad (3.7.45)$$

Using the substitution $u = w_1^2$ and changing variables, we can put Eq. (3.7.45) into the form Eq. (3.7.42). Indeed, the general form for an equation having elliptic function solutions is

$$(w')^2 = (w - a)(w - b)(w - c)(w - d) \qquad\qquad (3.7.46)$$

Equation (3.7.46) can also be transformed to either of the standard forms (Eqs. (3.7.42) or (3.7.45)). (The "bilinear" transformation $w = (\alpha + \beta w_1)/(\gamma + \delta w_1)$, $\alpha\delta - \beta\gamma \neq 0$, can be used to transform Eq. (3.7.40) to Eqs. (3.7.42) or (3.7.45).) We also note that the autonomous (i.e., the coefficients are independent of z) second-order differential equation

$$\frac{d^2 w}{dz^2} = w^3 + e w^2 + f w, \qquad e, f \text{ constant} \qquad\qquad (3.7.47)$$

can be solved by multiplying Eq. (3.7.47) by dw/dz and integrating. Then by factorization we may put the result in the form (3.7.46).

The six new equations that Painlevé discovered are not reducible to "known" differential equations. They are listed below, and are referred to as the six Painlevé transcendents listed as P_I through P_{VI}. It is understood that $w' \equiv dw/dz$.

$$P_I: \quad w'' = 6w^2 + z$$

$$P_{II}: \quad w'' = 2w^3 + zw + a$$

$$P_{III}: \quad w'' = \frac{(w')^2}{w} - \frac{w'}{z} + \frac{(aw^2 + b)}{z} + cw^3 + \frac{d}{w}$$

$$P_{IV}: \quad w'' = \frac{(w')^2}{2w} + \frac{3w^3}{2} + 4zw^2 + 2(z^2 - a)w + \frac{b}{w}$$

$$P_V: \quad w'' = \left(\frac{1}{2w} + \frac{1}{w-1}\right)(w')^2 - \frac{w'}{z} + \frac{(w-1)^2}{z^2}\left(aw + \frac{b}{w}\right)$$
$$+ \frac{cw}{z} + \frac{dw(w+1)}{w-1}$$

$$P_{VI}: \quad w'' = \frac{1}{2}\left(\frac{1}{w} + \frac{1}{w-1} + \frac{1}{w-z}\right)(w')^2$$
$$- \left(\frac{1}{z} + \frac{1}{z-1} + \frac{1}{w-z}\right)w'$$
$$+ \frac{w(w-1)(w-2)}{z^2(z-1)^2}\left[a + \frac{bz}{w^2} + \frac{c(z-1)}{(w-1)^2} + \frac{dz(z-1)}{(w-z)^2}\right]$$

where a, b, c, d are arbitrary constants.

It turns out that the sixth equation contains the first five by a limiting procedure, carried out by first transforming w and z appropriately in terms of a suitable (small) parameter, and then taking limits of the parameter to zero. Recent research has shown that these six equations can be linearized by transforming the equations via a somewhat complicated sequence of transformations into linear integral equations. The methods to understand these transformations and related solutions involve methods of complex analysis, to be discussed later in Chapter 7 on Riemann–Hilbert boundary value problems.

Second- and higher-order nonlinear equations need not have only poles or algebraic singularities. For example, the equation

$$\frac{d^2w}{dz^2} = \left(\frac{dw}{dz}\right)^2 \left(\frac{2w-1}{w^2+1}\right) \tag{3.7.48}$$

has the solution

$$w(z) = \tan(\log(az + b)) \tag{3.7.49}$$

where a and b are arbitrary constants. Hence the point $z = -b/a$ is a branch point, and the function $w(z)$ has no limit as z approaches this point. Similarly, the equation

$$\frac{d^2w}{dz^2} = \frac{\alpha - 1}{\alpha w} \left(\frac{dw}{dz}\right)^2 \tag{3.7.50}$$

has the solution

$$w(z) = c(z - d)^\alpha \tag{3.7.51}$$

where c and d are arbitrary constants. Equation (3.7.51) has an algebraic branch point only if $\alpha = m/n$ where m and n are integers, otherwise the point $z = d$ is a transcendental branch point.

Third-order equations may possess even more exotic movable singular points. Indeed, motivated by Painlevé's work, Chazy (1911) showed that the following equation

$$\frac{d^3w}{dz^3} = 2w\frac{d^2w}{dz^2} - 3\left(\frac{dw}{dz}\right)^2 \tag{3.7.52}$$

was solvable via a rather nontrivial transformation of coordinates. His solution shows that the general solution of Eq. (3.7.52) possesses a movable natural barrier. Indeed the barrier is a circle, whose center and radius depend on initial values. Interestingly enough, the solution w in Eq. (3.7.52) is related to the following system of equations (first considered in the case $\epsilon = -1$ by Darboux (1878) and then solved by Halphen (1881), which we refer to as the Darboux-Halphens system (when $\epsilon = -1$):

$$\frac{dw_1}{dz} = w_2w_3 + \epsilon w_1(w_2 + w_3)$$

$$\frac{dw_2}{dz} = w_3w_1 + \epsilon w_2(w_3 + w_1)$$

$$\frac{dw_3}{dz} = w_1w_2 + \epsilon w_3(w_1 + w_2) \tag{3.7.53}$$

In particular, when $\epsilon = -1$, Chazy's equation is related to the solutions of Eq. (3.7.53) by

$$w = -2(w_1 + w_2 + w_3)$$

When $\epsilon = 0$, Eqs. (3.7.53) are related (by scaling) to Eqs. (3.7.44a,b,c), and the solution may be written in terms of elliptic functions. Equations (3.7.53) with $\epsilon = -1$ arise in the study of relativity and integrable systems (Ablowitz and Clarkson, 1991).

In fact, Chazy's and the Darboux-Halphen system can be solved in terms of certain special functions that are generalizations of trigonometric and elliptic function, so-called automorphic functions that we will study further in Chapter 5 (Section 5.8). By direct calculation (whose details are outlined in the exercises) we can verify that the following (owing to Chazy) yields a solution to Eq. (3.7.52). Transform to a new independent variable

$$z(s) = \frac{\chi_2(s)}{\chi_1(s)} \tag{3.7.54a}$$

where χ_1 and χ_2 are two linearly independent solutions of the following hypergeometric equation (see Eq. (3.7.35c), where $a = b = \frac{1}{12}$ and $c = \frac{1}{2}$):

$$\frac{d^2\chi}{ds^2} = \alpha(s)\frac{d\chi}{ds} + \beta(s)\chi \tag{3.7.54b}$$

where

$$\alpha(s) = \left(\frac{\frac{7s}{6} - \frac{1}{2}}{s(1-s)}\right) \quad \text{and} \quad \beta(s) = \frac{1}{144s(1-s)}$$

that is

$$s(1-s)\frac{d^2\chi}{ds^2} + \left(\frac{1}{2} - \frac{7s}{6}\right)\frac{d\chi}{ds} - \frac{\chi}{144} = 0 \tag{3.7.54c}$$

Then the solution w of Chazy's equation can be expressed as follows:

$$w(s(z)) = 6\frac{d}{dz}\log\chi_1 = \frac{6}{\chi_1}\frac{d\chi_1}{ds}\frac{ds}{dz} = \frac{6\frac{d\chi_1}{ds}}{\chi_1(dz/ds)}$$

$$= \frac{6\chi_1}{W(\chi_1, \chi_2)}\frac{d\chi_1}{ds} \tag{3.7.55}$$

where $W(\chi_1, \chi_2)$ is the Wronskian of χ_1 and χ_2, which satisfies $W' = \alpha W$, or

$$W(\chi_1, \chi_2) = \chi_1\frac{d\chi_2}{ds} - \chi_2\frac{d\chi_1}{ds} = s^{-1/2}(1-s)^{-2/3}W_0 \tag{3.7.56}$$

where W_0 is an arbitrary constant. Although this yields, in principle, only a special solution to Eq. (3.7.52), the general solution can be obtained by making the transformation $\chi_1 \mapsto a\chi_1 + b\chi_2$, $\chi_2 \mapsto c\chi_1 + d\chi_2$, with a, b, c, and d as arbitrary constants normalized to $ad - bc = 1$.

However, to understand the properties of the solution $w(z)$, we really need to understand the conformal map $z = z(s)$ and its inverse $s = s(z)$. (From $s(z)$ and $\chi_1(s(z))$ we find the solution $w(s(z))$.) Usually, this map is denoted by $s = s(z; \alpha, \beta, \gamma)$, where α, β, γ are three parameters related to the hypergeometric equation (3.7.54c), which are in this case $\alpha = 0$, $\beta = \pi/2$, $\gamma = \pi/3$. This function is called a **Schwarzian triangle function**, and the map transforms the region defined inside a "circular triangle" (a triangle whose sides are either straight lines or circular arcs – at least one side being an arc) in the z plane to the upper half s plane. It turns out that by reflecting the triangle successively about any of its sides, and repeating this process infinitely, we can analytically continue the function $s(z, 0, \pi/2, \pi/3)$ everywhere inside a circle. For the solution normalized as in Eqs. (3.7.54a–c) this is a circle centered at the origin. The function $s = s(z; 0, \pi/2, \pi/3)$ is single valued and analytic inside the circle, but the circumference of the circle is a natural boundary – which in this case can be shown to consist of a dense set of essential singularities. The reader can find a further discussion of mappings of circular triangles and Schwarzian triangle functions in Section 5.8. Such functions are special cases of what are often called **automorphic functions**.   Automorphic functions have the property that $s(\gamma(z)) = s(z)$, where $\gamma(z) = \frac{az+b}{cz+d}$, $ad - bc = 1$, and as such are generalization of periodic functions, for example, elliptic functions.

It is worth remarking that the Darboux-Halphen system (3.7.53) can also be solved in terms of a Schwarzian triangle function. In fact, the solutions ω_1, ω_2, ω_3, are given by the formulae

$$\omega_1 = -\frac{1}{2}\frac{d}{dz}\log\frac{s'(z)}{s}, \qquad \omega_2 = -\frac{1}{2}\frac{d}{dz}\log\frac{s'(z)}{1-s},$$
$$\omega_3 = -\frac{1}{2}\frac{d}{dz}\log\frac{s'(z)}{s(1-s)} \tag{3.7.57}$$

where $s(z)$ satisfies the equation

$$\{s, z\} = -\left(\frac{1}{s^2} + \frac{1}{(1-s)^2} + \frac{1}{s(1-s)}\right)\frac{(s'(z))^2}{2} \tag{3.7.58a}$$

and the term $\{s, z\}$ is the **Schwarzian derivative** defined by

$$\{s, z\} = \frac{s'''}{s'} - \frac{3}{2}\left(\frac{s''}{s'}\right)^2 \tag{3.7.58b}$$

Equation (3.7.58a) is obtained when we substitute ω_1, ω_2, and ω_3 given by Eq. (3.7.57) into Eq. (3.7.53), with $\epsilon = -1$. The function $s(z)$ is the "zero angle" Schwarzian triangle function, $s(z) = s(z, 0, 0, 0)$, which is discussed in Section 5.8. In the exercises, the following transformation involving Schwarzian derivatives is established:

$$\{z, s\} = \{s, z\} \frac{(-1)}{(s'(z))^2} \tag{3.7.59}$$

Using Eqs. (3.7.58a)–(3.7.59), we obtain the equation

$$\{z, s\} = \frac{1}{2} \left(\frac{1}{s^2} + \frac{1}{(1-s)^2} + \frac{1}{s(1-s)} \right) \tag{3.7.60}$$

In Section 5.8 we show how to solve Eq. (3.7.60), and thereby find the inverse transformation $z = z(s)$ in terms of hypergeometric functions. We will not go further into these because it will take us too far outside the scope of this book.

Problems for Section 3.7

1. Discuss the nature of the singular points (location, fixed, or movable) of the following differential equations and solve the differential equations.

    (a) $z\dfrac{dw}{dz} = 2w + z$ (b) $z\dfrac{dw}{dz} = w^2$

    (c) $\dfrac{dw}{dz} = a(z)w^3$, $a(z)$ is an entire function of z.

    (d) $z^2\dfrac{d^2w}{dz^2} + z\dfrac{dw}{dz} + w = 0$

2. Solve the differential equation

$$\frac{dw}{dz} = w - w^2$$

    Show that it has poles as its only singularity.

3. Given the equation

$$\frac{dw}{dz} = p(z)w^2 + q(z)w + r(z)$$

    where $p(z)$, $q(z)$, $r(z)$ are (for convenience) entire functions of z

(a) Letting $w = \alpha(z)\phi'(z)/\phi(z)$, show that taking $\alpha(z) = -1/p(z)$ eliminates the term $(\phi'/\phi)^2$, and find that $\phi(z)$ satisfies

$$\phi'' - \left(q(z) + \frac{p'(z)}{p(z)}\right)\phi' + p(z)r(z)\phi = 0$$

(b) Explain why the function $w(z)$ has, as its only singular points, movable poles. Where are they located?

4. Determine the indicial equation and the basic form of expansion representing the solution to the following equations:

(a) $z^2\dfrac{d^2w}{dz^2} + z\dfrac{dw}{dz} + (z^2 - p^2)w = 0,$ p not integer,

(Bessel's Equation)

(b) $\left(1 - z^2\right)\dfrac{d^2w}{dz^2} + 2z\dfrac{dw}{dz} + p(p+1)w = 0,$ p not integer,

(Legendre's Equation)

(c) $z(1-z)\dfrac{d^2w}{dz^2} + [c - (a+b+1)z]$

$\dfrac{dw}{dz} - abw = 0,$ one solution is satisfactory,

(Hypergeometric Equation)

5. Suppose we are given the equation $d^2w/dz^2 = 2w^3$.

(a) Let us look for a solution of the form

$$w = \sum_{n=0}^{\infty} a_n(z - z_0)^{n-r} = a_0(z - z_0)^{-r} + a_1(z - z_0)^{1-r} + \cdots$$

for z near z_0. Substitute this into the equation to determine that $r = 1$ and $a_0 = \pm 1$.

(b) "Linearize" about the basic solution by letting $w = \pm 1/(z - z_0) + v$ and dropping quadratic terms in v to find $d^2v/dz^2 = 12v/(z - z_0)^2$. Solve this equation (Cauchy–Euler type) to find

$$v = A(z - z_0)^{-2} + B(z - z_0)^3$$

(c) Explain why this indicates that all coefficients of subsequent powers in the following expansion (save possibly a_4)

$$w = \frac{\pm 1}{(z - z_0)} + a_1 + a_2(z - z_0) + a_3(z - z_0)^2 + a_4(z - z_0)^3 + \cdots$$

can be solved uniquely. Substitute the expansion into the equation for w, and find a_1, a_2, and a_3, and establish the fact that a_4 is *arbitrary*. We obtain two arbitrary constants in this expansion: z_0 and a_4. The solution to $w'' = 2w^3$ can be expressed in terms of elliptic functions; its general solution is known to have only simple poles as its movable singular points.

(d) Show that a similar expansion works when we consider the equation

$$\frac{d^2 w}{dz^2} = z w^3 + z w$$

(this is the second Painlevé equation (Ince, 1956)), and hence that the formal analysis indicates that the only movable algebraic singular points are poles. (Painlevé proved that there are no other singular points for this equation.)

(e) Show that this expansion *fails* when we consider

$$\frac{d^2 w}{dz^2} = 2w^3 + z^2 w$$

because a_4 cannot be found. This indicates that a more general expansion is required. (In fact, another term of the form $b_4(z-z_0)^3 \log(z-z_0)$ must be added at this order, and further logarithmic terms must be added at all subsequent orders in order to obtain a consistent formal expansion.)

6. In this exercise we describe the verification that formulae (3.7.54a)–(3.7.56) indeed satisfy Chazy's equation.

(a) Use Eqs. (3.7.54a)–(3.7.55) to verify, by differentiation and resubstitution, the following formulae for the first three derivatives of w. (Use the linear equation (3.7.54b)

$$\frac{d^2 \chi}{ds^2} = \alpha(s) \frac{d\chi}{ds} + \beta(s)\chi$$

to resubstitute the second derivative $\chi''(s)$ in terms of the first derivative $\chi'(s)$ and the function $\chi(s)$ successively, thereby eliminating higher derivatives of $\chi(s)$.)

$$\frac{dw}{dz} = 6\left(\chi_1^4\beta + \chi_1^2(\chi_1')^2\right)/\mathcal{W}^2 \tag{i}$$

$$\frac{d^2w}{dz^2} = 6\left[\chi_1^6(\beta' - 2\alpha\beta) + \chi_1^5\chi_1'6\beta + 2\chi_1^3(\chi_1')^3\right]/\mathcal{W}^3 \tag{ii}$$

$$\frac{d^3w}{dz^3} = 6\left[\chi_1^8\left(\beta'' - 2\alpha'\beta - 5\alpha\beta' + 6\alpha^2\beta + 6\beta^2\right)\right.$$
$$\left. + \chi_1^7\chi_1'(2\beta' - 24\alpha\beta) + 6\chi_1^4(\chi_1')^4 + 36\chi_1^6(\chi_1')^2\right]/\mathcal{W}^4 \tag{iii}$$

where $\mathcal{W}$ is given by Eq. (3.7.56).

(b) By inserting (i)–(iii) into Chazy's equation (3.7.52) show that all terms cancel except for the following equation in α and β:

$$\beta'' - 2\alpha'\beta - 5\alpha\beta' + 6\alpha^2\beta + 24\beta^2 = 0 \tag{iv}$$

Show that the specific choices, as in Eq. (3.7.54b)

$$\alpha(s) = \left(\frac{\frac{7s}{6} - \frac{1}{2}}{s(1-s)}\right) \quad \text{and} \quad \beta(s) = \frac{1}{144s(1-s)}$$

satisfy (iv) and hence verify Chazy's solution.

7. Consider an invertible function $s = s(z)$.

(a) Show that the derivative d/dz transforms according to the relationship
$$\frac{d}{dz} = \frac{1}{z'(s)}\frac{d}{ds}.$$

(b) As in Eq. (3.7.58b), the Schwarzian derivative is defined as $\{s, z\} = (s''/s')' - \frac{1}{2}(s''/s')^2$. Show that

$$\{s, z\} = \frac{1}{z'(s)}\frac{d^2}{ds^2}\left(\frac{1}{z'(s)}\right) - \frac{1}{2}\left(\frac{d}{ds}\left(\frac{1}{z'(s)}\right)\right)^2$$

$$= -\frac{1}{(z'(s))^2}\{z, s\}$$

(c) Consequently, establish that

$$\{z, s\} = \{s, z\}\left(-\frac{1}{(s'(z))^2}\right)$$

8. In this exercise we derive a different representation for the solution of Chazy's equation.

(a) Show that

$$s''(z) = \frac{d}{dz}\left(s'(z)\right) = s'(z)\frac{d}{ds}\left(\frac{1}{z'(s)}\right)$$

(b) Use $z(s) = \chi_2(s)/\chi_1(s)$, where χ_1 and χ_2 satisfy the hypergeometric equation (3.7.54a), and the Wronskian relation $W(\chi_1, \chi_2) = (\chi_1\chi_2' - \chi_1'\chi_2) = W_0 s^{-1/2}(1-s)^{-2/3}$ in the above formulae, to show that

$$s''(z) = s'(z)\frac{d}{ds}\left(s^{\frac{1}{2}}(1-s)^{\frac{2}{3}}\chi_1^2(s)/W_0\right)$$

$$= s'(z)\left(\frac{1}{2s}s'(z) + \frac{2}{3(1-s)}s'(z) + \frac{2\chi_1'}{\chi_1}s'(z)\right)$$

(c) Use Chazy's solution (3.7.55), $w = \dfrac{6\chi_1'}{\chi_1}s'(z)$, to show that

$$w = z\frac{s''}{s} - \frac{3}{2}\frac{s'}{s} + \frac{2s'}{1-s}$$

$$= \frac{1}{2}\frac{d}{dz}\log\frac{(s')^6}{s^3(1-s)^4}$$

(d) Note that here $s(z)$ is the Schwarzian triangle function with angles 0, $\pi/2$, $\pi/3$; that is,

$$s(z) = s(z, 0, \pi/2, \pi/3)$$

The fact that Chazy's and the Darboux-Halphen system are related by the equation $w = -2(\omega_1 + \omega_2 + \omega_3)$ allows us to find a relation between the above Schwarzian $s(z, 0, \pi/2, \pi/3)$ (for Chazy's equation) and the one used in the text for the solution of the Darboux-Halphen system with zero angles, $s(z, 0, 0, 0)$. Call the latter Schwarzian $\hat{s}(z)$, that is, $\hat{s}(z) = s(z, 0, 0, 0)$. Show that Eq. (3.7.57) and $w = -2(w_1 + w_2 + w_3)$ yields the relationship

$$\frac{1}{2}\frac{d}{dz}\log\frac{(s')^6}{s^3(1-s)^4} = \frac{d}{dz}\log\frac{(\hat{s}')^3}{\hat{s}^2(1-\hat{s})^2}$$

or

$$\frac{(s')^6}{s^3(1-s)^4} = A\frac{(\hat{s}')^6}{\hat{s}^4(1-\hat{s})^4}$$

where A is a constant.

*3.8 Computational Methods

In this section we discuss some of the concrete aspects involving computation in the study of complex analysis. Our purpose here is not to be extensive in our discussion but rather to illustrate some basic ideas that can be readily implemented. We will discuss two topics: the evaluation of (a) Laurent series and (b) the solution of differential equations, both of which relate to our discussions in this chapter. We note that an extensive discussion of computational methods and theory can be found in Henrici (1977).

*3.8.1 Laurent Series

In Section 3.3 we derived the Laurent series representation of a function analytic in an annulus, $R_1 \leq |z - z_0| \leq R_2$. It is given by the formulae (3.3.1) and (3.3.2), which we repeat here for the convenience of the reader:

$$f(z) = \sum_{n=-\infty}^{\infty} c_n (z - z_0)^n \qquad (3.8.1)$$

where

$$c_n = \frac{1}{2\pi i} \oint_C \frac{f(z)\,dz}{(z - z_0)^{n+1}} \qquad (3.8.2)$$

and C is any simple closed contour in the annulus which encloses the inner boundary $|z - z_0| = R_1$. We shall take C to be a circle of radius r. Accordingly the change of variables

$$z = z_0 + re^{i\theta} \qquad (3.8.3)$$

where r is the radius of a circle with $R_1 \leq r \leq R_2$, allows us to rewrite Eq. (3.8.2) as

$$\hat{c}_n = \frac{1}{2\pi} \int_0^{2\pi} f(\theta) e^{-in\theta}\,d\theta \qquad (3.8.4)$$

where $c_n = \hat{c}_n / r^n$. In fact, Eq. (3.8.4) gives the Fourier coefficients of the function

$$f(\theta) = \sum_{n=-\infty}^{\infty} \hat{c}_n e^{in\theta} \qquad (3.8.5)$$

with period 2π defined on the circle (3.8.3). Equation (3.8.4) can be used as a computational tool after discretization. We consider $2N$ points equally spaced along the circle, with $\theta_j = -\pi + hj$, $j = 0, 1, \ldots, 2N - 1$, and $\int_0^{2\pi} \to \sum_{j=-N}^{N-1}$ with $d\theta \to \Delta\theta = h = 2\pi/(2N) = \pi/N$ (note that when $j = 2N$ then $\theta_{2N} = \pi$).

The following discretization corresponds to what is usually called the discrete Fourier transform:

$$f(\theta_j) = \sum_{n=-N}^{N-1} \hat{c}_n e^{in\theta_j} \tag{3.8.6}$$

where

$$\hat{c}_n = \frac{1}{2N} \sum_{j=-N}^{N-1} f(\theta_j) e^{-in\theta_j} \tag{3.8.7}$$

We note that the formulae (3.8.6) and (3.8.7) can be calculated directly, at a "cost" of $O(N^2)$ multiplications. (The notation $O(N^2)$ means proportional to N^2; a formal definition can be found in section 6.1.) Moreover, it is well known that in fact, the computational "cost" can be reduced significantly to $O(N \log N)$ multiplications by means of the fast fourier transform (FFT), (see, e.g., Henrici (1977)).

Given a function at $2N$ equally spaced points on a circle, one can readily compute the discrete Fourier coefficients, $\hat{c}_n$. The approximate Laurent coefficients are then given by $c_n = \hat{c}_n/r^n$. (For all the numerical examples below we use $r = 1$.) As N increases, the approximation improves rapidly if the continuous function is expressible as a Laurent series. However, if the function $f(z)$ were analytic, we would find that the coefficients with negative indices would be zero (to a very good approximation).

Example 3.8.1 Consider the functions (a) $f(z) = 1/z$ and (b) $f(z) = e^{1/z}$. Note that with $z_0 = 0$ the exact answers are (a) $c_{-1} = 1$ and $c_n = 0$ for $n \neq 0$, (b) $c_n = 1/(-n)!$ for $n \leq 0$, and $c_n = 0$ for $n \geq 1$. The magnitude of the numerically computed coefficients, using $N = 16$, are shown in Figure 3.8.1 ($*$ represents

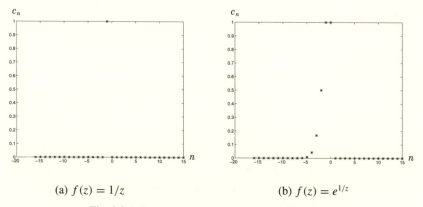

(a) $f(z) = 1/z$ (b) $f(z) = e^{1/z}$

Fig. 3.8.1. Laurent coefficients c_n for two functions

the coefficient). Note that for part (a) we obtain only one significantly nonzero coefficient ($c_{-1} \approx 1$ and $c_n \approx 0$ to high accuracy). In part (b) we find that $c_n \approx 0$ for $n \geq 1$; all the coefficients agree with the exact values to a high degree of accuracy. Note that the coefficients decay rapidly for large negative n.

*3.8.2 Differential Equations

The solution of differential equations in the complex plane can be approximated by many of the computational methods often studied in numerical analysis. We shall discuss "time-stepping" methods and series methods.

We consider the scalar differential equation

$$\frac{dy}{dz} = f(z, y) \tag{3.8.8}$$

with the initial condition $y(z_0) = y_0$, where f is analytic in both arguments in some domain D containing $z = z_0$. The key ideas are best illustrated by the explicit Euler method. Here dy/dz is approximated by the difference $(y(z + h_n) - y(z))/h_n$. Call $z_{n+1} = z_n + h_n$, $y(z_n) = y_n$, and note that h_n is *complex*; that is, h_n can take any direction in the complex plane. Also note that we allow the step size, h_n, to vary from one time step to the next, which is necessary if, for instance, we want to integrate around the unit circle. In this application we keep $|h_n|$ constant. Hence, at $z = z_n$, we have the approximation

$$y_{n+1} = y_n + h_n f(z_n, y_n) \tag{3.8.9}$$

with the initial condition $y(z_0) = y_0$. It can be shown that under suitable assumptions, Eq. (3.8.9) is an $O(h_n^2)$ approximation over every step and an $O(h_n)$ approximation if we integrate over a finite time T with $h_n \to 0$. Equation (3.8.9) is straightforward to apply as we now show.

Example 3.8.2 Approximate the solution of the equation

$$\frac{dy}{dz} = y^2, \qquad y(1) = -2$$

as z traverses along the contour C, where C is the unit circle in the complex plane. We discretize along the circle and take $z_n = e^{i\theta_n}$, where $\theta_n = 2\pi n/N$ and $h_n = z_{n+1} - z_n$. The exact solution of $dy/dz = y^2$ is $y = 1/(A - z)$, where A is an arbitrary complex constant and we see that it has a pole at the

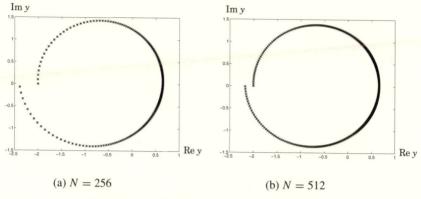

(a) $N = 256$ (b) $N = 512$

Fig. 3.8.2. Explicit Euler's method, Example 3.8.2

location $z = A$. For the initial value $y(1) = -2$, $A = \frac{1}{2}$, and the pole is located at $z = \frac{1}{2}$.

Because we are taking a circuit around the unit circle, we never get close to the singular point. Because the solution is single valued, we expect to return to the initial value after one circuit. We use the approximation (3.8.9) with $f(z_n, y_n) = y_n^2$, $y_0 = y(z_0) = -2$. The solutions using $N = 256$ and $N = 512$ are shown in Figure 3.8.2, where we plot the real part of y versus the imaginary part of y: $y(z_n) = y_R(z_n) + iy_I(z_n)$ for $n = 0, 1, \ldots, N$. Although the solution using $N = 512$ shows an improvement, the approximate solution is not single valued as it should be. This is due to the inaccuracy of the Euler method.

We could improve the solution by increasing N even further, but in practice one uses more accurate methods that we now quote. By using more accurate Taylor series expansions of $y(z_n + h_n)$ we can find the following second- and fourth-order accurate methods:

(a) second-order Runge–Kutta (RK2)

$$y_{n+1} = y_n + \frac{1}{2}h_n(k_{n_1} + k_{n_2}) \tag{3.8.10}$$

where $k_{n_1} = f(z_n, y_n)$ and $k_{n_2} = f(z_n + h_n, y_n + h_n k_{n_1})$
(b) fourth-order Runge–Kutta (RK4)

$$y_{n+1} = y_n + \frac{1}{6}h_n(k_{n_1} + 2k_{n_2} + 2k_{n_3} + k_{n_4}) \tag{3.8.11}$$

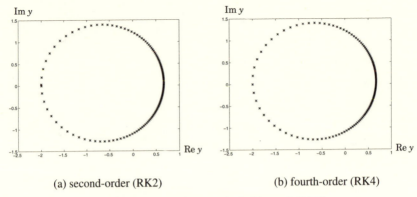

(a) second-order (RK2)   (b) fourth-order (RK4)

Fig. 3.8.3. Runge–Kutta methods (Example 3.8.3)

where

$$k_{n_1} = f(z_n, y_n), \qquad k_{n_2} = f\left(z_n + \frac{1}{2}h_n, y_n + \frac{1}{2}h_n k_{n_1}\right)$$

$$k_{n_3} = f\left(z_n + \frac{1}{2}h_n, y_n + \frac{1}{2}h_n k_{n_2}\right), \qquad k_{n_4} = f(z_n + h_n, y_n + h_n k_{n_3})$$

Example 3.8.3 We illustrate how the above methods, RK2 and RK4, work on the same problem as Example 3.8.2 above, choosing h_n in the same way as before. Using $N = 128$ we see in Figure 3.8.3 that the solution is indeed single valued (to numerical accuracy) as expected. It is clear that the solutions obtained by these methods are nearly single valued; they are a significant improvement over Euler's method. Moreover, RK4 is an improvement over RK2, although RK4 requires more function evaluations and more computer time.

Example 3.8.4 Consider the differential equation

$$\frac{dy}{dz} = \frac{1}{2}y^3$$

with initial values

   (a) $y(1) = 1$, the exact solution is $y(z) = 1/\sqrt{2 - z}$;

   (b) $y(1) = 2i$, the exact solution is $y(z) = 2i/\sqrt{4z - 3}$.

The general solution is $y(z) = (z_0 - z)^{-1/2}$, where the proper branch of the square root is chosen to agree with the initial value. We integrate around the unit circle (choosing h_n as in the previous examples) using RK2 and RK4 for $N = 128$; the results are shown in Figure 3.8.4. For the initial value in part (a) the singularity lies outside the unit circle and the numerical solutions are single

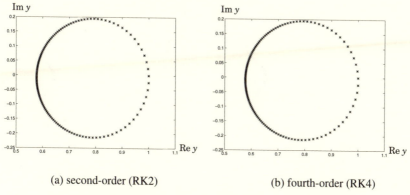

(a) second-order (RK2)   (b) fourth-order (RK4)

Fig. 3.8.4. Part (a) of Example 3.8.4, using $y(1) = 1$ and $N = 128$

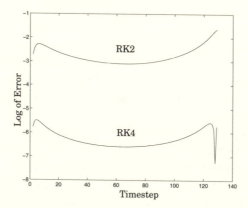

Fig. 3.8.5. The error in the numerical solutions shown in Figure 3.8.4

valued. Figure 3.8.5 shows the logarithm of the absolute value of the errors in the calculations graphed in Figure 3.8.4. Note that the error in RK4 is several orders of magnitude smaller than the error in RK2. For the initial value of part (b) the branch point is at $z = 3/4$ and thus lies inside the unit circle and the solutions are clearly not single valued. Numerically (see Fig. 3.8.6) we find that the jump in the function $y(z)$ is approximately $4i$ as we traverse the circle from $\theta = 0$ to 2π, as expected from the exact solution.

As long as there are no singular points on or close to the integration contour there will be no difficulty in implementing the above time-stepping algorithms. However, in practice one frequently has nearby singular points and the contour may need to be modified in order to analytically continue the solution. In this case, it is sometimes useful to use series methods to approximate the

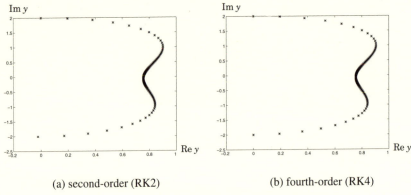

(a) second-order (RK2)   (b) fourth-order (RK4)

Fig. 3.8.6. Part (b) of Example 3.8.4, using $y(1) = 2i$ and $N = 128$

solution of the differential equation and estimate the radius of convergence as the calculation proceeds. This is discussed next.

Given Eq. (3.8.8) and noting Cauchys' Theorem for differential equations, Theorem 3.7.1, we can look for a series solution of the form $y = \sum_{n=0}^{\infty} A_n$ $(z - z_0)^n$. By inserting this series into the equation, we seek to develop a recursion relation between the coefficients; this can be difficult or unwieldy in complicated cases, but computationally speaking, it can almost always be accomplished. Having found such a recursion relation, we can evaluate the coefficients A_n and find an approximation to the radius of convergence: $R = |z - z_0| = \lim_{n \to \infty} |A_n/A_{n+1}|$. As we proceed in the calculation we estimate the radius of convergence (for large n). We may need to modify our contour if the radius of convergence begins to shrink and move in a direction where the radius of convergence enlarges or remains acceptably large.

Example 3.8.5 Evaluate the series solution to the equation

$$\frac{dy}{dz} = y^2 + 1 \qquad (3.8.12)$$

with $y(0) = 1$. The exact solution is obtained by integrating

$$\frac{dy}{1 + y^2} = dz$$

to find $y = \tan(z + \pi/4)$. In order to obtain a recursion relation associated with the coefficients of the series solution $y = \sum_{n=0}^{\infty} A_n(z - z_0)^n$ it is useful to use the series product formula:

$$\sum_{n=0}^{\infty} A_n(z - z_0)^n \sum_{m=0}^{\infty} B_m(z - z_0)^m = \sum_{n=0}^{\infty} C_n(z - z_0)^n$$

where $C_n = \sum_{p=0}^{n} A_p B_{n-p}$. The insertion of the series for y into Eq. (3.8.12) yields

$$\sum_{n=0}^{\infty} n A_n(z - z_0)^{n-1} = \left(\sum_{n=0}^{\infty} A_n(z - z_0)^n \right)^2 + 1$$

Using the product formula and the transformation

$$\sum_{n=0}^{\infty} n A_n(z - z_0)^{n-1} = \sum_{n=0}^{\infty} (n + 1) A_{n+1}(z - z_0)^n$$

we obtain the equation

$$\sum_{n=0}^{\infty} (n + 1) A_{n+1}(z - z_0)^n = \sum_{n=0}^{\infty} \left(\sum_{p=0}^{n} A_p A_{n-p} \right) (z - z_0)^n + 1$$

and hence the recursion relation

$$(n + 1) A_{n+1} = \sum_{p=0}^{n} A_p A_{n-p} + \delta_{n,0} \tag{3.8.13}$$

where $\delta_{n,0}$ is the Kronecker delta function; $\delta_{n,0} = 1$ if $n = 0$ and 0 otherwise. Because we have posed the differential equation at $z = 0$, we begin with $z_0 = 0$ and $A_0 = y_0 = 1$. It is straightforward to compute the coefficients from this formula. Computing the ratios up to $n = 12$ (for example) we find that the final terms yield $\lim_{n \to \infty} A_n/A_{n+1} \approx A_{11}/A_{12} = 0.78539816$. It is clear that the series converges inside a radius of convergence R of approximately $\pi/4$ as it should. Suppose we use this series up to $z = 0.1$ in steps of 0.01. This means we use the recursion relation (3.8.13), but we use it repeatedly after each time step; that is, for each of the values $A_0 = y(z_j)$, $z_j = 0, 0.01, 0.02, \ldots, 0.10$, we calculate the corresponding, successive coefficients, A_n, from Eq. (3.8.13) before we proceed to the next z_j. This means that in the series solution for y we are reexpanding about a new point $z_0 = z_j$. We obtain (still using $n = 12$ coefficients) $y(0.1) = 1.22305$ and an approximate radius of convergence $R = 0.6854$. Note that these values are very good approximations of the analytical values. We can also evaluate the series by moving into the complex plane. For example, if we expand around $z = 0.1$ and move in steps of $0.01i$ to $z = 0.1 + 0.1i$, we obtain $y(0.1 + 0.1i) = 1.1930 + 0.2457i$ and an approximate radius of convergence $R = 0.6967$. The series expansion is now seen to be valid in a larger region. This is true because we are now moving away from the singularity. The procedure can be repeated and we can analytically extend the solution by reexpanding the series about new points and employing the recursion relation to move into any region where the solution is analytic. In this way we can "internally" decide on how big a region of analyticity we wish to cover and always be sure to move into regions where the series solution is valid.

Problems for Section 3.8

1. Find the magnitude of the numerically computed Laurent coefficients with $z_0 = 0$ (using $N = 32$) for (i) $f(z) = e^z$, (ii) $f(z) = \sqrt{z}$, (iii) $f(z) = 1/\sqrt{z}$, (iv) $f(z) = \tan 1/z$, and show that they agree with those in Figure 3.8.7.

   (a) Do the Laurent coefficients in Figure 3.8.7 correspond to what you would expect from analytical considerations? What is the true behavior of each function; that is, what kind of singularities do these functions have?

   (b) Note that the coefficients decay at very different rates for the examples (i) to (iv). Explain why this is the case. (Hint: Relate it to the single-valuedness of the function.)

2. Consider the differential equation $\dfrac{dy}{dz} = y^2$, $y(z_0) = y_0$.

   (a) Show that the analytical solution is given by

   $$y(z) = \frac{y_0}{1 - y_0(z - z_0)}$$

   (b) Write down the position of the singularity of the solution (a) above. What is the nature of the singularity?

   (c) From (b) above note that the position of the singularity depends on the initial values, that is, z_0 and y_0. Choose $z_0 = 1$ and find the values y_0 for which the singularity lies inside the unit circle.

   (d) Use the time-stepping numerical techniques discussed in this section (Euler, RK2, and RK4) to compute the solution on the unit circle $z = e^{i\theta}$ as θ varies from $\theta = 0$ to 4π.

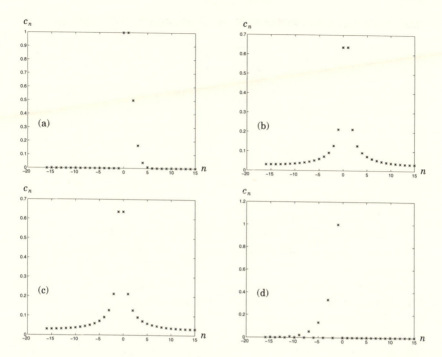

Fig. 3.8.7. Laurent coefficients c_n for Problem 1: (a) $f(z) = e^z$; (b) $f(z) = z^{1/2}$; (c) $f(z) = 1/z^{1/2}$; (d) $f(z) = \tan(1/z)$.

3. Repeat Problem 2 above for the differential equation $dy/dz = \frac{1}{2}y^3$, $y(z_0) = y_0$.

4. Consider the equation

$$\frac{dy}{dz} + 2zy = 1, \qquad y(1) = 1$$

(a) Solve this equation using the series method. Evaluate the solution as we traverse the unit circle. Show that the solution is single valued.

(b) Evaluate an approximation to $y(-1)$ from the series.

(c) Show that an exact representation of the solution in terms of integrals is

$$y(z) = \int_1^z e^{t^2 - z^2}\, dt + e^{1 - z^2}$$

and verify that, by evaluating $y(z)$ by a Taylor series (i.e., use $e^{t^2} = 1 + t^2 + t^4/2! + t^6/3! + t^8/4! + \cdots$), the answer obtained from this series is a good approximation to that obtained in part (b).

4

Residue Calculus and Applications of Contour Integration

In this chapter we extend Cauchy's Theorem to cases where the integrand is not analytic, for example, if the integrand possesses isolated singular points. Each isolated singular point contributes a term proportional to what is called the residue of the singularity. This extension, called the residue theorem, is very useful in applications such as the evaluation of definite integrals of various types. The residue theorem provides a straightforward and sometimes the only method to compute these integrals. We also show how to use contour integration to compute the solutions of certain partial differential equations by the techniques of Fourier and Laplace transforms.

4.1 Cauchy Residue Theorem

Let $f(z)$ be analytic in the region D, defined by $0 < |z - z_0| < \rho$, and let $z = z_0$ be an isolated singular point of $f(z)$. The Laurent expansion of $f(z)$ (discussed in Section 3.3) in D is given by

$$f(z) = \sum_{n=-\infty}^{\infty} C_n(z - z_0)^n \qquad (4.1.1)$$

with

$$C_n = \frac{1}{2\pi i} \oint_C \frac{f(z)\, dz}{(z - z_0)^{n+1}} \qquad (4.1.2)$$

where C is a simple closed contour lying in D. The negative part of series $\sum_{n=-\infty}^{-1} C_n(z - z_0)^n$ is referred to as the **principal part** of the series. The coefficient C_{-1} is called the **residue** of $f(z)$ at z_0, sometimes denoted as

206

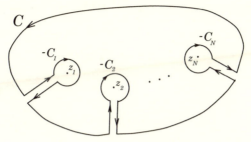

Fig. 4.1.1. Proof of Theorem 4.1.1

$C_{-1} = \text{Res}\,(f(z); z_0)$. We note when $n = -1$, Eq. (4.1.2) yields

$$\oint_C f(z)\,dz = 2\pi i C_{-1} \qquad (4.1.3)$$

Thus Cauchy's Theorem is now seen to suitably generalize to functions $f(z)$ with one isolated singular point. Namely, we had previously proven that for $f(z)$ analytic in D the integral $\oint_C f(z)\,dz = 0$, where C was a closed contour in D. Equation (4.1.3) shows that the correct modification of Cauchy's Theorem, when $f(z)$ contains *one* isolated singular point at $z_0 \in D$, is that the integral be proportional to the residue (C_{-1}) of $f(z)$ at z_0. In fact, this concept is easily extended to functions with a finite number of isolated singular points. The result is often referred to as the **Cauchy Residue Theorem**, which we now state.

Theorem 4.1.1 Let C be a simple closed contour inside and on which f is analytic, except for a finite number of isolated singular points $z_1, \ldots, z_N$. Then

$$\oint f(z)\,dz = 2\pi i \sum_{j=1}^{N} a_j \qquad (4.1.4)$$

where a_j is the residue of $f(z)$ at $z = z_j$, denoted by $a_j = \text{Res}\,(f(z); z_j)$.

Proof Consider Figure 4.1.1. We enclose each of the points z_j by small non-intersecting closed curves, each of which lies within C: $C_1, C_2, \ldots, C_N$ and is connected to the main closed contour by cross cuts. Because the integrals along the cross cuts vanish, we find that on the contour $\Gamma = C - C_1 - C_2 - \cdots - C_N$ (with each contour taken in the positive sense)

$$\int_\Gamma f(z)\,dz = 0$$

which follows from Cauchy's Theorem. Thus

$$\oint_C f(z)\,dz = \sum_{j=1}^{N} \oint_{C_j} f(z)\,dz \qquad (4.1.5)$$

We now use the result (4.1.3) about each singular point. Because $f(z)$ has a Laurent expansion in the neighborhood of each singular point, $z = z_j$, Eq. (4.1.4) follows. ∎

Some prototypical examples are described below.

Example 4.1.1 Evaluate

$$I_k = \frac{1}{2\pi i} \oint_{C_0} z^k \, dz, \qquad k \in \mathbf{Z}$$

where C_0 is the unit circle $|z| = 1$. Because z^k is analytic for $k = 0, 1, 2, \ldots$, we have $I_k = 0$ for $k = 0, 1, 2, \ldots$. Similarly, for $k = -2, -3, \ldots$ we find that the residue of z^k is zero; hence $I_k = 0$. For $k = -1$ the residue of z^{-1} is unity and thus $I_{-1} = 1$.

We write $I_k = \delta_{k,-1}$, where

$$\delta_{k,\ell} = \begin{cases} 1 & \text{when } k = \ell \\ 0 & \text{otherwise} \end{cases}$$

is referred to as the Kronecker delta function.

Example 4.1.2 Evaluate

$$I = \frac{1}{2\pi i} \oint_{C_0} z\, e^{1/z} \, dz$$

where C_0 is the unit circle $|z| = 1$. The function $f(z) = z e^{1/z}$ is analytic for all $z \neq 0$ inside C_0 and has the following Laurent expansion about $z = 0$:

$$z e^{1/z} = z\left(1 + \frac{1}{z} + \frac{1}{2! z^2} + \frac{1}{3! z^3} + \cdots\right)$$

Hence the residue Res $\left(z e^{1/z}; 0\right) = 1/2!$, and we have

$$I = \frac{1}{2}$$

Example 4.1.3 Evaluate

$$I = \oint_{C_2} \frac{z+2}{z(z+1)}\, dz$$

where C_2 is the circle $|z| = 2$.

We write the integrand as a partial fraction

$$\frac{z+2}{z(z+1)} = \frac{A}{z} + \frac{B}{z+1}$$

hence $z + 2 = A(z+1) + Bz$, and we deduce (taking $z = 0$, $z = -1$) that $A = 2$ and $B = -1$. (In fact, the coefficients $A = 2$ and $B = -1$ are the residues of the function $\frac{z+2}{z(z+1)}$ at $z = 0$ and $z = -1$, respectively.) Thus

$$I = \oint_C \left(\frac{2}{z} - \frac{1}{z+1} \right) dz = 2\pi i (2 - 1) = 2\pi i$$

where we note that the residue about $z = 0$ of $2/z$ is 2 and the residue of $1/(z+1)$ about $z = -1$ is 1.

So far we have evaluated the residue by expanding $f(z)$ in a Laurent expansion about the point $z = z_j$. Indeed, if $f(z)$ has an essential singular point at $z = z_0$, then expansion in terms of a Laurent expansion is the only general method to evaluate the residue. If, however, $f(z)$ has a pole in the neighborhood of z_0, then there is a simple formula, which we now give.

Let $f(z)$ be defined by

$$f(z) = \frac{\phi(z)}{(z - z_0)^m} \tag{4.1.6}$$

where $\phi(z)$ is analytic in the neighborhood of $z = z_0$, m is a positive integer, and f has a pole of order m. Then the residue of $f(z)$ at z_0 is given by

$$C_{-1} = \frac{1}{(m-1)!} \left(\frac{d^{m-1}}{dz^{m-1}} \phi \right)(z = z_0) \tag{4.1.7}$$

(This means that one first computes the $(m - 1)^{st}$ derivative of $\phi(z)$ and then evaluates it at $z = z_0$.)

The derivation of this formula follows from the fact that if $f(z)$ has a pole of order m at $z = z_0$, then it can be written in the form (4.1.6). Because $\phi(z)$ is analytic in the neighborhood of z_0

$$\phi(z) = \phi(z_0) + \phi'(z_0)(z - z_0) + \cdots + \frac{\phi^{(m-1)}(z_0)}{(m-1)!}(z - z_0)^{m-1} + \cdots$$

Dividing this expression by $(z - z_0)^m$, it follows that the coefficient of the $(z - z_0)^{-1}$ term, denoted by C_{-1}, is given by Eq. (4.1.7).

A simple pole has $m = 1$, hence the formula

$$C_{-1} = \phi(z_0) = \lim_{z \to z_0} ((z - z_0)f(z)) \tag{4.1.8}$$

$$\text{(simple pole)}$$

Suppose our function is given by a ratio of two functions $N(z)$ and $D(z)$, where both are analytic in the neighborhood of $z = z_0$

$$f(z) = \frac{N(z)}{D(z)} \tag{4.1.9}$$

Then if $D(z)$ has a zero of order m at z_0, we may write $D(z) = (z - z_0)^m \tilde{D}(z)$, where $\tilde{D}(z_0) \neq 0$ and $\tilde{D}(z)$ is analytic near $z = z_0$. Hence $f(z)$ takes the form (4.1.6) where $\phi(z) = N(z)/\tilde{D}(z)$ and Eq. (4.1.7) applies. In the special case of a simple pole, $m = 1$, from the Taylor series of $N(z)$ and $D(z)$, we have $N(z) = N(z_0) + (z - z_0)N'(z_0) + \cdots$, and $\tilde{D}(z) = D'(z_0) + (z - z_0)\frac{D''(z_0)}{2!}$ $+ \cdots$, whereupon $\phi(z_0) = \frac{N(z_0)}{D'(z_0)}$, and Eq. (4.1.8) yields

$$C_{-1} = \frac{N(z_0)}{D'(z_0)} \tag{4.1.10}$$

In the following problems we illustrate the use of formulae (4.1.7) and (4.1.10).

Example 4.1.4 Evaluate

$$I = \frac{1}{2\pi i} \oint_{C_2} \left(\frac{3z + 1}{z(z - 1)^3} \right) dz$$

where C_2 is the circle $|z| = 2$.

The function

$$f(z) = \frac{3z + 1}{z(z - 1)^3}$$

has the form (4.1.6) near $z = 0, z = 1$. We have

$$\mathrm{Res}\,(f(z); 0) = \left(\frac{3z+1}{(z-1)^3}\right)_{z=0} = -1$$

$$\mathrm{Res}\,(f(z); 1) = \frac{1}{2!}\left(\frac{d^2}{dz^2}\left(\frac{3z+1}{z}\right)\right)_{z=1}$$

$$= \frac{1}{2!}\left(\frac{d^2}{dz^2}\left(3 + \frac{1}{z}\right)\right)_{z=1} = +1$$

hence $I = 0$.

Example 4.1.5 Evaluate

$$I = \frac{1}{2\pi i}\oint_{C_0} \cot z \, dz$$

where C_0 is the unit circle $|z| = 1$.

The function $\cot z = \cos z / \sin z$ is a ratio of two analytic functions whose singularities occur at the zeroes of $\sin z : z = n\pi, n = 0, \pm 1, \pm 2, \ldots$. Because the contour C_0 encloses only the singularity $z = 0$, we can use formula (4.1.10) to find

$$I = \lim_{z\to 0}\frac{\cos z}{(\sin z)'} = 1$$

Sometimes it is useful to work with the residue at infinity. The residue at infinity, $\mathrm{Res}\,(f(z), \infty)$, in analogy with the case of finite isolated singular points (see Eq. (4.1.5)), is given by the formula

$$\mathrm{Res}\,(f(z); \infty) = \frac{1}{2\pi i}\oint_{C_\infty} f(z)\, dz \tag{4.1.11a}$$

where C_∞ denotes the limit $R \to \infty$ of a circle C_R with radius $|z| = R$. For example, if $f(z)$ is analytic at infinity with $f(\infty) = 0$, it has the expansion $f(z) = a_{-1}/z + a_{-2}/z^2 + \cdots$, hence we find that

$$\mathrm{Res}\,(f(z), \infty) = \frac{1}{2\pi i}\oint_{C_\infty} f(z)\, dz$$

$$= \lim_{R\to\infty}\frac{1}{2\pi i}\int_0^{2\pi}\left(\frac{a_{-1}}{Re^{i\theta}} + \frac{a_{-2}}{(Re^{i\theta})^2} + \cdots\right)i Re^{i\theta}\, d\theta$$

$$= a_{-1} \tag{4.1.11b}$$

In fact Eq. (4.1.11a,b) holds even when $f(\infty) \neq 0$, as long as $f(z)$ has a Laurent series in the neighborhood of $z = \infty$.

As mentioned earlier it is sometimes convenient, when analyzing the behavior of a function near infinity, to make the change of variables $z = 1/t$. Using $dz = -\frac{1}{t^2} dt$ and noting that the counterclockwise (positive direction) of C_R: $z = Re^{i\theta}$ transforms to a clockwise rotation (negative direction) in t: $t = 1/z = (1/R)e^{-i\theta} = \epsilon e^{-i\theta}$, $\epsilon = 1/R$, we have

$$\text{Res}(f(z); \infty) = \frac{1}{2\pi i} \oint_{C_\infty} f(z)\, dz = \frac{1}{2\pi i} \oint_{C_\epsilon} \left(\frac{1}{t^2}\right) f\left(\frac{1}{t}\right) dt \quad (4.1.12)$$

where C_ϵ is the limit as $\epsilon \to 0$ of a small circle ($\epsilon = 1/R$) around the origin in the t plane. Hence the residue at ∞ is given by

$$\text{Res}(f(z); \infty) = \text{Res}\left\{\frac{1}{t^2} f\left(\frac{1}{t}\right); 0\right\}$$

that is, the right-hand side is the coefficient of t^{-1} in the expansion of $f(1/t)/t^2$ near $t = 0$; the left-hand side is the coefficient of z^{-1} in the expansion of $f(z)$ at $z = \infty$. Sometimes we write

$$\text{Res}(f(z); \infty) = \lim_{z \to \infty} (zf(z)) \quad \text{when } f(\infty) = 0. \quad (4.1.13)$$

The concept of residue at infinity is quite useful when we integrate rational functions. Rational functions have only isolated singular points in the extended plane and are analytic elsewhere. Let $z_1, z_2, \ldots, z_N$ denote the finite singularities. Then *for every rational function,*

$$\sum_{j=1}^{N} \text{Res}(f(z); z_j) = \text{Res}(f(z); \infty) \quad (4.1.14)$$

This follows from an application of the residue theorem. We know that

$$\frac{1}{2\pi i} \lim_{R \to \infty} \oint_{C_R} f(z)\, dz = \sum_{j=1}^{N} \text{Res}(f(z); z_j)$$

because $f(z)$ has poles at $\{z_j\}_{j=1}^{N}$. On the other hand, because $f(z)$ is a rational function it has a Laurent series near infinity, hence we have $\text{Res}(f(z); \infty) = (1/2\pi i) \lim_{R \to \infty} \oint_{C_R} f(z)dz$.

We illustrate the use of the residue at infinity in the following examples.

Example 4.1.6 We consider the problem worked earlier, Example 4.1.4, but we now use $\text{Res}(f(z); \infty)$.

We note that all the singularities of $f(z)$ lie inside C_2, and the integrand is a rational function with $f(\infty) = 0$. Thus $I = \text{Res}(f(z); \infty)$. Because $f(z) = 3/z^3 + \cdots$ as $z \to \infty$, we use Eq. (4.1.13) to find

$$\text{Res}(f(z); \infty) = \lim_{z \to \infty} \frac{(3z + 1)}{(z - 1)^3} = 0$$

Hence $I = 0$, as we had already found by a somewhat longer calculation!

We illustrate this idea with another problem.

Example 4.1.7 Evaluate

$$I = \frac{1}{2\pi i} \oint_C \frac{a^2 - z^2}{a^2 + z^2} \frac{dz}{z}$$

where C is any simple closed contour enclosing the points $z = 0, z = \pm ia$.

The function

$$f(z) = \frac{a^2 - z^2}{a^2 + z^2} \frac{1}{z}$$

is a rational function with $f(\infty) = 0$, hence it has only isolated singular points, and note that $f(z) = -1/z + \cdots$ as $z \to \infty$.

$$I = \text{Res}(f(z); \infty)$$

We again use Eq. (4.1.13) to find

$$I = \lim_{z \to \infty} (zf(z)) = -1$$

The value $w(z_j)$, defined by

$$w(z_j) = \frac{1}{2\pi i} \oint_C \frac{dz}{z - z_j} = \frac{1}{2\pi i} [\log(z - z_j)]_C = \frac{\Delta\theta_j}{2\pi}, \qquad (4.1.15)$$

is called the **winding number** of the curve C around the point z_j. Here, $\Delta\theta_j$ is the total change in the argument of $z - z_j$ when z traverses the curve C around the point z_j. The value $w(z_j)$ represents the number of times (positive means counterclockwise) that C winds around z_j.

By the process of deformation of contours, including the introduction of cross cuts and the like, one can generalize the Cauchy Residue Theorem (4.1.1) to

$$\oint_C f(z) \, dz = 2\pi i \sum_{j=1}^{N} w(z_j) a_j, \qquad a_j = \text{Res}(f(z); z_j) \qquad (4.1.16)$$

where the hypothesis of Theorem 4.1.1 remains intact except for allowing the contour C to be nonsimple — hence the need for introducing the winding numbers $w(z_j)$ at every point $z = z_j$ with residue $a_j = \mathrm{Res}(f(z); z_j)$.

In applications it is usually clear how to break up a nonsimple contour into a series of simple contours; we shall not go through the formal proof in the general case. Rather than proving Eq. (4.1.16) in general, we illustrate the procedure of breaking up a nonsimple contour into a series of simple contours with an example.

Example 4.1.8 Use Eq. (4.1.16) to evaluate

$$I = \oint_C \frac{dz}{z^2 + a^2}$$

where C is the nonsimple contour of Figure 4.1.2.

The residue of $1/(z^2 + a^2)$ is

$$\mathrm{Res}\left(\frac{1}{z^2 + a^2}; \pm ia\right) = \left(\frac{1}{2z}\right)_{\pm ia} = \pm \frac{1}{2ai}$$

We see from Figure 4.1.2 that the winding numbers are $w(ia) = +2$ and $w(-ia) = +1$. Thus

$$I = 2\pi i \left[2\left(\frac{1}{2ai}\right) - \frac{1}{2ai}\right] = \frac{\pi}{a}$$

More generally, corresponding to any two closed curves C_1 and C_2 we have

$$\oint_{C_1} \frac{dz}{z^2 + a^2} = \oint_{C_2} \frac{dz}{z^2 + a^2} + \frac{N\pi}{a} \tag{4.1.17}$$

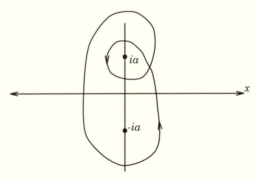

Fig. 4.1.2. Nonsimple curve for Example 4.1.8

where N is an appropriate integer related to the winding numbers of C_1 and C_2. Note that $N\pi/a$ is intimately related to the function

$$\Phi(z) = \int_{z_0}^{z} \frac{du}{u^2 + a^2} = \frac{1}{a} \tan^{-1} \frac{z}{a} + \Phi_0$$

where

$$\Phi_0 = \frac{-1}{a} \tan^{-1}\left(\frac{z_0}{a}\right)$$

or $z = a \tan a(\Phi - \Phi_0)$. Because z is periodic, with period $N\pi/a$, changing Φ by $N\pi/a$ yields the same value for z because the period of $\tan x$ is π.

Incorporating the winding numbers in Cauchy's Residue Theorem shows that, in the general case, the difference between two contours, C_1 and C_2, of a function $f(z)$ analytic inside these contours, save for a finite number of isolated singular points, is given by

$$\left(\oint_{C_1} - \oint_{C_2}\right) f(z)\, dz = 2\pi i \sum_{j=1}^{N} w_j a_j$$

The points $a_j = \text{Res}\,(f(z); z_j)$ are the periods of the inverse function $z = z(\Phi)$, defined by

$$\Phi(z) = \int_{z_0}^{z} f(z)\, dz; \qquad z = z(\Phi)$$

Problems for Section 4.1

1. Evaluate the integrals $\frac{1}{2\pi i} \oint_C f(z)\, dz$, where C is the unit circle centered at the origin and $f(z)$ is given below.

   (a) $\dfrac{z+1}{2z^3 - 3z^2 - 2z}$ (b) $\dfrac{\cosh(1/z)}{z}$ (c) $\dfrac{e^{-\cosh z}}{4z^2 + \pi^2}$

   (d) $\dfrac{\log(z+2)}{2z+1}$, principal branch   (e) $\dfrac{(z+1/z)}{z(2z - 1/2z)}$

2. Evaluate the integrals $\frac{1}{2\pi i} \oint_C f(z)\, dz$, where C is the unit circle centered at the origin with $f(z)$ given below. Do these problems by both (i) enclosing the singular points inside C and (ii) enclosing the singular points outside

C (by including the point at infinity). Show that you obtain the same result in both cases.

$$\text{(a)} \quad \frac{z^2 + 1}{z^2 - a^2}, \quad a^2 < 1 \qquad \text{(b)} \quad \frac{z^2 + 1}{z^3} \qquad \text{(c)} \quad z^2 e^{-1/z}$$

3. Determine the type of singular point each of the following functions has at $z = \infty$:

    (a) z^m, $m =$ positive integer   (b) $z^{1/3}$ (c) $(z^2 + a^2)^{1/2}$, $a^2 > 0$

        (d) $\log z$ (e) $\log(z^2 + a^2)$, $a^2 > 0$ (f) e^z

        (g) $z^2 \sin \dfrac{1}{z}$ (h) $\dfrac{z^2}{z^3 + 1}$ (i) $\sin^{-1} z$ (j) $\log(1 - e^{1/z})$

4. Let $f(z)$ be analytic outside a circle C_R enclosing the origin.
    (a) Show that

$$\frac{1}{2\pi i} \oint_{C_R} f(z)\, dz = \frac{1}{2\pi i} \oint_{C_\rho} f\left(\frac{1}{t}\right) \frac{dt}{t^2}$$

    where C_ρ is a circle of radius $1/R$ enclosing the origin. For $R \to \infty$ conclude that the integral can be computed to be Res $(f(1/t)/t^2; 0)$.

    (b) Suppose $f(z)$ has the convergent Laurent expansion

$$f(z) = \sum_{j=-\infty}^{-1} A_j z^j$$

    Show that the integral above equals A_{-1}.

5. (a) The following identity for Bessel functions is valid:

$$\exp\left(\frac{w}{2}(z - 1/z)\right) = \sum_{n=-\infty}^{\infty} J_n(w) z^n$$

    Show that

$$J_n(w) = \frac{1}{2\pi i} \oint_C \exp\left(\frac{w}{2}(z - 1/z)\right) \frac{dz}{z^{n+1}}$$

    where C is the unit circle centered at the origin.

(b) Use $\exp\left(\frac{w}{2}(z - 1/z)\right) = \exp\left(\frac{w}{2}z\right)\exp\left(-\frac{w}{2z}\right)$; multiply the two series for exponentials to compute the following series representation for the Bessel function of "0th" ($n = 0$) order:

$$J_0(w) = \sum_{k=0}^{\infty}(-1)^k\frac{w^{2k}}{(k!)^2 2^{2k}}$$

6. Consider the following integral

$$I_R = \oint_{C_R}\frac{dz}{z^2\cosh z}$$

where C_R is a square centered at the origin whose sides lie along the lines $x = \pm(R+1)\pi$ and $y = \pm(R+1)\pi$, where R is a positive integer. Evaluate this integral both by residues and by direct evaluation of the line integral and show that in either case $\lim_{R\to\infty} I_R = 0$, where the limit is taken over the integers. (In the direct evaluation, use estimates of the integrand.)

7. Suppose we know that everywhere outside the circle C_R, radius R centered at the origin, $f(z)$ and $g(z)$ are analytic with $\lim_{z\to\infty} f(z) = C_1$ and $\lim_{z\to\infty}(zg(z)) = C_2$, where C_1 and C_2 are constant. Show

$$\frac{1}{2\pi i}\oint_{C_R} g(z)e^{f(z)}\,dz = C_2 e^{C_1}$$

8. Suppose $f(z)$ is a meromorphic function (i.e., $f(z)$ is analytic everywhere in the finite z plane except at isolated points where it has poles) with N simple zeroes (i.e., $f(z_0) = 0$, $f'(z_0) \neq 0$) and M simple poles inside a circle C. Show

$$\frac{1}{2\pi i}\oint_C \frac{f'(z)}{f(z)}\,dz = N - M$$

4.2 Evaluation of Certain Definite Integrals

We begin this section by developing methods to evaluate integrals of the form

$$I = \int_{-\infty}^{\infty} f(x)\,dx \tag{4.2.1}$$

where $f(x)$ will be specified later. Integrals with infinite limits converge depending on the existence of a limit; namely, we say that I **converges** if the two limits in

$$I = \lim_{L \to \infty} \int_{-L}^{\alpha} f(x)\,dx + \lim_{R \to \infty} \int_{\alpha}^{R} f(x)\,dx, \qquad \alpha \text{ finite} \qquad (4.2.2)$$

exist. When evaluating integrals in complex analysis, it is useful (as we will see) to consider a more restrictive limit by taking $L = R$, and this is sometimes referred to as the **Cauchy Principal Value at Infinity**, I_p:

$$I_p = \lim_{R \to \infty} \int_{-R}^{R} f(x)\,dx \qquad (4.2.3)$$

If Eq. (4.2.2) is convergent, then $I = I_p$ by simply taking as a special case $L = R$. It is possible for I_p to exist but not the more general limit (4.2.2). For example, if $f(x)$ is odd and nonzero at infinity (e.g. $f(x) = x$), then $I_p = 0$ but I will not exist. In applications one frequently checks the convergence of I by using the usual tests of calculus and *then* one evaluates the integral via Eq. (4.2.3). In what follows, unless otherwise explicitly stated, we shall only consider integrals with infinite limits whose convergence can be established in the sense of Eq. (4.2.2).

We first show how to evaluate integrals of the form

$$I = \int_{-\infty}^{\infty} f(x)\,dx$$

where $f(x) = N(x)/D(x)$, where $N(x)$ and $D(x)$ are real polynomials (that is, $f(x)$ is a rational function), $D(x) \neq 0$ for $x \in \mathbb{R}$, and $D(x)$ is at least 2 degrees greater than the degree of $N(x)$; the latter hypothesis implies convergence of the integral. The method is to consider the integral

$$\oint_C f(z)\,dz = \int_{-R}^{R} f(x)\,dx + \int_{C_R} f(z)\,dz \qquad (4.2.4)$$

(see Figure 4.2.1) in which C_R is a large semicircle and the contour C encloses all the singularities of $f(z)$, namely, those locations where $D(z) = 0$, that is, $z_1, z_2, \ldots, z_N$. We use Cauchy's Residue Theorem and suitable analysis showing that $\lim_{R \to \infty} \int_{C_R} f(z)\,dz = 0$ (this is true owing to the assumptions on $f(x)$ and is proven in Theorem 4.2.1), in which case from (4.2.4) we have, as $R \to \infty$,

$$\int_{-\infty}^{\infty} f(x)\,dx = 2\pi i \sum_{j=1}^{N} \operatorname{Res}(f(z); z_j) \qquad (4.2.5)$$

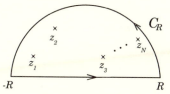

Fig. 4.2.1. Evaluating Eq. (4.2.5) with the contour in the upper half plane

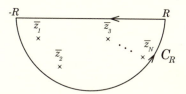

Fig. 4.2.2. Evaluating Eq. (4.2.5) with the contour in the lower half plane

The integral can also be evaluated by using the closed contour in the lower half plane, shown in Figure 4.2.2. Note that because $D(x)$ is a real polynomial, its complex zeroes come in complex conjugate pairs.

We illustrate the method first by an example.

Example 4.2.1 Evaluate

$$I = \int_{-\infty}^{\infty} \frac{x^2}{x^4 + 1} dx$$

We begin by establishing that the contour integral along the semicircular arc described in Eq. (4.2.4) vanishes as $R \to \infty$. Using $f(z) = z^2/(z^4 + 1)$, $z = Re^{i\theta}$, $dz = iRe^{i\theta} d\theta$, $|dz| = R d\theta$, we have

$$\left| \int_{C_R} f(z) \, dz \right| \leq \int_{\theta=0}^{\pi} \frac{|z|^2}{|z^4 + 1|} |dz| \leq \int_{\theta=0}^{\pi} \frac{|z|^2}{|z|^4 - 1} |dz|$$

$$= \frac{\pi R^3}{R^4 - 1} \xrightarrow[R \to \infty]{} 0$$

These inequalities follow from $|z^4 + 1| \geq |z|^4 - 1$, which implies $1/|z^4 + 1| \leq 1/(R^4 - 1)$; we have used the integral inequalities of Chapter 2 (see, for example, Theorem 2.4.2). Thus we have shown how Eq. (4.2.5) is arrived at in this example.

The residues of the function $f(z)$ are easily calculated from Eq. (4.1.10) of Section 4.1 by noting that all poles are simple; they may be found by solving $z^4 = -1 = e^{i\pi}$, and hence there is one pole located in each of the four quadrants. We shall use the contour in Figure 4.1.1 so we need only the zeroes in the first and second quadrants: $z_1 = e^{i\pi/4}$ and $z_2 = e^{i(\pi/4+\pi/2)} = e^{3i\pi/4}$. Thus Eq. (4.2.5) yields

$$
\begin{aligned}
I &= 2\pi i \left[\left(\frac{z^2}{4z^3} \right)_{z_1} + \left(\frac{z^2}{4z^3} \right)_{z_2} \right] \\
&= \frac{2\pi i}{4} \left(e^{-i\pi/4} + e^{-3i\pi/4} \right) = \frac{\pi}{2} \left(e^{i\pi/4} + e^{-i\pi/4} \right) \\
&= \pi \cos(\pi/4) = \pi/\sqrt{2}
\end{aligned}
$$

where we have used $i = e^{i\pi/2}$. We also note that if we used the contour depicted in Figure 4.2.2 and evaluated the residues in the third and fourth quadrants, we would arrive at the same result – as we must.

More generally, we have the following theorem.

Theorem 4.2.1 Let $f(z) = N(z)/D(z)$ be a rational function such that the degree of $D(z)$ exceeds the degree of $N(z)$ by at least two. Then

$$
\lim_{R \to \infty} \int_{C_R} f(z)\, dz = 0
$$

Proof We write

$$
f(z) = \frac{a_n z^n + a_{n-1} z^{n-1} + \cdots + a_1 z + a_0}{b_m z^m + b_{m-1} z^{m-1} + \cdots + b_1 z + b_0}
$$

then, using the same ideas as in Example 4.2.1

$$
\begin{aligned}
\left| \oint_{C_R} f(z)\, dz \right| &\leq \int_0^\pi (R\, d\theta) \frac{|a_n||z|^n + |a_{n-1}||z|^{n-1} + \cdots + |a_1||z| + |a_0|}{|b_m||z|^m - |b_{m-1}||z|^{m-1} - \cdots - |b_1||z| - |b_0|} \\
&= \frac{\pi R(|a_n|R^n + \cdots + |a_0|)}{|b_m|R^m - |b_{m-1}|R^{m-1} - \cdots - |b_0|} \xrightarrow[R \to \infty]{} 0
\end{aligned}
$$

since $m \geq n + 2$. ∎

Integrals that are closely related to the one described above are of the form

$$
I_1 = \int_{-\infty}^\infty f(x) \cos kx\, dx, \qquad I_2 = \int_{-\infty}^\infty f(x) \sin kx\, dx,
$$

$$
I_{3\pm} = \int_{-\infty}^\infty f(x) e^{\pm ikx}\, dx, \qquad (k > 0)
$$

where $f(x)$ is a rational function satisfying the conditions in Theorem 4.2.1. These integrals are evaluated by a method similar to the ones described earlier. When evaluating integrals such as I_1 or I_2, we first replace them by integrals of the form I_3. We evaluate, say I_{3+}, by using the contour in Figure 4.2.1. Again, we need to evaluate the integral along the upper semicircle. Because $e^{ikz} = e^{ikx}e^{-ky}$ $(z = x + iy)$, we have $|e^{ikz}| \le 1$ $(y > 0)$ and

$$\left| \int_{C_R} f(z)e^{ikz}\, dz \right| \le \int_0^R |f(z)|\, |dz| \xrightarrow[R\to\infty]{} 0$$

from the results of Theorem 4.2.1. Thus using

$$I_{3+} = \int_{-\infty}^{\infty} f(x)e^{ikx}\, dx$$

$$= \int_{-\infty}^{\infty} f(x)\cos kx\, dx + i \int_{-\infty}^{\infty} f(x)\sin kx\, dx,$$

we have from (4.2.5) suitably modified,

$$I_{3+} = I_1 + iI_2 = 2\pi i \sum_{j=1}^{\infty} \text{Res}\left(f(z)e^{ikz}; z_j \right) \tag{4.2.6}$$

and hence by taking real and imaginary parts of Eq. (4.2.6), we can compute I_1 and I_2.

It should be remarked that to evaluate I_{3-}, we use a semicircular contour in the lower half of the plane, that is, Figure 4.2.2. The calculations are similar to those before, save for the fact that we need to compute the residues in the lower half plane and we find that $I_{3-} = I_1 - iI_2 = -2\pi i \sum_{j=1}^{N} \text{Res}(f(z); z_j)$.

We note that in other applications one might need to consider integrals $\int_{C_R} e^{-kz} f(z)\, dz$ where C_R is a semicircle in the *right half plane*, and/or $\int_{C_L} e^{kz} f(z)\, dz$ where C_L is a semicircle in the *left half plane*. The methods to show such integrals are zero as $R \to \infty$ are similar to those presented above, hence there is no need to elaborate further.

Example 4.2.2 Evaluate

$$I = \int_{-\infty}^{\infty} \frac{\cos kx}{(x + b)^2 + a^2}\, dx, \qquad k > 0,\ a > 0,\ b \text{ real}$$

We consider

$$I_+ = \int_{-\infty}^{\infty} \frac{e^{ikx}}{(x + b)^2 + a^2}\, dx$$

and use the contour in Figure 4.2.1 to find

$$I_+ = 2\pi i \; \text{Res}\left(\frac{e^{ikz}}{(z+b)^2+a^2}; z_0 = ia - b\right) \qquad (a > 0)$$

$$= 2\pi i \left(\frac{e^{ikz}}{2(z+b)}\right)_{z_0 = ia - b} = \frac{\pi}{a} e^{-ka} e^{-ibk} \qquad (a, k > 0)$$

From

$$I_+ = \int_{-\infty}^{\infty} \frac{\cos kx}{(x+b)^2+a^2}\,dx + i \int_{-\infty}^{\infty} \frac{\sin kx}{(x+b)^2+a^2}\,dx$$

we have

$$I = \frac{-\pi}{a} e^{-ka} \cos bk$$

and

$$J = \int_{-\infty}^{\infty} \frac{\sin kx}{(x+b)^2+a^2}\,dx = \frac{\pi}{a} e^{-ka} \sin bk$$

If $b = 0$, the latter formula reduces to $J = 0$, which also follows directly from the fact that the integrand is odd. The reader can verify that

$$\left| \int_{C_R} \frac{e^{ikz}}{(z+b)^2+a^2}\,dz \right| \le \int_{C_R} \frac{|dz|}{|z|^2 - 2|b||z| - a^2 - b^2}$$

$$= \frac{\pi R}{R^2 - 2bR - (a^2+b^2)} \xrightarrow[R \to \infty]{} 0$$

In applications we frequently wish to evaluate integrals like $I_{3\pm}$ involving $f(x)$ for which all that is known is $f(x) \to 0$ as $|x| \to \infty$. From calculus we know that in these cases the integral still converges, conditionally, but our estimates leading to Eq. (4.2.6) must be made more carefully. We say that $f(z) \to 0$ **uniformly** as $R \to \infty$ in C_R if $|f(z)| \le K_R$, where K_R depends only on R (not on arg z) and $K_R \to 0$ as $R \to \infty$. We have the following lemma, called Jordan's Lemma.

Lemma 4.2.2 (Jordan) Suppose that on the circular arc C_R in Figure 4.2.1 we have $f(z) \to 0$ uniformly as $R \to \infty$. Then

$$\lim_{R \to \infty} \int_{C_R} e^{ikz} f(z)\,dz = 0 \qquad (k > 0)$$

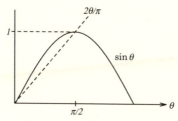

Fig. 4.2.3. Jordan's Lemma

Proof With $|f(z)| \leq K_R$, K_R constant,

$$I = \left| \int_{C_R} e^{ikz} f(z) \, dz \right| \leq \int_0^\pi e^{-ky} K_R R \, d\theta$$

Using $y = R \sin \theta$, and $\sin(\pi - \theta) = \sin \theta$

$$\int_0^\pi e^{-ky} \, d\theta = \int_0^\pi e^{-kR \sin \theta} \, d\theta = 2 \int_0^{\pi/2} e^{-kR \sin \theta} \, d\theta$$

But in the region $0 \leq \theta \leq \pi/2$ we also have the estimate $\sin \theta \geq 2\theta/\pi$ (see Figure 4.2.3).
Thus

$$I \leq 2K_R R \int_0^{\pi/2} e^{-2kR\theta/\pi} \, d\theta = \frac{2K_R R\pi}{2kR} \left(1 - e^{-kR} \right)$$

and $I \to 0$ as $R \to \infty$ because $K_R \to 0$.   ∎

We note that if $k < 0$, a similar result holds for the contour in Figure 4.2.2. *Moreover, by suitably rotating the contour, Jordan's Lemma applies to the cases $k = i\ell$ for $\ell \neq 0$.* Consequently, Eq. (4.2.6) follows whenever Jordan's Lemma applies. Jordan's Lemma is used in the following example.

Example 4.2.3 Evaluate

$$I = 2 \int_{-\infty}^\infty \frac{x \sin \alpha x \cos \beta x}{x^2 + \gamma^2} \, dx, \qquad \gamma > 0, \alpha, \beta \text{ real.}$$

The trigonometric formula

$$\sin \alpha x \cos \beta x = \frac{1}{2} [\sin(\alpha - \beta)x + \sin(\alpha + \beta)x]$$

motivates the introduction of the integrals

$$J = \int_{-\infty}^{\infty} \frac{xe^{i(\alpha-\beta)x}}{x^2+\gamma^2}\, dx + \int_{-\infty}^{\infty} \frac{xe^{i(\alpha+\beta)x}}{x^2+\gamma^2}\, dx$$

$$= J_1 + J_2$$

Jordan's Lemma applies because the function $f(z) = z/(z^2+\gamma^2) \to 0$ uniformly as $z \to \infty$ and we note that,

$$|f| \le \frac{R}{R^2-\gamma^2} \equiv K_R$$

We note that the denominator is only one degree higher than the numerator. If $\alpha - \beta > 0$, then we close our contour in the upper half plane and the only residue is $z = i\gamma$ ($\gamma > 0$), hence

$$J_1 = i\pi e^{-(\alpha-\beta)\gamma}$$

On the other hand, if $\alpha - \beta < 0$, we close in the lower half plane and find

$$J_1 = -i\pi e^{(\alpha-\beta)\gamma}$$

Combining these results

$$J_1 = i\pi\, \text{sgn}(\alpha - \beta)e^{-|\alpha-\beta|\gamma}$$

Similarly, for I_2 we find

$$J_2 = i\pi\, \text{sgn}(\alpha + \beta)e^{-|\alpha+\beta|\gamma}$$

Thus

$$J = i\pi\left[\text{sgn}(\alpha - \beta)e^{-|\alpha-\beta|\gamma} + \text{sgn}(\alpha + \beta)e^{-|\alpha+\beta|\gamma}\right]$$

and, by taking the imaginary part

$$I = \pi\left[\text{sgn}(\alpha - \beta)e^{-|\alpha-\beta|\gamma} + \text{sgn}(\alpha + \beta)e^{-|\alpha+\beta|\gamma}\right]$$

If we take $\text{sgn}(0) = 0$ then the case $\alpha = \beta$ is incorporated in this result. This could either be established directly using $\sin\alpha x \cos\alpha x = \frac{1}{2}\sin 2\alpha x$, or by noting that $J_1 = 0$ owing to the oddness of the integrand. This is a consequence of employing the Cauchy Principal Value integral. (Note that the integral I is convergent.)

We now consider a class of real integrals of the following type:

$$I = \int_0^{2\pi} f(\sin\theta, \cos\theta)\, d\theta$$

where $f(x, y)$ is a rational function of x, y. We make the substitution

$$z = e^{i\theta}, \qquad dz = ie^{i\theta}\, d\theta$$

Then, using $\cos\theta = (e^{i\theta} + e^{-i\theta})/2$ and $\sin\theta = (e^{i\theta} - e^{-i\theta})/2i$, we have

$$\cos\theta = (z + 1/z)/2, \qquad \sin\theta = (z - 1/z)/2i \qquad (4.2.7)$$

Thus

$$\int_0^{2\pi} d\theta\, f(\sin\theta, \cos\theta) = \oint_{C_0} \frac{dz}{iz} f\left(\frac{z - 1/z}{2i}, \frac{z + 1/z}{2}\right)$$

where C_0 is the unit circle $|z| = 1$. Using the residue theorem

$$
\begin{aligned}
I &= \oint_{C_0} f\left(\frac{z - 1/z}{2i}, \frac{z + 1/z}{2}\right)\left(\frac{dz}{iz}\right) \\
&= 2\pi i \sum_{j=1}^{N} \operatorname{Res}\left(\frac{f\left(\frac{z-1/z}{2i}, \frac{z+1/z}{2}\right)}{iz}; z_j\right)
\end{aligned}
$$

The fact that $f(x, y)$ is a rational function of x, y implies that the residue calculation amounts to finding the zeroes of a polynomial.

Example 4.2.4 Evaluate

$$I = \int_0^{2\pi} \frac{d\theta}{A + B\sin\theta} \qquad (A^2 > B^2, A > 0)$$

Employing the substitution (4.2.7) with C_0 the unit circle $|z| = 1$, and assuming, for now, that $B \neq 0$,

$$
\begin{aligned}
I &= \oint_{C_0} \frac{dz}{iz} \frac{1}{\left(A + B\left(\frac{z - 1/z}{2i}\right)\right)} = \oint_{C_0} \frac{2\, dz}{2i\, Az + B(z^2 - 1)} \\
&= \frac{2}{B} \oint_{C_0} \frac{dz}{z^2 + 2i\frac{A}{B}z - 1}
\end{aligned}
$$

The roots of the denominator z_1 and z_2 that satisfy $(z - z_1)(z - z_2) = z^2 + 2i\,Az/B - 1 = 0$ are given by

$$z_1 = -i\frac{A}{B} + i\sqrt{\left(\frac{A}{B}\right)^2 - 1} = \frac{-iA + i\sqrt{A^2 - B^2}}{B}$$

$$z_2 = -i\frac{A}{B} - i\sqrt{\left(\frac{A}{B}\right)^2 - 1} = \frac{-iA - i\sqrt{A^2 - B^2}}{B}$$

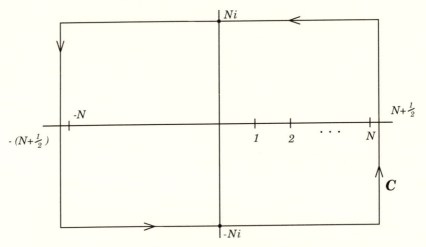

Fig. 4.2.4. Rectangular contour C

Because $z_1 z_2 = -1$, we find that $|z_1||z_2| = 1$; hence if one root is inside C_0, the other is outside. Because $A^2 - B^2 > 0$, and $A > 0$, it follows that $|z_1| < |z_2|$; hence z_1 lies inside. Thus, computing the residue of the integral, we have, from Eq. (4.1.8)

$$I = 2\pi i \left(\frac{2}{B}\right)\left(\frac{1}{z_1 - z_2}\right)$$

$$= \frac{4\pi i}{B}\frac{B}{2i\sqrt{A^2 - B^2}} = \frac{2\pi}{\sqrt{A^2 - B^2}}.$$

(The value of I when $B = 0$ is $2\pi/A$.) We note that we also have computed

$$I = \int_0^{2\pi} \frac{d\theta}{A + B\cos\theta}$$

simply by making the substitution $\theta = \pi/2 + \phi$ inside the original integral.

As another illustration of the residue theorem and calculation of integrals, we describe how to obtain a "pole" expansion of a function via a contour integral.

Example 4.2.5 Evaluate

$$I = \frac{1}{2\pi i}\oint_C \frac{\pi \cot \pi \zeta}{z^2 - \zeta^2}\,d\zeta, \qquad (z^2 \neq 0, 1, 2, 3, \ldots)$$

where C is the contour given by the rectangle $(-N - \frac{1}{2}) \le x \le (N + \frac{1}{2})$, $-N \le y \le N$ (see Figure 4.2.4). Show that it implies

$$\pi \cot \pi z = z \sum_{n=-\infty}^{\infty} \frac{1}{z^2 - n^2}$$

$$= z \left(\frac{1}{z^2} + \frac{2}{z^2 - 1^2} + \frac{2}{z^2 - 2^2} + \cdots \right) \qquad (4.2.8)$$

$$z \ne 0, \pm 1, \pm 2, \ldots$$

We take N sufficiently large so that z lies inside C. The poles are located at $\zeta = n = 0, \pm 1, \pm 2, \ldots, \pm N$, and at $\zeta = \pm z$; hence

$$I = \sum_{n=-N}^{N} \pi \left(\frac{\cos \pi \zeta}{\pi \cos \pi \zeta} \frac{1}{z^2 - \zeta^2} \right)_{\zeta = n}$$

$$+ \pi \left(\frac{\cot \pi \zeta}{-2\zeta} \right)_{\zeta = z} + \pi \left(\frac{\cot \pi \zeta}{-2\zeta} \right)_{\zeta = -z}$$

$$= \sum_{n=-N}^{N} \frac{1}{z^2 - n^2} - \pi \frac{\cot \pi z}{z}$$

Next we estimate the contour integral on the vertical sides, $\zeta = \pm(N + \frac{1}{2}) + i\eta$. Here the integrand satisfies

$$\left| \frac{\pi \cot \pi \zeta}{z^2 - \zeta^2} \right| \le \frac{\pi |\tanh \eta|}{|\zeta|^2 - |z|^2} \le \frac{\pi}{N^2 - |z|^2}$$

Because $|\zeta| > N$, $|\tanh \eta| \le 1$ and we used

$$|\cot \pi \zeta| = \left| \frac{\sin \left[\pi \left(N + \frac{1}{2} \right) \right] (\sinh \eta)(-i)}{\sin \left[\pi \left(N + \frac{1}{2} \right) \right] (\cosh \eta)} \right|$$

On the horizontal sides, $\zeta = \xi \pm iN$, and the integral satisfies

$$\left| \frac{\pi \cot \pi \zeta}{z^2 - \zeta^2} \right| \le \frac{\pi \coth \pi N}{|\zeta|^2 - |z|^2} \le \frac{\pi \coth \pi N}{N^2 - |z|^2},$$

because $|\zeta| > N$ and we used

$$|\cot \pi \zeta| = \left| \frac{e^{\mp \pi N} e^{i\pi\xi} + e^{\pm \pi N} e^{-i\pi\xi}}{e^{\mp \pi N} e^{i\pi\xi} - e^{\pm \pi N} e^{-i\pi\xi}} \right|$$

$$\leq \frac{e^{\pi N} + e^{-\pi N}}{e^{\pi N} - e^{-\pi N}} = \coth \pi N$$

Thus

$$I = \frac{1}{2\pi i} \oint_C \left| \frac{\pi \cot \pi \zeta}{z^2 - \zeta^2} \right| |d\zeta|$$

$$\leq \frac{1}{2\pi} \frac{2(2N)\pi}{N^2 - |z|^2} + \frac{1}{2\pi} \frac{2(2N+1)\pi \coth \pi N}{N^2 - |z|^2} \xrightarrow[N \to \infty]{} 0,$$

since $\coth \pi N \to 1$ as $N \to \infty$. Hence we recover Eq. (4.2.8) in the limit $N \to \infty$. Formula (4.2.8) is referred to as a Mittag–Leffler expansion of the function $\pi \cot \pi z$. (The interested reader will find a discussion of Mittag-Leffler expansions in Section 3.6 of Chapter 3.) Note that this kind of expansion takes a different form than does a Taylor series or Laurent series. It is an expansion based upon the poles of the function $\cot \pi z$.

The result (4.2.8) can be integrated to yield an infinite product representation of $\sin \pi z$. Namely, from

$$\frac{d}{dz} \log \sin \pi z = \pi \cot \pi z$$

it follows by integration (taking the principal branch for the logarithm) that

$$\log \sin \pi z = \log z + A_0 + \sum_{n=1}^{\infty} (\log(z^2 - n^2) - A_n)$$

where A_0 and A_n are constants. The constants are conveniently evaluated at $z = 0$ by noting that $\lim_{z \to 0} \log \frac{\sin \pi z}{z} = \log \pi$. Thus $A_0 = \log \pi$, and $A_n = \log(-n^2)$; hence taking the exponential yields

$$\frac{\sin \pi z}{\pi} = z \prod_{n=1}^{\infty} \left(1 - \frac{z^2}{n^2} \right) \tag{4.2.9}$$

This is an example of the so-called Weierstrass Factor Theorem, discussed in Section 3.6.

It turns out that Eq. (4.2.8) can also be obtained by evaluation of a different integral, a fact that is not immediately apparent. We illustrate this in the following example.

Example 4.2.6 Evaluate

$$I = \frac{1}{2\pi i} \oint_C \pi \cot \pi \zeta \left(\frac{1}{\zeta} - \frac{1}{\zeta - z} \right) d\zeta \qquad (z \neq 0, \pm 1, \pm 2, \ldots)$$

where C is the same contour as in Example 4.2.5 and is depicted in Figure 4.2.4. Residue calculation yields

$$I = ((-)\pi \cot \pi \zeta)_{\zeta = z} + \sum_{n=-N}^{N} {}' \left\{ \frac{\pi \cos \pi \zeta}{\pi \cos \pi \zeta} \left(\frac{1}{\zeta} - \frac{1}{\zeta - z} \right) \right\}_{\zeta = n, n \neq 0}$$

$$+ \left(\frac{\pi \cos \pi \zeta}{\pi \cos \pi \zeta} \frac{(-)}{\zeta - z} \right)_{\zeta = 0}$$

$$= -\pi \cot \pi z + \sum_{n=-N}^{N} {}' \left(\frac{1}{z - n} + \frac{1}{n} \right) + \frac{1}{z}$$

where $\sum_{n=-N}^{N} {}'$ means we omit the $n = 0$ contribution. We also note that the contribution from the double pole at $\zeta = 0$ vanishes because $\cot \pi \zeta / \zeta \sim 1/(\pi \zeta^2) - \pi/3 + \cdots$ as $\zeta \to 0$.

Finally, we estimate the integral I on the boundary in the same manner as in Example 4.2.5 to find

$$|I| \leq \frac{1}{2\pi} \oint_C |\pi \cot \pi \zeta| \frac{|z|}{|\zeta|(|\zeta| - |z|)} |d\zeta|$$

$$|I| \leq \frac{|z|}{2\pi} \left(\frac{4N\pi}{N(N - |z|)} + \frac{2(2N + 1)\pi \coth \pi N}{N(N - |z|)} \right)$$

$$\longrightarrow 0 \quad \text{as} \quad N \to \infty$$

Hence

$$\pi \cot \pi z = \sum_{n=-N}^{N} {}' \left(\frac{1}{z - n} + \frac{1}{n} \right) + \frac{1}{z} \qquad (4.2.10)$$

Note the expansion (4.2.10) has a suitable "convergence factor" $(1/n)$ inside the sum, otherwise it would diverge. When we combine the terms appropriately

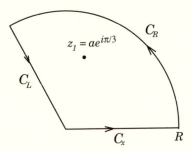

Fig. 4.2.5. Contour for Example 4.2.7

for $n = \pm 1, \pm 2, \ldots$, we find

$$\pi \cot \pi z = \frac{1}{z} + \frac{1}{z-1} + \frac{1}{1} + \frac{1}{z+1} - \frac{1}{1} + \frac{1}{z-2} + \frac{1}{2} + \frac{1}{z+2} - \frac{1}{2} + \cdots$$

$$= \frac{1}{z} + \frac{2z}{z^2 - 1} + \frac{2z}{z^2 - 2^2} + \cdots = \sum_{n=-\infty}^{\infty} \frac{z}{z^2 - n^2}$$

which is Eq. (4.2.8).

When employing contour integration, sometimes it is necessary to employ special properties of the integrand, as is illustrated below.

Example 4.2.7 Evaluate

$$I = \int_0^\infty \frac{dx}{x^3 + a^3}, \qquad a > 0$$

Because we have an integral on $(0, \infty)$, we cannot immediately use a contour like that of Figure 4.2.1. If the integral was $\int_0^\infty f(x)dx$ where $f(x)$ was an even function $f(x) = f(-x)$, then $\int_0^\infty f(x)dx = \frac{1}{2}\int_{-\infty}^\infty f(x)dx$. However, in this case the integrand is not even, and for $x < 0$ has a singularity. Nevertheless there is a symmetry that can be employed: namely, $(xe^{2\pi i/3})^3 = x^3$. This suggests using the contour of Figure 4.2.5, where C_R is the sector $R\,e^{i\theta} : 0 \leq \theta \leq 2\pi/3$.

We therefore have

$$\oint_C \frac{dz}{z^3 + a^3} = \left(\int_{C_L} + \int_{C_x} + \int_{C_R} \right) \frac{dz}{z^3 + a^3}$$

$$= 2\pi i \sum_j \text{Res}\left(\frac{1}{z^3 + a^3}; z_j \right)$$

The only pole inside C satisfies $z^3 = -a^3 = a^3 e^{i\pi}$ and is given by $z_1 = ae^{i\pi/3}$.

The residue is obtained from

$$\text{Res}\left(\frac{1}{z^3 + a^3}; z_1\right) = \left(\frac{1}{3z^2}\right)_{z_1} = \frac{1}{3a^2 e^{2\pi i/3}} = \frac{1}{3a^2} e^{-2\pi i/3}$$

The integral on C_R tends to zero because of Theorem 4.2.1. Alternatively, by direct calculation,

$$\left|\int_{C_R} \frac{dz}{z^3 + a^3}\right| \leq \frac{2\pi R}{3(R^3 - a^3)} \to 0, \qquad R \to \infty$$

The integral on C_L is evaluated by making the substitution $z = e^{2\pi i/3} r$ (where the orientation is taken into account)

$$\int_{C_L} \frac{dz}{z^3 + a^3} = \int_{r=R}^0 \frac{e^{2\pi i/3}}{r^3 + a^3} dr = e^{-2\pi i/3} I.$$

Thus taking into account the contributions from C_x ($0 \leq z = x \leq R$) and from C_L, we have

$$I(1 - e^{2\pi i/3}) = \lim_{R \to \infty} \int_0^R \frac{dr}{r^3 + a^3}(1 - e^{2\pi i/3}) = \frac{2\pi i}{3a^2} e^{-2\pi i/3}$$

Thus

$$I = \frac{2\pi i}{3a^2} \frac{e^{-2\pi i/3}}{1 - e^{2\pi i/3}} = \frac{\pi}{3a^2}\left(\frac{2i}{e^{-i\pi/3} - e^{i\pi/3}}\right) e^{-i\pi}$$

$$= \frac{\pi}{3a^2 \sin \pi/3} = \frac{2\pi}{3\sqrt{3}a^2}$$

The following example, similar in spirit to Eq. (4.2.7), allows us to calculate the following conditionally convergent integrals

$$C = \int_0^\infty \cos(tx^2)\, dx \tag{4.2.11}$$

$$S = \int_0^\infty \sin(tx^2)\, dx \tag{4.2.12}$$

Example 4.2.8 Evaluate

$$I = \int_0^\infty e^{itx^2}\, dx$$

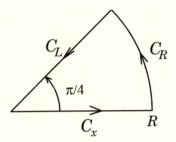

Fig. 4.2.6. Contour for Example 4.2.8

For convenience we take $t > 0$. Consider the contour depicted in Figure 4.2.6 where the contour C_R is the sector $Re^{i\theta} : 0 \le \theta \le \pi/4$. Because e^{itz^2} is analytic inside $C = C_x + C_R + C_L$, we have

$$\oint_C e^{itz^2} dz = \left(\int_{C_L} + \int_{C_x} + \int_{C_R} \right) e^{itz^2} dz = 0$$

The integral on C_R is estimated using the same idea as in Jordan's Lemma ($\sin \theta \ge 2\theta/\pi$ for $0 \le \theta \le \pi/2$)

$$\left| \int_{C_R} e^{itz^2} dz \right| = \left| \int_0^{\pi/4} e^{itR^2(\cos 2\theta + i \sin 2\theta)} Re^{i\theta} i \, d\theta \right|$$

$$\le \int_0^{\pi/4} Re^{-tR^2 \sin 2\theta} \, d\theta$$

$$\le \int_0^{\pi/4} Re^{-tR^2 \frac{4\theta}{\pi}} \, d\theta = \frac{\pi}{4tR}(1 - e^{-tR^2})$$

where we used $\sin x \ge \frac{2x}{\pi}$ for $0 < x < \frac{\pi}{2}$. Thus $| \int_{C_R} e^{itz^2} dz| \to 0$ as $R \to \infty$. Hence on C_x, $z = x$, and on C_L, $z = re^{i\pi/4}$;

$$\int_{C_x} e^{itz^2} dz = \int_0^R e^{itx^2} dx$$

$$\int_{C_L} e^{itz^2} dz = \int_R^0 e^{-tr^2} dr e^{i\pi/4}.$$

Thus

$$I = \int_0^\infty e^{itx^2} dx = e^{i\pi/4} \int_0^\infty e^{-tr^2} dr$$

and this transforms I to a well-known real definite integral that can be evaluated directly. We use polar coordinates

$$J^2 = \left(\int_0^\infty e^{-tx^2} \, dx \right)^2 = \int_0^\infty \int_0^\infty e^{-t(x^2+y^2)} \, dx \, dy$$

$$= \int_{\theta=0}^{\pi/2} \int_{\rho=0}^\infty e^{-t\rho^2} \rho \, d\rho \, d\theta = \frac{\pi}{4t}$$

Thus, taking $R \to \infty$,

$$J = \int_0^\infty e^{-tx^2} \, dx = \frac{1}{2} \sqrt{\frac{\pi}{t}} \qquad (4.2.13)$$

and

$$I = e^{i\pi/4} \frac{1}{2} \sqrt{\frac{\pi}{t}} = \left(\cos \frac{\pi}{4} + i \sin \frac{\pi}{4} \right) \frac{1}{2} \sqrt{\frac{\pi}{t}}$$

Hence Eqs. (4.2.11) and (4.2.12) are found to be

$$S = C = \frac{1}{2} \sqrt{\frac{\pi}{2t}} \qquad (4.2.14)$$

Incidentally, it should be noted that we cannot evaluate the integral I in the same way (via polar coordinates) we do on J because I is not absolutely convergent.

The following example exhibits still another variant of contour integration.

Example 4.2.9 Evaluate

$$I = \int_{-\infty}^\infty \frac{e^{px}}{1 + e^x} \, dx$$

for $0 < \text{Re } p < 1$. The condition on p is required for convergence of the integral. Consider the contour depicted in Figure 4.2.7.

$$\oint_C \frac{e^{pz}}{1 + e^z} \, dz = \left(\int_{C_x} + \int_{C_{SR}} + \int_{C_{SL}} + \int_{C_T} \right) \frac{e^{pz}}{1 + e^z} \, dz$$

$$= 2\pi i \sum_j \text{Res} \left(\frac{e^{pz}}{1 + e^z}; z_j \right)$$

The only poles of the function $e^{pz}/(1 + e^z)$ occur when $e^z = -1$ or by taking the logarithm $z = i(\pi + 2n\pi), n = 0, \pm 1, \pm 2, \ldots$. The contour is chosen such

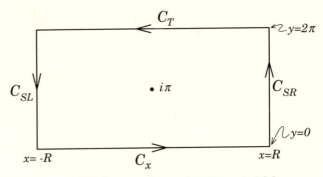

Fig. 4.2.7. Contour of integration, Example 4.2.9

that $z = x + iy, 0 \le y \le 2\pi$; hence the only pole inside the contour is $z = i\pi$, with the residue

$$\text{Res}\left(\frac{e^{pz}}{1+e^z}; i\pi\right) = \left(\frac{e^{pz}}{e^z}\right)_{z=i\pi} = e^{(p-1)i\pi}$$

The integrals along the sides are readily estimated and shown to vanish as $R \to \infty$. Indeed on C_{SR}: $z = R + iy, 0 \le y \le 2\pi$

$$\left|\int_{C_{SR}} \frac{e^{pz}}{1+e^z} dz\right| = \left|\int_0^{2\pi} \frac{e^{p(R+iy)}}{1+e^{R+iy}} i\, dy\right| \le \frac{e^{pR}}{e^R - 1} 2\pi \to 0,$$

$$R \to \infty, \qquad (\text{Re } p < 1)$$

On C_{SL} : $z = -R + iy, 0 \le y \le 2\pi$

$$\left|\int_{C_{SL}} \frac{e^{pz}}{1+e^z} dz\right| = \left|\int_{2\pi}^0 \frac{e^{p(-R+iy)}}{1+e^{-R+iy}} i\, dy\right| \le \frac{e^{-pR}}{1-e^{-R}} 2\pi \to 0,$$

$$R \to \infty, \qquad (\text{Re } p > 0)$$

The integral on the top has $z = x + 2\pi i, e^z = e^x$, so

$$\int_{C_T} \frac{e^{pz}}{1+e^z} dz = e^{2\pi i p} \int_{+R}^{-R} \frac{e^{px}}{1+e^x} dx$$

Hence, putting all of this together, we have, as $R \to \infty$,

$$\int_{-\infty}^{\infty} \frac{e^{px}}{1+e^x} dx \left(1 - e^{2\pi i p}\right) = 2\pi i\, e^{(p-1)i\pi}$$

or

$$\int_{-\infty}^{\infty} \frac{e^{px}}{1+e^x} dx = 2\pi i \frac{e^{-i\pi}}{e^{-ip\pi} - e^{ip\pi}} = \frac{\pi}{\sin p\pi}$$

Problems for Section 4.2

1. Evaluate the following real integrals.

(a) $\displaystyle\int_0^{\infty} \frac{dx}{x^2 + a^2}$, $a^2 > 0$

(Verify your answer by using usual antiderivatives.)

(b) $\displaystyle\int_0^{\infty} \frac{dx}{(x^2 + a^2)^2}$, $a^2 > 0$

(c) $\displaystyle\int_0^{\infty} \frac{dx}{(x^2 + a^2)(x^2 + b^2)}$, $a^2, b^2 > 0$

(d) $\displaystyle\int_0^{\infty} \frac{dx}{x^6 + 1}$

2. Evaluate the following real integrals by residue integration:

(a) $\displaystyle\int_{-\infty}^{\infty} \frac{x \sin x}{(x^2 + a^2)} dx$; $a^2 > 0$

(b) $\displaystyle\int_{-\infty}^{\infty} \frac{\cos kx}{(x^2 + a^2)(x^2 + b^2)} dx$; $a^2, b^2, k > 0$

(c) $\displaystyle\int_{-\infty}^{\infty} \frac{x \cos kx}{x^2 + 4x + 5} dx$; $k > 0$ (d) $\displaystyle\int_0^{\infty} \frac{\cos kx}{x^4 + 1} dx$

(e) $\displaystyle\int_0^{\infty} \frac{x^3 \sin kx}{x^4 + a^4} dx$; $a^4 > 0$ (f) $\displaystyle\int_0^{2\pi} \frac{d\theta}{1 + \cos^2 \theta}$

(g) $\displaystyle\int_0^{\pi/2} \sin^4 \theta \, d\theta$ (h) $\displaystyle\int_0^{2\pi} \frac{d\theta}{(5 - 3\sin\theta)^2}$

(i) $\displaystyle\int_{-\infty}^{\infty} \frac{\cos kx \cos mx}{(x^2 + a^2)} dx$, $a^2 > 0$.

3. Show

$$\int_0^{2\pi} \cos^{2n} \theta \, d\theta = \int_0^{2\pi} \sin^{2n} \theta \, d\theta = \frac{2\pi}{2^{2n}} B_n$$

where $B_n = 2^{2n}(1 \cdot 3 \cdot 5 \cdots (2n-1))/(2 \cdot 4 \cdot 6 \cdots (2n))$. (Hint: the fact that in the binomial expansion of $(1+w)^{2n}$ the coefficient of the term w^n is B_n.)

4. Show that
$$\int_0^\infty \frac{\cosh ax}{\cosh \pi x} \, dx = \frac{1}{2} \sec\left(\frac{a}{2}\right), \qquad |a| < \pi$$

Use a rectangular contour with corners at $\pm R$ and $\pm R + i$.

5. Consider a rectangular contour with corners at $b \pm iR$ and $b+1 \pm iR$. Use this contour to show that

$$\lim_{R \to \infty} \frac{1}{2\pi i} \int_{b-iR}^{b+iR} \frac{e^{az}}{\sin \pi z} \, dz = \frac{1}{\pi(1+e^{-a})}$$

where $0 < b < 1$, $|\operatorname{Im} a| < \pi$.

6. Consider a rectangular contour C_R with corners at $(\pm R, 0)$ and $(\pm R, a)$. Show that

$$\oint_{C_R} e^{-z^2} \, dz = \int_{-R}^R e^{-x^2} \, dx - \int_{-R}^R e^{-(x+ia)^2} \, dx + J_R = 0$$

where
$$J_R = \int_0^a e^{-(R+iy)^2} i \, dy - \int_0^a e^{-(-R+iy)^2} i \, dy$$

Show $\lim_{R \to \infty} J_R = 0$, whereupon we have $\int_{-\infty}^\infty e^{-(x+ia)^2} dx = \int_{-\infty}^\infty e^{-x^2} dx = \sqrt{\pi}$, and consequently, deduce that $\int_{-\infty}^\infty e^{-x^2} \cos 2ax \, dx = \sqrt{\pi} e^{-a^2}$.

7. Use a sector contour with radius R, as in Figure 4.2.6, centered at the origin with angle $0 \le \theta \le \frac{2\pi}{5}$ to find, for $a > 0$,

$$\int_0^\infty \frac{dx}{x^5 + a^5} = \frac{\pi}{5a^4 \sin \frac{\pi}{5}}$$

8. Consider the contour integral

$$I(N) = \frac{1}{2\pi i} \oint_{C(N)} \frac{\pi \csc \pi \zeta}{z^2 - \zeta^2} \, d\zeta$$

where the contour $C(N)$ is the rectangular contour depicted in Figure 4.2.4 (see also Example 4.2.5).

(a) Show that calculation of the residues implies that

$$I(N) = \sum_{n=-N}^{N} \frac{(-1)^n}{z^2 - n^2} - \frac{\pi \csc \pi z}{z}$$

(b) Estimate the line integral along the boundary and show that $\lim_{N \to \infty} I(N) = 0$ and consequently, that

$$\pi \csc \pi z = z \sum_{n=-\infty}^{\infty} \frac{(-1)^n}{z^2 - n^2}$$

(c) Use the result of part (b) to obtain the following representation of π:

$$\pi = 2 \sum_{n=-\infty}^{\infty} \frac{(-1)^n}{1 - 4n^2}$$

9. Consider a rectangular contour with corners $\left(N + \frac{1}{2}\right)(\pm 1 \pm i)$ to evaluate

$$\frac{1}{2\pi i} \oint_{C(N)} \frac{\pi \cot \pi z \coth \pi z}{z^3} \, dz$$

and in the limit as $N \to \infty$, show that

$$\sum_{n=1}^{\infty} \frac{\coth n\pi}{n^3} = \frac{7}{180} \pi^3$$

Hint: note

$$\text{Res}\left(\frac{\pi \cot z \coth z}{z^3}; 0\right) = -\frac{7\pi^3}{45}$$

4.3 Indented Contours, Principal Value Integrals, and Integrals With Branch Points

4.3.1 Principal Value Integrals

In Section 4.2 we introduced the notion of the Cauchy Principal Value integral at infinity (see Eq. (4.2.3)). Frequently in applications we are also interested in integrals with integrands that have singularities at a finite location. Consider the integral $\int_a^b f(x)\,dx$, where $f(x)$ has a singularity at x_0, $a < x_0 < b$. Convergence of such an integral depends on the existence of the following limit, where $f(x)$ has a singularity at $x = x_0$:

$$I = \lim_{\epsilon \to 0^+} \int_a^{x_0 - \epsilon} f(x)\,dx + \lim_{\delta \to 0^+} \int_{x_0 + \delta}^b f(x)\,dx \qquad (4.3.1)$$

We say the integral $\int_a^b f(x)\,dx$ is convergent if and only if Eq. (4.3.1) exists and is finite; otherwise we say it is divergent. The integral might exist even if the $\lim_{x \to x_0} f(x)$ is infinite or is divergent. For example, the integral $\int_0^2 dx/(x - 1)^{1/3}$ is convergent, whereas the integral $\int_0^2 dx/(x-1)^2$ is divergent. Sometimes by restricting the definition (4.3.1), we can make sense of a divergent integral. In this respect the so-called Cauchy Principal Value integral (where $\delta = \epsilon$ in Eq. (4.3.1))

$$\fint_a^b f(x)\,dx = \lim_{\epsilon \to 0^+} \left(\int_a^{x_0 - \epsilon} + \int_{x_0 + \epsilon}^b \right) f(x)\,dx \qquad (4.3.2)$$

is quite useful. We use the notation $\fint_a^b$ to denote the Cauchy Principal Value integral. (Here the Cauchy Principal Value integral is required because of the singularity at $x = x_0$. We usually do not explicitly refer to where the singularity occurs unless there is a special reason to do so, such as when the singularity is at infinity.) We say the Cauchy Principal Value integral exists if and only if the limit (4.3.2) exists. For example, the integral

$$\int_{-1}^2 \frac{1}{x}\,dx = \lim_{\epsilon \to 0^+} \int_{-1}^{-\epsilon} \frac{1}{x}\,dx + \lim_{\delta \to 0^+} \int_\delta^2 \frac{1}{x}\,dx$$

$$= \lim_{\epsilon \to 0^+} \ln|\epsilon| - \lim_{\delta \to 0^+} \ln|\delta| + \ln 2$$

does not exist, whereas

$$\fint_{-1}^2 \frac{1}{x}\,dx = \lim_{\epsilon \to 0}(\ln|\epsilon| - \ln|\epsilon|) + \ln 2 = \ln 2$$

does exist.

More generally, in applications we are sometimes interested in functions on an infinite interval with many points $\{x_i\}_{i=1}^N$ for which $\lim_{x \to x_i} f(x)$ is either infinite or does not exist. We say the following Cauchy Principal Value integral

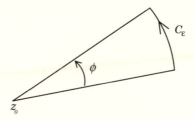

Fig. 4.3.1. Small circular arc C_ϵ

exists if and only if for $a < x_1 < x_2 < \cdots < x_N < b$

$$\fint_{-\infty}^{\infty} f(x)\, dx = \lim_{R \to \infty} \left(\int_{-R}^{a} + \int_{b}^{R} \right) f(x)\, dx + \lim_{\epsilon_1, \epsilon_2, \ldots, \epsilon_N \to 0^+} \left(\int_{a}^{x_1 - \epsilon_1} \right.$$

$$+ \int_{x_1 + \epsilon_1}^{x_2 - \epsilon_2} + \int_{x_2 + \epsilon_2}^{x_3 - \epsilon_3} + \cdots + \int_{x_{N-1} + \epsilon_{N-1}}^{x_N - \epsilon_N} + \left. \int_{x_N + \epsilon_N}^{b} \right) f(x)\, dx$$

$$(4.3.3)$$

exists. In practice we usually combine the integrals and consider the double limit $R \to \infty$ and $\epsilon_i \to 0^+$, for example, $\int_{-R}^{x_1 - \epsilon_1}$ in Eq. (4.3.3), and do not bother to partition the integrals into intermediate values with a, b inserted. Examples will serve to clarify this point. Hereafter we consider $\epsilon_i > 0$, and $\lim_{\epsilon_i \to 0}$ means $\lim_{\epsilon_i \to 0^+}$.

The following theorems will be useful in the sequel. We consider integrals on a small circular arc with radius ϵ, center $z = z_0$, and with the arc subtending an angle ϕ (see Figure 4.3.1). There are two important cases: (a) $(z - z_0) f(z) \to 0$ uniformly (independent of the angle θ along C_ϵ) as $\epsilon \to 0$, and (b) $f(z)$ possesses a simple pole at $z = z_0$.

Theorem 4.3.1 (a) Suppose that on the contour C_ϵ, depicted in Figure 4.3.1, we have $(z - z_0) f(z) \to 0$ uniformly as $\epsilon \to 0$.
Then

$$\lim_{\epsilon \to 0} \int_{C_\epsilon} f(z)\, dz = 0$$

(b) Suppose $f(z)$ has a simple pole at $z = z_0$ with residue $\mathrm{Res}\,(f(z); z_0) = C_{-1}$. Then for the contour C_ϵ

$$\lim_{\epsilon \to 0} \int_{C_\epsilon} f(z)\, dz = i\phi C_{-1} \qquad (4.3.4)$$

where the integration is carried out in the positive (counterclockwise) sense.

Proof (a) The hypothesis, $(z - z_0) f(z) \to 0$ uniformly as $\epsilon \to 0$, means that on C_ϵ, $|(z - z_0) f(z)| \le K_\epsilon$, where K_ϵ depends on ϵ, not on arg $(z - z_0)$, and $K_\epsilon \to 0$ as $\epsilon \to 0$. Estimating the integral $(z = z_0 + \epsilon\, e^{i\theta})$ using $|f(z)| \le K_\epsilon/\epsilon$

$$\left| \int_{C_\epsilon} f(z)\, dz \right| \le \int_{C_\epsilon} |f(z)|\, |dz|$$

$$\le \frac{K_\epsilon}{\epsilon} \int_0^\phi \epsilon\, d\theta = K_\epsilon \phi \to 0, \qquad \epsilon \to 0$$

(b) If $f(z)$ has a simple pole with Res $(f(z); z_0) = C_{-1}$, then from the Laurent expansion of $f(z)$ in the neighborhood of $z = z_0$

$$f(z) = \frac{C_{-1}}{z - z_0} + g(z)$$

where $g(z)$ is analytic in the neighborhood of $z = z_0$. Thus

$$\lim_{\epsilon \to 0} \int_{C_\epsilon} f(z)\, dz = \lim_{\epsilon \to 0} C_{-1} \int_{C_\epsilon} \frac{dz}{z - z_0} + \lim_{\epsilon \to 0} \int_{C_\epsilon} g(z)\, dz$$

The first integral on the right-hand side is evaluated, using $z = z_0 + \epsilon\, e^{i\theta}$, to find

$$\int_{C_\epsilon} \frac{dz}{z - z_0} = \int_0^\phi \frac{i\epsilon\, e^{i\theta}\, d\theta}{\epsilon\, e^{i\theta}} = i\phi$$

In the second integral, $|g(z)| \le M = $ constant in the neighborhood of $z = z_0$ because $g(z)$ is analytic there; hence we can apply part (a) of this theorem to find that the second integral vanishes in the limit of $\epsilon \to 0$, and we recover Eq. (4.3.4). ∎

As a first example we show

$$\int_{-\infty}^\infty \frac{\sin ax}{x}\, dx = \text{sgn}(a)\, \pi \tag{4.3.5}$$

where

$$\text{sgn}(a) = \begin{cases} -1 & a < 0 \\ 0 & a = 0 \\ 1 & a > 0 \end{cases} \tag{4.3.6}$$

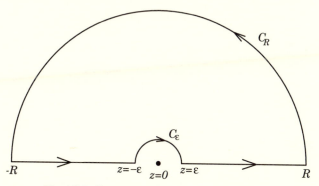

Fig. 4.3.2. Contour of integration, Example 4.3.1

Example 4.3.1 Evaluate

$$I = \fint_{-\infty}^{\infty} \frac{e^{iax}}{x} \, dx, \qquad a \text{ real} \tag{4.3.7}$$

Let us first consider $a > 0$ and the contour depicted in Figure 4.3.2. Because there are no poles enclosed by the contour, we have

$$\oint_C \frac{e^{iaz}}{z} \, dz = \left(\int_{-R}^{-\epsilon} + \int_{\epsilon}^{R} \right) \frac{e^{iax}}{x} \, dx + \int_{C_\epsilon} \frac{e^{iaz}}{z} \, dz + \int_{C_R} \frac{e^{iaz}}{z} \, dz = 0$$

Because $a > 0$, the integral $\int_{C_R} e^{iaz} \, dz/z$ satisfies Jordan's Lemma (i.e., Theorems 4.2.2); hence it vanishes as $R \to \infty$. Similarly, $\int_{C_\epsilon} e^{iaz} \, dz/z$ is calculated using Theorem 4.3.1(b) to find

$$\lim_{\epsilon \to 0} \int_{C_\epsilon} \frac{e^{iaz}}{z} \, dz = -i\pi$$

where we note that on C_ϵ the angle subtended is π. The minus sign is a result of the direction being *clockwise*. Taking the limit $R \to \infty$ we have,

$$I = \fint_{-\infty}^{\infty} \frac{\cos ax}{x} \, dx + i \int_{-\infty}^{\infty} \frac{\sin ax}{x} \, dx = i\pi$$

Thus by setting real and imaginary parts equal we obtain $\fint (\cos ax)/x \, dx = 0$ (which is consistent with the fact that $(\cos ax)/x$ is odd) and Eq. (4.3.5) with $a > 0$. The case $a < 0$ follows because $\frac{\sin ax}{x}$ is odd in a. The same result could be obtained by using the contour in Figure 4.3.3. We also note that there is no need for the principal value in the integral (4.3.5) because it is a (weakly) convergent integral.

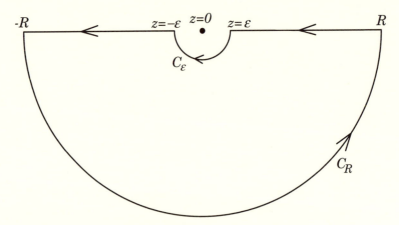

Fig. 4.3.3. Alternative contour, Example 4.3.1

The following example is similar except that there are two locations where the integral has principal value contributions.

Example 4.3.2 Evaluate

$$I = \int_{-\infty}^{\infty} \frac{\cos x - \cos a}{x^2 - a^2}\, dx, \qquad a \text{ real}$$

We note that the integral is convergent and well defined at $x = \pm a$ because l'Hospital's rule shows

$$\lim_{x \to \pm a} \frac{\cos x - \cos a}{x^2 - a^2} = \lim_{x \to \pm a} \frac{-\sin x}{2x} = \frac{-\sin a}{2a}$$

We evaluate I by considering

$$J = \oint_C \frac{e^{iz} - \cos a}{z^2 - a^2}\, dz$$

where the contour C is depicted in Figure 4.3.4. Because there are no poles enclosed by C, we have

$$0 = \oint_C \frac{e^{iz} - \cos a}{z^2 - a^2}\, dz$$
$$= \left\{ \int_{-R}^{-a-\epsilon_1} + \int_{-a+\epsilon_1}^{a-\epsilon_2} + \int_{a+\epsilon_2}^{R} + \int_{C_{\epsilon_1}} + \int_{C_{\epsilon_2}} + \int_{C_R} \right\} \frac{e^{iz} - \cos a}{z^2 - a^2}\, dz$$

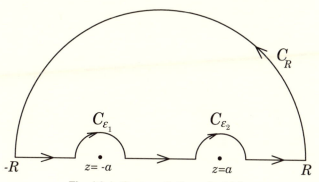

Fig. 4.3.4. Contour C, Example 4.3.2

Along C_R we find, by Theorems 4.2.1 and 4.2.2, that

$$\lim_{R \to \infty} \left| \int_{C_R} \frac{e^{iz} - \cos a}{z^2 - a^2} \, dz \right| = 0$$

Similarly, from Theorem 4.3.1 we find (note that the directions of C_{ϵ_1} and C_{ϵ_2} are clockwise; that is, in the negative direction)

$$\lim_{\epsilon_1 \to 0} \int_{C_{\epsilon_1}} \frac{e^{iz} - \cos a}{z^2 - a^2} \, dx = -i\pi \left(\frac{e^{iz} - \cos a}{2z} \right)_{z=-a} = \frac{\pi \sin a}{2a}$$

and

$$\lim_{\epsilon_2 \to 0} \int_{C_{\epsilon_2}} \frac{e^{iz} - \cos a}{z^2 - a^2} \, dx = -i\pi \left(\frac{e^{iz} - \cos a}{2z} \right)_{z=a} = \frac{\pi \sin a}{2a}$$

Thus as $R \to \infty$

$$\fint_{-\infty}^{\infty} \frac{e^{ix} - \cos a}{x^2 - a^2} \, dx = -\pi \frac{\sin a}{a}$$

and hence, by taking the real part, $I = -\pi (\sin a)/a$.

Again we note that the Cauchy Principal Value integral was only a device used to obtain a result for a well-defined integral. We also mention the fact that in practice one frequently calculates contributions along contours such as C_{ϵ_i} by carrying out the calculation directly without resorting to Theorem 4.3.1.

Our final illustration of Cauchy Principal Values is the evaluation of an integral similar to that of Example 4.2.9.

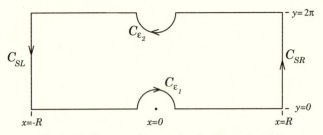

Fig. 4.3.5. Contour of integration for Example 4.3.3

Example 4.3.3 Evaluate

$$I = \int_{-\infty}^{\infty} \frac{e^{px} - e^{qx}}{1 - e^x} \, dx$$

where $0 < p, q < 1$.

We observe that this integral is convergent and well defined. We evaluate two separate integrals:

$$I_1 = \fint_{-\infty}^{\infty} \frac{e^{px}}{1 - e^x} \, dx$$

and

$$I_2 = \fint_{-\infty}^{\infty} \frac{e^{qx}}{1 - e^x} \, dx,$$

noting that $I = I_1 - I_2$.

In order to evaluate I_1, we consider the contour depicted in Figure 4.3.5

$$J = \oint_C \frac{e^{pz}}{1 - e^z} \, dz$$

$$= \left(\int_{-R}^{-\epsilon_1} + \int_{\epsilon_1}^{R} \right) \frac{e^{px}}{1 - e^x} \, dx + \left(\int_{C_{SR}} + \int_{C_{SL}} \right) \frac{e^{pz}}{1 - e^z} \, dz$$

$$+ \left(\int_{R}^{-\epsilon_2} + \int_{\epsilon_2}^{-R} \right) \frac{e^{2\pi i p} e^{px}}{1 - e^x} \, dx + \left(\int_{C_{\epsilon_1}} + \int_{C_{\epsilon_2}} \right) \frac{e^{pz}}{1 - e^z} \, dz$$

Along the top path line we take $z = x + 2\pi i$. The integral $J = 0$ because no singularities are enclosed. The estimates of Example 4.2.9 show that the integrals along the sides C_{SL} and C_{SR} vanish. From Theorem 4.3.1 we have

$$\lim_{\epsilon_1 \to 0} \int_{C_{\epsilon_1}} \frac{e^{pz}}{1 - e^z} \, dz = -i\pi \left(\frac{e^{pz}}{-e^z} \right)_{z=0} = i\pi$$

and

$$\lim_{\epsilon_2 \to 0} \int_{C_{\epsilon_2}} \frac{e^{pz}}{1 - e^z} dz = -i\pi \left(\frac{e^{pz}}{-e^z} \right)_{z=2\pi i} = i\pi e^{2\pi ip}$$

Hence, taking $R \to \infty$,

$$I_1 = \dashint_{-\infty}^{\infty} \frac{e^{px}}{1 - e^x} dx = -i\pi \frac{1 + e^{2\pi ip}}{1 - e^{2\pi ip}} = \pi \cot \pi p$$

Clearly, a similar analysis is valid for I_2 where p is replaced by q. Thus, putting all of this together, we find

$$I = \pi (\cot \pi p - \cot \pi q)$$

4.3.2 Integrals with Branch Points

In the remainder of this section we consider integrands that involve branch points. To evaluate the integrals, we introduce suitable branch cuts associated with the relevant multivalued functions. The procedure will be illustrated by a variety of examples.

Before working out examples we prove a theorem that will be useful in providing estimates for cases where Jordan's Lemma is not applicable.

Theorem 4.3.2 If on a circular arc C_R of radius R and center $z = 0$, $zf(z) \to 0$ uniformly as $R \to \infty$, then

$$\lim_{R \to \infty} \int_{C_R} f(z) \, dz = 0$$

Proof Let ϕ be the angle enclosed by the arc C_R. Then

$$\left| \int_{C_R} f(z) \, dz \right| \le \int_0^\phi |f(z)| R \, d\theta \le K_R \phi$$

Because $zf(z) \to 0$ *uniformly*, it follows that $|zf(z)| = R|f(z)| \le K_R$, $K_R \to 0$, as $R \to \infty$.   ∎

Example 4.3.4 Use contour integration to evaluate

$$I = \int_0^\infty \frac{dx}{(x + a)(x + b)}, \qquad a, b > 0$$

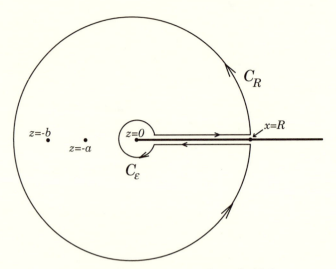

Fig. 4.3.6. "Keyhole" contour, Example 4.3.4

Because the integrand is not even, we cannot extend our integration region to the entire real line. Hence the methods of Section 4.2 will not work directly. Instead, we consider the contour integral

$$J = \oint_C \frac{\log z}{(z+a)(z+b)} \, dz$$

where C is the "keyhole" contour depicted in Figure 4.3.6. We take $\log z$ to be on its principal branch $z = re^{i\theta} : 0 \le \theta < 2\pi$, and choose a branch cut along the x-axis, $0 \le x < \infty$.

An essential ingredient of the method is that owing to the location of the branch cut, the sum of the integrals on each side of the cut do not cancel.

$$J = \int_\epsilon^R \frac{\log x}{(x+a)(x+b)} \, dx + \int_R^\epsilon \frac{\log(x e^{2\pi i})}{(x+a)(x+b)} \, dx$$

$$+ \left(\int_{C_\epsilon} + \int_{C_R} \right) \frac{\log z}{(z+a)(z+b)} \, dz$$

$$= 2\pi i \left\{ \left(\frac{\log z}{z+a} \right)_{z=-b} + \left(\frac{\log z}{z+b} \right)_{z=-a} \right\}$$

Theorems 4.3.1a and 4.3.2 show that

$$\lim_{\epsilon \to 0} \int_{C_\epsilon} \frac{\log z}{(z+a)(z+b)} \, dz = 0$$

and

$$\lim_{R \to \infty} \int_{C_R} \frac{\log z}{(z+a)(z+b)} \, dz = 0$$

Using $\log(x \, e^{2\pi i}) = \log x + 2\pi i$, $\log(-a) = \log |a| + i\pi$ (for $a > 0$) to simplify the expression for J, we have

$$-2\pi i \int_0^\infty \frac{dx}{(x+a)(x+b)} = 2\pi i \left(\frac{\log b - i\pi}{a - b} + \frac{\log a - i\pi}{b - a} \right)$$

hence

$$I = \left(\frac{\log b/a}{b - a} \right)$$

Of course, we could have evaluated this integral by elementary methods because (we worked this example only for illustrative purposes)

$$I = \left(\frac{1}{b - a} \right) \int_0^\infty \left(\frac{1}{x + a} - \frac{1}{x + b} \right) dx$$

$$= \left(\frac{1}{b - a} \right) \left[\ln \left(\frac{x + a}{x + b} \right) \right]_0^\infty = \frac{\log b/a}{b - a}$$

Another example is the following.

Example 4.3.5 Evaluate

$$I = \int_0^\infty \frac{\log^2 x}{x^2 + 1} \, dx$$

Consider

$$J = \oint_C \frac{\log^2 z}{z^2 + 1} \, dz$$

where C is the contour depicted in Figure 4.3.7, and we take the principal branch of $\log z : z = r \, e^{i\theta}, 0 \leq \theta < 2\pi$.

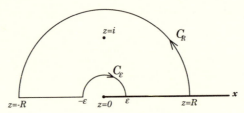

Fig. 4.3.7. Contour of integration, Example 4.3.5

We have

$$J = \int_R^\epsilon \frac{[\log(re^{i\pi})]^2}{(re^{i\pi})^2 + 1} e^{i\pi} \, dr + \int_\epsilon^R \frac{\log^2 x}{x^2 + 1} \, dx$$

$$+ \left(\int_{C_R} + \int_{C_\epsilon} \right) \frac{\log^2 z}{z^2 + 1} \, dz$$

$$= 2\pi i \left(\frac{\log^2 z}{2z} \right)_{z=i=e^{i\pi/2}}$$

Theorems 4.3.1a and 4.3.2 show that $\int_{C_\epsilon} \to 0$ as $\epsilon \to 0$ and $\int_{C_R} \to 0$ as $R \to \infty$. Thus the above equation simplifies, and

$$2 \int_0^\infty \frac{\log^2 x}{x^2 + 1} \, dx + 2i\pi \int_0^\infty \frac{\log x}{x^2 + 1} \, dx - \pi^2 \int_0^\infty \frac{dx}{x^2 + 1}$$

$$= 2\pi i \frac{(i\pi/2)^2}{2i} = -\frac{\pi^3}{4}$$

However, the last integral can also be evaluated by contour integration, using the method of Section 4.2 (with the contour C_R depicted in Figure 4.2.1) to find

$$\int_0^\infty \frac{dx}{x^2 + 1} = \frac{1}{2} \int_{-\infty}^\infty \frac{dx}{x^2 + 1} = \frac{1}{2} 2\pi i \left(\frac{1}{2z} \right)_{z=i} = \frac{\pi}{2}$$

Hence we have

$$2 \int_0^\infty \frac{\log^2 x \, dx}{x^2 + 1} + 2\pi i \int_0^\infty \frac{\log x \, dx}{x^2 + 1} = \frac{\pi^3}{4}$$

whereupon, by taking the real and imaginary parts

$$I = \frac{\pi^3}{8} \quad \text{and} \quad \int_0^\infty \frac{\log x}{x^2 + 1} \, dx = 0$$

An example that uses some of the ideas of Examples 4.3.4 and 4.3.5 is the following.

Example 4.3.6 Evaluate

$$I = \int_0^\infty \frac{x^{m-1}}{x^2 + 1} \, dx, \qquad 0 < m < 2$$

The condition on m is required for the convergence of the integral.
    Consider the contour integral

$$J = \oint_C \frac{z^{m-1}}{z^2 + 1} \, dz$$

where C is the keyhole contour in Figure 4.3.6. We take the principal branch of $z^m : z = re^{i\theta}, 0 \le \theta < 2\pi$. The residue theorem yields

$$J = \int_\epsilon^R \frac{x^{m-1}}{x^2 + 1} \, dx + \int_R^\epsilon \frac{(xe^{2\pi i})^{m-1}}{x^2 + 1} \, dx$$

$$+ \left(\int_{C_\epsilon} + \int_{C_R} \right) \frac{z^{m-1}}{z^2 + 1} \, dz$$

$$= 2\pi i \left[\left(\frac{z^{m-1}}{2z} \right)_{z=i=e^{i\pi/2}} + \left(\frac{z^{m-1}}{2z} \right)_{z=-i=e^{3i\pi/2}} \right]$$

Theorems 4.3.1a and 4.3.2 imply that $\int_{C_\epsilon} \to 0$ as $\epsilon \to 0$, and $\int_{C_R} \to 0$ as $R \to \infty$. Therefore the above equation simplifies to

$$\int_0^\infty \frac{x^{m-1}}{x^2 + 1} dx (1 - e^{2\pi i m}) = 2\pi i \left(\frac{e^{i\pi(m-1)/2}}{2i} - \frac{e^{3i\pi(m-1)/2}}{2i} \right)$$

$$= -\pi i \left(e^{im\pi/2} + e^{3im\pi/2} \right)$$

Hence

$$I = -\pi i \, e^{im\pi/2} \left(\frac{1 + e^{im\pi}}{1 - e^{2im\pi}} \right) = \pi \frac{\cos \frac{m\pi}{2}}{\sin m\pi} = \frac{\pi}{2 \sin \frac{m\pi}{2}}$$

The following example illustrates how we deal with more complicated multivalued functions and their branch cut structure. The reader is encouraged to review Section 2.3 before considering the next example.

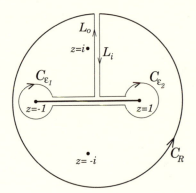

Fig. 4.3.8. Contour C for Example 4.3.7

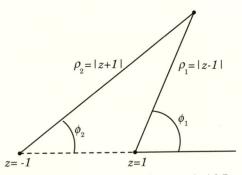

Fig. 4.3.9. Polar coordinates for example 4.3.7

Example 4.3.7 Evaluate

$$I = \int_{-1}^{1} \frac{\sqrt{1-x^2}}{1+x^2} \, dx$$

where the square root function takes on a positive value (i.e., $\sqrt{1} = +1$) in the range $-1 < x < 1$.

We consider the contour integral

$$J = \oint_{C} \frac{(z^2 - 1)^{1/2}}{1 + z^2} \, dz$$

where the contour C and the relevant branch cut structure are depicted in Figure 4.3.8.

Before calculating the various contributions to J we first discuss the multi-valued function $\sqrt{z^2 - 1}$ (see Figure 4.3.9). Using polar coordinates, we find

$$(z^2 - 1)^{\frac{1}{2}} = \sqrt{\rho_1 \rho_2} e^{i(\phi_1 + \phi_2)/2}, \qquad 0 \le \phi_1, \phi_2 < 2\pi$$

where

$$(z - 1)^{\frac{1}{2}} = \sqrt{\rho_1} e^{i\phi_1/2}, \qquad \rho_1 = |z - 1|$$

and

$$(z + 1)^{\frac{1}{2}} = \sqrt{\rho_2} e^{i\phi_2/2}, \qquad \rho_2 = |z + 1|$$

With this choice of branch, we find

$$(z^2 - 1)^{\frac{1}{2}} = \begin{cases} \sqrt{x^2 - 1}, & 1 < x < \infty \\ -\sqrt{x^2 - 1}, & -\infty < x < -1 \\ i\sqrt{1 - x^2}, & -1 < x < 1, \quad y \to 0^+ \\ -i\sqrt{1 - x^2}, & -1 < x < 1, \quad y \to 0^- \end{cases}$$

Using these expressions in the contour integral J, it follows that

$$J = \int_{-1+\epsilon_1}^{1-\epsilon_2} \frac{i\sqrt{1 - x^2}}{1 + x^2} dx + \int_{1-\epsilon_2}^{-1+\epsilon_1} \frac{-i\sqrt{1 - x^2}}{1 + x^2} dx$$

$$+ \left(\int_{C_{\epsilon_1}} + \int_{C_{\epsilon_2}} + \int_{C_R} \right) \frac{(z^2 - 1)^{1/2}}{1 + z^2} dz$$

$$= 2\pi i \left[\left(\frac{(z^2 - 1)^{1/2}}{2z} \right)_{z=e^{i\pi/2}} + \left(\frac{(z^2 - 1)^{1/2}}{2z} \right)_{z=e^{3i\pi/2}} \right]$$

We note that the crosscut integrals vanish

$$\left(\int_{L_0} + \int_{L_i} \right) \frac{(z^2 - 1)^{1/2}}{1 + z^2} dz = 0$$

because L_0 and L_i are chosen in a region where $(z^2 - 1)^{1/2}$ is continuous and single-valued, and L_0 and L_i are arbitrarily close to each other. Theorem 4.3.1a shows that $\int_{C_{\epsilon_i}} \to 0$ as $\epsilon_i \to 0$; that is

$$\left| \int_{C_{\epsilon_i}} \frac{(z^2 - 1)^{1/2}}{1 + z^2} dz \right| \leq \int_0^{2\pi} \frac{(|(z - 1)(z + 1)|^{1/2})}{|2 + \epsilon_i^2 e^{2i\theta} + 2\epsilon_i e^{i\theta}|} \epsilon_i \, d\theta$$

$$\leq \int_0^{2\pi} \frac{\sqrt{\epsilon_i^2 + 2\epsilon_i}}{2 - 2\epsilon_i - \epsilon_i^2} \epsilon_i \, d\theta \xrightarrow[\epsilon_i \to 0]{} 0$$

The contribution on C_R is calculated as follows:

$$\int_{C_R} \frac{(z^2-1)^{1/2}}{1+z^2} \, dz = \int_0^{2\pi} \frac{(R^2 e^{2i\theta}-1)^{1/2}}{1+R^2 e^{2i\theta}} i R e^{i\theta} \, d\theta$$

where we note that $(R^2 e^{2i\theta} - 1)^{1/2} \approx R e^{i\theta}$ as $R \to \infty$ because the chosen branch implies $\lim_{z \to \infty}(z^2 - 1)^{1/2} = z$. Hence

$$\lim_{R \to \infty} \int_{C_R} \frac{(z^2-1)^{1/2}}{1+z^2} \, dz = 2\pi i$$

Calculation of the residues requires computing the correct branch of $(z^2 - 1)^{1/2}$.

$$\left(\frac{(z^2-1)^{1/2}}{2z} \right)_{z=e^{i\pi/2}} = \frac{\sqrt{2} e^{i(3\pi/4+\pi/4)/2}}{2i} = \frac{1}{\sqrt{2}}$$

$$\left(\frac{(z^2-1)^{1/2}}{2z} \right)_{z=e^{3i\pi/2}} = \frac{\sqrt{2} e^{i(5\pi/4+7\pi/4)/2}}{-2i} = \frac{1}{\sqrt{2}}$$

Taking $\epsilon_i \to 0$, $R \to \infty$, and substituting the above results in the expression for J, we find

$$2i \int_{-1}^{1} \frac{\sqrt{1-x^2}}{1+x^2} dx = 2\pi i \left(\sqrt{2} - 1 \right)$$

hence

$$I = \pi(\sqrt{2} - 1)$$

It should be remarked that the contribution along C_R is proportional to the residue at infinity. Namely, calling

$$f(z) = \frac{(z^2-1)^{1/2}}{1+z^2}$$

it follows that

$$f(z) = \frac{(z^2(1-1/z^2))^{\frac{1}{2}}}{z^2 \left(1+\frac{1}{z^2}\right)} = \frac{1}{z}\left(1 - \frac{1}{2z^2} + \cdots\right)\left(1 - \frac{1}{z^2} + \cdots\right)$$

Thus the coefficient of $\frac{1}{z}$ is unity; hence from definition (4.1.11a,b)

$$\text{Res}(f(z); \infty) = 1$$

and

$$\lim_{R \to \infty} \int_{C_R} \frac{(z^2 - 1)^{1/2}}{1 + z^2} = 2\pi i \operatorname{Res}\left(\frac{(z^2 - 1)^{1/2}}{1 + z^2}; \infty \right) = 2\pi i$$

The above analysis shows that calculating I follows from

$$2I = 2\pi i \sum_{j=1}^{2} \operatorname{Res}(f(z); z_j) - z\pi i \operatorname{Res}(f(z); \infty)$$

where the three residues are calculated at

$$z_1 = e^{i\pi/2}, \qquad z_2 = e^{3i\pi/2}, \qquad z_\infty = \infty$$

the minus sign is due to the orientation of the contour C_R with respect to z_∞.

The calculation involving the residue at infinity can be carried out for a large class of functions. In practice one usually computes the contribution at infinity by evaluating an integral along a large circular contour, C_R, in the same manner as we have done here.

Problems for Section 4.3

1. (a) Use principal value integrals to show that

$$\int_0^\infty \frac{\cos kx - \cos mx}{x^2} \, dx = \frac{-\pi}{2}(|k| - |m|)$$

Hint: note that the function $f(z) = (e^{ikz} - e^{imz})/z^2$ has a simple pole at the origin.

    (b) Let $k = 2, m = 0$ to deduce that

$$\int_0^\infty \frac{\sin^2 x}{x^2} \, dx = \frac{\pi}{2}$$

2. Show that

$$\int_0^\infty \frac{\sin x}{x(x^2 + 1)} \, dx = \frac{\pi}{2} \left(1 - \frac{1}{e} \right)$$

3. Show that

$$\int_{-\infty}^\infty \frac{(\cos x - 1)}{x^2(x^2 + a^2)} \, dx = -\frac{\pi}{a^2} + \frac{\pi}{a^3}(1 - e^{-a}), \qquad a > 0$$

4. Use a rectangular contour with corners at $\pm R$ and $\pm R + i\pi/k$, with an appropriate indentation, to show that

$$\int_0^\infty \frac{x}{\sinh kx} \, dx = \frac{\pi^2}{4k|k|} \qquad \text{for } k \neq 0$$

5. Projection operators can be defined as follows. Consider a function $F(z)$

$$F(z) = \frac{1}{2\pi i} \int_C \frac{f(\zeta)}{\zeta - z} \, d\zeta$$

where C is a contour, typically infinite (e.g. the real axis) or closed (e.g. a circle) and z lies off the contour. Then the "plus" and "minus" projections of $F(z)$ at $z = \zeta_0$ are defined by the following limit:

$$F^\pm(\zeta_0) = \lim_{z \to \zeta_0^\pm} \left[\frac{1}{2\pi i} \int_C \frac{f(\zeta)}{\zeta - z} \, d\zeta \right]$$

where $\zeta_0^\pm$ are points just inside $(+)$ or outside $(-)$ of a closed contour (i.e., $\lim_{z \to \zeta_0^+}$ denotes the limit from points z inside the contour C) or to the left $(+)$ or right $(-)$ of an infinite contour. Note: the "+" region lies to the *left* of the contour; where we take the standard orientation for a contour, that is, the contour is taken with counterclockwise orientation. To simplify the analysis, we will assume that $f(x)$ can be analytically extended in the neighborhood of the real axis.

(a) Show that

$$F^\pm(\zeta_0) = \frac{1}{2\pi i} \fint_C \frac{f(\zeta)}{\zeta - \zeta_0} \, d\zeta \pm \frac{1}{2} f(\zeta_0).$$

where $\fint_C$ is the principal value integral that omits the point $\zeta = \zeta_0$.

(b) Suppose that $f(\zeta) = 1/(\zeta^2 + 1)$, and the contour C is the real axis (infinite), with orientation take from $-\infty$ to ∞; find $F^\pm(\zeta_0)$.

(c) Suppose that $f(\zeta) = 1/(\zeta^2 + a^2)$, $a^2 > 1$, and the contour C is the unit circle centered at the origin with counterclockwise orientation; find $F^\pm(\zeta_0)$.

6. An important application of complex variables is to solve equations for functions analytic in a certain region, given a relationship on a boundary (see Chapter 7). A simple example of this is the following. Solve for the function $\psi^+(z)$ analytic in the upper half plane and $\psi^-(z)$ analytic

in the lower half plane, given the following relationship on the real axis
($\operatorname{Re} z = x$):

$$\psi^+(x) - \psi^-(x) = f(x)$$

where, say, $f(x)$ is differentiable and absolutely integrable. A solution
to this problem for function $\psi^\pm(z) \to 0$ as $|z| \to \infty$ is given by

$$\psi^\pm(x) = \lim_{\epsilon \to 0^+} \frac{1}{2\pi i} \int_{-\infty}^\infty \frac{f(\zeta)}{\zeta - (x \pm i\epsilon)} \, d\zeta$$

To simplify the analysis, we will assume that $f(x)$ can be analytically
extended in the neighborhood of the real axis.

(a) Explain how this solution could be formally obtained by introducing
the projection operators

$$P^\pm = \lim_{\epsilon \to 0^+} \frac{1}{2\pi i} \int_{-\infty}^\infty \frac{d\zeta}{\zeta - (x \pm i\epsilon)}$$

and in particular why it should be true that

$$P^+\psi^+ = \psi^+, \quad P^-\psi^- = -\psi^-, \quad P^+\psi^- = P^-\psi^+ = 0$$

(b) Verify the results in part (a) for the example

$$\psi^+(x) - \psi^-(x) = \frac{1}{x^4 + 1}$$

and find $\psi^\pm(z)$ in this example.

(c) Show that

$$\psi^\pm(x) = \frac{1}{2\pi i} \fint_{-\infty}^\infty \frac{f(\zeta)}{\zeta - x} \, d\zeta \pm \frac{1}{2} f(x)$$

In operator form the first term is usually denoted as $H(f(x))/2i$,
where $H(f(x))$ is called the Hilbert transform. Then show that

$$\psi^\pm(x) = \frac{1}{2i} H(f(x)) \pm \frac{1}{2} f(x)$$

or, in operator form

$$\psi^\pm(x) = \frac{1}{2}(\pm 1 - i H) f(x)$$

7. Use the keyhole contour of Figure 4.3.6 to show that on the principal branch of x^k

(a) $I(a) = \int_0^\infty \dfrac{x^{k-1}}{(x+a)} \, dx = \dfrac{\pi}{\sin k\pi} a^{k-1}, \quad 0 < k < 1, \quad a > 0$

(b) $\displaystyle \int_0^\infty \dfrac{x^{k-1}}{(x+1)^2} \, dx = \dfrac{(1-k)\pi}{\sin k\pi}, \quad 0 < k < 2$

Verify this result by evaluating $I'(1)$ in part (a).

8. Use the technique described in Problem 7 above, using $\oint_C f(z)(\log z)^2 dz$, to establish that

(a) $I(a) = \displaystyle\int_0^\infty \dfrac{\log x}{x^2 + a^2} \, dx = \dfrac{\pi}{2a} \log a, \quad a > 0$

(b) $\displaystyle\int_0^\infty \dfrac{\log x}{(x^2+1)^2} dx = -\dfrac{\pi}{4}$

Verify this by computing $I'(1)$ in part (a).

9. Use the keyhole contour in Figure 4.3.6 to establish that

$$\int_0^\infty \frac{x^{-k}}{x^2 + 2x \cos \phi + 1} \, dx = \frac{\pi}{\sin k\pi} \frac{\sin(k\phi)}{\sin \phi}$$

for $0 < k < 1, 0 < \phi < \pi$.

10. By using a large semicircular contour, enclosing the left half plane with a suitable keyhole contour, show that

$$\frac{1}{2\pi i} \int_{a-i\infty}^{a+i\infty} \frac{e^{zt}}{\sqrt{z}} \, dz = \frac{1}{\sqrt{\pi t}} \quad \text{for } a, t > 0$$

This is the inverse Laplace transform of the function $1/\sqrt{z}$. (The Laplace transform and its inverse are discussed in Section 4.5.)

11. Consider the integral

$$I_R = \frac{1}{2\pi i} \oint_{C_R} \frac{dz}{(z^2 - a^2)^{1/2}}, \quad a^2 > 0$$

where C_R is a circle of radius R centered at the origin enclosing the points $z = \pm a$. Take the principal value of the square root.

(a) Evaluate the residue of the integrand at infinity and show that $I_R = 1$.

(b) Evaluate the integral by defining the contour around the branch points and along the branch cuts between $z = -a$ to $z = a$, to find (see Section 2.3) that $I_R = \dfrac{2}{\pi} \displaystyle\int_0^a \dfrac{dx}{\sqrt{a^2 - x^2}}$. Use the well-known indefinite integral $\displaystyle\int \dfrac{dx}{\sqrt{a^2 - x^2}} = \sin^{-1} x/a + \text{const}$ to obtain the same result as in part (a).

12. Use the transformation $t = (x - 1)/(x + 1)$ on the principal branch of the following functions to show that

(a)
$$\int_{-1}^1 \left(\frac{1+t}{1-t}\right)^{k-1} dt = \frac{2(1-k)\pi}{\sin k\pi}, \quad 0 < k < 2$$

(b)
$$\int_{-1}^1 \log\left(\frac{1+t}{1-t}\right) \frac{dt}{1-at} = \frac{1}{2a} \log^2\left(\frac{1+a}{1-a}\right), \quad 0 < a < 1$$

13. Use the keyhole contour of Figure 4.3.6 to show for the principal branch of $x^{1/2}$ and $\log x$

$$\int_0^\infty \frac{x^{1/2} \log x}{(1+x^2)} dx = \frac{\pi^2}{2\sqrt{2}}$$

and

$$\int_0^\infty \frac{x^{1/2}}{(1+x^2)} dx = \frac{\pi}{\sqrt{2}}$$

14. Consider the following integral with the principal branch of the square root, $\displaystyle\int_0^1 \sqrt{x(1-x)}\, dx$. Use the contour integral $I_R = \dfrac{1}{2\pi i} \displaystyle\oint_{C_R} \sqrt{z(z-1)}\, dz$, where C_R is the "two-keyhole" or "dogbone" contour similar to Figure 4.3.8, this time enclosing $z = 0$ and $z = 1$. Take the branch cut on the real axis between $z = 0$ and $z = 1$.

(a) Show that the behavior at $z = \infty$ implies that $\displaystyle\lim_{R\to\infty} I_R = \frac{-1}{8}$.

(b) Use the principal branch of this function to show

$$I_R = \frac{1}{\pi} \int_0^1 \sqrt{x(1-x)}\, dx$$

and conclude that $\displaystyle\int_0^1 \sqrt{x(1-x)}\, dx = \frac{\pi}{8}$.

(c) Use the same method to show

$$\int_0^1 x^n \sqrt{x(1-x)} = \pi b_{n+2}$$

where b_{n+2} is the coefficient of the term x^{n+2} in the binomial expansion of $(1-x)^{1/2}$; that is

$$(1-x)^{\frac{1}{2}} = 1 - \frac{1}{2}x - \frac{1}{8}x^2 - \cdots.$$

15. In Problem 10 of Section 2.6 we derived the formula

$$v(r, \varphi) = v(r = 0) + \frac{1}{2\pi} \int_0^{2\pi} u(\theta) \frac{2r\sin(\varphi - \theta)}{1 - 2r\cos(\varphi - \theta) + r^2} d\theta$$

where $u(\theta)$ is given on the unit circle and the harmonic conjugate to $u(r, \varphi)$, $v(r, \varphi)$ is determined by the formula above. (In that exercise $u(r, \varphi)$ was also derived.) Let $\zeta = re^{i\varphi}$. Show that as $r \to 1$, $\zeta \to e^{i\varphi}$ with $z = e^{i\theta}$, the above formula may be written as (use the trigonometric identity (**I**) below)

$$v(\varphi) = v(0) - \frac{1}{2\pi i}\oint_0^{2\pi} u(\theta) \frac{e^{i\varphi} + e^{i\theta}}{e^{i\varphi} - e^{i\theta}} d\theta$$

where the integral is taken as the Cauchy principal value. Show that

$$\textbf{(I)} \quad \frac{e^{i\varphi} + e^{i\theta}}{e^{i\varphi} - e^{i\theta}} = i\left(\frac{\sin(\theta - \varphi)}{1 - \cos(\theta - \varphi)}\right) = i\cot\left(\frac{\theta - \varphi}{2}\right)$$

using

$$\cos x = 2\cos^2 \frac{x}{2} - 1 = 1 - 2\sin^2 \frac{x}{2}, \qquad \sin x = 2\sin \frac{x}{2}\cos \frac{x}{2},$$

and therefore deduce that

$$v(\varphi) = v(0) + \frac{1}{2\pi}\oint_0^{2\pi} u(\theta)\cot\left(\frac{\varphi - \theta}{2}\right) d\theta.$$

This formula relates the boundary values, on the circle, between the real and imaginary parts of a function $f(z) = u + iv$, which is analytic inside the circle.

4.4 The Argument Principle, Rouché's Theorem

The Cauchy Residue Theorem can be used to obtain a useful result regarding the number of zeroes and poles of a meromorphic function. In what follows, we refer to second order, third order, ... poles as "poles of order 2,3,...."

Theorem 4.4.1 (*Argument Principle*) Let $f(z)$ be a meromorphic function defined inside and on a simple closed contour C, with no zeroes or poles on C. Then

$$I = \frac{1}{2\pi i} \oint_C \frac{f'(z)}{f(z)} \, dz = N - P = \frac{1}{2\pi} [\arg f(z)]_C \qquad (4.4.1)$$

where N and P are the number of zeroes and poles, respectively, of $f(z)$ inside C; where a multiple zero or pole is counted according to its multiplicity, and where $\arg f(z)$ is the argument of $f(z)$; that is, $f(z) = |f(z)| \exp(i \arg f(z))$ and $[\arg f(z)]_C$ denotes the change in the argument of $f(z)$ over C.

Proof Suppose $z = z_i$ is a zero/pole of order n_i. Then

$$\frac{f'(z)}{f(z)} = \frac{\pm n_i}{z - z_i} + \phi(z) \qquad (4.4.2)$$

where ϕ is analytic in the neighborhood of z and the plus/minus sign stands for the zero/pole case, respectively. Formula (4.4.2) follows from the fact that if $f(z)$ has a zero of order n_{iz}, we can write $f(z)$ as $f(z) = (z - z_i)^{n_{iz}} g(z)$, where $g(z_i) \neq 0$ and $g(z)$ is analytic in the neighborhood of z_i; whereas if $f(z)$ has a pole of order n_{ip}, then $f(z) = g(z)/(z - z_i)^{n_{ip}}$. Equation (4.4.2) then follows by differentiation with $\phi(z) = g'(z)/g(z)$. Applying the Cauchy Residue Theorem yields

$$I = \frac{1}{2\pi i} \oint_C \frac{f'(z)}{f(z)} \, dz = \sum_{i_z=1}^{M_z} n_{iz} - \sum_{i_p=1}^{M_p} n_{ip} = N - P$$

where M_z (M_p) is the number of zero (pole) locations z_{i_z} (z_{i_p}) with multiplicity n_{iz} (n_{ip}), respectively.

In order to show that $I = [\arg f(z)]_C/2\pi$, we parametrize C as in Section 2.4. Namely, we let $z = z(t)$ on $a \leq t \leq b$, where $z(a) = z(b)$. Thus we have, for I, the line integral

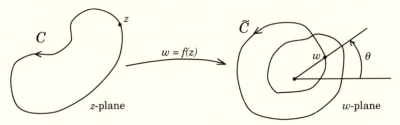

Fig. 4.4.1. Mapping w

$$I = \frac{1}{2\pi i} \int_a^b \frac{f'(z(t))}{f(z(t))} z'(t) \, dt = \frac{1}{2\pi i} [\log f(z(t))]_{t=a}^b$$

$$= \frac{1}{2\pi i} [\log |f(z(t))| + i \arg f(z(t))]_{t=a}^b = \frac{1}{2\pi i} [\arg f(z)]_C \quad (4.4.3)$$

where we have taken the principal branch of the logarithm. ∎

Geometrically, Eq. (4.4.3) corresponds to the following. Consider Figure 4.4.1.

Let $w = f(z)$ be the image of the point z under the mapping $w = f(z)$, and let $\theta = \arg f(z)$ be the angle that the ray from the origin to w makes with respect to the horizontal; then $w = |f(z)| \exp(i\theta)$. Equation (4.4.3) corresponds to the number of times the point w winds around the origin on the image curve $\tilde{C}$ when z moves around C. Under the transformation $w = f(z)$ we find

$$\frac{1}{2\pi i} \oint_C \frac{f'(z)}{f(z)} \, dz = \frac{1}{2\pi i} \oint_{\tilde{C}} \frac{dw}{w} = \frac{1}{2\pi i} [\arg w]_{\tilde{C}} \quad (4.4.4)$$

The quantity $\frac{1}{2\pi i} [\arg w]_{\tilde{C}}$ is called the **winding number** of $\tilde{C}$ about the origin.

The following extension of Theorem 4.4.1 is useful. Suppose that $f(z)$ and C satisfy the hypothesis of Theorem 4.4.1, and let $h(z)$ be analytic inside and on C. Then

$$J = \frac{1}{2\pi i} \oint_C \frac{f'(z)}{f(z)} h(z) \, dz$$

$$= \sum_{i_z=1}^{M_z} n_{iz} h(z_{iz}) - \sum_{i_p=1}^{M_p} n_{ip} h(z_{ip}) \quad (4.4.5)$$

where $h(z_{iz})$ and $h(z_{ip})$ are the values of $h(z)$ at the locations z_{iz} and z_{ip}. Formula (4.4.5) follows from Eq. (4.4.2) by evaluating the contour integral

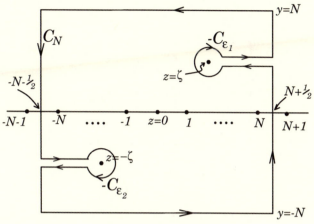

Fig. 4.4.2. Contour C_N

associated with

$$\frac{f'(z)}{f(z)}h(z) = \frac{\pm n_i}{z - z_i}h(z) + h(z)\phi(z)$$

This amounts to obtaining the residue

$$\mathrm{Res}\left(\frac{f'(z)}{f(z)}h(z); z_i\right) = \pm n_i\, h(z_i)$$

If the zeroes/poles are simple, then in Eq. (4.4.5) we take $n_{iz} = n_{ip} = 1$.

Example 4.4.1 Consider the following integral:

$$I(\zeta) = \frac{1}{2\pi i}\oint_{C_N}\frac{\pi\cot\pi z}{\zeta^2 - z^2}\,dz$$

where C_N is depicted in Figure 4.4.2. Deduce that $\pi\cot\pi\zeta = \zeta\sum_{n=-\infty}^{\infty}\frac{1}{\zeta^2 - n^2}$.

We have an integral of the form (4.4.5), where $f(z) = \sin\pi z$, $h(z) = \frac{1}{\zeta^2 - z^2}$, $z_i = n$, and $n = 0, \pm1, \pm2, \ldots, \pm N$. Deforming the contour and using Eq. (4.4.5) with $n_{iz} = 1$, $n_{ip} = 0$, we find

$$\frac{1}{2\pi i}\left(\oint_{C_{NR}} - \oint_{C_{\epsilon_1}} - \oint_{C_{\epsilon_2}}\right)\frac{\pi\cot\pi z}{\zeta^2 - z^2}\,dz = \sum_{n=-N}^{N}\frac{1}{\zeta^2 - n^2}$$

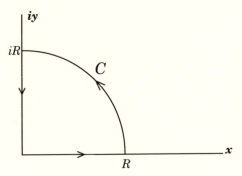

Fig. 4.4.3. Contour in first quadrant

where C_{NR} denotes the rectangular contour alone, without crosscuts and circles around $z = \pm\zeta$. Taking $N \to \infty$ and $\epsilon_i \to 0$, we have, after computing the residues about C_{ϵ_i}

$$\lim_{N \to \infty} \frac{1}{2\pi i} \oint_{C_{NR}} \frac{\pi \cot \pi z}{\zeta^2 - z^2} \, dz - \left[\left(\frac{\pi \cot \pi z}{-2z} \right)_{z=\zeta} + \left(\frac{\pi \cot \pi z}{-2z} \right)_{z=-\zeta} \right]$$

$$= \sum_{n=-\infty}^{\infty} \frac{1}{\zeta^2 - n^2}$$

It was shown in Section 4.2, Example 4.2.5, that $\oint_{C_{NR}} \to 0$ as $N \to \infty$; hence we have the series representation

$$\pi \cot \pi \zeta = \zeta \sum_{n=-\infty}^{\infty} \frac{1}{\zeta^2 - n^2} \tag{4.4.6}$$

The Argument Principle is also frequently used to determine the number of zeroes located in a given region of the plane.

Example 4.4.2 Determine the number of zeroes located inside the first quadrant of the function

$$f(z) = z^3 + 1$$

Consider the contour in Figure 4.4.3.

We begin at $z = 0$ where $f(z) = 1$. We take

$$\arg f(z) = \phi, \qquad \tan \phi = \frac{\operatorname{Im} f(x, y)}{\operatorname{Re} f(x, y)}$$

with the principal branch $\phi = 0$ corresponding to $f(z) = 1$. On the circle $|z| = R$,

$$z = R\,e^{i\theta}, \qquad f(z) = R^3 e^{3i\theta}\left(1 + \frac{1}{R^3 e^{3i\theta}}\right)$$

On the real axis $z = x$, $f(x, y) = x^3 + 1$ is positive and continuous; hence at $x = R$, $y = 0$, $\theta(R, 0) = 0$. Next we follow the arg $f(x, y)$ around the circular contour. For large R, $1 + 1/(R^3 e^{3i\theta})$ is near 1; hence arg $f(z) \approx 3\theta$, and at $z = iR$, $\theta = \pi/2$; hence arg $f(z) \approx 3\pi/2$. Now along the imaginary axis $z = iy$

$$f(x, y) = -iy^3 + 1$$

as y traverses from $y \to \infty$ to $y \to 0^+$, $\tan\phi$ goes from $-\infty$ to 0^- (note $\tan\phi \approx -y^3$ for $y \to \infty$); that is, ϕ goes from $\frac{3\pi}{2}$ to 2π. Note also that we are in the fourth quadrant because Im $f < 0$ and Re $f > 0$.

Thus the Argument Principle gives

$$N = \frac{1}{2\pi}[\arg f]_C = 1$$

because arg f changed by 2π over this circuit. ($P = 0$, because $f(z)$ is a polynomial, hence analytic in C.) Thus we have one zero located in the first quadrant. Because complex roots for a real polynomial must have a complex conjugate root, it follows that the other complex root occurs in the fourth quadrant. Cubic equations with real coefficients have at least one real root. Because it is not on the positive real axis ($x^3 + 1 > 0$ for $x > 0$), it is necessarily on the negative axis.

In this particular example we could have evaluated the roots directly, that is, $z^3 = -1 = e^{i\pi} = e^{i\pi + 2n\pi}$, $n = 0, 1, 2$, or $z_1 = e^{i\pi/3}$, $z_2 = e^{i\pi} = -1$, and $z_3 = e^{5i\pi/3}$. In more complicated examples an explicit calculation of the roots is usually impossible. In such examples the Argument Principle can be a very effective tool. However, our purpose here was only to give the reader the basic ideas of the method as applied to a simple example. As we have seen, one must calculate the number of changes of sign in Im $f(x, y)$ and Re $f(x, y)$ in order to compute arg $f(x, y)$.

A result that is essentially a corollary to the Argument Principle is the following, often termed Rouché's Theorem.

Theorem 4.4.2 (Rouché) Let $f(z)$ and $g(z)$ be analytic on and inside a simple closed contour C. If $|f(z)| > |g(z)|$ on C, then $f(z)$ and $[f(z) + g(z)]$ have the same number of zeroes inside the contour C.

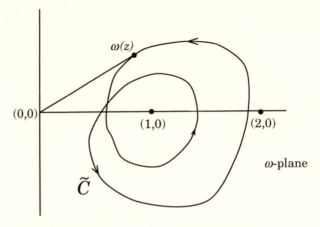

Fig. 4.4.4. Contour $\tilde{C}$, proof of Rouché's Theorem

In Theorem 4.4.2, multiple zeroes are enumerated in the same manner as in the Argument Principle.

Proof Because $|f(z)| > |g(z)| \geq 0$ on C, then $|f(z)| \neq 0$; hence $f(z) \neq 0$ on C. Thus, calling

$$w(z) = \frac{f(z) + g(z)}{f(z)}$$

it follows that the contour integral $\frac{1}{2\pi i} \oint_C \frac{w'(z)}{w(z)} \, dz$ is well defined (no poles on C). Moreover, $(w(z) - 1) = g/f$, whereupon

$$|w(z) - 1| < 1 \tag{4.4.7}$$

and hence all points $w(z)$ in the w plane lie within the circle of unit radius centered at $(1, 0)$. Thus we conclude that the origin $w = 0$ cannot be enclosed by $\tilde{C}$ (see Figure 4.4.4). $\tilde{C}$ is the image curve in the w-plane (if $w = 0$ were enclosed then $|w - 1| = 1$ somewhere on $\tilde{C}$). Hence $[\arg w(z)]_C = 0$ and $N = P$ for $w(z)$. Therefore the number of zeroes of $f(z)$ (poles of $w(z)$) equals the number of zeroes of $f(z) + g(z)$. ∎

Rouché's theorem can be used to prove the fundamental theorem of algebra (also discussed in Section 2.6). Namely, every polynomial

$$P(z) = z^n + a_{n-1}z^{n-1} + a_{n-2}z^{n-2} + \cdots + a_0$$

has n and only n roots counting multiplicities: $P(z_i) = 0, i = 1, 2, \ldots, n$. We call $f(z) = z^n$ and $g(z) = a_{n-1}z^{n-1} + a_{n-2}z^{n-2} + \cdots + a_0$. For $|z| > 1$ we find

$$|g(z)| \le |a_{n-1}||z|^{n-1} + |a_{n-2}||z|^{n-2} + \cdots + |a_0|$$
$$\le (|a_{n-1}| + |a_{n-2}| + \cdots + |a_0|)|z|^{n-1}$$

If our contour C is taken to be a circle with radius R greater than unity, $|f(z)| = R^n > |g(z)|$ whenever

$$R > \max(1, |a_{n-1}| + |a_{n-2}| + \cdots + |a_0|)$$

Hence $P(z) = f(z) + g(z)$ has the same number of roots as $f(z) = z^n = 0$, which is n. Moreover, all of the roots of $P(z)$ are contained inside the circle $|z| < R$ because by the above estimate for R

$$|P(z)| = |z^n + g(z)| \ge R^n - |g(z)| > 0$$

and therefore does not vanish for $|z| \ge R$.

Example 4.4.3 Show that all the roots of $P(z) = z^8 - 4z^3 + 10$ lie between $1 \le |z| \le 2$.

First we consider the circular contour C_1: $|z| = 1$. We take $f(z) = 10$, and $g(z) = z^8 - 4z^3$. Thus $|f(z)| = 10$, and $|g(z)| \le |z|^8 + 4|z|^3 = 5$. Hence $|f| > |g|$, which implies that $P(z)$ has no roots on C_1. Because f has no roots inside C_1, neither does $P(z) = (f + g)(z)$. Next we take $f(z) = z^8$ and $g(z) = -4z^3 + 10$. On the circular contour C_2: $|z| = 2$, $|f(z)| = 2^8 = 256$, and $|g(z)| \le 4|z|^3 + 10 = 42$, so $|f| > |g|$ and thus $P(z)$ has no roots on C_2. Hence the number of roots of $(f + g)(z)$ equals the number of zeroes of $f(z) = z^8$. Thus $z^8 = 0$ implies that there are eight roots inside C_2. Because they cannot be inside or on C_1, they lie in the region $1 < |z| < 2$.

Example 4.4.4 Show that there is exactly one root inside the contour C_1: $|z| = 1$, for

$$h(z) = e^z - 4z - 1$$

We take $f(z) = -4z$ and $g(z) = e^z - 1$ on C_1

$$|f(z)| = |4z| = 4, \quad |g(z)| = |e^z - 1| \le |e^z| + 1 < e + 1 < 4$$

Thus $|f| > |g|$ on C_1, and hence $h(z) = (f + g)(z)$ has the same number of roots as $f(z) = 0$, which is *one*.

Problems for Section 4.4

1. Verify the Argument Principle, Theorem 4.4.1, in the case of the following functions. Take the contour C to be a unit circle centered at the origin.

(a) z^n, n an integer (positive or negative)   (b) e^z (c) $\coth 4\pi z$

(d) $P(z)/Q(z)$, where $P(z)$ and $Q(z)$ are polynomials of degree N and M, respectively, and have all their zeroes inside C.

(e) What happens if we consider $f(z) = e^{1/z}$?

2. Show

$$\frac{1}{2\pi i} \oint_C \frac{f'(z)}{(f(z) - f_0)} \, dz = N$$

where N is the number of points z where $f(z) = f_0$ (a constant) inside C; $f'(z)$ and $f(z)$ are analytic inside and on C; and $f(z) \neq f_0$ on the boundary of C.

3. Use the Argument Principle to show that

(a) $f(z) = z^5 + 1$ has one zero in the first quadrant, and
(b) $f(z) = z^7 + 1$ has two zeroes in the first quadrant.

4. Show that there are no zeroes of $f(z) = z^4 + z^3 + 5z^2 + 2z + 4$ in the first quadrant. Use the fact that on the imaginary axis, $z = iy$, the argument of the function for large y starts with a certain value that corresponds to a quadrant of the argument of $f(z)$. Each change in sign of Re $f(iy)$ and Im $f(iy)$ corresponds to a suitable change of quadrant of the argument.

5. (a) Show that $e^z - (4z^2 + 1) = 0$ has exactly two roots for $|z| < 1$. Hint: in Rouché's Theorem use $f(z) = -4z^2$ and $g(z) = e^z - 1$, so that when C is the unit circle

$$|f(z)| = 4 \quad \text{and} \quad |g(z)| = |e^z - 1| \le e^z + 1$$

(b) Show that the improved estimate $|g(z)| \le e - 1$ can be deduced from $e^z - 1 = \int_0^z e^w \, dw$ and that this allows us to establish that $e^z - (2z+1) = 0$, thus there is exactly one root for $|z| < 1$.

6. Suppose that $f(z)$ is analytic in a region containing a simple contour C. Let $|f(z)| \le M$ on C, and show via Rouché's Theorem that $|f(z)| \le M$ inside C. (The maximum of an analytic function is attained on its boundary; this provides an alternate proof of the maximum modulus result in Section 2.6.) Hint: suppose there is a value of of $f(z)$, say f_0, such that $|f_0| > M$. Consider the two functions $-f_0$ and $f(z) - f_0$, and use $|f(z) - f_0| \ge |f_0| - |f(z)|$ in Rouché's Theorem to deduce that $f(z) \neq f_0$.

7. (a) Consider the mapping $w = z^3$. When we encircle the origin in the z plane one time, how many times do we encircle the origin in the w plane? Explain why this agrees with the Argument Principle.

   (b) Suppose we consider $w = z^3 + a_2 z^2 + a_1 z + a_0$ for three constants a_0, a_1, a_2. If we encircle the origin in the z plane once on a very large circle, how many times do we encircle the w plane?

   (c) Suppose we have a mapping $w = f(z)$ where $f(z)$ is analytic inside and on a simple closed contour C in the z plane. Let us define $\tilde{C}$ as the (nonsimple) image in the w plane of the contour C in the z plane. If we deduce that it encloses the origin ($w = 0$) N times, and encloses the point ($w = 1$) M times, what is the change in arg w over the contour $\tilde{C}$? If we had a computer available, what algorithm should be designed (if it is at all possible) to determine the change in the argument?

*4.5 Fourier and Laplace Transforms

One of the most valuable tools in mathematics, physics, and engineering is making use of the properties a function takes on in a so-called *transform* (or *dual*) space. In suitable function spaces, defined below, the Fourier transform pair is given by the following relations:

$$f(x) = \frac{1}{2\pi} \int_{-\infty}^{\infty} \hat{F}(k) e^{ikx} \, dk \qquad (4.5.1)$$

$$\hat{F}(k) = \int_{-\infty}^{\infty} f(x) e^{-ikx} \, dx \qquad (4.5.2)$$

$\hat{F}(k)$ is called the **Fourier transform** of $f(x)$. The integral in Eq. (4.5.1) is referred to as the **inverse Fourier transform.** In mathematics, the study of Fourier transforms is central in fields like harmonic analysis. In physics and engineering, applications of Fourier transforms are crucial, for example, in the study of quantum mechanics, wave propagation, and signal processing. In this section we introduce the basic notions and give a heuristic derivation of Eqs. (4.5.1)–(4.5.2). In the next section we apply these concepts to solve some of the classical partial differential equations.

In what follows we make some general remarks about the relevant function spaces for $f(x)$ and $\hat{F}(k)$. However, in our calculations we will apply complex variable techniques and will not use any deep knowledge of function spaces. Relations (4.5.1)–(4.5.2) always hold if $f(x) \in L^2$ and $\hat{F}(k) \in L^2$, where $f \in L^2$ refers to the function space of **square integrable functions**

$$\|f\|_2 = \left(\int_{-\infty}^{\infty} |f|^2(x) \, dx \right)^{1/2} < \infty \qquad (4.5.3)$$

In those cases where $f(x)$ and $f'(x)$ are continuous everywhere apart from a finite number of points for which $f(x)$ has integrable and bounded discontinuities (such a function $f(x)$ is said to be **piecewise smooth**), it turns out that at each point of discontinuity, call it x_0, the integral given by Eq. (4.5.1) converges to the mean: $\lim_{\epsilon \to 0^+} \frac{1}{2}[f(x_0 + \epsilon) + f(x_0 - \epsilon)]$. At points where $f(x)$ is continuous, the integral (4.5.1) converges to $f(x)$.

There are other function spaces for which Eqs. (4.5.1)–(4.5.2) hold. If $f(x) \in L^1$, the space of **absolutely integrable functions** (functions f satisfying $\int_{-\infty}^{\infty} |f(x)| \, dx < \infty$), then $\hat{F}(k)$ tends to zero for $|k| \to \infty$ and belongs to a certain function space of functions decaying at infinity (sometimes referred to as **bounded mean oscillations** (BMO)). Conversely, if we start with a suitably decaying function $\hat{F}(k)$ (i.e., in BMO), then $f(x) \in L^1$. In a sense, purely from a symmetry point of view, such function spaces may seem less natural than when $f(x)$, $\hat{F}(k)$ are both in L^2 for Eqs. (4.5.1)–(4.5.2) to be valid. Nevertheless, in some applications $f(x) \in L^1$ and not L^2. Applications sometimes require the use of Fourier transforms in spaces for which no general theory applies, but nevertheless, specific results can be attained. It is outside the scope of this text to examine L^p ($p = 1, 2, \ldots$) function spaces where the general results pertaining to Eqs. (4.5.1)–(4.5.2) can be proven. Interested readers can find such a discussion in various books on complex or Fourier analysis, such as Rudin (1966). Unless otherwise specified, we shall assume that our function $f(x) \in L^1 \cap L^2$, that is, $f(x)$ is both absolutely and square integrable. It follows that

$$|\hat{F}(k)| \le \|f\|_1 = \int_{-\infty}^{\infty} |f(x)| \, dx$$

and

$$\|\hat{F}\|_2 = \|f\|_2$$

The first relationship follows directly. The second is established later in this section. In those cases for which $f(x)$ is piecewise smooth, an elementary though tedious proof of Eqs. (4.5.1)–(4.5.2) can be constructed by suitably breaking up the interval $(-\infty, \infty)$ and using standard results of integration (Titchmarsh, 1948).

Statements analogous to Eqs. (4.5.1)–(4.5.2) hold for functions on finite intervals $(-L, L)$, which may be extended as periodic functions of period $2L$:

$$f(x) = \sum_{n=-\infty}^{\infty} \hat{F}_n \, e^{in\pi x/L} \tag{4.5.4}$$

$$\hat{F}_n = \frac{1}{2L} \int_{-L}^{L} f(x) \, e^{-in\pi x/L} \, dx \tag{4.5.5}$$

The values $\hat{F}_n$ are called the **Fourier coefficients** of the Fourier series representation of the function $f(x)$ given by Eq. (4.5.4).

A simple example of how Fourier transforms may be calculated is afforded by the following.

Example 4.5.1 Let $a > 0$, $b > 0$, and $f(x)$ be given by

$$f(x) = \begin{cases} e^{-ax} & x > 0 \\ e^{bx} & x < 0 \end{cases}$$

Then, from Eq. (4.5.2), we may compute $\hat{F}(k)$ to be

$$\hat{F}(k) = \int_0^\infty e^{-ax-ikx}\,dx + \int_{-\infty}^0 e^{bx-ikx}\,dx$$

$$= \frac{1}{a+ik} + \frac{1}{b-ik} \tag{4.5.6}$$

The inversion (i.e., reconstruction of $f(x)$) via Eq. (4.5.1) is ascertained by calculating the appropriate contour integrals. In this case we use closed semi-circles in the upper ($x > 0$) and lower ($x < 0$) half k-planes. The inversion can be done either by combining the two terms in Eq. (4.5.6) or by noting that

$$\frac{1}{2\pi}\int_{-\infty}^\infty \frac{e^{ikx}}{a+ik}\,dk = \begin{cases} e^{-ax} & x > 0 \\ 1/2 & x = 0 \\ 0 & x < 0 \end{cases}$$

and

$$\frac{1}{2\pi}\int_{-\infty}^\infty \frac{e^{ikx}}{b-ik}\,dk = \begin{cases} 0 & x > 0 \\ 1/2 & x = 0 \\ e^{bx} & x < 0 \end{cases}$$

The values at $x = 0$ take into account both the pole and the principal value contribution at infinity; that is, $\frac{1}{2\pi}\int_{C_R}\frac{dk}{a+ik} = 1/2$ where on $C_R : k = R\,e^{i\theta}$, $0 \le \theta \le \pi$. Thus Eq. (4.5.1) gives convergence to the mean value at $x = 0$: $\frac{1}{2} + \frac{1}{2} = 1 = (\lim_{x\to 0^+} f(x) + \lim_{x\to 0^-} f(x))/2$.

A function that will lead us to a useful result is the following:

$$\Delta(x; \epsilon) = \begin{cases} \dfrac{1}{2\epsilon} & |x| < \epsilon \\ 0 & |x| > \epsilon \end{cases} \tag{4.5.7}$$

Its Fourier transform is given by

$$\hat{\Delta}(k; \epsilon) = \frac{1}{2\epsilon}\int_{-\epsilon}^\epsilon e^{-ikx}\,dx = \frac{\sin k\epsilon}{k\epsilon} \tag{4.5.8}$$

Certainly, $\Delta(x; \epsilon)$ is both absolutely and square integrable; so it is in $L^1 \cap L^2$, and $\hat{\Delta}(k; \epsilon)$ is in L^2. It is natural to ask what happens as $\epsilon \to 0$. The function defined by Eq. (4.5.7) tends, as $\epsilon \to 0$, to a novel "function" called the **Dirac delta function**, denoted by $\delta(x)$ and having the following properties:

$$\delta(x) = \lim_{\epsilon \to 0} \Delta(x; \epsilon) \qquad (4.5.9)$$

$$\int_{-\infty}^{\infty} \delta(x - x_0)dx = \lim_{\epsilon \to 0} \int_{x=x_0-\epsilon}^{x=x_0+\epsilon} \delta(x - x_0)dx = 1 \qquad (4.5.10)$$

$$\int_{-\infty}^{\infty} \delta(x - x_0) f(x)\, dx = f(x_0) \qquad (4.5.11)$$

where $f(x)$ is continuous. Equations (4.5.10)–(4.5.11) can be ascertained by using the limit definitions (4.5.7) and (4.5.9). The function defined in Eq. (4.5.9) is often called a unit impulse "function"; it has an arbitrarily large value concentrated at the origin, whose integral is unity. The delta function, $\delta(x)$, is not a mathematical function in the conventional sense, as it has an arbitrarily large value at the origin. Nevertheless, there is a rigorous mathematical framework in which these new functions – called distributions – can be analyzed. Interested readers can find such a discussion in, for example, Lighthill (1959). For our purposes the device of the limit process $\epsilon \to 0$ is sufficient. We also note that Eq. (4.5.7) is not the only valid representation of a delta function; for example, others are given by

$$\delta(x) = \lim_{\epsilon \to 0} \left\{ \frac{1}{\sqrt{\pi \epsilon}} e^{-x^2/\epsilon} \right\} \qquad (4.5.12a)$$

$$\delta(x) = \lim_{\epsilon \to 0} \left\{ \frac{\epsilon}{\pi(\epsilon^2 + x^2)} \right\} \qquad (4.5.12b)$$

It should be noted that, formally speaking, the Fourier transform of a delta function is given by

$$\hat{\delta}(k) = \int_{-\infty}^{\infty} \delta(x) e^{-ikx}\, dx = 1 \qquad (4.5.13)$$

It does not vanish as $|k| \to \infty$; indeed, it is a constant (unity; note that $\delta(x)$ is not L^1). Similarly, it turns out from the theory of distributions (motivated by the inverse Fourier transform), that the following alternative definition of a delta function holds:

$$\delta(x) = \frac{1}{2\pi} \int_{-\infty}^{\infty} \hat{\delta}(k) e^{ikx}\, dk$$

$$= \frac{1}{2\pi} \int_{-\infty}^{\infty} e^{ikx}\, dk \qquad (4.5.14)$$

Formula (4.5.14) allows us a simple (but formal) way to verify Eqs. (4.5.1)–(4.5.2). Namely, by using Eqs. (4.5.1)–(4.5.2) and assuming that interchanging integrals is valid, we have

$$f(x) = \frac{1}{2\pi} \int_{-\infty}^{\infty} dk \, e^{ikx} \left(\int_{-\infty}^{\infty} dx' \, f(x') \, e^{-ikx'} \right)$$

$$= \int_{-\infty}^{\infty} dx' \, f(x') \left(\frac{1}{2\pi} \int_{-\infty}^{\infty} dk \, e^{ik(x-x')} \right)$$

$$= \int_{-\infty}^{\infty} dx' \, f(x') \delta(x - x') \tag{4.5.15}$$

In what follows, the Fourier transform of derivatives will be needed. The Fourier transform of a derivative is readily obtained via integration by parts.

$$\hat{F}_1(k) \equiv \int_{-\infty}^{\infty} f'(x) \, e^{-ikx} \, dx$$

$$= \left[f(x) \, e^{-ikx} \right]_{-\infty}^{\infty} + ik \int_{-\infty}^{\infty} f(x) \, e^{-ikx} \, dx$$

$$= ik \, \hat{F}(k) \tag{4.5.16a}$$

and by repeated integration by parts

$$\hat{F}_n(k) = \int_{-\infty}^{\infty} f^{(n)}(x) e^{-ikx} \, dx = (ik)^n \hat{F}(k) \tag{4.5.16b}$$

Formulae (4.5.16a,b) will be useful when we examine solutions of differential equations by transform methods.

It is natural to ask what is the Fourier transform of a product. The result is called the **convolution product**; it is *not* the product of the Fourier transforms. We can readily derive this formula. We use two transform pairs: one for a function $f(x)$ (Eqs. (4.5.1)–(4.5.2)) and another for a function $g(x)$, replacing $f(x)$ and $\hat{F}(k)$ in Eqs. (4.5.1)–(4.5.2) by $g(x)$ and $\hat{G}(k)$, respectively. We define the convolution product,

$$(f * g)(x) = \int_{-\infty}^{\infty} f(x - x')g(x') \, dx' = \int_{-\infty}^{\infty} f(x')g(x - x') \, dx.$$

(The latter equality follows by renaming variables.) We take the Fourier transform of $(f * g)(x)$ and interchange integrals (allowed since both f and g are absolutely integrable) to find

$$\int_{-\infty}^{\infty} (f * g)(x) e^{-ikx} \, dx = \int_{-\infty}^{\infty} dx \left(\int_{-\infty}^{\infty} f(x - x') g(x') dx' \right) e^{-ikx}$$

$$= \int_{-\infty}^{\infty} dx' e^{ikx'} g(x') \int_{-\infty}^{\infty} dx e^{-ik(x-x')} f(x - x')$$

$$= \hat{F}(k) \hat{G}(k). \qquad (4.5.17)$$

Hence, by taking the inverse transform of this result,

$$\frac{1}{2\pi i} \int_{-\infty}^{\infty} e^{ikx} \hat{F}(k) \hat{G}(k) \, dk = \int_{-\infty}^{\infty} g(x') f(x - x') \, dx'$$

$$= \int_{-\infty}^{\infty} g(x - x') f(x') dx'. \qquad (4.5.18)$$

The latter equality is accomplished by renaming the integration variables. Note that if $f(x) = \delta(x - x')$, then $\hat{F}(k) = 1$, and Eq. (4.5.18) reduces to the known transform pair for $g(x)$, that is, $g(x) = \frac{1}{2\pi} \int_{-\infty}^{\infty} \hat{G}(k) e^{ikx} \, dk$.

A special case of Eq. (4.5.18) is the so-called **Parseval formula**, obtained by taking $g(x) = \overline{f}(-x)$, where $\overline{f}(x)$ is the complex conjugate of $f(x)$, and evaluating Eq. (4.5.17) at $x = 0$:

$$\int_{-\infty}^{\infty} f(x) \overline{f}(x) \, dx = \frac{1}{2\pi} \int_{-\infty}^{\infty} \hat{F}(k) \hat{G}(k) \, dk.$$

Function $\hat{G}(k)$ is now the Fourier transform of $\overline{f}(-x)$:

$$\hat{G}(k) = \int_{-\infty}^{\infty} e^{-ikx} \overline{f}(-x) \, dx = \int_{-\infty}^{\infty} e^{ikx} \overline{f}(x) \, dx$$

$$= \overline{\left(\int_{-\infty}^{\infty} e^{-ikx} f(x) \, dx \right)} = \overline{(\hat{F}(k))}$$

Hence we have the Parseval formula:

$$\int_{-\infty}^{\infty} |f|^2(x) \, dx = \frac{1}{2\pi} \int_{-\infty}^{\infty} |\hat{F}(k)|^2 \, dk \qquad (4.5.19)$$

In some applications, $\int_{-\infty}^{\infty} |f|^2(x) \, dx$ refers to the energy of a signal. Frequently it is the term $\int_{-\infty}^{\infty} |\hat{F}(k)|^2 \, dk$ which is really measured (it is sometimes referred to as the **power spectrum**), which then gives the energy as per Eq. (4.5.19).

In those cases where we have $f(x)$ being an even or odd function, then the Fourier transform pair reduces to the so-called **cosine transform** or **sine transform** pair. Because $f(x)$ being even/odd in x means that $\hat{F}(k)$ will be even or odd in k, the pair (4.5.1)–(4.5.2) reduces to statements about semiinfinite functions in the space L^2 on $(0, \infty)$ (i.e., $\int_0^\infty |f(x)|^2 dx < \infty$) or in the space L^1 on $(0, \infty)$ (i.e., $\int_0^\infty |f(x)| dx < \infty$). For even functions, $f(x) = f(-x)$, the following definitions

$$f(x) = \frac{1}{\sqrt{2}} f_c(x), \qquad \hat{F}(k) = \sqrt{2}\hat{F}_c(k) \qquad (4.5.20)$$

(or more generally $\hat{F}(k) = a\hat{F}_c(k)$, $f(x) = bf_c(x)$, $a = 2b$, $b \neq 0$) yield the Fourier cosine transform pair

$$f_c(x) = \frac{2}{\pi} \int_0^\infty \hat{F}_c(k) \cos kx\, dk \qquad (4.5.21)$$

$$\hat{F}_c(k) = \int_0^\infty f_c(x) \cos kx\, dx \qquad (4.5.22)$$

For odd functions, $f(x) = -f(-x)$, the definitions

$$f(x) = \frac{1}{\sqrt{2}} f_s(x), \qquad \hat{F}(k) = -\sqrt{2}i\,\hat{F}_s(k)$$

yield the Fourier sine transform pair

$$f_s(x) = \frac{2}{\pi} \int_0^\infty \hat{F}_s(k) \sin kx\, dk \qquad (4.5.23)$$

$$\hat{F}_s(k) = \int_0^\infty f_s(x) \sin kx\, dx \qquad (4.5.24)$$

Obtaining the Fourier sine or cosine transform of a derivative employs integration by parts; for example

$$\hat{F}_{c,1}(k) = \int_0^\infty f'(x) \cos kx\, dk$$

$$= [f(x) \cos kx]_0^\infty + k \int_0^\infty f(x) \sin kx\, dk = k\hat{F}_s(k) - f(0) \quad (4.5.25)$$

and

$$\hat{F}_{s,1}(k) = \int_0^\infty f'(x) \sin kx\, dk$$

$$= [f(x) \sin kx]_0^\infty - k \int_0^\infty f(x) \cos kx\, dk = -k\hat{F}_c(k) \quad (4.5.26)$$

are formulae for the first derivative. Similar results obtain for higher derivatives.

It turns out to be useful to extend the notion of Fourier transforms. One way to do this is to consider functions that have support only on a semi-interval. We take $f(x) = 0$ on $x < 0$ and replace $f(x)$ by $e^{-cx} f(x) (c > 0)$ for $x > 0$. Then Eqs. (4.5.1)–(4.5.2) satisfy, using $\hat{F}(k)$ from Eq. (4.5.2) in Eq. (4.5.1):

$$e^{-cx} f(x) = \frac{1}{2\pi} \int_{-\infty}^{\infty} dk \, e^{ikx} \left[\int_{0}^{\infty} e^{-ikx'} e^{-cx'} f(x') \, dx' \right]$$

hence

$$f(x) = \frac{1}{2\pi} \int_{-\infty}^{\infty} dk \, e^{(c+ik)x} \left[\int_{0}^{\infty} e^{-(c+ik)x'} f(x') \, dx' \right]$$

Within the above integrals we define $s = c + ik$, where c is a fixed real constant, and make the indicated redefinition of the limits of integration to obtain

$$f(x) = \frac{1}{2\pi i} \int_{c-i\infty}^{c+i\infty} ds \, e^{sx} \left(\int_{0}^{\infty} e^{-sx'} f(x') \, dx' \right)$$

or, in a form analogous to Eqs. (4.5.1)–(4.5.2):

$$f(x) = \frac{1}{2\pi i} \int_{c-i\infty}^{c+i\infty} \hat{F}(s) e^{sx} \, ds \tag{4.5.27}$$

$$\hat{F}(s) = \int_{0}^{\infty} f(x) e^{-sx} \, dx \tag{4.5.28}$$

Formulae (4.5.27)–(4.5.28) are referred to as the **Laplace transform** ($\hat{F}(s)$) and the inverse Laplace transform of a function ($f(x)$), respectively. The usual function space for $f(x)$ in the Laplace transform (analogous to $L^1 \cap L^2$ for $f(x)$ in Eq. (4.5.1)) are those functions satisfying:

$$\int_{0}^{\infty} e^{-cx} |f(x)| \, dx < \infty. \tag{4.5.29}$$

Note $\text{Re } s = c$ in Eqs. (4.5.27)–(4.5.28). If Eq. (4.5.29) holds for some $c > 0$; if so, then $f(x)$ is said to be **of exponential order**.

The integral (4.5.27) is generally carried out by contour integration. The contour from $c - i\infty$ to $c + i\infty$ is referred to as the **Bromwich contour**, and c is taken to the right of all singularities in order to insure (4.5.29). Closing the contour to the right will yield $f(x) = 0$ for $x < 0$.

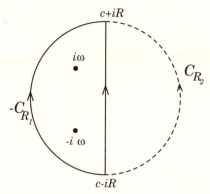

Fig. 4.5.1. Bromwich contour, Example 4.5.2

We give two examples.

Example 4.5.2 Evaluate the inverse Laplace transform of $\hat{F}(s) = \frac{1}{s^2+w^2}$, with w a real constant. See Figure 4.5.1.

$$f(x) = \frac{1}{2\pi i} \int_{c-i\infty}^{c+i\infty} \frac{e^{sx}}{s^2 + w^2} \, ds$$

For $x < 0$ we close the contour to the right of the Bromwich contour. Because no singularities are enclosed, we have, on C_{R_2}

$$s = c + Re^{i\theta}, \quad \frac{-\pi}{2} < \theta < \frac{\pi}{2}$$

thus $f(x) = 0, x < 0$ because $\int_{C_{R_2}} \to 0$ as $R \to \infty$. On the other hand, for $x > 0$, closing to the left yields (we note that $\int_{C_{R_1}} \to 0$ as $R \to \infty$ where on C_{R_1}: $s = c + Re^{i\theta}, \frac{\pi}{2} \le \theta \le \frac{3\pi}{2}$)

$$f(x) = \sum_{j=1}^{2} \text{Res}\left(\frac{e^{sx}}{s^2 + w^2}; s_j\right), \qquad s_1 = iw, \quad s_2 = -iw$$

$$f(x) = \frac{e^{iwx}}{2iw} - \frac{e^{-iwx}}{2iw} = \frac{\sin wx}{w} \qquad (x > 0)$$

Example 4.5.3 Evaluate the inverse Laplace transform of the function

$$\hat{F}(s) = s^{-a}, \qquad (0 < a < 1)$$

where we take the branch cut along the negative real axis (see Figure 4.5.2).

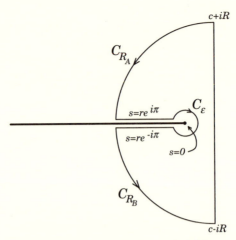

Fig. 4.5.2. Contour for Example 4.5.3

As is always the case, $f(x) = 0$ for $x < 0$ when we close to the right. Closing to the left yields (schematically)

$$f(x) + \left(\int_{C_{R_A}} + \int_{C_{R_B}} + \int_{s=re^{i\pi}} + \int_{s=re^{-i\pi}} + \int_{C_\epsilon} \right) \frac{\hat{F}(s)}{2\pi i} e^{sx} \, ds = 0$$

The right-hand side vanishes because there are no singularities enclosed by this contour. The integrals $\int_{C_{R_A}}, \int_{C_{R_B}}$, and $\int_{C_\epsilon}$ vanish as $R \to \infty$ (on C_{R_A} : $s = c + Re^{i\theta}, \frac{\pi}{2} < \theta < \pi$; on C_{R_B} : $s = c + Re^{i\theta}, -\pi < \theta < \frac{-\pi}{2}$), and $\epsilon \to 0$ (on C_ϵ : $s = \epsilon e^{i\theta}, -\pi < \theta < \pi$). On C_ϵ we have

$$\left| \int_{C_\epsilon} \frac{F(s)}{2\pi i} e^{sx} \, ds \right| \le \int_{-\pi}^{\pi} \epsilon^{1-a} \frac{e^{\epsilon x}}{2\pi} \, d\theta \xrightarrow[\epsilon \to 0]{} 0$$

Hence

$$f(x) - \frac{1}{2\pi i} \int_{r=\infty}^{0} r^{-a} e^{-ia\pi} e^{-rx} \, dr - \frac{1}{2\pi i} \int_{r=0}^{\infty} r^{-a} e^{ia\pi} e^{-rx} \, dr = 0$$

or

$$f(x) = \frac{(e^{ia\pi} - e^{-ia\pi})}{2\pi i} \int_0^\infty r^{-a} e^{-rx} \, dr \qquad (4.5.30)$$

The **gamma function**, or factorial function, is defined by

$$\Gamma(z) = \int_0^\infty u^{z-1} e^{-u} \, du \qquad (4.5.31)$$

This definition implies that $\Gamma(n) = (n-1)!$ when n is a positive integer: $n = 1, 2, \ldots$. Indeed, we have by integration by parts

$$\Gamma(n+1) = [-u^n e^{-u}]_0^\infty + n \int_0^\infty u^{n-1} e^{-u} \, du$$
$$= n\Gamma(n) \tag{4.5.32}$$

and when $z = 1$, Eq. (4.5.31) directly yields $\Gamma(1) = 1$. Equation (4.5.32) is a difference equation, which, when supplemented with the starting condition $\Gamma(1) = 1$, can be solved for all n. So when $n = 1$, Eq. (4.5.32) yields $\Gamma(2) = \Gamma(1) = 1$; when $n = 2$, $\Gamma(3) = 2\Gamma(2) = 2!$, ..., and by induction, $\Gamma(n) = (n-1)!$ for positive integer n. We often use Eq. (4.5.31) for general values of z, requiring only that Re $z > 0$ in order for there to be an integrable singularity at $u = 0$. With the definition (4.5.31) and rescaling $rx = u$ we have

$$f(x) = \left(\frac{\sin a\pi}{\pi}\right) x^{a-1} \Gamma(1-a)$$

The Laplace transform of a derivative is readily calculated.

$$\hat{F}_1(s) = \int_0^\infty f'(x) e^{-sx} \, dx$$
$$= [f(x)e^{-sx}]_0^\infty + s \int_0^\infty f(x)e^{-sx} \, dx$$
$$= s\hat{F}(s) - f(0)$$

By integration by parts n times it is found that

$$\hat{F}_n(s) = \int_0^\infty f^{(n)}(x)e^{-sx} dx$$
$$= s^n \hat{F}(s) - f^{(n-1)}(0) - sf^{(n-2)}(0) - \cdots$$
$$- s^{n-2} f'(0) - s^{n-1} f(0) \tag{4.5.33}$$

The Laplace transform analog of the convolution result for Fourier transforms (4.5.17) takes the following form. Define

$$h(x) = \int_0^x g(x')f(x-x') \, dx' \tag{4.5.34}$$

We show that the Laplace transform of h is the product of the Laplace transform of g and f.

$$\hat{H}(s) = \int_0^\infty h(x)\,e^{-sx}\,dx$$

$$= \int_0^\infty dx\,e^{-sx} \int_0^x g(x')f(x-x')\,dx'$$

By interchanging integrals, we find that

$$\hat{H}(s) = \int_0^\infty dx'\,g(x') \int_{x'}^\infty dx\,e^{-sx}\,f(x-x')$$

$$= \int_0^\infty dx'\,g(x')\,e^{-sx'} \int_{x'}^\infty dx\,e^{-s(x-x')}\,f(x-x')$$

$$= \int_0^\infty dx'\,g(x')\,e^{-sx'} \int_0^\infty du\,e^{-su}\,f(u)$$

hence

$$\hat{H}(s) = \hat{G}(s)\hat{F}(s) \tag{4.5.35}$$

The convolution formulae (4.5.34–4.5.35) can be used in a variety of ways. We note that if $\hat{F}(s) = 1/s$, then Eq. (4.5.27) implies $f(x) = 1$. We use this in the following example.

Example 4.5.4 Evaluate $h(x)$ where the Laplace transform of $h(x)$ is given by

$$\hat{H}(s) = \frac{1}{s(s^2+1)}$$

We have two functions:

$$\hat{F}(s) = \frac{1}{s}, \qquad \hat{G}(s) = \frac{1}{s^2+1}$$

Using the result of Example 4.5.2 shows that $g(x) = \sin x$; hence Eq. (4.5.35) implies that

$$h(x) = \int_0^x \sin x'\,dx' = 1 - \cos x$$

This may also be found by the partial fractions decomposition

$$\hat{H}(s) = \frac{1}{s} - \frac{s}{s^2+1}$$

and noting that $f(x) = \frac{s}{s^2+w^2}$ has, as its Laplace transform, $\hat{F}(s) = \cos wx$.

Problems for Section 4.5

1. Obtain the Fourier transforms of the following functions:

    (a) $e^{-|x|}$ (b) $\dfrac{1}{x^2 + a^2}$, $a^2 > 0$

    (c) $\dfrac{1}{(x^2 + a^2)^2}$, $a^2 > 0$ (d) $\dfrac{\sin ax}{(x + b)^2 + c^2}$

2. Obtain the inverse Fourier transform of the following functions:

    (a) $\dfrac{1}{k^2 + w^2}$, $w^2 > 0$ (b) $\dfrac{1}{(k^2 + w^2)^2}$, $w^2 > 0$

    Note the duality between direct and inverse Fourier transforms.

3. Show that the Fourier transform of the "Gaussian" $f(x) = \exp(-(x - x_0)^2/a^2)$ is also a Gaussian:

    $$\hat{F}(k) = a\sqrt{\pi}e^{-(ka/2)^2}$$

4. Obtain the Fourier transform of the following functions, and thereby show that the Fourier transforms of hyperbolic secant functions are also related to hyperbolic secant functions.

    (a) $\operatorname{sech} a(x - x_0)e^{i\omega x}$ (b) $\operatorname{sech}^2 a(x - x_0)$

5. (a) Obtain the Fourier transform of $f(x) = (\sin \omega x)/(x)$, $\omega > 0$.
    (b) Show that $f(x)$ in part (a) is not L_1, that is, $\int_{-\infty}^{\infty} |f(x)| dx$ does not exist. Despite this fact, we can obtain the Fourier transform, so $f(x) \in L_1$ is a sufficient condition, but is not necessary, for the Fourier transform to exist.

6. Suppose we are given the differential equation

    $$\frac{d^2 u}{dx^2} - \omega^2 u = -f(x)$$

    with $u(x = \pm\infty) = 0$, $\omega > 0$.

    (a) Take the Fourier transform of this equation to find (using Eq. (4.5.16))

    $$\hat{U}(k) = \hat{F}(k)/(k^2 + \omega^2)$$

    where $\hat{U}(k)$ and $\hat{F}(k)$ are the Fourier transform of $u(x)$ and $f(x)$, respectively.

(b) Use the convolution product (4.5.17) to deduce that

$$u(x) = \frac{1}{2\omega} \int_{-\infty}^{\infty} e^{-\omega|x-\zeta|} f(\zeta) \, d\zeta$$

and thereby obtain the solution of the differential equation.

7. Obtain the Fourier sine transform of the following functions:

(a) $e^{\omega x}$, $\omega > 0$ (b) $\dfrac{x}{x^2 + 1}$ (c) $\dfrac{\sin \omega x}{x^2 + 1}$, $\omega > 0$

8. Obtain the Fourier cosine transform of the following functions:

(a) $e^{-\omega x}$, $\omega > 0$ (b) $\dfrac{1}{x^2 + 1}$ (c) $\dfrac{\cos \omega x}{x^2 + 1}$, $\omega > 0$

9. (a) Assume that $u(\infty) = 0$ to establish that

$$\int_0^\infty \frac{d^2 u}{dx^2} \sin kx \, dx = k \, u(0) - k^2 \hat{U}_s(k)$$

where $\hat{U}_s(k) = \int_0^\infty u(x) \sin kx \, dx$ is the sine transform of $u(x)$ (Eq. (4.5.23)).

(b) Use this result to show that by taking the Fourier sine transform of

$$\frac{d^2 u}{dx^2} - \omega^2 u = -f(x)$$

with $u(0) = u_0$, $u(\infty) = 0$, $\omega > 0$, yields, for the Fourier sine transform of $u(x)$

$$\hat{U}_s(k) = \frac{u_0 k + \hat{F}_s(k)}{k^2 + \omega^2}$$

where $\hat{F}_s(k)$ is the Fourier sine transform of $f(x)$.

(c) Use the analog of the convolution product for the Fourier sine transform

$$\frac{1}{2} \int_0^\infty [f(|x - \zeta|) - f(x + \zeta)] g(\zeta) \, d\zeta$$

$$= \frac{2}{\pi} \int_0^\infty \sin kx \, \hat{F}_c(k) \hat{G}_s(k) \, dk,$$

where $\hat{F}_c(k)$ is the Fourier cosine transform of $f(x)$ and $\hat{G}_s(k)$ is the Fourier sine transform of $g(x)$, to show that the solution of the differential equation is given by

$$u(x) = u_0 e^{-\omega x} + \frac{1}{2\omega} \int_0^\infty (e^{-\omega|x-\zeta|} - e^{-\omega(x+\zeta)}) f(\zeta) \, d\zeta$$

(The convolution product for the sine transform can be deduced from the usual convolution product (4.5.17) by assuming in the latter formula that $f(x)$ is even and $g(x)$ is an odd function of x.)

10. (a) Assume that $u(\infty) = 0$ to establish that

$$\int_0^\infty \frac{d^2 u}{dx^2} \cos kx \, dx = -\frac{du}{dx}(0) - k^2 \hat{U}_c(k)$$

where $\hat{U}_c(k) = \int_0^\infty u(x) \cos kx \, dx$ is the cosine transform of $u(x)$.

(b) Use this result to show that taking the Fourier cosine transform of

$$\frac{d^2 u}{dx^2} - \omega^2 u = -f(x), \quad \text{with} \quad \frac{du}{dx}(0) = u_0', \quad u(\infty) = 0, \quad \omega > 0$$

yields, for the Fourier cosine transform of $u(x)$

$$\hat{U}_c(k) = \frac{\hat{F}_c(k) - u_0'}{k^2 + \omega^2}$$

where $\hat{F}_c(k)$ is the Fourier cosine transform of $f(x)$.

(c) Use the analog of the convolution product of the Fourier cosine transform

$$\frac{1}{2} \int_0^\infty (f(|x - \zeta|) + f(x + \zeta)) g(\zeta) \, d\zeta$$

$$= \frac{2}{\pi} \int_0^\infty \cos kx \, \hat{F}_c(k) \hat{G}_c(k) \, dk$$

to show that the solution of the differential equation is given by

$$u(x) = -\frac{u_0'}{\omega} e^{-\omega x} + \frac{1}{2\omega} \int_0^\infty \left(e^{-\omega|x-\zeta|} + e^{-\omega|x+\zeta|} \right) f(\zeta) \, d\zeta$$

(The convolution product for the cosine transform can be deduced from the usual convolution product (4.5.17) by assuming in the latter formula that $f(x)$ and $g(x)$ are even functions of x.)

11. Obtain the inverse Laplace transforms of the following functions, assuming $\omega, \omega_1, \omega_2 > 0$.

$$\text{(a)} \quad \frac{s}{s^2 + \omega^2} \qquad \text{(b)} \quad \frac{1}{(s + \omega)^2} \qquad \text{(c)} \quad \frac{1}{(s + \omega)^n}$$

$$\text{(d)} \quad \frac{s}{(s + \omega)^n} \qquad \text{(e)} \quad \frac{1}{(s + \omega_1)(s + \omega_2)} \qquad \text{(f)} \quad \frac{1}{s^2(s^2 + \omega^2)}$$

$$\text{(g)} \quad \frac{1}{(s + \omega_1)^2 + \omega_2^2} \qquad \text{(h)} \quad \frac{1}{(s^2 - \omega^2)^2}$$

12. Show explicitly that the Laplace transform of the second derivative of a function of x satisfies

$$\int_0^\infty f''(x)e^{-sx}\,dx = s^2 \hat{F}(s) - sf(0) - f'(0)$$

13. Establish the following relationships, where we use the notation $\mathcal{L}(f(x)) \equiv \hat{F}(s)$:

(a) $$\mathcal{L}(e^{ax} f(x)) = \hat{F}(s - a) \qquad a > 0$$

(b) $$\mathcal{L}(f(x - a)H(x - a)) = e^{-as}\hat{F}(s) \qquad a > 0,$$

where

$$H(x) = \begin{cases} 1 & x \geq 0 \\ 0 & x < 0 \end{cases}$$

(c) Use the convolution product formula for Laplace transforms to show that the inverse Laplace transform of $\hat{H}(s) = 1/(s^2(s^2 + \omega^2))$ satisfies

$$h(x) = \frac{1}{\omega} \int_0^x x' \sin\omega(x - x')\,dx'$$

$$= \frac{x}{\omega^2} - \frac{\sin\omega x}{\omega^3}$$

Verify this result by using partial fractions.

14. (a) Show that the inverse Laplace transform of $\hat{F}(s) = e^{-as^{1/2}}/s, a > 0$, is given by

$$f(x) = 1 - \frac{1}{\pi} \int_0^\infty \frac{\sin(ar^{1/2})}{r} e^{-rx}\,dr$$

Note that the integral converges at $r = 0$.

(b) Use the definition of the error function integral

$$\mathrm{erf}\, x = \frac{2}{\sqrt{\pi}} \int_0^x e^{-r^2}\, dr$$

to show that an alternative form for $f(x)$ is

$$f(x) = 1 - \mathrm{erf}\left(\frac{a}{2\sqrt{x}}\right)$$

(c) Show that the inverse Laplace transform of $\hat{F}(s) = e^{-as^{1/2}}/s^{1/2}, a > 0$, is given by

$$f(x) = \frac{1}{\pi} \int_0^\infty \frac{\cos(ar^{1/2})}{r^{1/2}} e^{-rx}\, dr$$

or the equivalent forms

$$f(x) = \frac{2}{\pi\sqrt{x}} \int_0^\infty \cos\frac{au}{\sqrt{x}} e^{-u^2}\, du = \frac{1}{\sqrt{\pi x}} e^{-a^2/4x}$$

Verify this result by taking the derivative with respect to a in the formula of part (a).

(d) Follow the procedure of part (c) and show that the inverse Laplace transform of $\hat{F}(s) = e^{-as^{1/2}}$ is given by

$$f(x) = \frac{a}{2\sqrt{\pi}x^{3/2}} e^{-a^2/4x}$$

15. Show that the inverse Laplace transform of the function

$$\hat{F}(s) = \frac{1}{\sqrt{s^2 + \omega^2}}$$

is given by

$$f(x) = \frac{1}{\pi} \int_{-\omega}^{\omega} \frac{e^{ixr}}{\sqrt{\omega^2 - r^2}}\, dr = \frac{2}{\pi} \int_0^1 \frac{\cos(\omega x \rho)}{\sqrt{1 - \rho^2}}\, d\rho$$

(The latter integral is a representation of $J_0(\omega x)$, the Bessel function of order zero.) Hint: Deform the contour around the branch points $s = \pm i\omega$, then show that the large contour at infinity and small contours encircling

$\pm i\omega$ are vanishingly small. It is convenient to use the polar representations $s+i\omega_1 = r_1 e^{i\theta_1}$ and $s-i\omega_2 = r_2 e^{i\theta_2}$, where $-3\pi/2 < \theta_i \le \pi/2, i = 1, 2$, and $\left(s^2 + \omega^2\right)^{1/2} = \sqrt{r_1 r_2}\, e^{i(\theta_1+\theta_2)/2}$. The contributions on both sides of the cut add to give the result.

16. Show that the inverse Laplace transform of the function $\hat{F}(s) = (\log s)/(s^2 + \omega^2)$ is given by

$$f(x) = \frac{\pi}{2\omega} \cos \omega x - \int_0^\infty \frac{e^{-rx}}{r^2 + \omega^2} dr \qquad \text{for } x > 0$$

Hint: Choose the branch $s = re^{i\theta}, -\pi \le \theta < \pi$. Show that the contour at infinity and around the branch point $s = 0$ are vanishingly small. There are two contributions along the branch cut that add to give the second (integral) term; the first is due to the poles at $s = \pm i\omega$.

17. (a) Show that the inverse Laplace transform of $\hat{F}_1(s) = \log s$ is given by $f_1(x) = -1/x$.
    (b) Do the same for $\hat{F}_2(s) = \log(s + 1)$ to obtain $f_2(x) = -e^{-x}/x$.
    (c) Find the inverse Laplace transform $\hat{F}(s) = \log((s + 1)/s)$ to obtain $f(x) = (1 - e^{-x})/x$, by subtracting the results of parts (a) and (b).
    (d) Show that we can get this result directly, by encircling both the $s = 0$ and $s = -1$ branch points and using the polar representations $s+1 = r_1 e^{i\theta_1}, s = r_2 e^{i\theta_2}, -\pi \le \theta_i < \pi, i = 1, 2$.

18. Establish the following results by formally inverting the Laplace transform.

$$\hat{F}(s) = \frac{1}{s} \frac{1 - e^{-\ell s}}{1 + e^{-\ell s}}, \qquad \ell > 0,$$

$$f(x) = \sum_{n=1,3,5,\dots} \left(\frac{4}{n\pi}\right) \sin \frac{n\pi x}{\ell}$$

Note that there are an infinite number of poles present in $\hat{F}(s)$; consequently, a straightforward continuous limit as $R \to \infty$ on a large semicircle will pass through one of these poles. Consider a large semicircle C_{R_N}, where R_N encloses N poles (e.g. $R_N = \frac{\pi i}{\ell}(N + \frac{1}{2})$) and show that as $N \to \infty$, $R_N \to \infty$, and the integral along C_{R_N} will vanish. Choosing appropriate sequences such as in this example, the inverse Laplace transform containing an infinite number of poles can be calculated.

19. Establish the following result by formally inverting the Laplace transform

$$\hat{F}(s) = \frac{1}{s} \frac{\sinh(sy)}{\sinh(s\ell)} \quad \ell > 0$$

$$f(x) = \frac{y}{\ell} + \sum_{n=1}^{\infty} \frac{2(-1)^n}{n\pi} \sin\left(\frac{n\pi y}{\ell}\right) \cos\left(\frac{n\pi x}{\ell}\right)$$

See the remark at the end of Problem 18, which explains how to show how the inverse Laplace transform can be proven to be valid in a situation such as this where there are an infinite number of poles.

*4.6 Applications of Transforms to Differential Equations

A particularly valuable technique available to solve differential equations in infinite and semiinfinite domains is the use of Fourier and Laplace transforms. In this section we describe some typical examples. The discussion is not intended to be complete. The aim of this section is to elucidate the transform technique, not to detail theoretical aspects regarding differential equations. The reader only needs basic training in the calculus of several variables to be able to follow the analysis. We shall use various classical partial differential equations (PDEs) as vehicles to illustrate methodology. Herein we will consider well-posed problems that will yield unique solutions. More general PDEs and the notion of well-posedness are investigated in considerable detail in courses on PDEs.

Example 4.6.1 Steady state heat flow in a semiinfinite domain obeys Laplace's equation. Solve for the bounded solution of Laplace's equation

$$\frac{\partial^2 \phi(x, y)}{\partial x^2} + \frac{\partial^2 \phi(x, y)}{\partial y^2} = 0 \tag{4.6.1}$$

in the region $-\infty < x < \infty$, $y > 0$, where on $y = 0$ we are given $\phi(x, 0) = h(x)$ with $h(x) \in L^1 \cap L^2$ (i.e., $\int_{-\infty}^{\infty} |h(x)| \, dx < \infty$ and $\int_{-\infty}^{\infty} |h(x)|^2 \, dx < \infty$).

This example will allow us to solve Laplace's Equation (4.6.1) by Fourier transforms. Denoting the Fourier transform in x of $\phi(x, y)$ as $\hat{\Phi}(k, y)$:

$$\hat{\Phi}(k, y) = \int_{-\infty}^{\infty} e^{-ikx} \phi(x, y) \, dx$$

taking the Fourier transform of Eq. (4.6.1), and using the result from Section 4.5 for the Fourier transform of derivatives, Eqs. (4.5.16a,b) (assuming the validity of interchanging y-derivatives and integrating over k, which can be verified *a posteriori*), we have

$$\frac{\partial^2 \hat{\Phi}}{\partial y^2} - k^2 \hat{\Phi} = 0 \qquad (4.6.2)$$

Hence

$$\hat{\Phi}(k, y) = A(k)e^{ky} + B(k)e^{-ky}$$

where $A(k)$ and $B(k)$ are arbitrary functions of k, to be specified by the boundary conditions. We require that $\hat{\Phi}(k, y)$ be bounded for all $y > 0$. In order that $\hat{\Phi}(k, y)$ yield a bounded function $\phi(x, y)$, we need

$$\hat{\Phi}(k, y) = C(k)e^{-|k|y} \qquad (4.6.3)$$

Denoting the Fourier transform of $\phi(x, 0) = h(x)$ by $\hat{H}(k)$ fixes $C(k) = \hat{H}(k)$, so that

$$\hat{\Phi}(k, y) = \hat{H}(k)e^{-|k|y} \qquad (4.6.4)$$

From Eq. (4.5.1) by direct integration (contour integration is not necessary) we find that $\hat{F}(k, y) = e^{-|k|y}$ is the Fourier transform of $f(x, y) = \frac{1}{\pi} \frac{y}{x^2+y^2}$, thus from the convolution formula Eq. (4.5.17) the solution to Eq. (4.6.1) is given by

$$\phi(x, y) = \frac{1}{\pi} \int_{-\infty}^{\infty} \frac{y\, h(x')}{(x - x')^2 + y^2}\, dx' \qquad (4.6.5)$$

If $h(x)$ were taken to be a Dirac delta function concentrated at $x = \zeta$, $h(x) = h_s(x - \zeta) = \delta(x - \zeta)$, then $\hat{H}(k) = e^{-ik\zeta}$, and from Eq. (4.6.4) directly (or Eq. (4.6.5)) a special solution to Eq. (4.6.1), $\phi_s(x, y)$ is

$$\phi_s(x, y) = G(x - \zeta, y) = \frac{1}{\pi}\left(\frac{y}{(x - \zeta)^2 + y^2}\right) \qquad (4.6.6)$$

Function $G(x - \zeta, y)$ is called a **Green's function**; it is a fundamental solution to Laplace's equation in this region. Green's functions have the property of solving a given equation with delta function inhomogeneity. From the boundary values $h_s(x - \zeta, 0) = \delta(x - \zeta)$ we may construct arbitrary initial values

$$\phi(x, 0) = \int_{-\infty}^{\infty} h(\zeta)\, \delta(x - \zeta)\, d\zeta = h(x) \qquad (4.6.7a)$$

and, because Laplace's equation is linear, we find by superposition that the general solution satisfies

$$\phi(x, y) = \int_{-\infty}^{\infty} h(\zeta) \, G(x - \zeta, y) \, d\zeta \qquad (4.6.7b)$$

which is Eq. (4.6.5), noting that ζ or x' are dummy integration variables. In many applications it is sufficient to obtain the Green's function of the underlying differential equation.

The formula (4.6.5) is sometimes referred to as the **Poisson formula for a half plane**. Although we derived it via transform methods, it is worth noting that a pair of such formulae can be derived from Cauchy's integral formula. We describe this alternative method now. Let $f(z)$ be analytic on the real axis and in the upper half plane and assume $\zeta f(\zeta) \to 0$ as $\zeta \to \infty$. Using a large closed semicircular contour such as that depicted in Figure 4.2.1 we have

$$f(z) = \frac{1}{2\pi i} \oint_C \frac{f(\zeta)}{\zeta - z} d\zeta$$

$$0 = \frac{1}{2\pi i} \oint_C \frac{f(\zeta)}{\zeta - \bar{z}} d\zeta$$

where $\text{Im } z > 0$ (in the second formula there is no singularity because the contour closes in the upper half plane and $\zeta = \bar{z}$ in the lower half plane). Adding and subtracting yields

$$f(z) = \frac{1}{2\pi i} \oint_C f(\zeta) \left(\frac{1}{\zeta - z} \pm \frac{1}{\zeta - \bar{z}} \right) d\zeta$$

The semicircular portion of the contour C_R vanishes as $R \to \infty$, implying the following on $\text{Im } \zeta = 0$ for the plus and minus parts of the above integral, respectively: calling $z = x + iy$ and $\zeta = x' + iy'$,

$$f(x, y) = \frac{1}{\pi i} \int_{-\infty}^{\infty} f(x', y' = 0) \left(\frac{x' - x}{(x - x')^2 + y^2} \right) dx'$$

$$f(x, y) = \frac{1}{\pi} \int_{-\infty}^{\infty} f(x', y' = 0) \left(\frac{y}{(x - x')^2 + y^2} \right) dx'$$

Calling

$$f(z) = f(x, y) = u(x, y) + i \, v(x, y), \qquad \text{Re } f(x, y = 0) = u(x, 0) = h(x)$$

and taking the imaginary part of the first and the real part of the second of the above formulae, yields the conjugate Poisson formulae for a half plane:

$$v(x, y) = \frac{1}{\pi} \int_{-\infty}^{\infty} h(x') \left(\frac{x' - x}{(x - x')^2 + y^2} \right) dx'$$

$$u(x, y) = \frac{1}{\pi} \int_{-\infty}^{\infty} h(x') \left(\frac{y}{(x - x')^2 + y^2} \right) dx'$$

Identifying $u(x, y)$ as $\phi(x, y)$, we see that the harmonic function $u(x, y)$ (because $f(z)$ is analytic its real and imaginary parts satisfy Laplace's equation) is given by the same formula as Eq. (4.6.5). Moreover, we note that the imaginary part of $f(z)$, $v(x, y)$, is determined by the real part of $f(z)$ on the boundary. We see that we cannot arbitrarily prescribe both the real and imaginary parts of $f(z)$ on the boundary. These formulae are valid for a half plane. Similar formulae can be obtained by this method for a circle (see also Example 10, Section 2.6).

Laplace's equation, (4.6.1), is typical of a steady state situation, for example, as mentioned earlier, steady state heat flow in a uniform metal plate. If we have time-dependent heat flow, the diffusion equation

$$\frac{\partial \phi}{\partial t} = k \nabla^2 \phi \tag{4.6.8}$$

is a relevant equation with k the diffusion coefficient. In Eq. (4.6.8), ∇^2 is the Laplacian operator, which in two dimensions is given by $\nabla^2 = \frac{\partial^2}{\partial x^2} + \frac{\partial^2}{\partial y^2}$. In one dimension, taking $k = 1$ for convenience, we have the following initial value problem:

$$\frac{\partial \phi(x, t)}{\partial t} = \frac{\partial^2 \phi(x, t)}{\partial x^2} \tag{4.6.9}$$

The Green's function for the problem on the line $-\infty < x < \infty$ is obtained by solving Eq. (4.6.9) subject to

$$\phi(x, 0) = \delta(x - \zeta)$$

Example 4.6.2 Solve for the Green's function of Eq. (4.6.9). Define

$$\hat{\Phi}(k, t) = \int_{-\infty}^{\infty} e^{-ikx} \phi(x, t) \, dx$$

whereupon the Fourier transform of Eq. (4.6.9) satisfies

$$\frac{\partial \hat{\Phi}(k, t)}{\partial t} = -k^2 \hat{\Phi}(k, t) \tag{4.6.10}$$

hence

$$\hat{\Phi}(k, t) = \hat{\Phi}(k, 0)e^{-k^2 t} = e^{-ik\zeta - k^2 t} \tag{4.6.11}$$

where $\hat{\Phi}(k, 0) = e^{-ik\zeta}$ is the Fourier transform of $\phi(x, 0) = \delta(x - \zeta)$. Thus, by the inverse Fourier transform, and calling $G(x - \zeta, t)$ the inverse transform of (4.6.11),

$$\begin{aligned}
G(x - \zeta, t) &= \frac{1}{2\pi} \int_{-\infty}^{\infty} e^{ik(x-\zeta) - k^2 t} \, dk \\
&= e^{-(x-\zeta)^2/4t} \cdot \frac{1}{2\pi} \int_{-\infty}^{\infty} e^{-(k - i\frac{x-\zeta}{2t})^2 t} \, dk \\
&= \frac{e^{-\frac{(x-\zeta)^2}{4t}}}{2\sqrt{\pi t}}
\end{aligned} \tag{4.6.12}$$

where we use $\int_{-\infty}^{\infty} e^{-u^2} \, du = \sqrt{\pi}$. Arbitrary initial values are included by again observing that

$$\phi(x, 0) = h(x) = \int_{-\infty}^{\infty} h(\zeta)\delta(x - \zeta) \, d\zeta$$

which implies

$$\phi(x, t) = \int_{-\infty}^{\infty} G(x - \zeta, t)h(\zeta) \, d\zeta = \frac{1}{2\sqrt{\pi t}} \int_{-\infty}^{\infty} h(\zeta) e^{-\frac{(x-\zeta)^2}{4t}} \, d\zeta \tag{4.6.13}$$

The above solution to Eq. (4.6.9) could also be obtained by using Laplace transforms. It is instructive to show how the method proceeds in this case. We begin by introducing the Laplace transform of $\phi(x, t)$ with respect to t:

$$\hat{\Phi}(x, s) = \int_0^{\infty} e^{-st} \phi(x, t) \, dt \tag{4.6.14}$$

Taking the Laplace transform in t of Eq. (4.6.8), with $\phi(x, 0) = \delta(x - \zeta)$, yields

$$\frac{\partial^2 \hat{\Phi}}{\partial x^2}(x, s) - s\hat{\Phi}(x, s) = -\delta(x - \zeta) \tag{4.6.15}$$

Hence the Laplace transform of the Green's function to Eq. (4.6.9) satisfies Eq. (4.6.15). We remark that generally speaking, any function $G(x - \zeta)$ satisfying

$$LG(x - \zeta) = -\delta(x - \zeta)$$

where L is a linear differential operator, is referred to as a Green's function. The general solution corresponding to $\phi(x, 0) = h(x)$ is obtained by superposition: $\phi(x, t) = \int_{-\infty}^{\infty} G(x - \zeta)h(\zeta) \, d\zeta$. Equation (4.6.15) is solved by first finding the bounded homogeneous solutions on $-\infty < x < \infty$, for $(x - \zeta) > 0$ and $(x - \zeta) < 0$:

$$\hat{\Phi}_+(x - \zeta, s) = A(s)e^{-s^{1/2}(x-\zeta)} \qquad \text{for } x - \zeta > 0$$

$$\hat{\Phi}_-(x - \zeta, s) = B(s)e^{s^{1/2}(x-\zeta)} \qquad \text{for } x - \zeta < 0 \qquad (4.6.16)$$

where we take $s^{1/2}$ to have a branch cut on the negative real axis; that is, $s = re^{i\theta}$, $-\pi \leq \theta < \pi$. This will allow us to readily invert the Laplace transform (Re $s > 0$).

The coefficients $A(s)$ and $B(s)$ in Eq. (4.6.16) are found by (a) requiring continuity of $\hat{\Phi}(x - \zeta, s)$ at $x = \zeta$ and by (b) integrating Eq. (4.6.15) from $x = \zeta - \epsilon$, to $x = \zeta + \epsilon$, and taking the limit as $\epsilon \to 0^+$. This yields a jump condition on $\frac{\partial \hat{\Phi}}{\partial x}(x - \zeta, s)$:

$$\left[\frac{\partial \hat{\Phi}}{\partial x}(x - \zeta, s) \right]_{x-\zeta=0^-}^{x-\zeta=0^+} = -1 \qquad (4.6.17)$$

Continuity yields $A(s) = B(s)$, and Eq. (4.6.17) gives

$$-s^{1/2}A(s) - s^{1/2}B(s) = -1 \qquad (4.6.18a)$$

hence

$$A(s) = B(s) = \frac{1}{2s^{1/2}} \qquad (4.6.18b)$$

Using Eq. (4.6.16), $\Phi(x - \zeta, s)$ is written in the compact form:

$$\Phi(x - \zeta, s) = \frac{e^{-s^{1/2}|x-\zeta|}}{2s^{1/2}} \qquad (4.6.19)$$

The solution $\phi(x, t)$ is found from the inverse Laplace transform:

$$\phi(x, t) = \frac{1}{2\pi i} \int_{c-i\infty}^{c+i\infty} \frac{e^{-s^{1/2}|x-\zeta|} e^{st}}{2s^{1/2}} \, ds \qquad (4.6.20)$$

for $c > 0$. To evaluate Eq. (4.6.20), we employ the same keyhole contour as in Example 4.5.3 in Section 4.5 (see Figure 4.5.2). There are no singularities enclosed, and the contours C_R and C_ϵ at infinity and at the origin vanish in the limit $R \to \infty$, $\epsilon \to 0$, respectively. We only obtain contributions along the top and bottom of the branch cut to find

$$\phi(x, t) = \frac{-1}{2\pi i} \int_\infty^0 \frac{e^{-ir^{1/2}|x-\zeta|} e^{-rt}}{2r^{1/2} e^{i\pi/2}} e^{i\pi} \, dr$$

$$+ \frac{-1}{2\pi i} \int_0^\infty \frac{e^{ir^{1/2}|x-\zeta|} e^{-rt}}{2r^{1/2} e^{-i\pi/2}} e^{-i\pi} \, dr \qquad (4.6.21)$$

In the second integral we put $r^{1/2} = u$; in the first we put $r^{1/2} = u$ and then take $u \to -u$, whereupon we find the same answer as before (see Eq. (4.6.12)):

$$\phi(x, t) = \frac{1}{2\pi} \int_{-\infty}^\infty e^{-u^2 t + iu|x-\zeta|} \, du$$

$$= \frac{1}{2\pi} \int_{-\infty}^\infty e^{-(u - i\frac{|x-\zeta|}{2t})^2 t} \, e^{-\frac{(x-\zeta)^2}{4t}} \, du$$

$$= \frac{e^{-(x-\zeta)^2/4t}}{2\sqrt{\pi t}} \qquad (4.6.22)$$

The Laplace transform method can also be applied to problems in which the spatial variable is on the semiinfinite domain. However, rather than use Laplace transforms, for variety and illustration, we show below how the sine transform can be used on Eq. (4.6.9) with the following boundary conditions:

$$\phi(x, 0) = 0, \qquad \phi(x = 0, t) = h(t), \ \lim_{x \to \infty} \frac{\partial \phi}{\partial x}(x, t) = 0,$$

$$\lim_{x \to \infty} \frac{\partial \phi}{\partial x}(x, t) = 0 \qquad (4.6.23)$$

Define, following Section 4.5

$$\phi(x, t) = \frac{2}{\pi} \int_0^\infty \hat{\Phi}_s(k, t) \sin kx \, dk \qquad (4.6.24a)$$

$$\hat{\Phi}_s(k, t) = \int_0^\infty \phi(x, t) \sin kx \, dk \qquad (4.6.24b)$$

We now operate on Eq. (4.6.9) with the integral $\int_0^\infty dx \sin kx$, and via integration by parts, find

$$\int_0^\infty \frac{\partial \phi^2}{\partial x^2} \sin kx \, dx = \left[\frac{\partial \phi}{\partial x}(x, t) \sin kx \right]_{x=0}^\infty - k \int_0^\infty \frac{\partial \phi}{\partial x} \cos kx \, dx$$

$$= k\phi(0, t) - k^2 \hat{\Phi}_s(k, t) \tag{4.6.25}$$

whereupon the transformed version of Eq. (4.6.9) is

$$\frac{\partial \hat{\Phi}_s}{\partial t}(k, t) + k^2 \hat{\Phi}_s(k, t) = k \, h(t) \tag{4.6.26}$$

The solution of Eq. (4.6.26) with $\phi(x, 0) = 0$ is given by

$$\hat{\Phi}_s(k, t) = \int_0^t h(t') k \, e^{-k^2(t-t')} \, dt' \tag{4.6.27}$$

If $\phi(x, 0)$ were nonzero, then Eq. (4.6.27) would have another term. For simplicity we only consider the case $\phi(x, 0) = 0$. Therefore

$$\phi(x, t) = \frac{2}{\pi} \int_0^\infty dk \sin kx \left\{ \int_0^t h(t') e^{-k^2(t-t')} k \, dt' \right\} \tag{4.6.28}$$

By integration we can show that (use $\sin kx = (e^{ikx} - e^{-ikx})/2$ and integrate by parts to obtain integrals such as those in (4.6.12))

$$J(x, t - t') = \int_0^\infty k \, e^{-k^2(t-t')} \sin kx \, dk$$

$$= \frac{\sqrt{\pi} x e^{-x^2/4(t-t')}}{4(t - t')^{3/2}} \tag{4.6.29}$$

hence by interchanging integrals in Eq. (4.6.28), we have

$$\phi(x, t) = \frac{2}{\sqrt{\pi}} \int_0^t h(t') J(x, t - t') \, dt'$$

When $h(t) = 1$, if we call $\eta = \frac{x}{2(t-t')^{1/2}}$, then $d\eta = \frac{x}{4(t-t')^{3/2}} dt'$, and we have

$$\phi(x, t) = \frac{2}{\pi} \int_{\frac{x}{2\sqrt{t}}}^\infty e^{-\eta^2} \, d\eta$$

$$\equiv \text{erfc}\left(\frac{x}{2\sqrt{t}} \right) \tag{4.6.30}$$

We note that $\mathrm{erfc}(x)$ is a well-known function, called the **complementary error function**: $\mathrm{erfc}(x) \equiv 1 - \mathrm{erf}(x)$, where $\mathrm{erf}(x) \equiv \frac{2}{\sqrt{\pi}} \int_0^x e^{-y^2} dy$.

It should be mentioned that the Fourier sine transform applies to problems such as Eq. (4.6.23) with fixed conditions on ϕ at the origin. Such solutions can be extended to the interval $-\infty < x < \infty$ where the initial values $\phi(x, 0)$ are themselves extended as an odd function on $(-\infty, 0)$. However, if we should replace $\phi(x = 0, t) = h(t)$ by a derivative condition, at the origin, say, $\frac{\partial \phi}{\partial x}(x = 0, t) = h(t)$, then the appropriate transform to use is a cosine transform.

Another type of partial differential equation that is encountered frequently in applications is the wave equation

$$\frac{\partial^2 \phi}{\partial x^2} - \frac{1}{c^2} \frac{\partial^2 \phi}{\partial t^2} = F(x, t) \tag{4.6.31}$$

where the constant $c, c > 0$, is the speed of propagation of the unforced wave. The wave equation governs vibrations of many types of continuous media with external forcing $F(x, t)$. If $F(x, t)$ vibrates periodically in time with constant frequency $\omega > 0$, say, $F(x, t) = f(x)e^{i\omega t}$, then it is natural to look for special solutions to Eq. (4.6.31) of the form $\phi(x, t) = \Phi(x)e^{i\omega t}$. Then $\Phi(x)$ satisfies

$$\frac{\partial^2 \Phi}{\partial x^2} + \left(\frac{\omega}{c}\right)^2 \Phi = f(x) \tag{4.6.32}$$

A real solution to Eq. (4.6.32) is obtained by taking the real part; this would correspond to forcing of $\phi(x, t) = \phi(x) \cos \omega t$. If we simply look for a Fourier transform solution of Eq. (4.6.32) we arrive at

$$\Phi(x) = \frac{-1}{2\pi} \int_C \frac{\hat{F}(k)}{k^2 - (\omega/c)^2} e^{ikx} \, dk \tag{4.6.33}$$

Unfortunately, for the standard contour C, k real, $-\infty < k < \infty$, Eq. (4.6.33) is not well defined because there are singularities in the denominator of the integrand in Eq. (4.6.33) when $k = \pm\omega/c$. Without further specification the problem is not well posed. The standard acceptable solution is found by specifying a contour C that is indented below $k = -\omega/c$ and above $k = +\omega/c$ (see Figure 4.6.1); this removes the singularities in the denominator.

This choice of contour turns out to yield solutions with outgoing waves at large distances from the source $F(x, t)$. A choice of contour reflects an imposed boundary condition. In this problem it is well known and is referred to as the

Fig. 4.6.1. Indented contour C

Sommerfeld radiation condition. An outgoing wave has the form $e^{i\omega(t-|x|/c)}$ (as t increases, x increases for a given choice of phase, i.e., on a fixed point on a wave crest). An incoming wave has the form $e^{i\omega(t+|x|/c)}$. Using the Fourier representation $\hat{F}(k) = \int_{-\infty}^{\infty} f(\zeta)e^{-i\zeta k}d\zeta$ in Eq. (4.6.33), we can write the function in the form

$$\Phi(x) = \int_{-\infty}^{\infty} f(\zeta)H(x-\zeta,\omega/c)\,d\zeta \qquad (4.6.34a)$$

where

$$H(x-\zeta,\omega/c) = \frac{-1}{2\pi} \int_C \frac{e^{ik(x-\zeta)}}{k^2 - (\omega/c)^2}\,dk \qquad (4.6.34b)$$

and the contour C is specified as in Figure 4.6.1. Contour integration of Eq. (4.6.34b) yields

$$H(x-\zeta,\omega/c) = \frac{i\,e^{-i|x-\zeta|(\omega/c)}}{2(\omega/c)} \qquad (4.6.34c)$$

At large distances from the source, $|x| \to \infty$, we have outgoing waves for the solution $\phi(x,t)$:

$$\phi(x,t) = \mathrm{Re}\left\{ \frac{i}{2(\omega/c)} \int_{-\infty}^{\infty} f(\zeta)e^{i\omega(t-|x-\zeta|/c)}\,d\zeta \right\}. \qquad (4.6.34d)$$

Thus, for example, if $f(\zeta)$ is a point source: $f(\zeta) = \delta(\zeta-x_0)$ where $\delta(\zeta-x_0)$ is a Dirac delta function concentrated at x_0, then (4.6.34d) yields

$$\phi(x,t) = -\frac{1}{2(\omega/c)} \sin\omega(t-|x-x_0|/c). \qquad (4.6.34e)$$

An alternative method to find this result is to add a damping mechanism to the original equation. Namely, if we add the term $-\epsilon\frac{\partial\phi}{\partial t}$ to the left-hand side of Eq. (4.6.31), then Eq. (4.6.33) is modified by adding the term $i\epsilon\omega$ to the denominator of the integrand. This has the desired effect of moving the poles off the real axis ($k_1 = -\omega/c + i\epsilon\alpha$, $k_2 = +\omega/c - i\epsilon\alpha$, where $\alpha = $ constant) in the same manner as indicated by Figure 4.6.1. By using Fourier transforms, and then taking the limit $\epsilon \to 0$ (small damping), the above results could have been obtained.

In practice, wave propagation problems such as the following one

$$\frac{\partial u}{\partial t} + \frac{\partial^3 u}{\partial x^3} = 0, \qquad -\infty < x < \infty, \qquad u(x, 0) = f(x), \qquad (4.6.35)$$

where again $f(x) \in L^1 \cap L^2$, are solved by Fourier transforms. Function $u(x, t)$ typically represents the small amplitude vibrations of a continuous medium such as water waves. One looks for a solution to Eq. (4.6.35) of the form

$$u(x, t) = \frac{1}{2\pi} \int_{-\infty}^{\infty} b(k, t) e^{ikx} \, dk \qquad (4.6.36)$$

Taking the Fourier transform of (4.6.35) and using (4.5.16) or alternatively, substitution of Eq. (4.6.36) into Eq. (4.6.35) – assuming interchanges of derivative and integrand are valid (a fact that can be shown to follow from rapid enough decay of $f(x)$ at infinity, that is, $f \in L^1 \cap L^2$) yields

$$\frac{\partial b}{\partial t} - ik^3 b = 0 \qquad (4.6.37a)$$

hence

$$b(k, t) = b(k, 0) e^{ik^3 t} \qquad (4.6.37b)$$

where

$$b(k, 0) = \int_{-\infty}^{\infty} f(x) e^{-ikx} \, dx \qquad (4.6.37c)$$

The solution (4.6.36) can be viewed as a superposition of waves of the form $e^{ikx - i\omega(k)t}$, $\omega(k) = -k^3$. Function $\omega(k)$ is referred to as the **dispersion relation**. The above integral representation, for general $f(x)$, is the "best" one can do, because we cannot evaluate it in closed form. However, as $t \to \infty$, the integral can be approximated by asymptotic methods, which will be discussed in Chapter 6, that is, the methods of stationary phase and steepest descents. Suffice it to say that the solution $u(x, t) \to 0$ as $t \to \infty$ (the initial values are said to disperse as $t \to \infty$) and the major contribution to the integral is found near the location where $\omega'(k) = x/t$; that is, $x/t = -3k^2$ in the integrand (where the phase $\Psi = kx - \omega(k)t$ is stationary: $\frac{\partial \Psi}{\partial k} = 0$). The quantity $\omega'(k)$ is called the **group velocity**, and it represents the speed of a packet of waves centered around wave number k. Using asymptotic methods for $x/t < 0$, as $t \to \infty$, $u(x, t)$ can be shown to have the following approximate form

$$u(x, t) \approx \frac{c}{\sqrt{t}} \left(\sum_{i=1}^{2} \frac{b(k_i)}{\sqrt{|k_i|}} e^{i(k_i x + k_i^3 t + \phi_i)} \right) \qquad (4.6.38)$$

$$k_1 = \sqrt{-x/3t}, \qquad k_2 = -\sqrt{-x/3t}, \qquad c, \phi_i \text{ constant}$$

When $x/t > 0$ the solution decays exponentially. As $x/t \to 0$, Eq. (4.6.38) may be rearranged and put into the following self-similar form

$$u(x, t) \approx \frac{d}{(3t)^{1/3}} A(x/(3t)^{1/3}) \qquad (4.6.39)$$

where d is constant and $A(\eta)$ satisfies (by substitution of (4.6.39) into (4.6.35))

$$A_{\eta\eta\eta} - \eta A_\eta - A = 0$$

or

$$A_{\eta\eta} - \eta A = 0 \qquad (4.6.40a)$$

with the boundary condition $A \to 0$ as $\eta \to \infty$. Equation (4.6.40a) is called Airy's equation. The integral representation of the solution to Airy's equation with $A \to 0$, $\eta \to \infty$, is given by

$$A(\eta) = \frac{1}{2\pi} \int_{-\infty}^{\infty} e^{i(k\eta + k^3/3)} \, dk \qquad (4.6.40b)$$

(See also the end of this section, Eq. (4.6.57).) Its wave form is depicted in Figure 4.6.2. Function $A(\eta)$ acts like a "matching" or "turning" of solutions from one type of behavior to another: i.e., from exponential decay as $\eta \to +\infty$ to oscillation as $\eta \to -\infty$ (see also Section 6.7).

Sometimes there is a need to use multiple transforms. For example, consider finding the solution to the following problem:

$$\frac{\partial^2 \phi}{\partial x^2} + \frac{\partial^2 \phi}{\partial y^2} - m^2 \phi = f(x, y), \qquad \phi(x, y) \to 0 \text{ as } x^2 + y^2 \to \infty \quad (4.6.41)$$

A simple transform in x satisfies

$$\phi(x, y) = \frac{1}{2\pi} \int_{-\infty}^{\infty} \Phi(k_1, y) e^{ik_1 x} \, dk_1$$

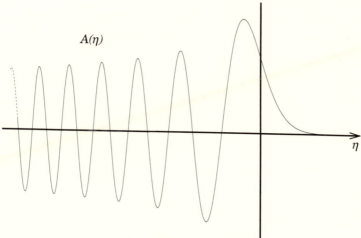

$A(\eta)$

η

Fig. 4.6.2. Airy function

We can take another transform in y to obtain

$$\phi(x, y) = \frac{1}{(2\pi)^2} \int_{-\infty}^{\infty} \int_{-\infty}^{\infty} \hat{\Phi}(k_1, k_2) e^{ik_1 x + ik_2 y} \, dk_1 \, dk_2 \qquad (4.6.42)$$

Using a similar formula for $f(x, y)$ in terms of its transform $\hat{F}(k_1, k_2)$, we find by substitution into Eq. (4.6.41)

$$\phi(x, y) = \frac{-1}{(2\pi)^2} \int\int \frac{\hat{F}(k_1, k_2) e^{ik_1 x + ik_2 y}}{k_1^2 + k_2^2 + m^2} \, dk_1 \, dk_2 \qquad (4.6.43)$$

Rewriting Eq. (4.6.43) using

$$\hat{F}(k_1, k_2) = \int_{-\infty}^{\infty} \int_{-\infty}^{\infty} f(x', y') e^{-ik_1 x' - ik_2 y'} \, dx' \, dy'$$

and interchanging integrals yields

$$\phi(x, y) = \int_{-\infty}^{\infty} \int_{-\infty}^{\infty} f(x', y') G(x - x', y - y') \, dx' \, dy' \qquad (4.6.44a)$$

where

$$G(x, y) = -\frac{1}{(2\pi)^2} \int_{-\infty}^{\infty} \int_{-\infty}^{\infty} \frac{e^{ik_1 x + ik_2 y}}{k_1^2 + k_2^2 + m^2} \, dk_1 \, dk_2 \qquad (4.6.44b)$$

By clever manipulation, one can evaluate Eq. (4.6.44b). Using the methods of Section 4.3, contour integration with respect to k_1 yields

$$G(x, y) = -\frac{1}{4\pi} \int_{-\infty}^{\infty} \frac{e^{ik_2 y - \sqrt{k_2^2 + m^2}|x|}}{\sqrt{k_2^2 + m^2}} \, dk_2 \qquad (4.6.45a)$$

Thus, for $x \neq 0$

$$\frac{\partial G}{\partial x}(x, y) = \frac{\text{sgn}(x)}{4\pi} \int_{-\infty}^{\infty} e^{ik_2 y - \sqrt{k_2^2 + m^2}|x|} \, dk_2 \qquad (4.6.45b)$$

where $\text{sgn} x = 1$ for $x > 0$, and $\text{sgn} x = -1$ for $x < 0$. Equation (4.6.45b) takes on an elementary form for $m = 0$ ($\sqrt{k_2^2} = |k_2|$):

$$\frac{\partial G}{\partial x}(x, y) = \frac{x}{2\pi(x^2 + y^2)}$$

and we have

$$G(x, y) = \frac{1}{4\pi} \ln(x^2 + y^2) \qquad (4.6.46)$$

The constant of integration is immaterial, because to have a vanishing solution $\phi(x, y)$ as $x^2 + y^2 \to \infty$, Eq. (4.6.44a) necessarily requires that $\int_{-\infty}^{\infty} \int_{-\infty}^{\infty} f(x, y) dx \, dy = 0$, which follows directly from Eq. (4.6.41) by integration with $m = 0$. Note that Eq. (4.6.43) implies that when $m = 0$, for the integral to be well defined, $\hat{F}(k_1 = 0, k_2 = 0) = 0$, which in turn implies the need for the vanishing of the double integral of $f(x, y)$. Finally, if $m \neq 0$, we only remark that Eq. (4.6.44b) or (4.6.45a) is transformable to an integral representation of a modified Bessel function of order zero:

$$G(x, y) = -\frac{1}{2\pi} K_0\left(m\sqrt{x^2 + y^2}\right) \qquad (4.6.47)$$

Interested readers can find contour integral representations of Bessel functions in many books on special functions.

Frequently, in the study of differential equations, integral representations can be found for the solution. Integral representations supplement series methods discussed in Chapter 3 and provide an alternative representation of a class of solutions. We give one example in what follows. Consider Airy's equation in the form (see also Eq. (4.6.40a))

$$\frac{d^2 y}{dz^2} - zy = 0 \qquad (4.6.48)$$

and look for an integral representation of the form

$$y(z) = \int_C e^{z\zeta} v(\zeta) \, d\zeta \qquad (4.6.49)$$

where the contour C and the function $v(\zeta)$ are to be determined. Formula (4.6.49) is frequently referred to as a generalized Laplace transform and the method as the generalized Laplace transform method. (Here C is generally not the Bromwich contour.) Equation (4.6.49) is a special case of the more general integral representation $\int_C K(z, \zeta) v(\zeta) d\zeta$. Substitution of Eq. (4.6.49) into Eq. (4.6.48), and assuming the interchange of differentiation and integration, which is verified *a posteriori* gives

$$\int_C (\zeta^2 - z) v(\zeta) e^{z\zeta} \, d\zeta = 0 \qquad (4.6.50)$$

Using

$$z e^{z\zeta} v = \frac{d}{d\zeta} (e^{z\zeta} v) - e^{z\zeta} \frac{dv}{d\zeta}$$

rearranging and integrating yields

$$- \left[e^{z\zeta} v(\zeta) \right]_C + \int_C \left(\zeta^2 v + \frac{dv}{d\zeta} \right) e^{z\zeta} \, d\zeta = 0 \qquad (4.6.51)$$

where the term in brackets, $[\cdot]_C$, stands for evaluation at the endpoints of the contour. The essence of the method is to choose C and $v(\zeta)$ such that both terms in Eq. (4.6.51) vanish. Taking

$$\frac{dv}{d\zeta} + \zeta^2 v = 0 \qquad (4.6.52)$$

implies that

$$v(\zeta) = A e^{-\zeta^3/3}, \qquad A = \text{constant} \qquad (4.6.53)$$

Thinking of an infinite contour, and calling $\zeta = R e^{i\theta}$, we find that the dominant term as $R \to \infty$ in $[\cdot]_C$ is due to $v(\zeta)$, which in magnitude is given by

$$|v(\zeta)| = |A| e^{-R^3 (\cos 3\theta)/3} \qquad (4.6.54)$$

Vanishing of this contribution for large values of ζ will occur for $\cos 3\theta > 0$, that is, for

$$-\frac{\pi}{2} + 2n\pi \le 3\theta < \frac{\pi}{2} + 2n\pi, \qquad n = 0, 1, 2 \qquad (4.6.55)$$

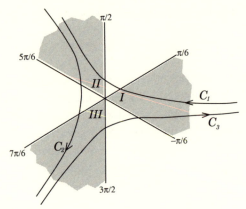

Fig. 4.6.3. Three standard contours

So we have three regions in which there is decay:

$$\begin{cases} -\frac{\pi}{6} < \theta < \frac{\pi}{6} & (I) \\ \frac{\pi}{2} < \theta < \frac{5\pi}{6} & (II) \\ \frac{7\pi}{6} < \theta < \frac{3\pi}{2} & (III) \end{cases} \qquad (4.6.56)$$

There are three standard contours C_i, $i = 1, 2$ and 3, in which the integrated term $[\cdot]_C$ vanishes as $R \to \infty$, depicted in Figure 4.6.3. The shaded region refers to regions of decay of $v(\zeta)$.

The three solutions, of which only two are linearly independent (because the equation is of second order), are denoted by

$$y_i(z) = \alpha_i \int_{C_i} e^{z\zeta - \zeta^3/3} \, d\zeta \qquad (4.6.57)$$

α_i being a convenient normalizing factor, $i = 1, 2, 3$. In order not to have a trivial solution, one must take the contour C between any two of the decaying regions. If we change variables $\zeta = ik$ then the solution corresponding to $i = 2$, $y_2(z)$, is proportional to the Airy function solution $A(\eta)$ discussed earlier (see Eq. (4.6.40b)).

Finally, we remark that this method applies to linear differential equations with coefficients depending linearly on the independent variable. Generalizations to other kernels $K(z, \zeta)$ (mentioned below Eq. (4.6.49)) can be made and is employed to solve other linear differential equations such as Bessel functions, Legendre functions, etc. The interested reader may wish to consult a reference such as Jeffreys and Jeffreys (1962).

Problems for Section 4.6

1. Use Laplace transform methods to solve the ODE

$$L\frac{dy}{dt} + Ry = f(t), \quad y(0) = y_0, \quad \text{constants } L, R > 0$$

(a) Let $f(t) = \sin \omega_0 t$, $\omega_0 > 0$, so that the Laplace transform of $f(t)$ is $\hat{F}(s) = \omega_0/(s^2 + \omega_0^2)$. Find

$$y(t) = y_0 e^{-\frac{R}{L}t} + \frac{\omega_0}{L\left((R/L)^2 + \omega_0^2\right)} e^{-\frac{R}{L}t} + \frac{R}{L^2} \frac{\sin \omega_0 t}{\left((R/L)^2 + \omega_0^2\right)}$$

$$- \frac{\omega_0}{L} \frac{\cos \omega_0 t}{\left((R/L)^2 + \omega_0^2\right)}$$

(b) Suppose $f(t)$ is an arbitrary continuous function that possesses a Laplace transform. Use the convolution product for Laplace transforms (Section 4.5) to find

$$y(t) = y_0 e^{-\frac{R}{L}t} + \int_0^t f(t')e^{-\frac{R}{L}(t-t')} \, dt'$$

(c) Let $f(t) = \sin \omega_0 t$ in (b) to obtain the result of part (a), and thereby verify your answer.

This is an example of an "L,R circuit" with impressed voltage $f(t)$ arising in basic electric circuit theory.

2. Use Laplace transform methods to solve the ODE

$$\frac{d^2 y}{dt^2} - k^2 y = f(t), \quad k > 0, \quad y(0) = y_0, \quad y'(0) = y_0'$$

(a) Let $f(t) = e^{-k_0 t}$, $k_0 \neq k$, $k_0 > 0$, so that the Laplace transform of $f(t)$ is $\hat{F}(s) = 1/(s + k_0)$, and find

$$y(t) = y_0 \cosh kt + \frac{y_0'}{k} \sinh kt + \frac{e^{-k_0 t}}{k_0^2 - k^2}$$

$$- \frac{\cosh kt}{k_0^2 - k^2} + \frac{(k_0/k)}{k_0^2 - k^2} \sinh kt$$

(b) Suppose $f(t)$ is an arbitrary continuous function that possesses a Laplace transform. Use the convolution product for Laplace transforms (Section 4.5) to find

$$y(t) = y_0 \cosh kt + \frac{y_0'}{k} \sinh kt + \int_0^t f(t') \frac{\sinh k(t - t')}{k} \, dt'$$

(c) Let $f(t) = e^{-k_0 t}$ in (b) to obtain the result in part (a). What happens when $k_0 = k$?

3. Consider the differential equation

$$\frac{d^3 y}{dt^3} + \omega_0{}^3 y = f(t), \quad \omega_0 > 0, \quad y(0) = y'(0) = y''(0) = 0$$

(a) Find that the Laplace transform of the solution, $\hat{Y}(s)$, satisfies (assuming that $f(t)$ has a Laplace transform $\hat{F}(s)$)

$$\hat{Y}(s) = \frac{\hat{F}(s)}{s^3 + \omega_0{}^3}$$

(b) Deduce that the inverse Laplace transform of $1/(s^3 + \omega_0{}^3)$ is given by

$$h(t) = \frac{e^{-\omega_0 t}}{3\omega_0{}^2} - \frac{2}{3\omega_0{}^2} e^{\omega_0 t / 2} \cos\left(\frac{\omega_0}{2}\sqrt{3}t - \frac{\pi}{3}\right)$$

and show that

$$y(t) = \int_0^t h(t') f(t - t') \, dt'$$

by using the convolution product for Laplace transforms.

4. Let us consider Laplace's equation $(\partial^2 \phi)/(\partial x^2) + (\partial^2 \phi)/(\partial y^2) = 0$, for $-\infty < x < \infty$ and $y > 0$, with the boundary conditions $(\partial \phi/\partial y)(x, 0) = h(x)$ and $\phi(x, y) \to 0$ as $x^2 + y^2 R \to \infty$. Find the Fourier transform solution. Is there a constraint on the data $h(x)$ for a solution to exist? If so, can this be explained another way?

5. Given the linear "free" Schrödinger equation (without a potential)

$$i\frac{\partial u}{\partial t} + \frac{\partial^2 u}{\partial x^2} = 0, \quad \text{with } u(x, 0) = f(x)$$

(a) solve this problem by Fourier transforms, by obtaining the Green's function in closed form, and using superposition. Recall that $\int_{-\infty}^{\infty} e^{iu^2} du = e^{i\pi/4} \sqrt{\pi}$.

(b) Obtain the above solution by Laplace transforms.

6. Given the heat equation

$$\frac{\partial \phi}{\partial t} = \frac{\partial^2 \phi}{\partial x^2}$$

with the following initial and boundary conditions

$$\phi(x, 0) = 0, \quad \frac{\partial \phi}{\partial x}(x = 0, t) = g(t),$$

$$\lim_{x \to \infty} \phi(x, t) = \lim_{x \to \infty} \frac{\partial \phi}{\partial x} = 0$$

(a) solve this problem by Fourier cosine transforms.
(b) Solve this problem by Laplace transforms.
(c) Show that the representations of (a) and (b) are equivalent.

7. Given the wave equation (with wave speed being unity)

$$\frac{\partial^2 \phi}{\partial t^2} - \frac{\partial^2 \phi}{\partial x^2} = 0$$

and the boundary conditions

$$\phi(x, t = 0) = 0, \quad \frac{\partial \phi}{\partial t}(x, t = 0) = 0,$$

$$\phi(x = 0, t) = 0, \quad \phi(x = \ell, t) = 1$$

(a) obtain the Laplace transform of the solution $\hat{\Phi}(x, s)$

$$\hat{\Phi}(x, s) = \frac{\sinh sx}{s \sinh s\ell}$$

(b) Obtain the solution $\phi(x, t)$ by inverting the Laplace transform to find

$$\phi(x, t) = \frac{x}{\ell} + \sum_{n=1}^{\infty} \frac{2(-1)^n}{n\pi} \sin\left(\frac{n\pi x}{\ell}\right) \cos\left(\frac{n\pi t}{\ell}\right)$$

(see also Problem (**19**), Section 4.5).

8. Given the wave equation

$$\frac{\partial^2 \phi}{\partial t^2} - \frac{\partial^2 \phi}{\partial x^2} = 0$$

and the boundary conditions

$$\phi(x, t = 0) = 0, \qquad \frac{\partial \phi}{\partial t}(x, t = 0) = 0,$$

$$\phi(x = 0, t) = 0, \qquad \phi(x = \ell, t) = f(t)$$

(a) show that the Laplace transform of the solution is given by

$$\hat{\Phi}(x, s) = \frac{\hat{F}(s) \sinh sx}{\sinh s\ell}$$

where $\hat{F}(s)$ is the Laplace transform of $f(t)$.

(b) Call the solution of the problem when $f(t) = 1$ (so that $\hat{F}(s) = 1/s$) to be $\phi_s(x, t)$. Show that the general solution is given by

$$\phi(x, t) = \int_0^t \frac{\partial \phi_s}{\partial t'}(x, t') f(t - t') \, dt'$$

9. Use multiple Fourier transforms to solve

$$\frac{\partial \phi}{\partial t} - \left(\frac{\partial^2 \phi}{\partial x^2} + \frac{\partial^2 \phi}{\partial y^2} \right) = 0$$

on the infinite domain $-\infty < x < \infty$, $-\infty < y < \infty$, $t > 0$, with $\phi(x, y) \to 0$ as $x^2 + y^2 \to \infty$, and $\phi(x, y, 0) = f(x, y)$. How does the solution simplify if $f(x, y)$ is a function of $x^2 + y^2$? What is the Green's function in this case?

10. Given the forced heat equation

$$\frac{\partial \phi}{\partial t} - \frac{\partial^2 \phi}{\partial x^2} = f(x, t), \qquad \phi(x, 0) = g(x)$$

on $-\infty < x < \infty$, $t > 0$, with $\phi, g, f \to 0$ as $|x| \to \infty$

(a) use Fourier transforms to solve the equation. How does the solution compare with the case $f = 0$?

(b) Use Laplace transforms to solve the equation. How does the method compare with that described in this section for the case $f = 0$?

11. Given the ODE

$$zy'' + (2r + 1)y' + zy = 0$$

look for a contour representation of the form $y = \int_C e^{z\zeta} v(\zeta)\, d\zeta$.

   (a) Show that if C is a *closed* contour and $v(\zeta)$ is single valued on this contour, then it follows that $v(\zeta) = A(\zeta^2 + 1)^{r-1/2}$.

   (b) Show that if $y = z^{-s}w$, then when $s = r$, w satisfies Bessel's equation $z^2 w'' + zw' + (z^2 - r^2)w = 0$, and a contour representation of the solution is given by

$$w = Az^r \oint_C e^{z\zeta} (\zeta^2 + 1)^{r-1/2}\, d\zeta$$

   Note that if $r = n + 1/2$ for integer n, then this representation yields the trivial solution. We take the branch cut to be inside the circle C when $(r - 1/2) \neq$ integer.

12. The hypergeometric equation

$$zy'' + (a - z)y' - by = 0$$

has a contour integral representation of the form $y = \int_C e^{z\zeta} v(\zeta)d\zeta$.

   (a) Show that one solution is given by

$$y = \int_0^1 e^{z\zeta} \zeta^{b-1}(1 - \zeta)^{a-b-1}\, d\zeta$$

   where $\operatorname{Re} b > 0$ and $\operatorname{Re}(a - b) > 0$.

   (b) Let $b = 1$, and $a = 2$; show that this solution is $y = (e^z - 1)/(z)$, and verify that it satisfies the equation.

   (c) Show that a second solution, $y_2 = vy_1$ (where the first solution is denoted as y_1) obeys

$$zy_1 v'' + (2zy_1' + (a - z)y_1)v' = 0$$

   Integrate this equation to find v, and thereby obtain a formal representation for y_2. What can be said about the analytic behavior of y_2 near $z = 0$?

13. Suppose we are given the following damped wave equation:

$$\frac{\partial^2 \phi}{\partial x^2} - \frac{1}{c^2}\frac{\partial^2 \phi}{\partial t^2} - \epsilon\frac{\partial \phi}{\partial t} = e^{i\omega t}\delta(x - \zeta), \quad \omega, \epsilon > 0$$

(a) Show that $\psi(x)$ where $\phi(x, t) = e^{i\omega t}\psi(x)$ satisfies

$$\psi'' + \left(\left(\frac{\omega}{c}\right)^2 - i\omega\epsilon\right)\psi = \delta(x - \zeta)$$

(b) Show that $\hat{\Psi}(k)$, the Fourier transform of $\psi(x)$, is given by

$$\hat{\Psi}(k) = \frac{-e^{-ik\zeta}}{k^2 - \left(\frac{\omega}{c}\right)^2 + i\omega\epsilon}$$

(c) Invert $\hat{\Psi}(k)$ to obtain $\psi(x)$, and in particular show that as $\epsilon \to 0^+$ we have

$$\psi(x) = \frac{ie^{i\frac{\omega}{c}|x-\zeta|}}{2\left(\frac{\omega}{c}\right)}$$

and that this has the effect of deforming the contour as described in Figure 4.6.1.

14. In this problem we obtain the Green's function of Laplace's equation in the upper half plane, $-\infty < x < \infty, 0 < y < \infty$, by solving

$$\frac{\partial^2 G}{\partial x^2} + \frac{\partial^2 G}{\partial y^2} = \delta(x - \zeta)\delta(y - \eta),$$

$$G(x, y = 0) = 0, \qquad G(x, y) \to 0 \quad \text{as} \quad r^2 = x^2 + y^2 \to \infty$$

(a) Take the Fourier transform of the equation with respect to x and find that the Fourier transform, $\hat{G}(k, y) = \int_{-\infty}^{\infty} G(x, y)e^{-ikx}\, dk$, satisfies

$$\frac{\partial^2 \hat{G}}{\partial y^2} - k^2\hat{G} = \delta(y - \eta)e^{-ik\zeta} \quad \text{with} \quad \hat{G}(x, y = 0) = 0$$

(b) Take the Fourier sine transform of $\hat{G}(k, y)$ with respect to y and find, for

$$\hat{G}_s(k, l) = \int_0^{\infty} \hat{G}(k, y)\sin ly\, dy$$

that it satisfies

$$\hat{G}_s(k, l) = -\frac{\sin l\eta e^{-ik\zeta}}{l^2 + k^2}$$

(c) Invert this expression with respect to k and find

$$\mathcal{G}(x, l) = -\frac{e^{-l|x-\zeta|} \sin l\eta}{2l}$$

whereupon

$$G(x, y) = -\frac{1}{\pi} \int_0^\infty \frac{e^{-l|x-\zeta|} \sin l\eta \sin ly}{l} \, dl$$

(d) Evaluate $G(x, y)$ to find

$$G(x, y) = \frac{1}{4\pi} \log \left(\frac{(x - \zeta)^2 + (y - \eta)^2}{(x - \zeta)^2 + (y + \eta)^2} \right)$$

Hint: Note that taking the derivative of $G(x, y)$ (of part (c) above) with respect to y yields an integral for $(\partial G)/(\partial y)$ that is elementary. Then one can integrate this result using $G(x, y = 0) = 0$ to obtain $G(x, y)$.

Part II

Applications of Complex Function Theory

The second portion of this text aims to acquaint the reader with examples of current practical applications of the theory of complex functions. Each of the chapters 5, 6 and 7 in Part II can be read independently.

5

Conformal Mappings and Applications

5.1 Introduction

A large number of problems arising in fluid mechanics, electrostatics, heat conduction, and many other physical situations can be mathematically formulated in terms of Laplace's equation (see also the discussion in Section 2.1). That is, all these physical problems reduce to solving the equation

$$\Phi_{xx} + \Phi_{yy} = 0 \tag{5.1.1}$$

in a certain region D of the z plane. The function $\Phi(x, y)$, in addition to satisfying Eq. (5.1.1), also satisfies certain boundary conditions on the boundary C of the region D. Recalling that the real and the imaginary parts of an analytic function satisfy Eq. (5.1.1), it follows that solving the above problem reduces to finding a function that is analytic in D and that satisfies certain boundary conditions on C. It turns out that the solution of this problem can be greatly simplified if the region D is either the upper half of the z plane or the unit disk. This suggests that instead of solving Eq. (5.1.1) in D, one should first perform a change of variables from the complex variable z to the complex variable $w = f(z)$, such that the region D of the z plane is mapped to the upper half plane of the w plane. Generally speaking, such transformations are called conformal, and their study is the main content of this chapter.

General properties of conformal transformations are studied in Sections 5.2 and 5.3. In Section 5.3 a number of theorems are stated, which are quite natural and motivated by heuristic considerations. The rigorous proofs are deferred to Section 5.5, which deals with more theoretical issues. We have denoted Section 5.5 as an optional (more difficult) section. In Section 5.4 a number of basic physical applications of conformal mapping are discussed, including problems from ideal fluid flow, steady state heat conduction, and electrostatics. Physical applications that require more advanced methods of conformal mapping are also included in later sections.

According to a celebrated theorem first discussed by Riemann, if D is a simply connected region D, which is not the entire complex z plane, then there exists an analytic function $f(z)$ such that $w = f(z)$ transforms D onto the upper half w plane. Unfortunately, this theorem does not provide a constructive approach for finding $f(z)$. However, for certain simple domains, such as domains bounded by polygons, it is possible to find an explicit formula (in terms of quadratures) for $f(z)$. Transformations of polygonal domains to the upper half plane are called Schwarz–Christoffel transformations and are studied in Section 5.6. A classically important case is the transformation of a rectangle to the upper half plane, which leads to elliptic integrals and elliptic functions. An important class of conformal transformations, called bilinear transformations, is studied in Section 5.7. Another interesting class of transformations involves a "circular polygon" (i.e., a polygon whose sides are circular arcs), which is studied in Section 5.8. The case of a circular triangle is discussed in some detail and relevant classes of functions such as Schwarzian functions and elliptic modular functions arise naturally. Some further interesting mathematical problems related to conformal transformations are discussed in Section 5.9.

5.2 Conformal Transformations

Let C be a curve in the complex z plane. Let $w = f(z)$, where $f(z)$ is some analytic function of z; define a change of variables from the complex variable z to the complex variable w. Under this transformation, the curve C is mapped to some curve C^* in the complex w plane. The precise form of C^* will depend on the precise form of C. However, there exists a geometrical property of C^* that is independent of the particular choice of C: Let z_0 be a point of the curve C, and assume that $f'(z_0) \neq 0$; under the transformation $w = f(z)$ the tangent to the curve C at the point z_0 is rotated counterclockwise by $\arg f'(z_0)$ (see Figure 5.2.1), $w_0 = f(z_0)$.

Before proving this statement, let us first consider the particular case that the transformation $f(z)$ is linear, that is, $f(z) = az + b$, $a, b \in \mathbb{C}$, and the curve C is a straight ray going through the origin. The mathematical description of such a curve is given by $z(s) = se^{i\varphi}$, where φ is constant, and the notation $z(s)$ indicates that for points on this curve, z is a function of s only. Under the transformation $w = f(z)$, this curve is mapped to $w(s) = az(s) + b = |a|s \exp[i(\varphi + \arg(a))] + b$, that is, to a ray rotated by $\arg(a) = \arg(f'(z))$; see Figure 5.2.2.

Let us now consider the general case. Points on a continuous curve C are characterized by the fact that their x and y coordinates are related. It turns out that, rather than describing this relationship directly, it is more convenient to describe it parametrically through the equations $x = x(s)$, $y = y(s)$, where $x(s)$ and $y(s)$ are real differentiable functions of the real parameter s. For

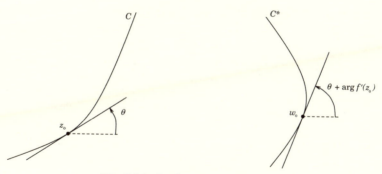

Fig. 5.2.1. Conformal transformation

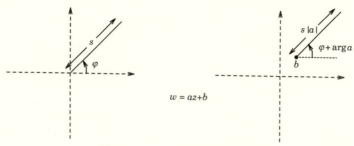

Fig. 5.2.2. Ray rotated by arg (a)

example, for the straight ray of Figure 5.2.2, $x = s \cos \varphi$, $y = s \sin \varphi$; for a circle with center at the origin and radius R, $x = R \cos s$, $y = R \sin s$, etc. More generally, the mathematical description of a curve C can be given by $z(s) = x(s) + iy(s)$. Suppose that $f(z)$ is analytic for z in some domain of the complex z plane denoted by D. Our considerations are applicable to that part of C that is contained in D. We shall refer to this part as an **arc** in order to emphasize that our analysis is local. For convenience of notation we shall denote it also by C. For such an arc, s belongs to some real interval $[a, b]$.

$$C : z(s) = x(s) + iy(s), \qquad s \in [a, b] \qquad (5.2.1)$$

We note that the image of a continuous curve is also continuous. Indeed, if we write $w = u(x, y) + iv(x, y)$, $u, v \in \mathbb{R}$, the image of the arc (5.2.1) is the arc C^* given by $w(s) = u(x(s), y(s)) + iv(x(s), y(s))$. Because x and y are continuous functions of s, it follows that u and v are also continuous functions of s, which establishes the continuity of C^*. Similarly, the image of a differentiable arc is a differentiable arc. However, the image of an arc that does not intersect itself is not necessarily nonintersecting. In fact, if $f(z_1) = f(z_2)$, $z_1, z_2 \in D$, any nonintersecting continuous arc passing through z_1 and z_2 will be mapped onto an arc that does intersect itself. Of course, one can avoid this if $f(z)$ takes no value more than once in D. We define $dz(s)/ds$ by

$$\frac{dz(s)}{ds} = \frac{dx(s)}{ds} + i\frac{dy(s)}{ds}$$

Let $f(z)$ be analytic in a domain containing the open neighborhood of $z_0 \equiv z(s_0)$. The image of C is $w(s) = f(z(s))$; thus by the chain rule

$$\left.\frac{dw(s)}{ds}\right|_{s=s_0} = f'(z_0)\left.\frac{dz(s)}{ds}\right|_{s=s_0} \tag{5.2.2}$$

If $f'(z_0) \neq 0$ and $z'(s_0) \neq 0$, it follows that $w'(s_0) \neq 0$ and

$$\arg(w'(s_0)) = \arg(z'(s_0)) + \arg(f'(z_0)) \tag{5.2.3}$$

or $\arg dw = \arg dz + \arg f'(z_0)$, where dw, dz are interpreted as infinitesimal line segments. This concludes the proof that, under the analytic transformation $f(z)$, the directed tangent to any curve through z_0 is rotated by an angle $\arg(f'(z_0))$.

An immediate consequence of the above geometrical property is that, for points where $f'(z) \neq 0$, analytic transformations preserve angles. Indeed, if two curves intersect at z_0, because the tangent of each curve is rotated by $\arg f'(z_0)$, it follows that the angle of intersection (in both magnitude and orientation), being the *difference* of the angles of the tangents, is preserved by such transformations. A transformation with this property is called **conformal**. We state this as a theorem; this theorem is enhanced in Sections 5.3 and 5.5.

Theorem 5.2.1 Assume that $f(z)$ is analytic and not constant in a domain D of the complex z plane. For any point $z \in D$ for which $f'(z) \neq 0$, this mapping is **conformal**, that is, it preserves the angle between two differentiable arcs.

Remark A conformal mapping, in addition to preserving angles, has the property of magnifying distances near z_0 by the factor $|f'(z_0)|$. Indeed, suppose that z is near z_0, and let w_0 be the image of z_0. Then the equation

$$|f'(z_0)| = \lim_{z \to z_0} \frac{|f(z) - f(z_0)|}{|z - z_0|}$$

implies that $|w - w_0|$ is approximately equal to $|f'(z_0)||z - z_0|$.

Example 5.2.1 Let D be the rectangular region in the z plane bounded by $x = 0$, $y = 0$, $x = 2$ and $y = 1$. The image of D under the transformation

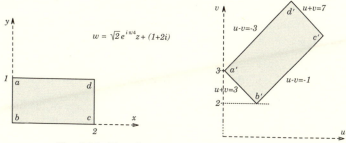

$$w = \sqrt{2}\,e^{i\pi/4}z + (1+2i)$$

Fig. 5.2.3. Transformation $w = (1 + i)z + (1 + 2i)$

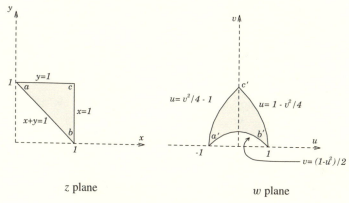

z plane   w plane

Fig. 5.2.4. Transformation $w = z^2$

$w = (1 + i)z + (1 + 2i)$ is given by the rectangular region D' of the w plane bounded by $u + v = 3$, $u - v = -1$, $u + v = 7$ and $u - v = -3$.

If $w = u + iv$, where $u, v \in \mathbb{R}$, then $u = x - y + 1$, $v = x + y + 2$. Thus the points a, b, c, and d are mapped to the points $(0,3)$, $(1,2)$, $(3,4)$, and $(2,5)$, respectively. The line $x = 0$ is mapped to $u = -y + 1$, $v = y + 2$, or $u + v = 3$ and is similar for the other sides of the rectangle.

The rectangle D is translated by $(1 + 2i)$, rotated by an angle $\pi/4$ in the counterclockwise direction, and dilated by a factor $\sqrt{2}$. In general, a linear transformation $f(z) = \alpha z + \beta$, translates by β, rotates by $\arg(\alpha)$, and dilates (or contracts) by $|\alpha|$. Because $f'(z) = \alpha \neq 0$, a linear transformation is always conformal. In this example, $\alpha = \sqrt{2}\exp(i\pi/4)$, $\beta = 1 + 2i$.

Example 5.2.2 Let D be the triangular region bounded by $x = 1$, $y = 1$, and $x + y = 1$. The image of D under the transformation $w = z^2$ is given by the curvilinear triangle $a'b'c'$ shown in Figure 5.2.3.

In this example, $u = x^2 - y^2$, $v = 2xy$. The line $x = 1$ is mapped to $u = 1 - y^2$, $v = 2y$, or $u = 1 - \frac{v^2}{4}$ and is similar for the other sides of the triangle. Because $f'(z) = 2z$ and the point $z = 0$ is outside D, it follows that this mapping is conformal; hence the angles of the triangle abc are equal to the respective angles of the curvilinear triangle $a'b'c'$.

Problems for Section 5.2

1. Show that under the transformation $w = 1/z$ the image of the lines $x = c_1 \neq 0$ and $y = c_2 \neq 0$ are the circles that are tangent at the origin to the v axis and to the u axis, respectively.

2. Find the image of the region R_z, bounded by $y = 0, x = 2$, and $x^2 - y^2 = 1$, for $x \geq 0$ and $y \geq 0$ (see Figure 5.2.5), under the transformation $w = z^2$.

3. Find a linear transformation that maps the circle C_1: $|z - 1| = 1$ onto the circle C_2: $|w - 3i/2| = 2$.

4. Show that the function $w = e^z$ maps the interior of the rectangle, R_z, $0 < x < 1, 0 < y < 2\pi$ $(z = x + iy)$, onto the interior of the annulus, R_w, $1 < |w| < e$, which has a jump along the positive real axis (see Figure 5.2.6).

5. Show that the mapping $w = \sqrt{1 - z^2}$ maps the hyperbola $2x^2 - 2y^2 = 1$ onto itself.

6. (a) Show that transformation $w = 2z + 1/z$ maps the exterior of the unit circle conformally onto the exterior of the ellipse:

$$\left(\frac{u}{3}\right)^2 + v^2 = 1$$

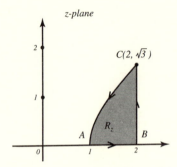

Fig. 5.2.5. Region in Problem 5.2.2

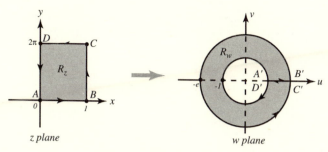

Fig. 5.2.6. Mapping of Problem 5.2.4

(b) Show that the transformation $w = \frac{1}{2}(ze^{-\alpha} + e^{\alpha}/z)$, for real constant α, maps the interior of the unit circle in the z plane onto the exterior of the ellipse $(u/\cosh\alpha)^2 + (v/\sinh\alpha)^2 = 1$ in the w plane.

5.3 Critical Points and Inverse Mappings

If $f'(z_0) = 0$, then the analytic transformation $f(z)$ ceases to be conformal. Such a point is called a **critical point** of f. Because critical points are zeroes of the analytic function f', they are isolated. In order to find what happens geometrically at a critical point, we use the following heuristic argument. Let $\delta z = z - z_0$, where z is a point near z_0. If the first nonvanishing derivative of $f(z)$ at z_0 is of the nth order, then representing δw by the Taylor series, it follows that

$$\delta w = \frac{1}{n!} f^{(n)}(z_0)(\delta z)^n + \frac{1}{(n+1)!} f^{(n+1)}(z_0)(\delta z)^{n+1} + \cdots \qquad (5.3.1)$$

where $f^{(n)}(z_0)$ denotes the nth derivative of $f(z)$ at $z = z_0$. Thus as $\delta z \to 0$

$$\arg(\delta w) \to n \arg(\delta z) + \arg\left(f^{(n)}(z_0)\right), \qquad (5.3.2)$$

This equation, which is the analog of Eq. (5.2.3), implies that the angle between any two infinitesimal line elements at the point z_0 is increased by the factor n. This suggests the following result.

Theorem 5.3.1 Assume that $f(z)$ is analytic and not constant in a domain D of the complex z plane. Suppose that $f'(z_0) = f''(z_0) = \cdots = f^{(n-1)}(z_0) = 0$, while $f^{(n)}(z_0) \neq 0$, $z_0 \in D$. Then the mapping $z \to f(z)$ magnifies n times the angle between two intersecting differentiable arcs that meet at z_0.

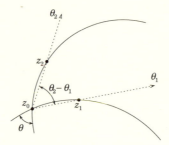

Fig. 5.3.1. Angle between line segments $(\theta_2 - \theta_1)$ tends to angle between arcs (θ) as $r \to 0$

Proof We now give a proof of this result. Let $z_1(s)$ and $z_2(s)$ be the equations describing the two arcs intersecting at z_0 (see Figure 5.3.1). If z_1 and z_2 are points on these arcs that have a distance r from z_0, it follows that

$$z_1 - z_0 = r e^{i\theta_1}, \qquad z_2 - z_0 = r e^{i\theta_2}, \qquad \text{or} \qquad \frac{z_2 - z_0}{z_1 - z_0} = e^{i(\theta_2 - \theta_1)}$$

The angle $\theta_2 - \theta_1$ is the angle formed by the linear segments connecting the points $z_1 - z_0$ and $z_2 - z_0$. As $r \to 0$, this angle tends to the angle formed by the two intersecting arcs in the complex z plane. Similar considerations apply for the complex w plane. Hence if θ and φ denote the angles formed by the intersecting arcs in the complex z plane and w plane, respectively, it follows that

$$\theta = \lim_{r \to 0} \arg \left(\frac{z_2 - z_0}{z_1 - z_0} \right), \qquad \varphi = \lim_{r \to 0} \arg \left(\frac{f(z_2) - f(z_0)}{f(z_1) - f(z_0)} \right) \qquad (5.3.3)$$

Hence

$$\varphi = \lim_{r \to 0} \arg \left\{ \left(\frac{\frac{f(z_2) - f(z_0)}{(z_2 - z_0)^n}}{\frac{f(z_1) - f(z_0)}{(z_1 - z_0)^n}} \right) \left(\frac{z_2 - z_0}{z_1 - z_0} \right)^n \right\} \qquad (5.3.4)$$

Using

$$f(z) = f(z_0) + \frac{f^{(n)}(z_0)}{n!}(z - z_0)^n + \frac{f^{(n+1)}(z_0)}{(n+1)!}(z - z_0)^{n+1} + \cdots \qquad (5.3.5)$$

it follows that

$$\lim_{r \to 0} \frac{f(z_2) - f(z_0)}{(z_2 - z_0)^n} = \lim_{r \to 0} \frac{f(z_1) - f(z_0)}{(z_1 - z_0)^n} = \frac{f^{(n)}(z_0)}{n!}$$

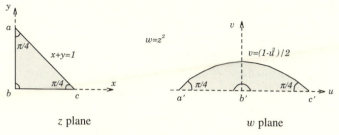

Fig. 5.3.2. Transformation $w = z^2$

Hence, Eqs. (5.3.4) and (5.3.3) imply

$$\varphi = \lim_{r \to 0} \arg \left(\frac{z_2 - z_0}{z_1 - z_0} \right)^n = n \lim_{r \to 0} \arg \left(\frac{z_2 - z_0}{z_1 - z_0} \right) = n\theta \qquad \blacksquare$$

Example 5.3.1 Let D be the triangular region bounded by $x = 0$, $y = 0$ and $x + y = 1$. The image of D under the transformation $w = z^2$ is given by the curvilinear triangle $a'b'c'$ shown in Figure 5.3.2 (note the difference from example 5.2.2).

In this example, $u = x^2 - y^2$, $v = 2xy$. The lines $x = 0$; $y = 0$; and $x + y = 1$ are mapped to $v = 0$ with $u \le 0$; $v = 0$ with $u \ge 0$; and $v = \frac{1}{2}(1 - u^2)$, respectively. The transformation $f(z) = z^2$ ceases to be conformal at $z = 0$. Because the second derivative of $f(z)$ at $z = 0$ is the first nonvanishing derivative, it follows that the angle at b (which is $\pi/2$ in the z plane) should be multiplied by 2. This is indeed the case, as the angle at b' in the w plane is π.

Critical points are also important in determining whether the function $f(z)$ has an inverse. Finding the inverse of $f(z)$ means solving the equation $w = f(z)$ for z in terms of w.

The following terminology will be useful. An analytic function $f(z)$ is called **univalent** in a domain D if it takes no value more than once in D. It is clear that a univalent function $f(z)$ provides a one-to-one map of D onto $f(D)$; it has a single-valued inverse on $f(D)$.

There are a number of basic properties of conformal maps that are useful and that we now point out to the reader. In this section we only state the relevant theorems; they are proven in the optional Section 5.5.

Theorem 5.3.2 Let $f(z)$ be analytic and not constant in a domain D of the complex z plane. The transformation $w = f(z)$ can be interpreted as a mapping of the domain D onto the domain $D^* = f(D)$ of the complex w plane.

The proof of this theorem can be found in Section 5.5. Because a domain is an open connected set, this theorem implies that open sets in the domain D of the z plane map to open sets D^* in the w plane. A consequence of this fact is that $|f(z)|$ cannot attain a maximum in D^* because any point $w = f(z)$ must be an interior point in the w plane. This theorem is useful because in practice we first find where the boundaries map. Then, since an open region is mapped to an open region, we need only find how one point is mapped if the boundary is a simple closed curve.

Suppose we try to construct the inverse in the neighborhood of some point z_0. If z_0 is not a critical point, then $w - w_0$ is given approximately by $f'(z_0)(z - z_0)$. Hence it is plausible that in this case, for every w there exists a unique z, that is, $f(z)$ is locally invertible. However, if z_0 is a critical point, and the first nonvanishing derivative at z_0 is $f^{(n)}(z_0)$, then $w - w_0$ is given approximately by $f^{(n)}(z_0)(z - z_0)^n/n!$. Hence now it is natural to expect that for every w there exist n different z's; that is, the inverse transformation is not single valued but it has a branch point of order n. These plausible arguments can actually be made rigorous (see Section 5.5).

Theorem 5.3.3 (1) Assume that $f(z)$ is analytic at z_0 and that $f'(z_0) \neq 0$. Then $f(z)$ is univalent in the neighborhood of z_0. More precisely, f has a unique analytic inverse F in the neighborhood of $w_0 \equiv f(z_0)$; that is, if z is sufficiently near z_0, then $z = F(w)$, where $w \equiv f(z)$. Similarly, if w is sufficiently near w_0 and $z \equiv F(w)$, then $w = f(z)$. Furthermore, $f'(z)F'(w) = 1$, which implies that the inverse map is conformal.

(2) Assume that $f(z)$ is analytic at z_0 and that it has a zero of order n; that is, the first nonvanishing derivative of $f(z)$ at z_0 is $f^{(n)}(z_0)$. Then to each w sufficiently close to $w_0 = f(z_0)$, there correspond n distinct points z in the neighborhood of z_0, each of which has w as its image under the mapping $w = f(z)$. Actually, this mapping can be decomposed in the form $w - w_0 = \zeta^n$, $\zeta = g(z - z_0)$, $g(0) = 0$, where $g(z)$ is univalent near z_0 and $g(z) = zH(z)$ with $H(0) \neq 0$.

The proof of this theorem can be found in Section 5.5.

Remark We recall that $w = z^n$ provides a one-to-one mapping of the z plane onto an n-sheeted Riemann surface in the w plane (see Section 2.2). If a complex number $w \neq 0$ is given without specification as to the sheet in which it lies, there are n possible values of z that give this w, and so $w = z^n$ has an n-valued inverse. However, when the Riemann surface is introduced, the correspondence becomes one-to-one, and $w = z^n$ has a single-valued inverse.

Theorem 5.3.4 Let C be a simple closed contour enclosing a domain D, and let $f(z)$ be analytic on C and in D. Suppose $f(z)$ takes no value more than once on C. Then (a) the map $w = f(z)$ takes C enclosing D to a simple closed contour C^* enclosing a domain D^* in the w plane; (b) $w = f(z)$ is a one-to-one map from D to D^*; and (c) if z traverses C in the positive direction, then $w = f(z)$ traverses C^* in the positive direction.

The proof of this theorem can be found in Section 5.5.

Remark By examining the mapping of simple closed contours it can be established that conformal maps preserve the connectivity of a domain. For example, the conformal map $w = f(z)$ of a simply connected domain in the z plane maps into a simply connected domain in the w plane. Indeed, a simple closed contour in the z plane can be continuously shrunk to a point, which must also be the case in the w plane – otherwise, we would violate Theorem 5.3.2.

Problems for Section 5.3

1. Find the families of curves on which $\operatorname{Re} z^2 = C_1$ for constant C_1, and $\operatorname{Im} z^2 = C_2$, for constant C_2. Show that these two families are orthogonal to each other.

2. Let D be the triangular region of Figure 5.3.2a, that is, the region bounded by $x = 0$, $y = 0$, and $x + y = 1$. Find the image of D under the mapping $w = z^3$. (It is sufficient to find a parameterization that describes the mapping of any of the sides.)

3. Express the transformations

   $$\text{(a)} \quad u = 4x^2 - 8y, \quad v = 8x - 4y^2$$
   $$\text{(b)} \quad u = x^3 - 3xy^2, \quad v = 3x^2y - y^3$$

   in the form $w = F(z, \bar{z})$, $z = x + iy$, $\bar{z} = x - iy$. Which of these transformations can be used to define a conformal mapping?

4. Show that the transformation $w = 2z^{-1/2} - 1$ maps the (infinite) domain exterior of the parabola $y^2 = 4(1 - x)$ conformally onto the domain $|w| < 1$. Explain why this transformation does not map the (infinite) domain interior of the parabola conformally onto the domain $|w| > 1$.

5. Let D denote the domain enclosed by the parabolae $v^2 = 4a(a - u)$ and $v^2 = 4a(a + u)$, $a > 0$, $w = u + iv$. Show that the function

$$w = c^2 \left[\int_0^z \frac{dt}{\sqrt{t(1 + t^2)}} \right]^2$$

where

$$\sqrt{a} = c \int_0^1 \frac{dt}{\sqrt{t(1 + t^2)}}$$

maps the unit circle conformally onto D.

5.4 Physical Applications

It was shown in Section 2.1 that the real and the imaginary parts of an analytic function satisfy Laplace's equation. This and the fact that the occurrence of Laplace's equation in physics is ubiquitous constitute one of the main reasons for the usefulness of complex analysis in applications. In what follows we first mention a few physical situations that lead to Laplace's equation. Then we discuss how conformal mappings can be effectively used to study the associated physical problems. Some of these ideas were introduced in Chapter 2.

A twice differentiable function $\Phi(x, y)$ satisfying Laplace's equation

$$\nabla^2 \Phi = \Phi_{xx} + \Phi_{yy} = 0 \tag{5.4.1}$$

in a region R is called harmonic in R. Let $V(z)$, $z = x + iy$, be analytic in R. If $V(z) = u(x, y) + iv(x, y)$, where $u, v \in \mathbb{R}$ and are twice differentiable, then both u and v are harmonic in R. Such functions are called conjugate functions. Given one of them (u or v), the other can be determined uniquely within an arbitrary additive constant (see Section 2.1).

Let u_1 and u_2 be the components of the vector $\mathbf{u}$ along the positive x and y axis, respectively. Suppose that the components of the vector $\mathbf{u}$ (where $\mathbf{u} = (u_1, u_2)$) satisfies the equation

$$\frac{\partial u_1}{\partial x} + \frac{\partial u_2}{\partial y} = 0 \tag{5.4.2}$$

Suppose further that the vector $\mathbf{u}$ can be derived from a potential, that is, there exists a scalar function Φ such that

$$u_1 = \frac{\partial \Phi}{\partial x}, \qquad u_2 = \frac{\partial \Phi}{\partial y} \tag{5.4.3}$$

Then Eqs. (5.4.2) and (5.4.3) imply that Φ is harmonic. These equations arise naturally in applications, as shown in the following examples.

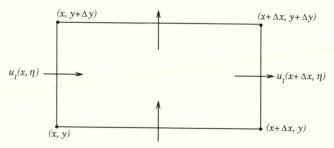

Fig. 5.4.1. Flow through a rectangle of sides Δx, Δy

Example 5.4.1 (Ideal Fluid Flow) A two-dimensional, steady, incompressible, irrotational fluid flow (see also the discussion in Section 2.1).

If a flow is two dimensional, it means that the fluid motion in any plane is identical to that in any other parallel plane. This allows one to study flow in a single plane that can be taken as the z plane. A flow pattern depicted in this plane can be interpreted as a cross section of an infinite cylinder perpendicular to this plane. If a flow is steady, it means that the velocity of the fluid at any point depends only on the position (x, y) and not on time. If the flow is incompressible, we take it to mean that the density (i.e., the mass per unit volume) of the fluid is constant. Let ρ and $\mathbf{u}$ denote the density and the velocity of the fluid. The law of conservation of mass implies Eq. (5.4.2). Indeed, consider a rectangle of sides Δx, Δy. See Figure 5.4.1.

The rate of accumulation of fluid in this rectangle is given by $\frac{d}{dt} \int_x^{x+\Delta x} \int_y^{y+\Delta y} \rho \, dx \, dy$. The rate of fluid entering along the side located between the points (x, y) and $(x, y + \Delta y)$ is given by $\int_y^{y+\Delta y} (\rho u_1)(x, \eta) \, d\eta$. A similar integral gives the rate of fluid entering the side between (x, y) and $(x + \Delta x, y)$. Letting ρ be a function of x, y, and (for the moment) t, conservation of mass implies

$$\frac{d}{dt} \int_x^{x+\Delta x} \int_y^{y+\Delta y} \rho \, dx \, dy = \int_y^{y+\Delta y} [(\rho u_1)(x, \eta) - (\rho u_1)(x + \Delta x, \eta)] \, d\eta$$
$$+ \int_x^{x+\Delta x} [(\rho u_2)(\xi, y) - (\rho u_2)(\xi, y + \Delta y)] \, d\xi$$

Dividing this equation by $\Delta x \Delta y$ and taking the limit as Δx, Δy tend to zero, and assuming that ρ, u_1, and u_2 are smooth functions of x, y, and t, it follows from calculus that

$$\frac{\partial \rho}{\partial t} + \frac{\partial (\rho u_1)}{\partial x} + \frac{\partial (\rho u_2)}{\partial y} = 0$$

Because the flow is steady $(\partial \rho)/(\partial t) = 0$, and because the flow is incompressible, ρ is constant. Hence this equation yields Eq. (5.4.2).

If the flow is irrotational, it means that the circulation of the fluid along any closed contour C is zero. The circulation around C is given by $\oint_C \mathbf{u} \cdot \mathbf{ds}$, where $\mathbf{ds}$ is the vector element of arc length along C. We could use a derivation similar to the above, or we could use Green's Theorem (see Section 2.5, Theorem 2.5.1 with (u, v) replaced by $\mathbf{u} = (u_1, u_2)$ and $\mathbf{ds} = (dx, dy)$), to deduce

$$\frac{\partial u_2}{\partial x} = \frac{\partial u_1}{\partial y} \tag{5.4.4}$$

This equation is a necessary and sufficient condition for the existence of a potential Φ, that is, Eq. (5.4.3). Therefore Eqs. (5.4.2) and (5.4.4) imply that Φ is harmonic.

Because the function Φ is harmonic, there must exist a conjugate harmonic function, say, $\Psi(x, y)$, such that

$$\Omega(z) = \Phi(x, y) + i\Psi(x, y) \tag{5.4.5}$$

is analytic. Differentiating $\Omega(z)$ and using the Cauchy–Riemann conditions (Eq. (2.1.4)), it follows that

$$\frac{d\Omega}{dz} = \frac{\partial \Phi}{\partial x} + i\frac{\partial \Psi}{\partial x} = \frac{\partial \Phi}{\partial x} - i\frac{\partial \Phi}{\partial y} = u_1 - iu_2 = \bar{u} \tag{5.4.6}$$

where $u = u_1 + iu_2$ is the velocity of the fluid. Thus the "complex velocity" of the fluid is given by

$$u = \overline{\left(\frac{d\Omega}{dz}\right)} \tag{5.4.7}$$

The function $\Psi(x, y)$ is called the stream function, while $\Omega(z)$ is called the complex velocity potential (see also the discussion in Section 2.1). The families of the curves $\Psi(x, y) = \text{const}$ are called streamlines of the flow. These lines represent the actual paths of fluid particles. Indeed, if C is the curve of the path of fluid particles, then the tangent to C has components $(u_1, u_2) = (\Phi_x, \Phi_y)$. Using the Cauchy–Riemann equations (2.1.4) we have

$$\Phi_x \Psi_x + \Phi_y \Psi_y = 0$$

it follows that as vectors $(\Phi_x, \Phi_y) \cdot (\Psi_x, \Psi_y) = 0$, that is, the vector perpendicular to C has components (Ψ_x, Ψ_y), which is, the gradient of Ψ. Hence we know from vector calculus that the curve C is given by $\Psi = \text{const}$.

Example 5.4.2 (Heat Flow) A two-dimensional, steady heat flow.

The quantity of heat conducted per unit area per unit time across a surface of a given solid is called heat flux. In many applications the heat flux, denoted by the vector $\mathbf{Q}$, is given by $\mathbf{Q} = -k\nabla T$, where T denotes the temperature of the solid and k is called the thermal conductivity, which is taken to be constant. Conductivity k depends on the material of the solid. Conservation of energy, in steady state, implies that there is no accumulation of heat inside a given simple closed curve C. Hence if we denote $Q_n = \mathbf{Q} \cdot \hat{n}$ where $\hat{n}$ is the unit outward normal

$$\oint_C Q_n \, ds = \oint_C (Q_1 \, dy - Q_2 \, dx) = 0$$

This equation, together with $\mathbf{Q} = -k\nabla T$, that is

$$Q_1 = -k\frac{\partial T}{\partial x}, \qquad Q_2 = -k\frac{\partial T}{\partial y}$$

and Green's Theorem 2.5.1 from vector calculus (see Section 2.5), imply that T satisfies Laplace's equation. Let Ψ be the harmonic conjugate function of T, then the function

$$\Omega(z) = T(x, y) + i\Psi(x, y) \tag{5.4.8}$$

is analytic. This function is called the complex temperature. The curves of the family $T(x, y) = \mathsf{const}$ are called isothermal lines.

Example 5.4.3 (Electrostatics) We have seen that the appearance of Laplace's equation in fluid flow is a consequence of the conservation of mass and of the assumption that the circulation of the flow along a closed contour equals zero (irrotationality). Furthermore, conservation of mass is equivalent to the condition that the flux of the fluid across any closed surface equals zero. The situation in electrostatics is similar: If $\mathbf{E}$ denotes the electric field, then the following two laws (consequences of the governing equations of time-independent electromagnetics) are valid. (a) The flux of $\mathbf{E}$ through any closed surface enclosing zero charge equals zero. This is a special case of what is known as Gauss's law; that is, $\oint_C E_n \, ds = q/\epsilon_0$, where E_n is the normal component of the electric field, ϵ_0 is the dielectric constant of the medium, and q is the net charge enclosed within C. (b) The electric field is derivable from a potential, or stated differently, the circulation of $\mathbf{E}$ around a simple closed contour equals zero. If the electric field vector is denoted by $\mathbf{E} = (E_1, E_2)$, then these two conditions imply

$$\frac{\partial E_1}{\partial x} + \frac{\partial E_2}{\partial y} = 0, \qquad \frac{\partial E_2}{\partial x} = \frac{\partial E_1}{\partial y} \qquad (5.4.9)$$

From Eq. (5.4.9) we have

$$E_1 = \frac{-\partial \Phi}{\partial x}, \qquad E_2 = \frac{-\partial \Phi}{\partial y} \qquad (5.4.10)$$

(the minus signs are standard convention) and thus from Eq. (5.4.9) the function Φ is harmonic, that is, it satisfies Laplace's equation. Let Ψ denote the function that is conjugate to Φ. Then the function

$$\Omega(z) = \Phi(x, y) + i\Psi(x, y) \qquad (5.4.11)$$

is analytic in any region not occupied by charge. This function is called the complex electrostatic potential. Differentiating $\Omega(z)$, and using the Cauchy–Riemann conditions, it follows that

$$\frac{d\Omega}{dz} = \frac{\partial \Phi}{\partial x} + i\frac{\partial \Psi}{\partial x} = \frac{\partial \Phi}{\partial x} - i\frac{\partial \Phi}{\partial y} = -\overline{E} \qquad (5.4.12)$$

where $E = E_1 + iE_2$ is the complex electric field ($\overline{E} = E_1 - iE_2$). The curves of the families $\Phi(x, y) = \text{const}$ and $\Psi(x, y) = \text{const}$ are called equipotential and flux lines, respectively. From Eq. (5.4.12) Gauss's law is equivalent to

$$\text{Im} \oint_C \overline{E}\, dz = \oint_C (E_1 dy - E_2 dx) = \oint_C E_n ds = q/\epsilon_0 \qquad (5.4.13)$$

We also note that integrals of the form $\int \overline{E} dz$ are invariant under a conformal transformation. More specifically, using Eq. (5.4.12), a conformal transformation $w = f(z)$ transforms the analytic function $\Omega(z)$ to $\Omega(w)$

$$\int \overline{E}\, dz = -\int \frac{d\Omega}{dz}\, dz = -\int \frac{d\Omega}{dw}\, dw = -\int d\Omega \qquad (5.4.14)$$

In order to find the unique solution Φ of Laplace's equation (5.4.1), one needs to specify appropriate boundary conditions. Let R be a simply connected region bounded by a simple closed curve C. There are two types of boundary-value problems that arise frequently in applications: (a) In the **Dirichlet problem** one specifies Φ on the boundary C. (b) In the **Neumann's problem** one specifies the normal derivative of Φ on the boundary C. (There is a third case, the "mixed case" where a combination of Φ and the normal derivative are given on the boundary. We will not discuss this possibility here.)

If a solution exists for a Dirichlet problem, then it must be unique. Indeed if Φ_1 and Φ_2 are two such solutions then $\Phi = \Phi_1 - \Phi_2$ is harmonic in R and $\Phi = 0$ on C. The well-known vector identity (derivable from Green's Theorem (2.5.1))

$$\oint_C \Phi \left(\frac{\partial \Phi}{\partial x} dy - \frac{\partial \Phi}{\partial y} dx \right) = \iint_R \left[\Phi \nabla^2 \Phi + \left(\frac{\partial \Phi}{\partial x} \right)^2 + \left(\frac{\partial \Phi}{\partial y} \right)^2 \right] dx\, dy$$

(5.4.15)

implies

$$\iint_R \left[\left(\frac{\partial \Phi}{\partial x} \right)^2 + \left(\frac{\partial \Phi}{\partial y} \right)^2 \right] dx\, dy = 0 \qquad (5.4.16)$$

Therefore Φ must be a constant in R, and because $\Phi = 0$ on C, we find that $\Phi = 0$ everywhere. Thus $\Phi_1 = \Phi_2$; that is, the solution is unique. The same analysis implies that if a solution exists for a Neumann problem ($\partial \Phi / \partial n = 0$ on C), then it is unique to within an arbitrary additive constant.

It is possible to obtain the solution of the Dirichlet and Neumann problems using conformal mappings. This involves the following steps:

(a) Use a conformal mapping to transform the region R of the z plane onto a simple region such as the unit circle or a half plane of the w plane.
(b) Solve the corresponding problem in the w plane.
(c) Use this solution and the inverse mapping function to solve the original problem (recall that if $f(z)$ is conformal ($f'(z) \neq 0$), then according to Theorem 5.3.3, $f(z)$ has a unique inverse).

This procedure is justified because of the following fact. Let $\Phi(x, y)$ be harmonic in the region R of the z plane. Assume that the region R is mapped onto the region R' of the w plane by the conformal transformation $w = f(z)$, where $w = u + iv$. Then $\Phi(x, y) = \Phi(x(u, v), y(u, v))$ is harmonic in R'. Indeed, by differentiation and use of the Cauchy–Riemann conditions (2.1.4) we can verify (see also Problem 7 in Section 2.1) that

$$\frac{\partial^2 \Phi}{\partial x^2} + \frac{\partial^2 \Phi}{\partial y^2} = \left| \frac{df}{dz} \right|^2 \left(\frac{\partial^2 \Phi}{\partial u^2} + \frac{\partial^2 \Phi}{\partial v^2} \right) \qquad (5.4.17)$$

which, because $df/dz \neq 0$, proves the above assertion. We use these ideas in the following example.

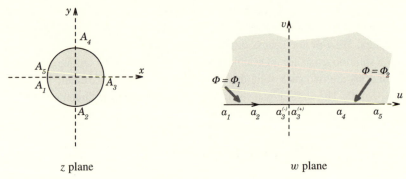

z plane   w plane

Fig. 5.4.2. Transformation of the unit circle

Example 5.4.4 Solve Laplace's equation for a function Φ inside the unit circle that on its circumference takes the value Φ_2 for $0 \le \theta < \pi$, and the value Φ_1 for $\pi \le \theta < 2\pi$. This problem can be interpreted as finding the steady state heat distribution inside a disk with a prescribed temperature Φ on the boundary.

An important class of conformal transformations are of the form $w = f(z)$ where

$$f(z) = \frac{az + b}{cz + d}, \qquad ad - bc \ne 0. \tag{5.4.18}$$

These transformations are called **bilinear transformations**. They will be studied in detail in Section 5.7. In this problem we can verify that the bilinear transformation (see also the discussion in Section 5.7, especially Eq. (5.7.18))

$$w = i\left(\frac{1 - z}{1 + z}\right) \tag{5.4.19}$$

that is,

$$u = \frac{2y}{(1 + x)^2 + y^2}, \qquad v = \frac{1 - (x^2 + y^2)}{(1 + x)^2 + y^2}$$

maps the unit circle onto the upper half of the w plane. (When z is on the unit circle $z = e^{i\theta}$, then $w(z) = u = \frac{\sin\theta}{1 + \cos\theta}$.) The arcs $A_1 A_2 A_3$ and $A_3 A_4 A_5$ are mapped onto the negative and positive real axis, respectively, of the w plane. Let $w = \rho e^{i\psi}$. The function $a\psi + b$, where a and b are real constants, is the real part of the analytic function $-ai \log w + b$ in the upper half plane and therefore is harmonic. Hence a solution of Laplace's equation in the upper half of the w plane, satisfying $\Phi = \Phi_1$ for $u < 0$, $v = 0$ (i.e., $\psi = \pi$) and $\Phi = \Phi_2$

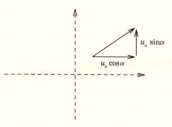

Fig. 5.4.3. Flow velocity

for $u > 0$, $v = 0$ (i.e., $\psi = 0$), is given by

$$\Phi = \Phi_2 - (\Phi_2 - \Phi_1)\frac{\psi}{\pi} = \Phi_2 - \frac{\Phi_2 - \Phi_1}{\pi} \tan^{-1}\left(\frac{v}{u}\right)$$

Owing to the uniqueness of solutions to the Dirichlet problem, this is the only solution. Using the expressions for u and v given by Eq. (5.4.19), it follows that in the x, y plane the solution to the problem posed in the unit circle is given by

$$\Phi(x, y) = \Phi_2 - \frac{\Phi_2 - \Phi_1}{\pi} \tan^{-1}\left[\frac{1 - (x^2 + y^2)}{2y}\right]$$

(See also Problem 9 in Section 2.2.)

Example 5.4.5 Find the complex potential and the streamlines of a fluid moving with a constant speed $u_0 \in \mathbb{R}$ in a direction making an angle α with the positive x axis. (See also Example 2.1.7.)

The x and y component of the fluid velocity are given by $u_0 \cos \alpha$ and $u_0 \sin \alpha$. The complex velocity is given by

$$u = u_0 \cos \alpha + i u_0 \sin \alpha = u_0 e^{i\alpha}$$

thus

$$\frac{d\Omega}{dz} = \bar{u} = u_0 e^{-i\alpha}, \quad \text{or} \quad \Omega = u_0 e^{-i\alpha} z$$

where we have equated the constant of integration to zero. Letting $\Omega = \Phi + i\Psi$, it follows that

$$\Psi(x, y) = u_0(y \cos \alpha - x \sin \alpha) = u_0 r \sin(\theta - \alpha)$$

The streamlines are given by the family of the curves $\Psi = \text{const}$, which are straight lines making an angle α with the positive x axis.

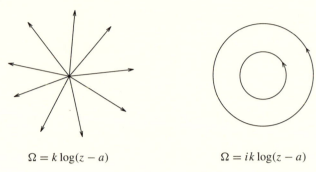

$$\Omega = k \log(z - a) \qquad\qquad \Omega = ik \log(z - a)$$

Fig. 5.4.4. Streamlines

Example 5.4.6 Analyze the flow pattern of a fluid emanating at a constant rate from an infinite line source perpendicular to the z plane at $z = 0$.

Let ρ and u_r denote the density (constant) and the radial velocity of the fluid, respectively. Let q denote the mass of fluid per unit time emanating from a line source of unit length. Then

$$q = (\text{density})(\text{flux}) = \rho(2\pi r u_r)$$

Thus

$$u_r = \frac{q}{2\pi\rho}\frac{1}{r} \equiv \frac{k}{r}, \qquad k > 0$$

where the constant $k = q/2\pi\rho$ is called the strength of the source. Integrating the equation $u_r = \partial\Phi/\partial r$ and equating the constant of integration to zero (Φ is cylindrically symmetric), it follows that $\Phi = k \log r$, and hence with $z = re^{i\theta}$

$$\Omega(z) = k \log z$$

The streamlines of this flow are given by $\Psi = \text{Im}\,\Omega(z) = \text{const}$, that is, $\theta = \text{const}$. These curves are rays emanating from the origin.

The complex potential $\Omega(z) = k \log(z - a)$ represents a "source" located at $z = a$. Similarly, $\Omega(z) = -k \log(z - a)$ represents a "sink" located at $z = a$ (because of the minus sign the velocity is directed toward $z = 0$).

It is clear that if $\Omega(z) = \Phi + i\Psi$ is associated with a flow pattern of streamlines $\Psi = \text{const}$, the function $i\Omega(z)$ is associated with a flow pattern of streamlines $\Phi = \text{const}$. These curves are orthogonal to the curves $\Psi = \text{const}$; that is, the flows associated with $\Omega(z)$ and $i\Omega(z)$ have orthogonal streamlines.

This discussion implies that in the particular case of the above example the streamlines of $\Omega(z) = ik \log z$ are concentric circles. Because $d\Omega/dz = ikz^{-1}$, it follows that the complex velocity is given by

$$\overline{\left(\frac{d\Omega}{dz}\right)} = \frac{k\sin\theta}{r} - \frac{ik\cos\theta}{r}$$

This represents the flow of a fluid rotating with a clockwise speed k/r around $z = 0$. This flow is usually referred to as a vortex flow, generated by a vortex of strength k localized at $z = 0$. If the vortex is localized at $z = a$, then the associated complex potential is given by $\Omega = ik\log(z-a)$. (See also Problems 7 and 8 of Section 2.2.)

Example 5.4.7 (*The force due to fluid pressure*) In the physical circumstances we are dealing with, one neglects viscosity, that is, the internal friction of a fluid. It can be shown from basic fluid equations that in this situation the pressure P of the fluid and the speed $|u|$ of the fluid are related by the so-called Bernoulli equation

$$P + \frac{1}{2}\rho|u|^2 = \alpha \tag{5.4.20}$$

where α is a constant along each streamline. Let $\Omega(z)$ be the complex potential of some flow and let the simple closed curve C denote the boundary of a cylindrical obstacle of unit length that is perpendicular to the z plane. We shall show that the force $F = X + iY$ exerted on this obstacle is given by

$$\bar{F} = \frac{1}{2}i\rho \oint_C \left(\frac{d\Omega}{dz}\right)^2 dz$$

Let ds denote an infinitesimal element around some point of the curve C, and let θ be the angle of the tangent to C at this point. The infinitesimal force

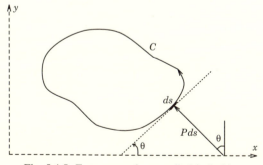

Fig. 5.4.5. Force exerted on a cylindrical object

exerted on the part of the cylinder corresponding to ds is perpendicular to ds and has magnitude $P ds$. (Recall that force equals pressure times area, and area equals ds times 1 because the cylinder has unit length). Hence

$$dF = dX + i\, dY = -P\, ds\, \sin\theta + i\, P\, ds\, \cos\theta = i\, Pe^{i\theta} ds$$

Also

$$dz = dx + i\, dy = ds\, \cos\theta + i\, ds\, \sin\theta = ds\, e^{i\theta}$$

Without friction, the curve C is a streamline of the flow. The velocity is tangent to this curve, where we denote the complex velocity as $u = |u|e^{i\theta}$; hence

$$\frac{d\Omega}{dz} = |u|\, e^{-i\theta} \tag{5.4.21}$$

The expression for dF and Bernoulli's equation (5.4.20) imply

$$F = X + iY = \oint_C i\left(\alpha - \frac{1}{2}\rho|u|^2\right)e^{i\theta} ds$$

The first term in the right-hand side of this equation equals zero because $\oint e^{i\theta} ds = \oint dz = 0$. Thus

$$\bar{F} = \frac{1}{2}i\rho \oint_C |u|^2 e^{-i\theta} ds = \frac{1}{2}i\rho \oint_C \left(\frac{d\Omega}{dz}\right)^2 e^{i\theta} ds = \frac{1}{2}i\rho \oint_C \left(\frac{d\Omega}{dz}\right)^2 dz$$

where we have used Eq. (5.4.21) to replace $|u|$ with $d\Omega/dz$, as well as dz with $ds\, e^{i\theta}$.

Example 5.4.8 Discuss the flow pattern associated with the complex potential

$$\Omega(z) = u_0\left(z + \frac{a^2}{z}\right) + \frac{i\gamma}{2\pi}\log z$$

This complex potential represents the superposition of a vortex of circulation of strength γ with a flow generated by the complex potential $u_0(z + a^2 z^{-1})$. (The latter flow was also discussed in Example 2.1.8.) Let $z = re^{i\theta}$; then if $\Omega = \Phi + i\Psi$

$$\Psi(x, y) = u_0\left(r - \frac{a^2}{r}\right)\sin\theta + \frac{\gamma}{2\pi}\log r, \quad a > 0, \quad a, \gamma \text{ real constants}$$

If $r = a$, then $\Psi(x, y) = \gamma \log a/2\pi = \text{const}$, therefore $r = a$ is a streamline.

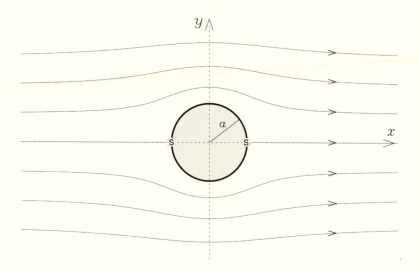

(Stagnation points marked with Ⓢ)

Fig. 5.4.6. Flow around a circular obstacle ($\gamma = 0$)

Furthermore

$$\frac{d\Omega}{dz} = u_0\left(1 - \frac{a^2}{z^2}\right) + \frac{i\gamma}{2\pi z}$$

which shows that as $z \to \infty$, the velocity tends to u_0. This discussion shows that the flow associated with $\Omega(z)$ can be considered as a flow with circulation about a circular obstacle. In the special case that $\gamma = 0$, this flow is depicted in Figure 5.4.6.

Note that when $\gamma = 0$, $d\Omega/dz = 0$ for $z = \pm a$; that is, there exist two points for which the velocity vanishes. Such points are called stagnation points see Figures (5.4.6–5.4.9); the streamline going through these points is given by $\Psi = 0$. In the general case of $\gamma \neq 0$, there also exist two stagnation points given by $d\Omega/dz = 0$, or

$$z = -\frac{i\gamma}{4\pi u_0} \pm \sqrt{a^2 - \frac{\gamma^2}{16\pi^2 u_0^2}}$$

If $0 \leq \gamma < 4\pi a u_0$, there are two stagnation points on the circle (see Figure 5.4.7). If $\gamma = 4\pi a u_0$, these two points coincide (see Figure 5.4.8) at $z = -ia$. If $\gamma > 4\pi a u_0$, then one of the stagnation points lies outside the circle, and one inside (see Figure 5.4.9).

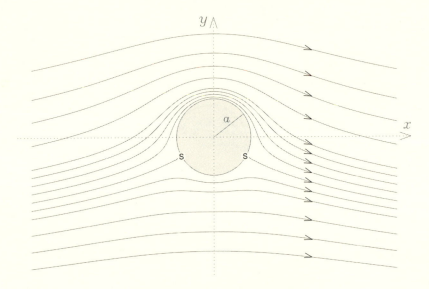

(Stagnation points marked with Ⓢ)

Fig. 5.4.7. Separate stagnation points ($\gamma < 4\pi a u_0$)

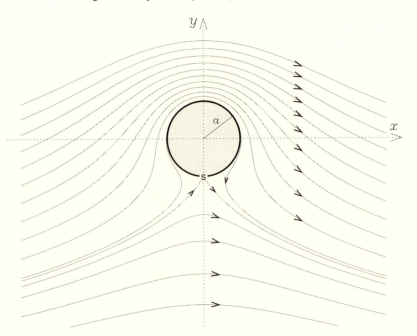

(Stagnation point marked with Ⓢ)

Fig. 5.4.8. Coinciding stagnation points ($\gamma = 4\pi a u_0$)

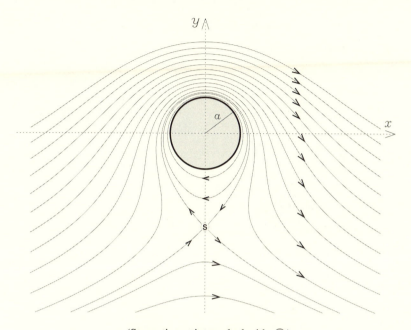

(Stagnation point marked with ⑤)

Fig. 5.4.9. Streamlines and stagnation point ($\gamma > 4\pi a u_0$)

Using the result of Example 5.4.7, it is possible to compute the force exerted on this obstacle

$$\bar{F} = \frac{1}{2} i \rho \oint_C \left[u_0 \left(1 - \frac{a^2}{z^2} \right) + \frac{i\gamma}{2\pi z} \right]^2 dz = -i\rho u_0 \gamma$$

(Recall that $\oint z^n dz = 2\pi i \delta_{n,-1}$, where $\delta_{n,-1}$ is the Kronecker delta function.) This shows that there exists a net force in the positive y direction of magnitude $\rho u_0 \gamma$. Such a force is known in aerodynamics as **lift**.

Example 5.4.9 Find the complex electrostatic potential due to a line of constant charge q per unit length perpendicular to the z plane at $z = 0$.

The relevant electric field is radial and has magnitude E_r. If C is the circular basis of a cylinder of unit length located at $z = 0$, it follows from Gauss's law (see Example 5.4.3) that

$$\oint_C E_r \, ds = E_r 2\pi r = 4\pi q, \quad \text{or} \quad E_r = \frac{2q}{r}$$

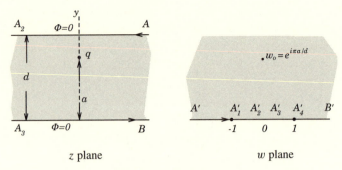

Fig. 5.4.10. Electrostatic potential between parallel plates

where q is the charge enclosed by the circle C, and here we have normalized to $\epsilon_0 = 1/4\pi$. Hence the potential satisfies

$$\frac{\partial \Phi}{\partial r} = -\frac{2q}{r}, \quad \text{or} \quad \Phi = -2q \log r, \quad \text{or} \quad \Omega(z) = -2q \log z$$

This is identical to the complex potential associated with a line source of strength $k = -2q$. From Eq. (5.4.13) we see that $\mathrm{Im}(\oint \overline{E}\, dz) = \mathrm{Im}(\oint -\Omega'(z)\, dz) = 4\pi q$, as it should.

Example 5.4.10 Consider two infinite parallel flat plates, separated by a distance d and maintained at zero potential. A line of charge q per unit length is located between the two planes at a distance a from the lower plate (see Figure 5.4.10). Find the electrostatic potential in the shaded region of the z plane.

The conformal mapping $w = \exp(\pi z/d)$ maps the shaded strip of the z plane onto the upper half of the w plane. So the point $z = ia$ is mapped to the point $w_0 = \exp(i\pi a/d)$; the points on the lower plate, $z = x$, and on the upper plate, $z = x + id$, map to the real axis $w = u$ for $u > 0$ and $u < 0$, respectively. Let us consider a line of charge q at w_0 and a line of charge $-q$ at $\overline{w_0}$. Consider the associated complex potential (see also the previous Example 5.4.9)

$$\Omega(w) = -2q \log(w - w_0) + 2q \log(w - \overline{w_0}) = 2q \log \left(\frac{w - \overline{w_0}}{w - w_0} \right)$$

Calling C_q a closed contour around the charge q, we see that Gauss' law is satisfied,

$$\oint_{C_q} E_n\, ds = \mathrm{Im} \oint_{C_q} \overline{E}\, dz = \mathrm{Im} \oint_{\tilde{C}_q} -\Omega'(w)\, dw = 4\pi q$$

where $\tilde{C}_q$ is the image of C_q in the w-plane. (Again, see Example 5.4.3, with $\epsilon_0 = 1/4\pi$.) Then, calling $\Omega = \Phi + i\Psi$, we see that Φ is zero on the real axis of the w plane (because $\log A/A^*$ is purely imaginary). Consequently, we have satisfied the boundary condition $\Phi = 0$ on the plates, and hence the electrostatic potential at any point of the shaded region of the z plane is given by

$$\Phi = 2q \, \mathrm{Re} \log \left[\frac{w - e^{-iv}}{w - e^{iv}} \right] = 2q \, \mathrm{Re} \log \left[\frac{e^{\frac{\pi z}{d}} - e^{-iv}}{e^{\frac{\pi z}{d}} - e^{iv}} \right], \qquad v \equiv \frac{\pi a}{d}$$

Problems for Section 5.4

1. Consider a source at $z = -a$ and a sink at $z = a$ of equal strengths k.

   (a) Show that the associated complex potential is $\Omega(z) = k \log[(z + a)/(z - a)]$.

   (b) Show that the flow speed is $2ka/\sqrt{a^4 - 2a^2 r^2 \cos 2\theta + r^4}$, where $z = re^{i\theta}$.

2. Use Bernoulli's equation (5.4.20) to determine the pressure at any point of the fluid of the flow studied in Example 5.4.6.

3. Consider a flow with the complex potential $\Omega(z) = u_0(z + a^2/z)$, the particular case $\gamma = 0$ of Example 5.4.8. Let p and p_∞ denote the pressure at a point on the cylinder and far from the cylinder, respectively.

   (a) Use Eq. (5.4.20) to establish that $p - p_\infty = \frac{1}{2}\rho u_0^2(1 - 4\sin^2\theta)$.

   (b) Show that a vacuum is created at the points $\pm ia$ the fluid is equal to or greater than $u_0 = \sqrt{2p_\infty/(3\rho)}$. This phenomenon is usually called **cavitation**.

4. Discuss the fluid flow associated with the complex velocity potential $\Omega(z) = Q_0 z + \frac{\bar{Q}_0 a^2}{z} + \frac{i\gamma}{2\pi} \log z$, $a > 0$, γ real, $Q_0 = U_0 + i V_0$. Show that the force exerted on the cylindrical obstacle defined by the flow field is given by $\mathbf{F} = i\rho \bar{Q}_0 \gamma$. This force is often referred to as the **lift**.

5. Show that the steady-state temperature at any point of the region given in Figure 5.4.11, where the temperatures are maintained as indicated in the figure, is given by

$$T(r, \theta) = \frac{10}{\pi} \tan^{-1} \left\{ \frac{(r^2 - 1)\sin\theta}{(r^2 + 1)\cos\theta - 2r} \right\}$$

$$- \frac{10}{\pi} \tan^{-1} \left\{ \frac{(r^2 - 1)\sin\theta}{(r^2 + 1)\cos\theta + 2r} \right\}$$

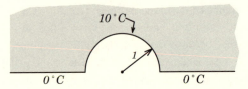

Fig. 5.4.11. Temperature distribution for Problem 5.4.5

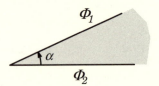

Fig. 5.4.12. Electrostatic potential for Problem 5.4.7

Hint: use the transformation $w = z + 1/z$ to map the above shaded region onto the upper half plane.

6. Let $\Omega(z) = z^{\alpha}$, where α is a real constant and $\alpha > \frac{1}{2}$. If $z = re^{i\theta}$ show that the rays $\theta = 0$ and $\theta = \pi/\alpha$ are streamlines and hence can be replaced by walls. Show that the speed of the flow is $\alpha r^{\alpha-1}$, where r is the distance from the corner.

7. Two semiinfinite plane conductors meet at an angle $0 < \alpha < \pi/2$ and are charged at constant potentials Φ_1 and Φ_2.

    Show that the potential Φ and the electric field $\mathbf{E} = (E_r, E_\theta)$ in the region between the conductors are given by

    $$\Phi = \Phi_2 + \left(\frac{\Phi_1 - \Phi_2}{\alpha}\right)\theta, \quad E_\theta = \frac{\Phi_2 - \Phi_1}{\alpha r}, \quad E_r = 0,$$

    where $z = re^{i\theta}, 0 \leq \theta \leq \alpha$.

8. Two semiinfinite plane conductors intersect at an angle $\alpha, 0 < \alpha < \pi$, and are kept at zero potential.

    A line of charge q per unit length is located at the point z_1, which is equidistant from both planes. Show that the potential in the shaded region is given by

    $$\mathrm{Re}\left\{-2q \log\left(\frac{z^{\frac{\pi}{\alpha}} - z_1^{\frac{\pi}{\alpha}}}{z^{\frac{\pi}{\alpha}} - \bar{z}_1^{\frac{\pi}{\alpha}}}\right)\right\}$$

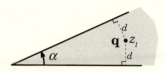

Fig. 5.4.13. Electrostatics, Problem 5.4.8.

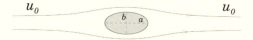

Fig. 5.4.14.

9. Consider the flow past an elliptic cylinder indicated in Figure 5.4.14.

    (a) Show that the complex potential is given by

    $$\Omega(z) = u_0 \left(\zeta + \frac{(a+b)^2}{4\zeta} \right)$$

    where

    $$\zeta \equiv \frac{1}{2} \left(z + \sqrt{z^2 - c^2} \right), \quad c^2 = a^2 - b^2$$

    (b) Show that the fluid speed at the top and bottom of the cylinder is $u_0(1 + \frac{b}{a})$.

10. Two infinitely long cylindrical conductors having cross sections that are confocal ellipses with foci at $(-c, 0)$ and $(c, 0)$ (see Figure 5.4.15) are kept at constant potentials Φ_1 and Φ_2.

    (a) Show that the mapping $z - i\eta_1 = c \sin \zeta = c \sin(\xi + i\eta)$ transforms the confocal ellipses in Figure 5.4.15 onto two parallel plates such as those depicted in Figure 5.4.10, where $\Phi = \Phi_j$ on $\eta = \eta_j$, with $\cosh \eta_j = R_j/c$, $j = 1, 2$. Use the transformation $w = \exp(\pi \zeta/d - i\eta_1)$, where $d = \eta_2 - \eta_1$ (see Example 5.4.10) to show that the complex potential is given by

    $$\Omega(w) = \Phi_1 + \frac{\Phi_2 - \Phi_1}{i\pi} \log w = \Phi_1 + \frac{\Phi_2 - \Phi_1}{id} \sin^{-1}\left(\frac{z}{c} - i\eta_1 \right).$$

    (b) If the capacitance of two perfect conductors is defined by $C = q/(\Phi_1 - \Phi_2)$, where q is the charge on the inside ellipse, use Gauss'

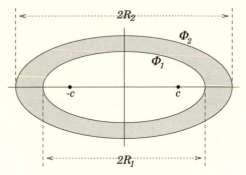

Fig. 5.4.15. Confocal ellipses, Problem 5.4.10

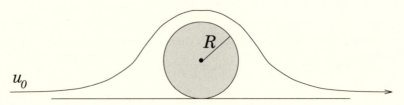

Fig. 5.4.16. Ideal flow, Problem 5.4.11

law to show that the capacitance per unit length is given by

$$C = \frac{1}{2d} = \frac{2\pi}{2\left(\cosh^{-1}\left(\frac{R_2}{c}\right) - \cosh^{-1}\left(\frac{R_1}{c}\right)\right)}$$

(c) Establish that as $c \to 0$ (two concentric circles):

$$C \to \frac{1}{2\log(R_2/R_1)}$$

11. A circular cylinder of radius R lies at the bottom of a channel of fluid that, at large distance from the cylinder, has constant velocity u_0.

(a) Show that the complex potential is given by

$$\Omega(z) = \pi R u_0 \coth\left(\frac{\pi R}{z}\right)$$

(b) Show that the difference in pressure between the top and the bottom points of the cylinder is $\rho\pi^4 u_0^2/32$, where ρ is the density of the fluid (see Eqs. (5.4.20)–(5.4.21)).

*5.5 Theoretical Considerations – Mapping Theorems

In Section 5.3, various mapping theorems were stated, but their proofs were postponed to this optional section.

Theorem 5.5.1 (originally stated as Theorem 5.3.2) Let $f(z)$ be analytic and not constant in a domain D of the complex z plane. The transformation $w = f(z)$ can be interpreted as a mapping of the domain D onto the domain $D^* = f(D)$ of the complex w plane. (Sometimes this theorem is summarized as "open sets map to open sets.")

Proof A point set is a domain if it is open and connected (see Section 1.2). An open set is connected if every two points of this set can be joined by a contour lying in this set. If we can prove that D^* is an open set, its connectivity is an immediate consequence of the fact that, because $f(z)$ is analytic, every continuous arc in D is mapped onto a continuous arc in D^*. The proof that D^* is an open set follows from an application of Rouche's Theorem (see Section 4.4), which states: if the functions $g(z)$ and $\tilde{g}(z)$ are analytic in a domain and on the boundary of this domain, and if on the boundary $|g(z)| > |\tilde{g}(z)|$, then in this domain the functions $g(z) - \tilde{g}(z)$ and $g(z)$ have exactly the same number of zeroes. Because $f(z)$ is analytic in D, then $f(z)$ has a Taylor expansion at a point $z_0 \in D$. Assume that $f'(z_0) \neq 0$. Then $g(z) \equiv f(z) - f(z_0)$ vanishes (it has a zero of order 1) at z_0. Because $f(z)$ is analytic, this zero is isolated (see Theorem 3.2.7). That is, there exists a constant $\varepsilon > 0$ such that $g(z) \neq 0$ for $0 < |z - z_0| \leq \varepsilon$. On the circle $|z - z_0| = \varepsilon$, $g(z)$ is continuous; hence there exists a positive constant A such that $A = \min |f(z) - f(z_0)|$ on $|z - z_0| = \varepsilon$. If $\tilde{g}(z) \equiv a$ is a complex constant such that $|a| < A$, then $|g(z)| > |a| = |\tilde{g}(z)|$ on $|z - z_0| = \varepsilon$, and Rouche's Theorem implies that $g(z) - \tilde{g}(z)$ vanishes in $|z - z_0| < \varepsilon$.

Hence, for every complex number $a = |a|e^{i\phi}$, $|a| < A$, we find that there is exactly one value for $g(z) = w - w_0 = a$ corresponding to every z inside $|z - z_0| < \varepsilon$. Therefore, if $z_0 \in D$, $f'(z_0) \neq 0$, and $w_0 = f(z_0)$, then for sufficiently small $\varepsilon > 0$ there exists a $\delta > 0$ such that the image of $|z - z_0| < \varepsilon$ contains the disk $|w - w_0| < \delta$ (here $\delta = A$), and therefore D^* is open. If $f'(z_0) = 0$, a slight modification of the above argument is required. If the first nonvanishing derivative of $f(z)$ at z_0 is of the nth order, then $g(z)$ has a zero of the nth order at $z = z_0$. The rest of the argument goes through as above, but in this case one obtains from Rouche's Theorem the additional information that in $|z - z_0| < \varepsilon$ the values w, for which $|w - w_0| < A$, will now be taken n times. ∎

Theorem 5.5.2 (originally stated as Theorem 5.3.3) (1) Assume that $f(z)$ is analytic at z_0 and that $f'(z_0) \neq 0$. Then $f(z)$ is univalent in the neighborhood of z_0. More precisely, f has a unique analytic inverse F in the neighborhood of $w_0 \equiv f(z_0)$; that is, if z is sufficiently near z_0, then $z = F(w)$, where $w \equiv f(z)$. Similarly, if w is sufficiently near w_0 and $z \equiv F(w)$, then $w = f(z)$. Furthermore, $f'(z)F'(w) = 1$, which implies that the inverse map is conformal.

(2) Assume that $f(z)$ is analytic at z_0 and that it has a zero of order n; that is, the first nonvanishing derivative of $f(z)$ at z_0 is $f^{(n)}(z_0)$. Then to each w sufficiently close to $w_0 \equiv f(z_0)$, there correspond n distinct points z in the neighborhood of z_0, each of which has w as its image under the mapping $w = f(z)$. Actually, this mapping can be decomposed in the form $w - w_0 = \zeta^n$, $\zeta = g(z - z_0)$, $g(0) = 0$ where $g(z)$ is univalent near z_0 and $g(z) = zH(z)$ with $H(0) \neq 0$.

Proof (1) The first part of the proof follows from Theorem 5.5.1, where it was shown that each w in the disk $|w - w_0| < A$ denoted by P is the image of a unique point z in the disk $|z - z_0| < \varepsilon$, where $w = f(z)$. This uniqueness implies $z = F(w)$ and $z_0 = F(w_0)$. The equations $w = f(z)$ and $z = F(w)$ imply the usual equation $w = f(F(w))$ satisfied by a function and its inverse. First we show that $F(w)$ is continuous in P and then show that this implies that $F(w)$ is analytic.

Let $w_1 \in P$ be the image of a unique point z_1 in $|z - z_0| < \varepsilon$. From Theorem 5.5.1, the image of $|z - z_1| < \varepsilon_1$ contains $|w - w_1| < \delta_1$, so for sufficiently small δ_1 we have $|z - z_1| = |F(w) - F(w_1)| < \varepsilon_1$, and therefore $F(w)$ is continuous.

Next assume that w_1 is near w. Then w and w_1 are the images corresponding to $z = F(w)$ and $z_1 = F(w_1)$, respectively. If w is fixed, the continuity of F implies that if $|w_1 - w|$ is small, then $|z_1 - z|$ is also small. Thus

$$\frac{F(w_1) - F(w)}{w_1 - w} = \frac{z_1 - z}{w_1 - w} = \frac{z_1 - z}{f(z_1) - f(z)} \longrightarrow \frac{1}{f'(z)} \tag{5.5.1}$$

as $|w_1 - w| \longrightarrow 0$.

Because $f(z) = w$ has only one solution for $|z - z_0| < \varepsilon$, it follows that $f'(z) \neq 0$. Thus Eq. (5.5.1) implies that $F'(w)$ exists and equals $1/f'(z)$. We also see, by the continuity of $f(z)$, that every z near z_0 has as its image a point near w_0. So if $|z - z_0|$ is sufficiently small, $w = f(z)$ is a point in P and $z = F(w)$. Thus $z = F(f(z))$ near z_0, which by the chain rule implies $1 = f'(z)F'(w)$, is consistent with Eq. (5.5.1).

(2) Assume for convenience, without loss of generality, that $z_0 = w_0 = 0$. Using the fact that the first $(n - 1)$ derivatives of $f(z)$ vanish at $z = z_0$, we see from its Taylor series that $w = z^n h(z)$, where $h(z)$ is analytic at $z = 0$ and $h(0) \neq 0$. Because $h(0) \neq 0$ there exists an analytic function $H(z)$ such that $h(z) = [H(z)]^n$, with $H(0) \neq 0$. (The function $H(z)$ can be found by taking the logarithm.) Thus $w = (g(z))^n$, where $g(z) = zH(z)$. The function $g(z)$ satisfies $g(0) = 0$ and $g'(0) \neq 0$, thus it is univalent near 0. The properties of $w = \zeta^n$ together with the fact that $g(z)$ is univalent imply the assertions of part (2) of Theorem 5.3.3. ∎

Theorem 5.5.3 (originally stated as Theorem 5.3.4) Let C be a simple closed contour enclosing a domain D, and let $f(z)$ be analytic on C and in D. Suppose $f(z)$ takes no value more than once on C. Then (a) the map $w = f(z)$ takes C enclosing D to a simple closed contour C^* enclosing a region D^* in the w plane; (b) $w = f(z)$ is a one-to-one map from D to D^*; and (c) if z traverses C in the positive direction, then $w = f(z)$ traverses C^* in the positive direction.

Proof (a) The image of C is a simple closed contour C^* because $f(z)$ is analytic and because $f(z)$ takes on no value more than once for z on C.

(b) Consider the following integral with the transformation $w = f(z)$, where w_0 corresponds to an arbitrary point $z_0 \in D$ and is not a point on C^*:

$$I = \frac{1}{2\pi i} \oint_C \frac{f'(z)\, dz}{f(z) - w_0} = \frac{1}{2\pi i} \oint_{C^*} \frac{dw}{w - w_0} \tag{5.5.2}$$

From the argument principal in Section 4.4 (Theorem 4.4.1) we find that $I = N - P$, where N and P are the number of zeroes and poles (respectively) of $f(z) - w_0$ enclosed within C. However, because $f(z)$ is analytic, $P = 0$ and $I = N$.

If w_0 lies outside C^*, the right-hand side of Eq. (5.5.2) is 0, and therefore $N = 0$ so that $f(z) \neq w_0$ inside C. If w_0 lies inside C^*, then the right-hand side of Eq. (5.5.2) is 1 (assuming for now, the usual positive convention in $\oint$), and therefore $f(z) = w_0$ once inside C. Finally, w_0 could not lie on C^* because it is an image of some point $z_0 \in D$, and, from Theorem 5.5.1 (open sets map to open sets), some point in the neighborhood of w_0 would need to be mapped to the exterior of C^*, which we have just seen is not possible.

Consequently, each value w_0 inside C^* is attained once and only once, and the transformation $w = f(z)$ is a one-to-one map.

(c) The above proof assumes that both C and C^* are traversed in the positive direction. If C^* is traversed in the negative direction, then the right-hand side of Eq. (5.5.2) would yield -1, which contradicts the fact that N must be positive. Clearly, C and C^* can both be traversed in the negative directions.

Finally, we conclude this section with a statement of the Riemann Mapping Theorem. First we remark that the entire finite plane, $|z| < \infty$, is simply connected. However, there exists no conformal map that maps the entire finite plane onto the unit disk. This is a consequence of Liouville's Theorem because an analytic function $w = f(z)$ such that $|f(z)| < 1$ for all finite $z \in \mathbb{C}$ would have to be constant. Similar reasoning shows that there exists no conformal map that maps the extended plane $|z| \leq \infty$ onto the unit disc. By Riemann's Mapping Theorem, these are the only simply connected domains that cannot be mapped onto the unit disk.

Theorem 5.5.4 (Riemann Mapping Theorem) Let D be a simply connected domain in the z plane, which is neither the z plane or the extended z plane. Then there exists a univalent function $f(z)$, such that $w = f(z)$ maps D onto the disk $|w| < 1$.

The proof of this theorem requires knowledge of the topological concepts of completeness and compactness. It involves considering families of mappings and solving a certain maximum problem for a family of bounded continuous functionals. This proof, which is nonconstructive, can be found in advanced textbooks (see, for example, Nehari (1952)). In the case of a simply connected domain bounded by a smooth Jordan curve, a simpler proof has been given (Garabedian, 1991).

Remarks It should be emphasized that the Riemann Mapping Theorem is a statement about simply connected open sets. It says nothing about the behavior of the mapping function on the boundary. However, for many applications of conformal mappings, such as the solution of boundary value problems, it is essential that one is able to define the mapping function on the boundary. For this reason it is important to identify those bounded regions for which the mapping function can be extended continuously to the boundary. It can be shown (Osgood–Carathéodory Theorem) that if D is bounded by a simple closed contour, then it is possible to extend the function f mapping D conformally onto the open unit disk in such a way that f is continuous on the boundaries.

A further consequence of all this is that fixing any three points of the mapping $f(z)$ uniquely determines the map. The essential reason for this is that two different maps onto the unit circle can be transformed to one another by a bilinear transformation, which can be shown to be fixed by three points (see Section 5.7).

We also note that there is a bilinear transformation (e.g. Eq. (5.4.19) and see also Eq. (5.7.18)) that maps the unit circle onto the upper half plane, so in the theorem we could equally well state that $w = f(z)$ maps D onto the upper half w plane.

5.6 The Schwarz–Christoffel Transformation

One of the most remarkable results in the theory of complex analysis is Riemann's Mapping Theorem, Theorem 5.5.4. This theorem states that any simply connected domain of the complex z plane, with the exception of the entire z plane and the extended entire z plane, can be mapped with a univalent transformation $w = f(z)$ onto the disk $|w| < 1$ or onto the upper half of the complex w plane. Unfortunately, the proof of this celebrated theorem is not constructive, that is, given a specific domain in the z plane, there is no general constructive approach for finding $f(z)$. Nevertheless, as we have already seen, there are many particular domains for which $f(z)$ can be constructed explicitly. One such domain is the interior of a polygon. Let us first consider an example of a very simple polygon.

Example 5.6.1 The interior of an open triangle of angle $\pi\alpha$, with vertex at the origin of the w plane is mapped to the upper half z plane by $w = z^\alpha, 0 < \alpha < 2$; see Figure 5.6.1.

If $z = re^{i\theta}$, $w = \rho e^{i\varphi}$, then the rays $\varphi = 0$ and $\varphi = \pi\alpha$ of the w plane are mapped to the rays $\theta = 0$ and $\theta = \pi$ of the z plane. We note that the

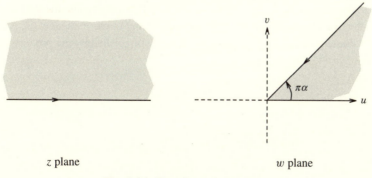

z plane $\qquad\qquad\qquad$ w plane

Fig. 5.6.1. Transformation $w = z^\alpha$

conformal property, that is that angles are preserved under the transformation $w = f(z) = z^{\alpha}$, doesn't hold at $z = 0$ since $f(z)$ is not analytic there when $\alpha \neq 1$.

The transformation $w = f(z)$ associated with a general polygon is called the Schwarz–Christoffel transformation. In deriving this transformation we will make use of the so-called Schwarz reflection principle. The most basic version of this principle is really based on the following elementary fact. Suppose that $f(z)$ is analytic in a domain D that lies in the upper half of the complex z plane. Let $\tilde{D}$ denote the domain obtained from D by reflection with respect to the real axis (obviously $\tilde{D}$ lies on the lower half of the complex z plane). Then corresponding to every point $z \in \tilde{D}$, the function $\tilde{f}(z) = \overline{f(\bar{z})}$ is analytic in $\tilde{D}$.

Indeed, if $f(z) = u(x, y) + iv(x, y)$, then $\overline{f(\bar{z})} = u(x, -y) - iv(x, -y)$. This shows that the real and imaginary parts of the function $\tilde{f}(z) \equiv \overline{f(\bar{z})}$ have continuous partial derivatives, and that the Cauchy–Riemann equations (see Section 2.1) for f imply the Cauchy–Riemann equations for $\tilde{f}$, and hence $\tilde{f}$ is analytic. Call $\tilde{u}(x, y') = u(x, -y)$, $\tilde{v}(x, y') = -v(x, -y)$, $y' = -y$, then the Cauchy Riemann conditions for $\tilde{u}$ and $\tilde{v}$ in terms of x and y' follow.

Example 5.6.2 The function $f(z) = 1/(z + i)$ is analytic in the upper half z plane. Use the Schwarz reflection principle to construct a function analytic in the lower half z plane.

The function $\overline{f(\bar{z})} = 1/(z - i)$ has a pole at $z = i$ as its only singularity; therefore it is analytic for $\text{Im} z \leq 0$.

The above idea not only applies to reflection about straight lines, but it also applies to reflections about circular arcs. (This is discussed more fully in Section 5.7.) In a special case, it implies that if $f(z)$ is analytic inside the unit circle, then the function $\overline{f(1/\bar{z})}$ is analytic outside the unit circle. Note that the points in the domains D and $\tilde{D}$ of Figure 5.6.2 are distinguished by the property that

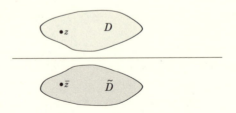

Fig. 5.6.2. Reflection principle

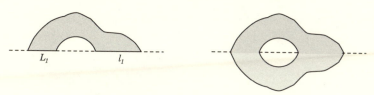

Fig. 5.6.3. Analytic continuation across the real axis

they are inverse points with respect to the real axis. If one uses a bilinear transformation (see, e.g., Eqs. (5.4.19) or (5.7.18)) to map the real axis onto the unit circle, the corresponding points z and $1/\bar{z}$ will be the "inverse" points with respect to this circle. (Note that the inverse points with respect to a circle are defined in property (viii) of Section 5.7).

The Schwarz reflection principle, across real line segments is the following. Suppose that the domain D has part of the real axis as part of its boundary. Assume that $f(z)$ is analytic in D and is continuous as z approaches the line segments $L_1, \ldots, L_n$ of the real axis and that $f(z)$ is real on these segments. Then $f(z)$ can be analytically continued across $L_1, \ldots, L_n$ into $\tilde{D}$ (see also Theorems 3.2.6, 3.2.7, and 3.5.2). Indeed, $f(z)$ is analytic in D, which implies that $\tilde{f}(z) \equiv \overline{f(\bar{z})}$ is analytic in $\tilde{D}$ because $f(z) = \tilde{f}(z)$ on the line segments (due to the reality condition). These facts, together with the continuity of $f(z)$ as z approaches $L_1, \ldots, L_n$, imply that the function $F(z)$ defined by $f(z)$ in D, by $\tilde{f}(z)$ in $\tilde{D}$, and by $f(z)$ on $L_1, \ldots, L_n$ is also analytic on these segments.

A particular case of such a situation is shown in Figure 5.6.3. Although D has two line segments in common with the real axis, let us assume that the conditions of continuity and reality are satisfied only on L_1. Then there exists a function analytic everywhere in the shaded region except on l_1.

The important assumption in deriving the above result was $\operatorname{Im} f = 0$ on $\operatorname{Im} z = 0$. If we think of $f(z)$ as a transformation from the z plane to the w plane, this means that a line segment of the boundary of D of the z plane is mapped into a line segment of the boundary of $f(D)$ in the w plane which are portions of the real w axis. By using linear transformations (rotations and translations) in the z and w plane, one can extend this result to the case that these transformed line segments are not necessarily on the real axis. In other words, the reality condition is modified to the requirement that $f(z)$ maps line segments in the z plane into line segments in the w plane. Therefore, if $w = f(z)$ is analytic in D and continuous in the region consisting of D together with the segments $L_1, \ldots, L_n$, and if these segments are mapped into line segments in the w plane, then $f(z)$ can be analytically continued across $L_1, \ldots, L_n$.

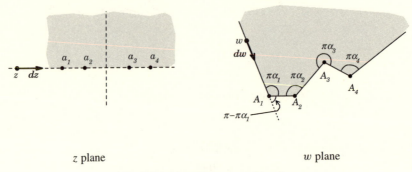

z plane   w plane

Fig. 5.6.4. Transformation of a polygon

As mentioned above the Schwarz reflection principle can be generalized to the case that line segments are replaced by circular arcs (see also, Nehari (1952)). We discuss this further in Section 5.7.

Theorem 5.6.1 *(Schwarz–Christoffel)* Let Γ be the piecewise linear boundary of a polygon in the w plane, and let the interior angles at successive vertices be $\alpha_1\pi, \ldots, \alpha_n\pi$. The transformation defined by the equation

$$\frac{dw}{dz} = \gamma(z - a_1)^{\alpha_1 - 1}(z - a_2)^{\alpha_2 - 1} \cdots (z - a_n)^{\alpha_n - 1} \qquad (5.6.1)$$

where γ is a complex number and $a_1, \ldots, a_n$ are real numbers, maps Γ into the real axis of the z plane and the interior of the polygon to the upper half of the z plane. The vertices of the polygon, A_1, A_2, ..., A_n, are mapped to the points $a_1, \ldots, a_n$ on the real axis. The map is an analytic one-to-one conformal transformation between the upper half z plane and the interior of the polygon.

In this application we consider both the map and its inverse, that is, mapping the w plane to the z plane; $z = F(w)$, or the z plane to the w plane; $w = f(z)$.

We first give a heuristic argument of how to derive Eq. (5.6.1). Our goal is to find an analytic function $f(z)$ in the upper half z plane such that $w = f(z)$ maps the real axis of the z plane onto the boundary of the polygon. We do this by considering the derivative of the mapping $\frac{dw}{dz} = f'(z)$, or $dw = f'(z)dz$. Begin with a point w on the polygon, say, to the left of the first vertex A_1 (see Figure 5.6.4) with its corresponding point z to the left of a_1 in the z plane. If we think of dw and dz as vectors on these contours, then arg $(dz) = 0$ (always) and arg $(dw) = 0$ (always, since this "vector" maintains a fixed direction) until

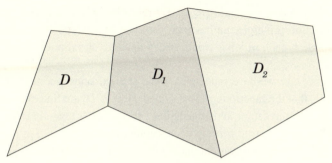

Fig. 5.6.5. Continuation of D

we traverse the first vertex. In fact arg (dw) only changes when we traverse the vertices. Thus arg $(f'(z)) =$ arg $(dw) -$ arg $(dz) =$ arg (dw). We see from Figure 5.6.4 that the *change* in arg $(f'(z))$ as we traverse (from left to right) the first vertex is $\pi - \pi\alpha_1$, and more generally, through any vertex A_ℓ the *change* is arg $(f'(z)) = \pi(1 - \alpha_\ell)$. This is precisely the behavior of the arguments of the function $(z - a_\ell)^{\alpha_\ell - 1}$: as we traverse a point $z = z_\ell$ we find that arg $(z - a_\ell) = \pi$ if z is real and on the left of a_ℓ and that arg $(z - a_\ell) = 0$ if z is real and on the right of a_ℓ. Thus arg $(z - a_\ell)$ *changes* by $-\pi$ as we traverse the point a_ℓ, and $(\alpha_\ell - 1)$ arg $(z - a_\ell)$ *changes* by $\pi(1 - \alpha_\ell)$. Because we have a similar situation at each vertex, this suggests that $\frac{dw}{dz} = f'(z)$ is given by the right-hand side of Eq. (5.6.1). Many readers may wish to skip the proof of this theorem, peruse the remarks that follow it, and proceed to the worked examples.

Proof We outline the essential ideas behind the proof. Riemann's mapping theorem, mentioned at the beginning of this section (see also Section 5.5) guarantees that such a univalent map $w = f(z)$ exists. (Actually, one could proceed on the assumption that the mapping function $f(z)$ exists and then verify that the function $f(z)$ defined by (5.6.1) satisfies the conditions of the theorem.) However we prefer to give a constructive proof. We now discuss its construction.

Let us consider the function $w = f(z)$ analytically continued across one of the sides of the polygon D in the w plane to obtain a function $f_1(z)$ in an adjacent polygon D_1; every point $w \in D$ corresponds to a point in the upper half z plane, and every point $w \in D_1$ corresponds to a symmetrical point, $\bar{z}$, in the lower half z plane. Doing this again along a side of D_1, we obtain a function $f_2(z)$ in the polygon D_2, etc., as indicated in Figure 5.6.5.

Each reflection of a polygon in the w plane across, say, the segment $A_k A_{k+1}$, corresponds (by the Schwarz reflection principle) to an analytic continuation of $f(z)$ across the line segment $a_k a_{k+1}$. By repeating this over and over, the Schwarz reflection principle implies that $f(z)$ can be analytically continued to form a single branch of what would be, in general, an infinitely branched function.

However, because we have reflections about straight sides, geometrical arguments imply that the functions $f(z) \in D$ and $f_2(z) \in D_2$ are linearly related to each other via a rotation and translation: that is, $f_2(z) = Af(z) + B$, $A = e^{i\alpha}$. The same is true for any even number of reflections, f_4, f_6, $\ldots$. But $g(z) = f''(z)/f'(z)$ is invariant under such linear transformations, so any point in the upper half z plane will correspond to a unique value of $g(z)$; because $g(z) = (f_2''(z))/(f_2'(z)) = (f_4''(z))/(f_4'(z)) = \cdots$, any even number of reflections returns us to the same value. Similarly, any odd number of reflections returns us a unique value of $g(z)$ corresponding to a point z in the lower half plane. The only possible locations for singularities correspond to the vertices of the polygon. On the real z axis, $g(z)$ is real (when $y = 0$, $g(z) = u_{xx}(x, 0)/u_x(x, 0)$, since $v(x, 0) = 0$) and may be continued by reflection to the lower half plane by $g(z) = \overline{g(\bar{z})}$. In this way, all points z (upper and lower half planes) are determined uniquely and the function $g(z)$ is therefore single valued.

Next, let us consider the points $z = a_\ell$ corresponding to the polygonal vertices A_ℓ. In the neighborhood of a vertex $z = a_\ell$ an argument such as that preceding this theorem shows that the mapping has the form

$$w = w_0 = f(z) - f(z_0) = (z - a_\ell)^{\alpha_\ell} \left[c_\ell^{(0)} + c_\ell^{(1)}(z - a_\ell) + c_\ell^{(2)}(z - a_\ell)^2 + \cdots \right]$$

Consequently, $g(z) = f''(z)/f'(z)$ is analytic in the extended z plane except for poles at the points $a_1, \ldots, a_n$ with residues $(\alpha_1 - 1), \ldots, (\alpha_n - 1)$. It follows from Liouville's Theorem that

$$\frac{f''(z)}{f'(z)} - \sum_{l=1}^{n} \frac{(\alpha_l - 1)}{z - a_l} = c \tag{5.6.2}$$

where c is some complex constant. But $f(z)$ is analytic at $z = \infty$ (assuming no vertex, $z = a_\ell$ is at infinity), so $f(z) = f(\infty) + b_1/z + b_2/z^2 + \ldots$; hence $f''(z)/f'(z) \to 0$ as as $z \to \infty$, which implies that $c = 0$. Integration of Eq. (5.6.2) yields Eq. (5.6.1). ∎

Remarks (1) For a closed polygon, $\sum_{l=1}^{n}(1 - \alpha_l) = 2$, and hence $\sum_{l=1}^{n} \alpha_l = (n-2)$ where n is the number of sides. This is a consequence of the well known geometrical property that the sum of the exterior angles of any closed polygon is 2π.

(2) It is shown in Section 5.7 that for bilinear transformations, the correspondence of three (and only three) points on the boundaries of two domains can be prescribed arbitrarily. Actually, it can be shown that this is true for any univalent transformation between the boundary of two simply-connected domains. In particular, any of the three vertices of the polygon, say A_1, A_2, and A_3, can be associated with any three points on the real axis a_1, a_2, a_3 (of course preserving order and orientation). More than three of the vertices a_ℓ cannot be prescribed arbitrarily, and the actual determination of $a_4, a_5, \ldots, a_n$ (sometimes called accessory parameters) might be difficult. In application, symmetry or other considerations usually are helpful, though numerical computation is usually the only means to evaluate the constants $a_4, a_5, \ldots, a_n$. Sometimes it is useful to fix more independent real conditions instead of fixing three points, (e.g. map a point $w_0 \in D$ to a fixed point z_0 in the z plane, and fix a direction of $f'(z_0)$, that is, fix arg $f'(z_0)$).

(3) The integration of Eq. (5.6.1) usually leads to multivalued funcions. A single branch is chosen by the requirement that $0 < \arg(z - a_l) < \pi, l = 1, \ldots, n$. The function $f(z)$ is analytic in the semiplane Im $z > 0$; it has branch points at $z = a_\ell$.

(4) Formula (5.6.1) holds when none of the points coincide with the point at infinity. However, using the transformation $z = a_n - 1/\zeta$, which transforms the point $z = a_n$ to $\zeta = \infty$ but transforms all other points a_ℓ to finite points $\zeta_\ell = 1/(a_n - a_\ell)$, we see that Eq. (5.6.1) yields

$$\frac{df}{d\zeta}(\zeta^2) = \gamma \left(a_n - a_1 - \frac{1}{\zeta} \right)^{\alpha_1 - 1} \cdots \left(a_n - a_{n-1} - \frac{1}{\zeta} \right)^{\alpha_{n-1} - 1} \left(-\frac{1}{\zeta} \right)^{\alpha_n - 1}$$

Using Remark (1), we have

$$\frac{df}{d\zeta} = \hat{\gamma}(\zeta - \zeta_1)^{\alpha_1 - 1} \cdots (\zeta - \zeta_{n-1})^{\alpha_{n-1} - 1} \tag{5.6.3a}$$

where $\zeta_\ell = 1/(a_n - a_\ell)$ and $\hat{\gamma}$ is a new constant. Thus, formula (5.6.1) holds with the point at ∞ removed. If the point $z = a_n$ is mapped to $\zeta = \infty$, then, by virtue of Remark (2), only two other vertices can be arbitrarily prescribed. Using Remark (1), we find that as $\zeta \to \infty$

$$\frac{df}{d\zeta} = \hat{\gamma} \zeta^{-\alpha_n - 1} \left[1 + \frac{c_1}{\zeta} + \cdots \right]$$

(5) Using the bilinear transformation

$$z = i\left(\frac{1+\zeta}{1-\zeta}\right), \qquad \zeta = \frac{z-i}{z+i}, \qquad \frac{dz}{d\zeta} = \frac{2i}{(1-\zeta)^2}$$

which transforms the upper half z plane onto the unit circle $|\zeta| < 1$, it follows that Eq. (5.6.1) also with z replaced by ζ and with suitable constants $\gamma \to \hat{\gamma}$, $a_\ell \to \zeta_\ell$, ζ_ℓ being on the unit circle, $\ell = 1, 2, \ldots, n$, which can be found by using the above bilinear transformation and Remark (1).

(6) These ideas can be used to map the complete exterior of a closed polygon (with n vertices) in the w plane to the upper half z plane. We note that at first glance one might not expect this to be possible because an annular region (not simply connected) cannot be mapped onto a half plane. In fact, the exterior of a polygon, which contains the point at infinity, is simply connected. A simple closed curve surrounding the closed polygon can be continuously deformed to the point at infinity. In order to obtain the formula in this case, we note that all of the interior angles $\pi\alpha_\ell$, $\ell = 1, 2, \ldots, n$, must be transformed to exterior angles $2\pi - \pi\alpha_\ell$ because we traverse the polygon in the opposite direction, keeping the exterior of the polygon to our left. Thus the change in arg $f'(z)$ at a vertex A_ℓ is $-(\pi - \pi\alpha_\ell)$ and therefore in Eq. (5.6.1) $(\alpha_\ell - 1) \to (1 - \alpha_\ell)$. We write the transformation in the form

$$\frac{dw}{dz} = g(z)(z - a_1)^{1-\alpha_1}(z - a_2)^{1-\alpha_2} \cdots (z - a_n)^{1-\alpha_n}$$

The function $g(z)$ is determined by properly mapping the point $w = \infty$, which is now an *interior* point of the domain to be mapped. Let us map $w = \infty$ to a point in the upper half plane, say, $z = ia_0$, $a_0 > 0$. We require $w(z)$ to be a conformal transformation at infinity, so near $z = ia_0$, $g(z)$ must be single valued and $w(z)$ should transform like

$$w(\zeta) = \gamma_1\zeta + \gamma_0 + \cdots, \qquad \zeta = \frac{1}{z - ia_0} \qquad \text{as } \zeta \to \infty, \text{ or } z \to ia_0$$

Similar arguments pertain to the mapping of the lower half plane by using the symmetry principle. Using the fact that the polygon is closed, $\sum_{\ell=1}^{n}(1 - \alpha_\ell) = 2$, and conformal at $z = \infty$, we deduce that

$$g(z) = \frac{\gamma}{(z - ia_0)^2(z + ia_0)^2}$$

(Note that $g(z)$ is real for real z.) Thus, the Schwarz–Christoffel formula mapping the exterior of a closed polygon to the upper half z plane is given by

$$\frac{dw}{dz} = \frac{\gamma}{(z - ia_0)^2(z + ia_0)^2}(z - a_1)^{1-\alpha_1}(z - a_2)^{1-\alpha_2} \cdots (z - a_n)^{1-\alpha_n} \quad (5.6.3b)$$

We also note that using the bilinear transformation of Remark (5) with $a_0 = 1$, we find that the Schwarz–Christoffel transformation from the *exterior* of a polygon to the *interior* of a unit circle $|\zeta| < 1$ is given by

$$\frac{dw}{d\zeta} = \frac{\hat{\gamma}}{\zeta^2}(\zeta - \zeta_1)^{1-\alpha_1}(\zeta - \zeta_2)^{1-\alpha_2} \cdots (\zeta - \zeta_n)^{1-\alpha_n} \quad (5.6.3c)$$

where the points $\zeta_i, i = 1, 2, \ldots, n$ lie on the unit circle.

Example 5.6.3 Determine the function that maps the half strip indicated in Figure 5.6.6 onto the upper half of the z plane.

We associate $A(\infty)$ with $a(\infty)$, $A_1(-k)$ with $a_1(-1)$, and $A_2(k)$ with $a_2(1)$. Then, from symmetry, we find that $B(\infty)$ is associated with $b(\infty)$. Equation (5.6.1) with $\alpha_1 = \alpha_2 = \frac{1}{2}$, $a_1 = -1$, and $a_2 = 1$ yields

$$\frac{dw}{dz} = \gamma(z + 1)^{-\frac{1}{2}}(z - 1)^{-\frac{1}{2}} = \frac{\tilde{\gamma}}{\sqrt{1 - z^2}}$$

Integration implies

$$w = \tilde{\gamma}\sin^{-1}z + c$$

When $z = 1$, $w = k$, and when $z = -1$, $w = -k$. Thus

$$k = \tilde{\gamma}\sin^{-1}(1) + c, \qquad -k = \tilde{\gamma}\sin^{-1}(-1) + c$$

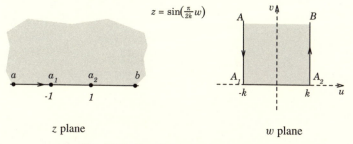

$$z = \sin\left(\frac{\pi}{2k}w\right)$$

z plane   w plane

Fig. 5.6.6. Transformation of a half strip

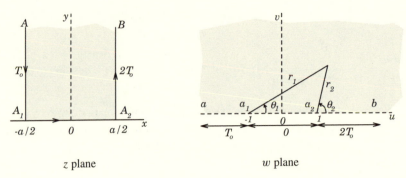

z plane   w plane

Fig. 5.6.7. Constant temperature boundary conditions

Using $\sin^{-1}(1) = \pi/2$ and $\sin^{-1}(-1) = -\pi/2$, these equations yield $c = 0$, $\tilde{\gamma} = 2k/\pi$. Thus $w = (2k/\pi)\sin^{-1} z$, and $z = \sin(\pi w/2k)$.

Example 5.6.4 A semiinfinite slab has its vertical boundaries maintained at temperature T_0 and $2T_0$ and its horizontal boundary at a temperature 0 (see Figure 5.6.7). Find the steady state temperature distribution inside the slab.

We shall use the result of Example 5.6.3, with $k = a/2$ (interchanging w and z). It follows that the transformation $w = \sin(\pi z/a)$ maps the semiinfinite slab onto the upper half w plane. The function $T = \alpha_1\theta_1 + \alpha_2\theta_2 + \alpha$ (see also Section 2.3, especially Eqs. (2.3.13)–(2.3.17)), where α_1, α_2, and α are real constants and is the imaginary part of $\alpha_1 \log(w + 1) + \alpha_2 \log(w - 1) + i\alpha$, and is thereby harmonic (i.e., because it is the imaginary part of an analytic function, it satisfies Laplace's equation) in the upper half strip. In this strip, $w + 1 = r_1 e^{i\theta_1}$ and $w - 1 = r_2 e^{i\theta_2}$, where $0 \leq \theta_1, \theta_2 \leq \pi$. To determine α_1, α_2, and α, we use the boundary conditions. If $\theta_1 = \theta_2 = 0$, then $T = 2T_0$; if $\theta_1 = 0$ and $\theta_2 = \pi$, then $T = 0$; if $\theta_1 = \theta_2 = \pi$, then $T = T_0$. Hence

$$T = \frac{T_0}{\pi}\theta_1 - \frac{2T_0}{\pi}\theta_2 + 2T_0 = \frac{T_0}{\pi}\tan^{-1}\frac{v}{u+1} - \frac{2T_0}{\pi}\tan^{-1}\frac{v}{u-1} + 2T_0$$

Using

$$w = u + iv = \sin\left(\frac{\pi z}{a}\right)$$

that is,

$$u = \sin\left(\frac{\pi x}{a}\right)\cosh\left(\frac{\pi y}{a}\right) \quad \text{and} \quad v = \cos\left(\frac{\pi x}{a}\right)\sinh\left(\frac{\pi y}{a}\right)$$

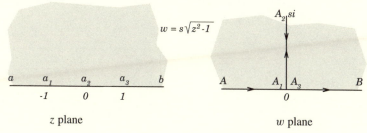

$w = s\sqrt{z^2 - 1}$

z plane

w plane

Fig. 5.6.8. Transformation of the cut half plane onto a slit

we find

$$T = \frac{T_0}{\pi} \tan^{-1}\left[\frac{\cos\left(\frac{\pi x}{a}\right)\sinh\left(\frac{\pi y}{a}\right)}{\sin\left(\frac{\pi x}{a}\right)\cosh\left(\frac{\pi y}{a}\right) + 1}\right] - \frac{2T_0}{\pi} \tan^{-1}\left[\frac{\cos\left(\frac{\pi x}{a}\right)\sinh\left(\frac{\pi y}{a}\right)}{\sin\left(\frac{\pi x}{a}\right)\cosh\left(\frac{\pi y}{a}\right) - 1}\right]$$
$$+ 2T_0$$

Example 5.6.5 Determine the function that maps the "slit" of height s depicted in Figure 5.6.8 onto the upper half of the z plane.

We associate $A_1(0-)$ with $a_1(-1)$, $A_2(si)$ with $a_2(0)$, and $A_3(0+)$ with $a_3(1)$. Then Eq. (5.6.1) with $\alpha_1 = \frac{1}{2}$, $a_1 = -1$, $\alpha_2 = 2$, $a_2 = 0$, $\alpha_3 = \frac{1}{2}$, and $a_3 = 1$ yields

$$\frac{dw}{dz} = \gamma(z+1)^{-\frac{1}{2}}z(z-1)^{-\frac{1}{2}} = \frac{\tilde{\gamma}z}{\sqrt{1-z^2}}$$

Thus

$$w = \delta\sqrt{z^2 - 1} + c$$

When $z = 0$, $w = si$, and when $z = 1$, $w = 0$. Thus

$$w = s\sqrt{z^2 - 1} \qquad (5.6.4)$$

Example 5.6.6 Find the flow past a vertical slit of height s, which far away from this slit is moving with a constant velocity U_0 in the horizontal direction.

It was shown in Example 5.6.5 that the transformation $w = s\sqrt{z^2 - 1}$ maps the vertical slit of height s in the w plane onto the real axis of the z plane. The flow field over a slit in the w plane is therefore transformed into a uniform flow in the z plane with complex velocity $\Omega(z) = U_0 z = U_0(x + iy)$, and with constant velocity U_0. The streamlines of the uniform flow in the z plane correspond to

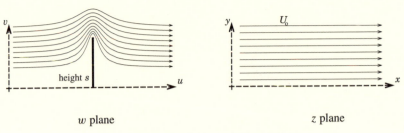

w plane   z plane

Fig. 5.6.9. Flow over vertical slit of height s

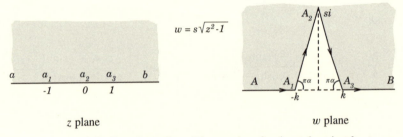

z plane   w plane

Fig. 5.6.10. Transformation of the exterior of an isosceles triangle

$y = c$ (recall that $\Omega = \Phi + i\Psi$ where Φ and Ψ are the velocity potential and stream function, respectively) where c is a positive constant and the flow field in the w plane is obtained from the complex potential $\Omega(w) = U_0((\frac{w}{s})^2+1)^{\frac{1}{2}}$. The image of each of the streamlines $y = c$ in the w plane is $w = s\sqrt{(x + ic)^2 - 1}$, $-\infty < x < \infty$. Note that $c = 0$ and $c \to \infty$ correspond to $v = 0$ and $v \to \infty$. Alternatively, from the complex potential $\Omega = \Phi + i\Psi$, $\Psi = 0$ when $v = 0$ and $\Omega'(w) \to U_0/s$ as $|w| \to \infty$. Thus one gets the flow past the vertical barrier depicted in Figure 5.6.9.

Example 5.6.7 Determine the function that maps the exterior of an isosceles triangle located in the upper half of the w plane onto the upper half of the z plane.

We note that this is not a mapping of a complete exterior of a closed polygon; in fact, this problem is really a modification of Example 5.6.5. We will show that a limit of this example as $k \to 0$ (see Figure 5.6.10) reduces to the previous one.

We associate $A_1(-k)$ with $a_1(-1)$, $A_2(si)$ with $a_2(0)$, and $A_3(k)$ with $a_3(1)$. The angles at A_1, A_2, and A_3 are given by $\pi - \pi\alpha$, $2\pi - (\pi - 2\pi\alpha)$, and $\pi - \pi\alpha$, respectively. Thus $\alpha_1 = 1 - \alpha$, $\alpha_2 = 1 + 2\alpha$, $\alpha_3 = 1 - \alpha$, and

Eq. (5.6.1) yields

$$\frac{dw}{dz} = \gamma(z+1)^{-\alpha}z^{2\alpha}(z-1)^{-\alpha} = \tilde{\gamma}\frac{z^{2\alpha}}{(1-z^2)^\alpha}$$

Thus

$$w = \tilde{\gamma}\int_0^z \frac{\zeta^{2\alpha}}{(1-\zeta^2)^\alpha}d\zeta + c \tag{5.6.5}$$

When $z = 0$, $w = si$, and when $z = 1$, $w = k$. Hence $c = si$, and

$$k = \tilde{\gamma}\int_0^1 \frac{\zeta^{2\alpha}}{(1-\zeta^2)^\alpha}d\zeta + si$$

The integral $\int_0^1$ can be expressed in terms of gamma functions by calling $t = \zeta^2$ and using a well-known result for integrals

$$B(p,q) = \int_0^1 t^{p-1}(1-t)^{q-1}dt = \frac{\Gamma(p)\Gamma(q)}{\Gamma(p+q)}$$

$B(p,q)$ is called the beta function. (See Eq. (4.5.30) for the definition of the gamma function, $\Gamma(z)$.) Using this equation with $p = \alpha + 1/2$, $q = 1 - \alpha$, and $\Gamma(\frac{3}{2}) = \frac{1}{2}\Gamma(\frac{1}{2})$, where $\Gamma(\frac{1}{2}) = \sqrt{\pi}$, it follows that $\tilde{\gamma}\Gamma(\alpha + 1/2)\Gamma(1 - \alpha) = (k - si)\sqrt{\pi}$. Thus

$$w = \frac{(k-si)\sqrt{\pi}}{\Gamma(\alpha+1/2)\Gamma(1-\alpha)}\int_0^z \frac{\zeta^{2\alpha}}{(1-\zeta^2)^\alpha}d\zeta + si \tag{5.6.6}$$

We note that Example 5.6.5 corresponds to the limit $k \to 0$, $\alpha \to \frac{1}{2}$ in Eq. (5.6.6), that is, under this limit, Eq. (5.6.6) reduces to Eq. (5.6.4)

$$w = si - si\int_0^z \frac{\zeta}{\sqrt{1-\zeta^2}}d\zeta = si\sqrt{1-z^2} = s\sqrt{z^2-1}$$

An interesting application of the Schwarz–Christoffel construction is the mapping of a rectangle. Despite the fact that it is a simple closed polygon, the function defined by the Schwarz–Christoffel transformation is not elementary. (Neither is the function elementary in the case of triangles.) In the case of a rectangle, we find that the mapping functions involve elliptic integrals and elliptic functions.

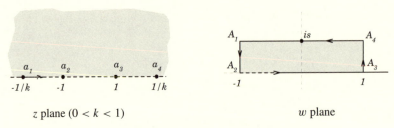

z plane $(0 < k < 1)$ w plane

Fig. 5.6.11. Transformation of a rectangle

Example 5.6.8 Find the function that maps the interior of a rectangle onto the upper half of the z plane. See Figure 5.6.11.

We associate $A_1(-1 + si)$ with $a_1(-1/k)$, $A_2(-1)$ with $a_2(-1)$, and $z = 0$ with $w = 0$. Then by symmetry, A_3, A_4 are associated with a_3, a_4 respectively. In this example we regard k as given and assume that $0 < k < 1$. Our goal is to determine both the transformation $w = f(z)$ and the constant s as a functions of k. In this case, $\alpha_1 = \alpha_2 = \alpha_3 = \alpha_4 = \frac{1}{2}$, $a_1 = -\frac{1}{k}$, $a_2 = -1$, $a_3 = 1$, and $a_4 = 1/k$. Furthermore, because $f(0) = 0$ (symmetry), the constant of integration is zero; thus Eq. (5.6.1) yields

$$\frac{dw}{dz} = \gamma(z - 1)^{-1/2}(z + 1)^{-1/2}(z - 1/k)^{-1/2}(z + 1/k)^{-1/2}$$

Then by integration manipulation and by redefining γ (to $\tilde{\gamma}$)

$$w = \tilde{\gamma} \int_0^z \frac{d\zeta}{\sqrt{(1 - \zeta^2)(1 - k^2\zeta^2)}} = \tilde{\gamma} F(z, k) \qquad (5.6.7)$$

The integral appearing in Eq. (5.6.7), with the choice of the branch defined earlier, in Remark 3 (w is real for real z, $|z| > 1/k$), is the so-called elliptic integral of the first kind; it is usually denoted by $F(z, k)$. (Note, from the integral in Eq. (5.6.7), that $F(z, k)$ is an odd function; i.e, $F(-z, k) = -F(z, k)$.) When $z = 1$, this becomes $F(1, k)$ which is referred to as the complete elliptic integral, usually denoted by $K(k) \equiv F(1, k) = \int_0^1 d\zeta/\sqrt{(1 - \zeta^2)(1 - k^2\zeta^2)}$. The association of $z = 1$ with $w = 1$ implies that $\tilde{\gamma} = 1/K(k)$. The association of $z = 1/k$ with $w = 1 + is$ yields

$$1 + is = \frac{1}{K} \int_0^{\frac{1}{k}} \frac{d\zeta}{\sqrt{(1 - \zeta^2)(1 - k^2\zeta^2)}}$$

$$= \frac{1}{K}\left(K + \int_1^{\frac{1}{k}} \frac{d\zeta}{\sqrt{(1 - \zeta^2)(1 - k^2\zeta^2)}}\right)$$

or $Ks = K'$, where K' denotes the associated elliptic integral (not the derivative), which is defined by

$$K'(k) = \int_1^{\frac{1}{k}} \frac{d\xi}{\sqrt{(\xi^2 - 1)(1 - k^2\xi^2)}}$$

This expression takes an alternative, standard form if one uses the substitution $\xi = (1 - k'^2\xi'^2)^{-1/2}$, where $k' = \sqrt{1 - k^2}$ (see Eq. (5.6.9)).

In summary, the transformation $f(z)$ and the constant s are given by

$$w = \frac{F(z, k)}{K(k)}, \qquad s = \frac{K'(k)}{K(k)} \tag{5.6.8}$$

where (the symbol $\equiv$ denotes "by definition")

$$F(z, k) \equiv \int_0^z \frac{d\zeta}{\sqrt{(1 - \zeta^2)(1 - k^2\zeta^2)}}, \qquad K(k) \equiv F(1, k)$$

$$K'(k) \equiv \int_0^1 \frac{d\xi}{\sqrt{(1 - \xi^2)[1 - (1 - k^2)\xi^2]}} \tag{5.6.9}$$

The parameter k is called the modulus of the elliptic integral. The inverse of Eq. (5.6.8) gives z as a function of w via one of the so-called Jacobian elliptic functions (see, e.g., Nehari (1952))

$$w = \frac{F(z, k)}{K(k)} \quad \Rightarrow \quad z = sn(wK, k) \tag{5.6.10}$$

We note that Example 5.6.3 (with $k = 1$ in that example) corresponds to the following limit in this example: $s \to \infty$, which implies that $k \to 0$ because $\lim_{k \to 0} K'(k) = \infty$, and $\lim_{k \to 0} K(k) = \pi/2$. Then the rectangle becomes an infinite strip, and Eq. (5.6.8a) reduces to equation

$$w = \frac{2}{\pi} \int_0^z \frac{d\zeta}{\sqrt{1 - \zeta^2}} = \frac{2}{\pi} \sin^{-1} z$$

Remark The fundamental properties of elliptic functions are their "double periodicity" and single valuedness. We illustrate this for one of the elliptic functions, the Jacobian "sn" function, which we have already seen in Eq. (5.6.10). A standard "normalized" definition is (replacing wK by w in Eq. (5.6.10))

$$w = F(z, k) \quad \Rightarrow \quad z = F^{-1}(w, k) = sn(w, k) \tag{5.6.11}$$

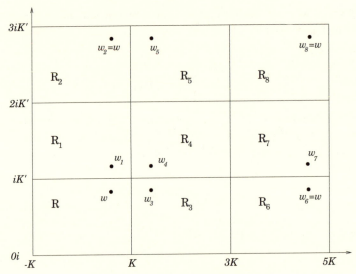

Fig. 5.6.12. Reflecting in the w plane

where again

$$F(z, k) \equiv \int_0^z \frac{d\zeta}{\sqrt{(1 - \zeta^2)(1 - k^2\zeta^2)}}$$

so that $\mathrm{sn}(0, k) = 0$ and $\mathrm{sn}'(0, k) = 1$ (in the latter we used $\frac{dz}{dw} = 1/\frac{dw}{dz}$).
  We shall show the double periodicity

$$\mathrm{sn}(w + n\omega_1 + im\omega_2, k) = \mathrm{sn}(w, k) \qquad (5.6.12)$$

where m and n are integers, $\omega_1 = 4K(k)$, and $\omega_2 = 2K'(k)$. Given the normalization of Eq. (5.6.11) it follows that the "fundamental" rectangle in the w plane corresponding to the upper half z plane is $A_1 = -K + iK'$, $a_1 = -1/k$, $A_2 = -K$, and $a_2 = -1$, with A_3, a_3, A_4, and a_4 as the points symmetric to these (in Figure 5.6.11, all points on the w plane are multiplied by K; this is now the rectangle R in Figure 5.6.12). The function $z = \mathrm{sn}(w, k)$ can be analytically continued by the symmetry principle.

  Beginning with any point w in the fundamental rectangle R we must obtain the same point w by symmetrically reflecting twice about a horizontal side of the rectangle, or twice about a vertical side of the rectangle, etc., which corresponds to returning to the same point in the upper half z plane each time. This yields the double periodicity relationship (5.6.12).

  These symmetry relationships also imply that the function $z = \mathrm{sn}(w, k)$ is single valued. Any point z in the upper half plane is uniquely determined and corresponds to an even number of reflections, and similarly, any symmetric point $\bar{z}$ in the lower half plane is found uniquely by an odd number of reflections. Analytic continuation of $z = \text{sn}(w, k)$ across any boundary therefore uniquely determines a value in the z plane. The only singularities of the map $w = f(z)$ are at the vertices of the rectangle, and near the vertices of the rectangle we have

$$w - A_i = C_i (z - a_i)^{\frac{1}{2}} \left[1 + c_i^{(1)}(z - a_i) + \cdots \right]$$

and we see that $z = \text{sn}(w, k)$ is analytic there as well. The "period rectangle" consists of any four rectangles meeting at a corner, such as R, R_1, R_3, R_4 in Figure 5.6.12. All other such period rectangles are periodic extensions of the fundamental rectangle. Two of the rectangles map to the upper half z plane and two map to the lower half z plane. Thus a period rectangle covers the z plane twice, that is, for $z = \text{sn}(w, k)$ there are two values of w that correspond to a fixed value of z.

For example, the zeroes of $\text{sn}(w, k)$ are located at $w = 2nK + 2miK'$ for integers m and n. From the definition of $F(z, k)$ we see that $F(0, k) = 0$. If we reflect the rectangle R to R_1, this zero is transformed to the location $w = 2iK'$, while reflecting to R_3 transforms the zero to $w = 2K$, etc. Hence two zeroes are located in each period rectangle. From the definition we also find the pairs $w = -K$, $z = -1$; $w = -K + iK'$, $z = -\frac{1}{k}$; and $w = iK'$, $z = \infty$. The latter is a simple pole.

Problems for Section 5.6

1. Use the Schwarz–Christoffel transformation to obtain a function that maps each of the indicated regions below in the w plane onto the upper half of the z plane in Figure 5.6.13a,b.

2. Find a function that maps the indicated region of the w plane in Figure 5.6.14 onto the upper half of the z plane, such that $(P, Q, R) \mapsto (-\infty, 0, \infty)$.

3. Show that the function $w = \displaystyle\int_0^z (dt/(1 - t^6)^{\frac{1}{3}})$ maps a regular hexagon into the unit circle.

4. Derive the Schwarz–Christoffel transformation that maps the upper half plane onto the triangle with vertices $(0, 0)$, $(0, 1)$, $(1, 0)$.

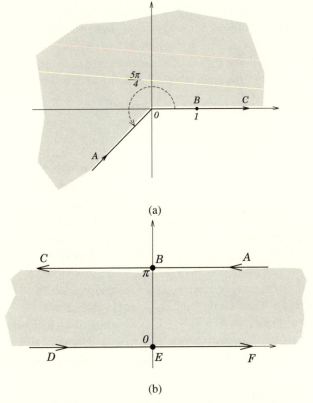

(a)

(b)

Fig. 5.6.13. Schwarz–Christoffel transformations–Problem 5.6.1

5. A fluid flows with initial velocity u_0 through a semiinfinite channel of
    width d and emerges through the opening AB of the channel (see Figure
    5.6.16). Find the speed of the fluid.

    Hint: First show that the conformal mapping $w = z + e^{2\pi z/d}$ maps the
    semiinfinite channel to an infinite channel, as illustrated in Figure 5.6.17.

6. The shaded region of Figure 5.6.18 represents a semiinfinite conductor
    with a vertical slit of height h in which the boundaries AD, DE (of height
    h) and DB are maintained at temperatures T_1, T_2, and T_3, respectively.

    Find the temperature everywhere. Hint: Use the conformal mapping
    studied in Example 5.6.5.

7. Utilize the Schwarz–Christoffel transformation in order to find the com-
    plex potential $F(w)$ governing the flow of a fluid over a step with velocity

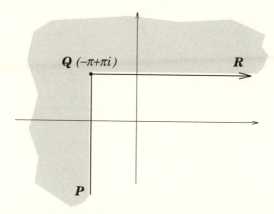

Fig. 5.6.14. Schwarz–Christoffel transformations–Problem 5.6.2

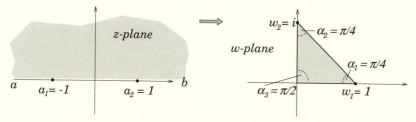

Fig. 5.6.15. Schwarz–Christoffel transformations–Problem 5.6.4

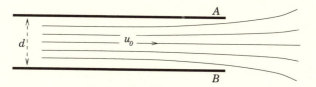

Fig. 5.6.16. Fluid flow–Problem 5.6.5

at infinity equal to q, where q is real. The step $A_1(-\infty)A_2(ih)A_3(0)$ is shown in Figure 5.6.19. The step is taken to be a streamline.

8. By using the Schwarz reflection principle, map the domain exterior to a T-shaped cut, shown in Figure 5.6.20, onto the half plane.

9. Find a conformal mapping onto the half plane, Im $w > 0$, of the z domain of the region illustrated in Figure 5.6.21 inside the strip $-\frac{\pi}{2} < \text{Re}\, z < \frac{\pi}{2}$, by taking cuts along the segments $[-\pi/2, -\pi/6]$ and $[\pi/6, \pi/2]$ of the real axis.

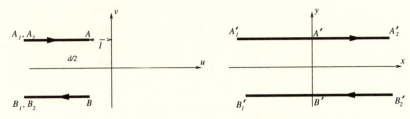

Fig. 5.6.17. Mapping Problem 5.6.5

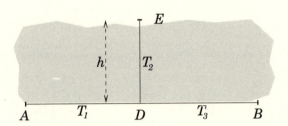

Fig. 5.6.18. Temperature distribution–Problem 5.6.6

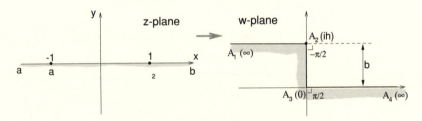

Fig. 5.6.19. Fluid flow–Problem 5.6.7

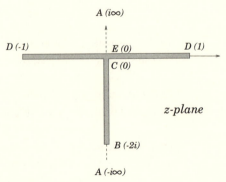

Fig. 5.6.20. "T-Shaped" region–Problem 5.6.8

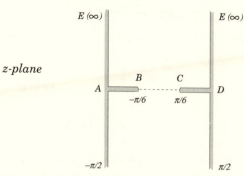

Fig. 5.6.21. Schwarz–Christoffel transformation–Problem 5.6.9

10. Show that the mapping $w = \int_0^z (dt/(t^{1/2}(t^2 - 1)^{1/2}))$ maps the upper half plane conformally onto the interior of a square.

11. Use the Schwarz–Christoffel transformation to show that

$$w = \log(2(z^2 + z)^{1/2} + 2z + 1)$$

maps the upper half plane conformally onto the interior of a semi infinite strip.

12. Show that the function $w = \int_1^z ((1 - \zeta^4)^{1/2}/\zeta^2)\, d\zeta$ maps the exterior of a square conformally onto the interior of the unit circle.

13. Show that a necessary (but not sufficient) condition for $z = z(w)$ appearing in the Schwarz–Christoffel formula to be single valued (i.e., for the inverse mapping to be defined and single valued) is that $\alpha_\ell = 1/n_\ell$, where n_ℓ is an integer.

14. Find the domain onto which the function

$$w = \int_0^z \frac{(1 + t^3)^\alpha}{(1 - t^3)^{\frac{2}{3} + \alpha}}\, dt, \qquad -1 < \alpha < \frac{1}{3}$$

maps the unit disk.

5.7 Bilinear Transformations

An important class of conformal mappings is given by the particular choice of $f(z)$

$$w = f(z) = \frac{az + b}{cz + d}, \qquad ad - bc \neq 0 \qquad (5.7.1)$$

where a, b, c, and d are complex numbers. This transformation is called **bilinear**, or **Möbius**, or sometimes linear fractional.

Bilinear transformations have a number of remarkable properties. Furthermore, these properties are global. Namely, they are valid for any z including $z = \infty$, that is, valid in the entire extended complex z plane. Throughout this section, we derive the most important properties of bilinear transformations (labeled (i)–(viii)):

(i) Conformality

Bilinear transformations are conformal.

Indeed, if $c = 0$, $f'(z) = \frac{a}{d} \neq 0$. If $c \neq 0$

$$f'(z) = \frac{ad - bc}{(cz + d)^2}$$

which shows that $f'(z)$ is well defined for $z \neq -d/c$, z finite. In order to analyze the point $z = \infty$, we let $\tilde{z} = 1/z$, then $w = f(z)$ becomes $w = (b\tilde{z} + a)/(d\tilde{z} + c)$, which is well behaved at $\tilde{z} = 0$. The image of the point $z = -d/c$ is $w = \infty$, which motivates the transformation $\tilde{w} = 1/w$. Then $\tilde{w} = (cz + d)/(az + b)$, and because the derivative of the right-hand side is not zero at $z = -d/c$, it follows that $w = f(z)$ is conformal at $\tilde{w} = 0$.

Example 5.7.1 Find the image of $x^2 - y^2 = 1$ under inversion, that is, under the transformation $w = 1/z$.

Using $z = x + iy$ and $\bar{z} = x - iy$, it follows that the hyperbola $x^2 - y^2 = 1$ can be written as

$$(z + \bar{z})^2 + (z - \bar{z})^2 = 4, \quad \text{or} \quad z^2 + \bar{z}^2 = 2$$

This becomes

$$\frac{1}{w^2} + \frac{1}{\bar{w}^2} = 2, \quad \text{or} \quad w^2 + \bar{w}^2 = 2|w|^4, \quad \text{or} \quad u^2 - v^2 = (u^2 + v^2)^2 \quad (5.7.2)$$

under $z = 1/w$, $w = u + iv$. For small u and v, Eq. (5.7.2) behaves like $u \approx \pm v$. This together with conformality at the points A and B suggests

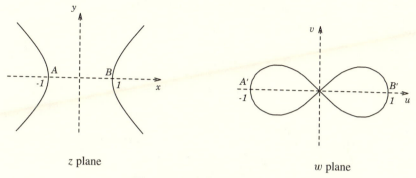

z plane

w plane

Fig. 5.7.1. Inversion ($w = 1/z$)

the lemniscate graph depicted in Figure 5.7.1. (In polar coordinates $w = \rho e^{i\phi}$, the lemniscate above is $\rho^2 = \cos 2\phi$.) Note that the left (right) branch of the hyperbola transforms to the left (right) lobe of the lemniscate.

(ii) Decomposition

Any bilinear transformation, which is not linear, can be decomposed into two linear transformations and an inversion.

Indeed, if $c = 0$ the transformation is linear. If $c \neq 0$, then it can be written in the form

$$w = \frac{a}{c} + \frac{bc - ad}{c(cz + d)}$$

This shows that a general bilinear transformation can be decomposed into the following three successive transformations:

$$z_1 = cz + d, \qquad z_2 = \frac{1}{z_1}, \qquad w = \frac{a}{c} + \frac{bc - ad}{c} z_2 \qquad (5.7.3)$$

(iii) Bilinear Transformations Form a Group

This means that bilinear transformations contain the identity transformation and that the inverse as well as the "product" of bilinear transformations are also bilinear transformations (i.e., closure under the operation product).

Indeed, the choice $b = c = 0$, $a = d = 1$, reduces $w = f(z)$ to $w = z$, that is, to the identity transformation. The inverse of the transformation $f(z)$ is obtained by solving the equation $w = f(z)$ for z in terms of w. Hence the inverse of $f(z)$ is given by

$$\frac{dz - b}{-cz + a} \tag{5.7.4}$$

This corresponds to Eq. (5.7.1) with $a \to d$, $b \to -b$, $c \to -c$, and $d \to a$, therefore $ad - bc \to ad - bc$, which shows that the inverse of $f(z)$ is also bilinear. In order to derive the product of two transformations, we let $z_2 = f_2(z)$ and $w = f_1(z_2)$, where f_1 and f_2 are defined by Eq. (5.7.1) with all the constants replaced by constants with subscripts 1 and 2, respectively. Computing $w = f_1(f_2(z)) = f_3(z)$, one finds

$$f_3(z) = \frac{a_3 z + b_3;}{c_3 z + d_3} \quad \begin{matrix} a_3 = a_1 a_2 + b_1 c_2, & b_3 = a_1 b_2 + b_1 d_2 \\ c_3 = c_1 a_2 + d_1 c_2, & d_3 = c_1 b_2 + d_1 d_2 \end{matrix} \tag{5.7.5}$$

It can be verified that $(a_3 d_3 - b_3 c_3) = (a_1 d_1 - b_1 c_1)(a_2 d_2 - b_2 c_2) \neq 0$, which together with Eq. (5.7.5) establishes that $f_3(z)$ is a bilinear transformation. Actually the operation product in this case is composition.

There exists an alternative, somewhat more elegant formulation of bilinear transformations: Associate with $f(z)$ the matrix

$$T = \begin{pmatrix} a & b \\ c & d \end{pmatrix} \tag{5.7.6}$$

The condition $ad - bc \neq 0$ implies that $\det T \neq 0$; that is, the matrix T is nonsingular. The inverse of T is denoted by

$$T^{-1} = \begin{pmatrix} d & -b \\ -c & a \end{pmatrix}$$

Equation (5.7.4) implies that the inverse of f is associated with the inverse of T. (If $ad - bc = 1$, then T^{-1} is exactly the matrix inverse.) Furthermore, Eq. (5.7.5) can be used to show that the product (composition) of two bilinear transformations $f_3(z) \equiv f_2(f_1(z))$ is associated with $T_1 T_2$. Using this notation and Property (ii), it follows that any bilinear transformation is either linear or can be decomposed in the form $T_1 T_2 T_3$ where T_1 and T_3 are associated with linear transformations and T_2 is associated with inversion, that is,

$$T_1 = \begin{pmatrix} a_1 & b_1 \\ 0 & 1 \end{pmatrix}, \quad T_2 = \begin{pmatrix} 0 & 1 \\ 1 & 0 \end{pmatrix}, \quad T_3 = \begin{pmatrix} a_3 & b_3 \\ 0 & 1 \end{pmatrix}$$

(iv) The Cross Ratio of Four Points Is an Invariant

This means that

$$\frac{(w_1 - w_4)(w_3 - w_2)}{(w_1 - w_2)(w_3 - w_4)} = \frac{(z_1 - z_4)(z_3 - z_2)}{(z_1 - z_2)(z_3 - z_4)} \tag{5.7.7}$$

where w_i is associated with z_i in Eq. (5.7.1). This property can be established by manipulating expressions of the form:

$$w_1 - w_2 = \frac{(ad - bc)(z_1 - z_2)}{(cz_1 + d)(cz_2 + d)} \tag{5.7.8}$$

An alternative proof is as follows. Let $X(z_1, z_2, z_3, z_4)$ denote the right-hand side of Eq. (5.7.7). Because every bilinear transformation can be decomposed into linear transformations and an inversion, it suffices to show that these transformations leave X invariant. Indeed, if z_l is replaced by $az_l + b$, then both the numerator and the denominator of X are multiplied by a^2, and hence X is unchanged. Similarly, if z_l is replaced by $1/z_l$, X is again unchanged.

If one of the points w_l, say w_1, is ∞, then the left-hand side of Eq. (5.7.7) becomes $(w_3 - w_2)/(w_3 - w_4)$. Therefore this ratio should be regarded as the ratio of the points ∞, w_2, w_3, w_4.

Equation (5.7.7) has the following important consequence: letting $w_4 = w$, $z_4 = z$, Eq. (5.7.7) becomes

$$\frac{(w_1 - w)(w_3 - w_2)}{(w_1 - w_2)(w_3 - w)} = \frac{(z_1 - z)(z_3 - z_2)}{(z_1 - z_2)(z_3 - z)} \tag{5.7.9}$$

Equation (5.7.9) can be written in the form $w = f(z)$, where $f(z)$ is some bilinear transformation uniquely defined in terms of the points z_l and w_l. This allows the interpretation that bilinear transformations take any three distinct points z_l, into any three distinct point w_l. Furthermore, a bilinear transformation is uniquely determined by these three associations.

The fact that a bilinear transformation is completely determined by how it transforms three points is consistent with the fact that $f(z)$ depends at most on three complex parameters (because one of c, d is different than zero and hence can be divided out).

Example 5.7.2 Find the bilinear transformation that takes the three points $z_1 = 0$, $z_2 = 1$, and $z_3 = -i$ into $w_1 = 1$, $w_2 = 0$, and $w_3 = i$.

Equation (5.7.9) implies

$$\frac{(1-w)i}{i-w} = \frac{(-z)(-i-1)}{(-1)(-i-z)}, \qquad \text{or} \qquad w = \frac{1-z}{1+z}$$

(v) Bilinear Transformations Have One or Two Fixed Points

By fixed points, we mean those points of the complex z plane that do not change their position if $z \to f(z)$. In other words, now we interpret the transformation $f(z)$ not as a transformation of one plane onto another, but as a transformation of the plane onto itself. The fixed points are the invariant points of this transformation. We can find these fixed points by solving the equation $z = (az + b)/(cz + d)$. It follows that we have a quadratic equation and, except for the trivial cases ($c = 0$ and $a = d$, or $b = c = 0$), there exist two fixed points. If $(d - a)^2 + 4bc = 0$, these two points coincide. We exclude these cases and denote the two fixed points by α and β. Equation (5.7.9) with $w_1 = z_1 = \alpha$, $w_3 = z_3 = \beta$ implies

$$\frac{w-\alpha}{w-\beta} = \lambda\left(\frac{z-\alpha}{z-\beta}\right) \qquad (5.7.10)$$

where λ is a constant depending on w_2, z_2, α, and β. Thus the general form of a bilinear transformation with two given fixed points α and β depends only on one extra constant λ.

Example 5.7.3 Find all bilinear transformations that map 0 and 1 to 0 and 1, respectively.

Equation (5.7.10) implies

$$\frac{w-1}{w} = \lambda\left(\frac{z-1}{z}\right) \qquad \text{or} \qquad w = \frac{z}{(1-\lambda)z + \lambda}$$

(vi) Bilinear Transformations Map Circles and Lines Into Circles or Lines

In particular, we will show that under inversion a line through the origin goes into a line through the origin, a line not through the origin goes into a circle, a circle through the origin goes into a line, and a circle not through the origin goes into a circle.

The derivation of these results follows. The mathematical expression of a line is

$$ax + by + c = 0, \qquad a, b, c \in \mathbb{R}$$

Using $z = x + iy$, $\bar{z} = x - iy$, this becomes

$$Bz + \bar{B}\bar{z} + c = 0, \qquad B = \frac{a}{2} - \frac{ib}{2} \tag{5.7.11}$$

The mathematical expression of a circle with center z_0 and radius ρ is $|z - z_0| = \rho$ or $(z - z_0)(\bar{z} - \bar{z}_0) = \rho^2$, or

$$z\bar{z} + \bar{B}z + B\bar{z} + c = 0; \qquad B = -z_0, \qquad c = |B|^2 - \rho^2 \tag{5.7.12}$$

Under the inversion transformation $w = 1/z$, the line given by Eq. (5.7.11), replacing $z = 1/w$ and $\bar{z} = 1/\bar{w}$, becomes

$$cw\bar{w} + \bar{B}w + B\bar{w} = 0$$

If $c = 0$ (which corresponds to the line (5.7.11) going through the origin), this equation defines a line going through the origin. If $c \neq 0$, using Eq. (5.7.12), the above equation is a circle of radius $|B|/|c|$ and with center at $-B/c$.

Under the inversion transformation, the circle defined by Eq. (5.7.12) becomes

$$cw\bar{w} + Bw + \bar{B}\bar{w} + 1 = 0$$

If $c = 0$ (which corresponds to the circle (5.7.12) going through the origin), this equation defines a line. If $c \neq 0$, the above equation is the equation of a circle.

Remark It is natural to group circles and lines together because a line in the extended complex plane can be thought of as a circle through the point $z = \infty$. Indeed, consider the line (5.7.11), and let $z = 1/\zeta$. Then Eq. (5.7.11) reduces to the equation $c\zeta\bar{\zeta} + B\bar{\zeta} + \bar{B}\zeta = 0$, which is the equation of a circle going through the point $\zeta = 0$, that is, $z = \infty$. Using the terminology of this remark, we say that the inversion transformation maps circles into circles.

The linear transformation $az + b$ translates by b, rotates by $\arg a$, and dilates (or contracts) by $|a|$ (see Example 5.2.1). Therefore a linear transformation also maps circles into circles. These properties of the linear and inversion transformations, together with Property (ii), imply that a general bilinear transformation maps circles into circles.

The interior and the exterior of a circle are called the complementary domains of the circle. Similarly, the complementary domains of a line are the two half planes, one on each side of the line. Let K, K_c denote the complementary domains of a circle in the z plane and K^*, K_c^* denote the complementary domains of the corresponding circle in the w plane (in this terminology for a circle we include the degenerate case of a line).

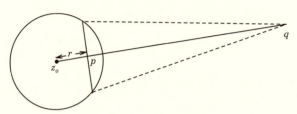

Fig. 5.7.2. Inverse points p, q

(vii) If K, K_c Denote the Complementary Domains of a Circle, Then Either
$$K^* = K \text{ and } K_c^* = K_c \text{ or } K^* = K_c \text{ and } K_c^* = K$$

The derivation of this property could be derived in a manner similar to the derivation of Properties (vi), where one uses inequalities instead of the equality (5.7.12). These inequalities follow from the fact that the interior and exterior of the circle are defined by $|z - z_0| < \rho$ and $|z - z_0| > \rho$, respectively. We will not go through the details here.

Example 5.7.4 The inversion $w = 1/z$ maps the interior (exterior) of the unit circle in the z plane to the exterior (interior) of the unit circle in the w plane.
   Indeed, if $|z| < 1$, then $|w| = 1/|z| > 1$.

(viii) Bilinear Transformations Map Inverse Points (With Respect to a Circle)
to Inverse Points

The points p and q are called inverse with respect to the circle of radius ρ and center z_0 if z_0, p, and q lie, in that order, on the same line and the distances $|z_0 - p|$ and $|z_0 - q|$ satisfy $|z_0 - p||z_0 - q| = \rho^2$ (see Figure 5.7.2).
   If the points z_0, p, and q lie on the same line, it follows that $p = z_0 + r_1 \exp(i\varphi)$, $q = z_0 + r_2 \exp(i\varphi)$. If they are inverse, then $r_1 r_2 = \rho^2$, or $(p - z_0)(\bar{q} - \bar{z}_0) = \rho^2$. Thus the mathematical description of two inverse points is

$$p = z_0 + re^{i\varphi}, \qquad q = z_0 + \frac{\rho^2}{r}e^{i\varphi}, \qquad r \neq 0 \qquad (5.7.13)$$

As $r \to 0$, $p = z_0$ and $q = \infty$. This is consistent with the geometrical description of the inverse points that show that as q recedes to ∞, p tends to the center. When a circle degenerates into a line, then the inverse points, with respect to the line, may be viewed as the points that are perpendicular to the line and are at equal distances from it.
   Using Eq. (5.7.13) and $z = z_0 + \rho e^{i\theta}$ (the equation for points on a circle), we have

$$\frac{z - p}{z - q} = \frac{\rho e^{i\theta} - re^{i\phi}}{\rho e^{i\theta} - \frac{\rho^2}{r}e^{i\phi}} = \frac{r}{\rho} \cdot \frac{re^{-i\theta} - \rho e^{-i\phi}}{re^{i\theta} - \rho e^{i\phi}} \cdot (-e^{i\phi}e^{i\theta}) \qquad (5.7.14)$$

whereupon

$$\left|\frac{z-p}{z-q}\right| = \frac{r}{\rho} \tag{5.7.15}$$

We also have the following. Let p and q be distinct complex numbers, and consider the equation

$$\left|\frac{z-p}{z-q}\right| = k, \qquad 0 < k \le 1 \tag{5.7.16}$$

It will be shown below that if $k = 1$, this equation represents a line and the points p and q are inverse points with respect to this line. If $k \ne 1$, this equation represents a circle with center at z_0 and radius ρ, given by

$$z_0 = \frac{p - k^2 q}{1 - k^2}, \qquad \rho = \frac{k|p-q|}{1-k^2} \tag{5.7.17}$$

the points p and q are inverse points with respect to this circle, and the point p is inside the circle.

Indeed, if $k = 1$, then $|z-p| = |z-q|$, which states that z is equidistant from p and q; the locus of such points is a straight line, which is the perpendicular bisector of the segment pq. If $k \ne 1$, then Eq. (5.7.16) yields

$$(z - p)(\bar{z} - \bar{p}) = k^2 (z - q)(\bar{z} - \bar{q})$$

This equation simplifies to Eq. (5.7.12), describing a circle with

$$B = \frac{k^2 q - p}{1 - k^2}, \qquad c = \frac{|p|^2 - k^2|q|^2}{1 - k^2}$$

These equations together with $z_0 = -B$, $\rho^2 = |B|^2 - c$ imply Eqs. (5.7.17). Equations (5.7.17) can be written as $|p - z_0| = k\rho$ and $|q - z_0| = \rho/k$, which shows that p and q are inverse points. Furthermore, because $k < 1$, the points p and q are inside and outside the circle, respectively.

Using Eq. (5.7.16), we demonstrate that bilinear transformations map inverse points to inverse points. Recall that bilinear transformations are compositions of linear transformations and inversion. If $w = az + b$, we must show that the points $\tilde{p} = ap + b$, $\tilde{q} = aq + b$ are inverse points with respect to the circle in the complex w plane. But

$$\left|\frac{w - \tilde{p}}{w - \tilde{q}}\right| = \frac{|z-p|}{|z-q|} = k$$

which shows that indeed $\tilde{p}$ and $\tilde{q}$ are inverse points in the w plane. Similarly, if $w = 1/z$, $\tilde{p} = 1/p$, and $\tilde{q} = 1/q$

$$\left| \frac{w - \tilde{p}}{w - \tilde{q}} \right| = \left| \frac{\frac{p-z}{pz}}{\frac{q-z}{qz}} \right| = \frac{|q|}{|p|} k$$

which shows that again $\tilde{p}$ and $\tilde{q}$ are inverse points, though $\tilde{p}$ might now lie inside or outside the circle. (Here we consider the generic case $p \neq 0, q \neq \infty$; the particular cases of $p = 0$ or $q = \infty$ are handled in a similar way.)

Example 5.7.5 A necessary and sufficient condition for a bilinear transformation to map the upper half plane $\text{Im} z > 0$ onto the unit disk $|w| < 1$, is that it be of the form

$$w = \beta \frac{z - \alpha}{z - \bar{\alpha}}, \qquad |\beta| = 1, \qquad \text{Im} \alpha > 0 \qquad (5.7.18)$$

Sufficiency: We first show that this transformation maps the upper half z plane onto $|w| < 1$. If z is on the real axis, then $|x - \alpha| = |x - \bar{\alpha}|$. Thus the real axis is mapped to $|w| = 1$; hence $y > 0$ is mapped onto one of the complementary domains of $|w| = 1$. Because $z = \alpha$ is mapped into $w = 0$, this domain is $|w| < 1$.

Necessity: We now show that the most general bilinear transformation mapping $y > 0$ onto $|w| < 1$ is given by Eq. (5.7.18). Because $y > 0$ is mapped onto one of the complementary domains of either a circle or of a line, $y = 0$ is to be mapped onto $|w| = 1$. Let α be a point in the upper half z plane that is mapped to the center of the unit circle in the w plane (i.e., to $w = 0$). Then $\bar{\alpha}$, which is the inverse point of α with respect to the real axis, must be mapped to $w = \infty$ (which is the inverse point of $w = 0$ with respect to the unit circle). Hence

$$w = \frac{a}{c} \frac{z - \alpha}{z - \bar{\alpha}}$$

Because the image of the real axis is $|w| = 1$, it follows that $|\frac{a}{c}| = 1$, and the above equation reduces to Eq. (5.7.18), where $\beta = a/c$.

Example 5.7.6 A necessary and sufficient condition for a bilinear transformation to map the disk $|z| < 1$ onto $|w| < 1$ is that it be of the form

$$w = \beta \frac{z - \alpha}{\bar{\alpha} z - 1}, \qquad |\beta| = 1, \qquad |\alpha| < 1 \qquad (5.7.19)$$

Sufficiency: We first show that this transformation maps $|z| < 1$ onto $|w| < 1$. If z is on the unit circle $z = e^{i\theta}$, then

$$|w| = |\beta| \left| \frac{e^{i\theta} - \alpha}{\bar{\alpha} e^{i\theta} - 1} \right| = \frac{|\alpha - e^{i\theta}|}{|\bar{\alpha} - e^{-i\theta}|} = 1$$

Hence $|z| < 1$ is mapped onto one of the complementary domains of $|w| = 1$. Because $z = 0$ is mapped into $\beta\alpha$, and $|\beta\alpha| < 1$, this domain is $|w| < 1$.

Necessity: We now show that the most general bilinear transformation mapping $|z| < 1$ onto $|w| < 1$ is given by Eq. (5.7.19). Because $|z| < 1$ is mapped onto $|w| < 1$, then $|z| = 1$ is to be mapped onto $|w| = 1$. Let α be the point in the unit circle that is mapped to $w = 0$. Then, from Eq. (5.7.13), $1/\bar{\alpha}$ (which is the inverse point of α with respect to $|z| = 1$) must be mapped to $w = \infty$. Hence, if $\alpha \neq 0$

$$w = \frac{a}{c} \frac{z - \alpha}{z - \frac{1}{\bar{\alpha}}} = \frac{a\bar{\alpha}}{c} \frac{z - \alpha}{\bar{\alpha} z - 1}$$

Because the image of $|z| = 1$ is $|w| = 1$, it follows that $|\frac{a\bar{\alpha}}{c}| = 1$, and the above equation reduces to Eq. (5.7.19), with $\beta = \frac{a\bar{\alpha}}{c}$. If $\alpha = 0$ and $\beta \neq 0$, then the points $0, \infty$ map into the points $0, \infty$, respectively, and $w = \beta z$, $|\beta| = 1$. Thus Eq. (5.7.19) is still valid.

It is worth noting that the process of successive inversions about an even number of circles is expressible as a bilinear transformation, as the following example illustrates.

Example 5.7.7 Consider a point z inside a circle C_1 of radius r and centered at the origin and another circle C_2 of radius R centered at z_0 (see Figure 5.7.3) containing the inverse point to z with respect to C_1. Show that two successive inversions of the point z about C_1 and C_2, respectively, can be expressed as a bilinear transformation.

The point $\tilde{z}$ is the inverse of z about the circle C_1 and is given by

$$z\bar{\tilde{z}} = r^2 \qquad \text{or} \qquad \tilde{z} = \frac{r^2}{\bar{z}}$$

The second inversion satisfies, for the point $\tilde{\tilde{z}}$

$$(\bar{\tilde{\tilde{z}}} - \bar{z_0})(\tilde{\tilde{z}} - z_0) = R^2$$

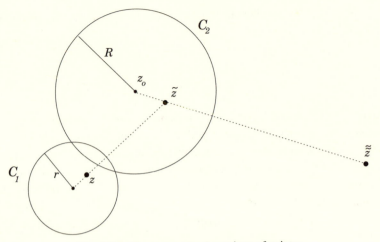

Fig. 5.7.3. Two successive inversions of point z

or

$$\tilde{\tilde{z}} = z_0 + \frac{R^2}{(\bar{\tilde{z}} - \overline{z_0})} = z_0 + \frac{R^2}{\frac{r^2}{z} - \overline{z_0}} = \frac{r^2 z_0 + (R^2 - |z_0|^2)z}{r^2 - \overline{z_0}z}$$

which is a bilinear transformation.

In addition to yielding a conformal map of the entire extended z plane, the bilinear transformation is also distinguished by the interesting fact that it is the only univalent function in the entire extended z plane.

Theorem 5.7.1 The bilinear transformation (5.7.1) is the only univalent function that maps $|z| \le \infty$ onto $|w| \le \infty$.

Proof Equation (5.7.8) shows that if $z_1 \ne z_2$, then $w_1 \ne w_2$; that is, a bilinear transformation is univalent. We shall now prove that a univalent function that maps $|z| \le \infty$ onto $|w| \le \infty$ must necessarily be bilinear.

To achieve this, we shall first prove that a univalent function that maps the finite complex z plane onto the finite complex w plane must be necessarily linear. We first note that if $f(z)$ is univalent in some domain D, then $f'(z) \ne 0$ in D. This is a direct consequence of Theorem 5.3.3, because if $f'(z_0) = 0$, $z_0 \in D$, then $f(z) - f(z_0)$ has a zero of order $n \ge 2$, and hence equation $f(z) = w$ has at least two distinct roots near z_0 for w near $f(z_0)$. It was shown in Theorem 5.3.2 (i.e., Theorem 5.5.1) that the image of $|z| < 1$ contains some disk $|w - w_0| < A$. This implies that ∞ is not an essential singularity of $f(z)$.

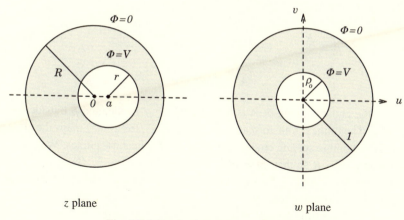

z plane   w plane

Fig. 5.7.4. Region between two cylinders

Because if ∞ is an essential singularity, then as $z \to \infty$, $f(z)$ comes arbitrarily close to w_0 (see Theorem 3.5.1); hence some values of f corresponding to $|z| > 1$ would also lie in the disk $|w - w_0| < A$, which would contradict the fact that f is univalent. It cannot have a branch point; therefore, $z = \infty$ is at worst a pole of f; that is, f is polynomial. But because $f'(z) \neq 0$ for $z \in D$, this polynomial must be linear. Having established the relevant result in the finite plane, we can now include infinities. Indeed, if $z = \infty$ is mapped into $w = \infty$, $f(z)$ being linear is satisfactory, and the theorem is proved. If $z_0 \neq \infty$ is mapped to $w = \infty$, then the transformation $\zeta = 1/(z - z_0)$ reduces this case to the case of the finite plane discussed above, in which case $w(\zeta)$ being linear (i.e., $w(\zeta) = a\zeta + b$) corresponds to $w(z)$ being bilinear. ∎

Example 5.7.8 Consider the region bounded by two cylinders perpendicular to the z plane; the bases of these cylinders are the discs bounded by the two circles $|z| = R$ and $|z - a| = r$, $0 < a < R - r$ ($R, r, a \in \mathbb{R}$). The inner cylinder is maintained at a potential V, while the outer cylinder is maintained at a potential zero. Find the electrostatic potential in the region between these two cylinders.

Recall (Eq. (5.7.16)) that the equation $|z - \alpha| = k|z - \beta|$, $k > 0$ is the equation of a circle with respect to which the points α and β are inverse to one another. If α and β are fixed, while k is allowed to vary, the above equation describes a family of nonintersecting circles. The two circles in the z plane can be thought of as members of this family, provided that α and β are chosen so that they are inverse points with respect to both of these circles (by symmetry considerations, we take them to be real), that is, $\alpha\beta = R^2$ and $(\alpha - a)(\beta - a) = r^2$. Solving for α and β we find

$$\beta = \frac{R^2}{\alpha} \quad \text{and} \quad \alpha = \frac{1}{2a}(R^2 + a^2 - r^2 - A),$$

$$A^2 \equiv [(R^2 + a^2 - r^2)^2 - 4a^2 R^2]$$

where the choice of sign of A is fixed by taking α inside, and β outside both circles. The bilinear transformation $w = \kappa(z - \alpha)/(\alpha z - R^2)$ maps the above family of nonintersecting circles into a family of concentric circles. We choose constant $\kappa = -R$ so that $|z| = R$ is mapped onto $|w| = 1$. (Here $z = Re^{i\theta}$ maps onto $w = (Be^{i\theta})/(\bar{B})$, where $B = R - \alpha e^{-i\theta}$, so that $|w| = 1$.) By using Eq. (5.7.17), for the circle $|z| = R$, $k_1^2 = \alpha/\beta$, and for the circle $|z - a| = r$, $k_2^2 = \frac{\alpha - a}{\beta - a}$. Thus from Eqs. (5.7.16)–(5.7.17) we see that the transformation

$$w = R\frac{z - \alpha}{R^2 - \alpha z} = \left(\frac{-R}{\alpha}\right)\left(\frac{z - \alpha}{z - \beta}\right)$$

maps $|z| = R$ onto $|w| = 1$ and maps $|z - a| = r$ onto $|w| = \rho_0$, where ρ_0 is given by

$$\rho_0 = k_2\left|\frac{R}{\alpha}\right| = \left|\frac{R}{\alpha}\right|\sqrt{\frac{\alpha - a}{\beta - a}}$$

From this information we can now find the solution of the Laplace equation that satisfies the boundary conditions. Calling $w = \rho e^{i\phi}$, this solution is $\Phi = V \log \rho / \log \rho_0$. Thus

$$\Phi = \frac{V}{\log \rho_0} \log|w| = \frac{V}{\log \rho_0} \log\left|R\frac{z - \alpha}{R^2 - \alpha z}\right|$$

From the mapping, we conclude that when $|z| = R$, $|w| = 1$; hence $\Phi = 0$, and when $|z - a| = r$, $|w| = \rho_0$, and therefore $\Phi = V$. Hence the real part of the analytic function $\Omega(w) = \frac{V}{\log \rho} \log w$ leads to a solution Φ ($\Omega = \Phi + i\Psi$) of Laplace's equation with the requisite boundary conditions.

In conclusion, we mention without proof, the Schwarz reflection principle pertaining to analytic continuation across arcs of circles. This is a generalization of the reflection principle mentioned in conjunction with the Schwarz–Christoffel transformation in Section 5.6, which required the analytic continuation of a function across straight line segments, for example, the real axis.

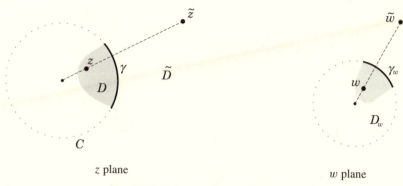

z plane   w plane

Fig. 5.7.5. Schwarz symmetry principle

Theorem 5.7.2 (Schwarz Symmetry Principle) Let $z \in D$ and $w \in D_w$ be points in the domains D and D_w, which contain circular arcs γ and γ_w respectively. (These arcs could degenerate into straight lines.) Let $f(z)$ be analytic in D and continuous in $D \cup \gamma$. If $w = f(z)$ maps D onto D_w so that the arc γ is mapped to γ_w, then $f(z)$ can be analytically continued across γ into the domain $\tilde{D}$ obtained from D by inversion with respect to the circle C of which γ is a part. Consequently, if z and $\tilde{z}$ are inverse points with respect to C, where $z \in D$ and $\tilde{z} \in \tilde{D}$, then the analytic continuation is given by $f(\tilde{z}) = \tilde{f}(z)$, where $\tilde{f}(z) = \tilde{w}$ is the inverse point to w with respect to circle C_w.

In fact, the proof of the symmetry principle can be reduced to that of symmetry across the real axis by transforming the circles C and C_w to the real axis, by bilinear transformations. We will not go into further detail here.

Thus, for example, let γ be the unit circle centered at the origin in the z plane, and let $f(z)$ be analytic within γ and continuous on γ. Then if $|f(z)| = R$ (i.e., γ_w is a circle of radius R in the w plane centered at the origin) on γ, then $f(z)$ can be analytically continued across γ by means of the formula $f(z) = R/\overline{f(1/\bar{z})}$ because $R/\bar{f}$ is the inverse point of f with respect to the circle of radius R centered at the origin and $1/\bar{z}$ is the inverse point to the point z inside the unit circle γ. On the other hand, suppose $f(z)$ maps to a real function on γ, then the formula for analytic continuation is given by $f(z) = \bar{f}(\frac{1}{\bar{z}})$ because $\bar{f}$ is the inverse point of f with respect to the real axis.

We note the "symmetry" in this continuation formula; that is, $\tilde{w} = \tilde{f}(z)$ is the inverse point to $w = f(z)$ with respect to the circle C_w of which γ_w is a part, and $\tilde{z}$ is the inverse point to z with respect to the circle C of which γ is a part. As indicated in Section 5.6, in the case where γ and γ_w degenerate into the real axis, this formula yields continuation of a function $f(z)$ where $f(z)$ is real for real z from the upper half plane to the lower half plane: $f(\bar{z}) = \bar{f}(z)$. Similar specializations obtain when the circles reduce to arbitrary straight lines. We also note that in Section 5.8 the symmetry principle across circular arcs is used in a crucial way.

Problems for Section 5.7

1. Show that the "cross ratios" associated with the points $(z, 0, 1, -1)$ and $(w, i, 2, 4)$ are $2z/(z+1)$ and $-2(w-i)/[(w-4)/(2-i)]$, respectively. Use these to find the bilinear transformation that maps $0, 1, -1$ to $i, 2, 4$, respectively.

2. Show that the transformation $w_1 = ((z+2)/(z-2))^{1/2}$ maps the z plane with a cut $-2 \leq \mathrm{Re}z \leq 2$ to the right half plane. Show that the latter is mapped onto the interior of the unit circle by the transformation $w = (w_1 - 1)/(w_1 + 1)$. Thus deduce the overall transformation that maps the simply connected region containing all points of the plane (including ∞) except the real points z in $-2 \leq z \leq 2$ onto the interior of the unit circle.

3. Show that the transformation $w = (z-a)/(z+a)$, $a = \sqrt{c^2 - \rho^2}$ (where c and ρ are real, $0 < \rho < c$), maps the domain bounded by the circle $|z - c| = \rho$ and the imaginary axis onto the annular domain bounded by $|w| = 1$ and an inner concentric circle (see Figure 5.7.6). Find the radius, δ, of the inner circle.

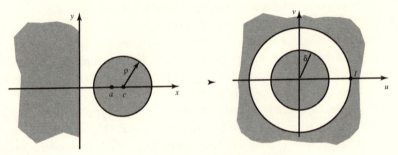

Fig. 5.7.6. Mapping of Problem 5.7.3

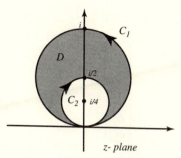

Fig. 5.7.7. Mapping of Problem 5.7.5

4. Show that the transformation $w_1 = [(1 + z)/(1 - z)]^2$ maps the upper half unit circle to the upper half plane and that $w_2 = (w_1 - i)/(w_1 + i)$ maps the latter to the interior of the unit circle. Use these results to find an elementary conformal mapping that maps a semicircular disk onto a full disk.

5. Let C_1 be the circle with center $i/2$ passing through 0, and let C_2 be the circle with center $i/4$ passing through 0 (see Figure 5.7.7). Let D be the region enclosed by C_1 and C_2. Show that the inversion $w_1 = 1/z$ maps D onto the strip $-2 < \operatorname{Im} w_1 < -1$ and the transformation $w_2 = e^{\pi w_1}$ maps this strip to the upper half plane. Use these results to find a conformal mapping that maps D onto the unit disk.

6. Find a conformal map f that maps the region between two circles $|z| = 1$ and $|z - \frac{1}{4}| = \frac{1}{4}$ onto an annulus $\rho_0 < |z| < 1$, and find ρ_0.

7. Find the function ϕ that is harmonic in the lens-shaped domain of Figure 5.7.8 and takes the values 0 and 1 on the bounding circular arcs. Hint: It is useful to note that the transformation $w = z/(z - (1 + i))$ maps the lens-shaped domain into the region $R_w : \frac{3\pi}{4} \le \arg w \le \frac{5\pi}{4}$ with $\phi = 1$ on $\arg w = 3\pi/4$ and $\phi = 0$ on $\arg w = 5\pi/4$. Then use the ideas introduced in Section 5.4 (c.f. Example 5.4.4) to find the corresponding harmonic function $\phi(w)$.

Note: ϕ can be interpreted as the steady state temperature inside an infinitely long strip (perpendicular to the plane) of material having this lens-shaped region as its cross section, with its sides maintained at the given temperatures.

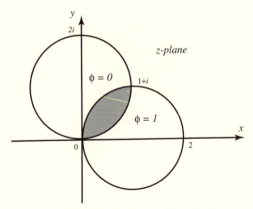

Fig. 5.7.8. Mapping of Problem 5.7.7

*5.8 Mappings Involving Circular Arcs

In Section 5.6 we showed that the mapping of special polygonal regions to the upper half plane involved trigonometric and elliptic functions. In this section we investigate the mapping of a region whose boundary consists of a **curvilinear polygon**, that is, a polygon whose sides are made up of circular arcs. We outline the main ideas, and in certain important special cases we will be led to an interesting class of functions called automorphic functions, which can be considered generalizations of elliptic functions. We will study a class of automorphic functions known as Schwarzian triangle functions, of which the best known (with zero angles) is the so-called elliptic modular function.

Consider a domain of the w plane bounded by circular arcs. Our aim is to find the transformation $w = f(z)$ that maps this domain onto the upper half of the z plane (see Figure 5.8.1).

The relevant construction is conceptually similar to the one used for linear polygons (i.e., the Schwarz–Christoffel transformation). We remind the reader that the crucial step in that construction is the introduction of the ratio f''/f'. The Schwarz reflection principle implies that this ratio is analytic in the entire z plane except at the points corresponding to the vertices of the polygon; near these vertices in the z plane, that is, near $z = a_\ell$

$$f(z) = (z - a_\ell)^{\alpha_\ell} \left[c_\ell^{(0)} + c_\ell^{(1)}(z - a_\ell) + \cdots \right]$$

therefore f''/f' has simple poles. These two facts and Liouville's Theorem imply the Schwarz–Christoffel transformation. The distinguished property of f''/f' is that it is invariant under a linear transformation; that is, if we transform $f = A\hat{f} + B$, where A and B are constant, then $f''/f' = \hat{f}''/\hat{f}'$. The fact that

z plane

w plane

Fig. 5.8.1. Mapping of a region whose boundary contains circular arcs

the mapping is constructed from a given polygon through an even number of Schwarz reflections implies that the most general form of the mapping is given by $f(z) = A\hat{f}(z) + B$ where A and B are constants.

The generalization of the above construction to the case of circular arcs is as follows. In Section 5.7 the Schwarz symmetry principle across circular arcs was discussed. We also mentioned in Section 5.7 that an even succession of inversions across circles can be expressed as a bilinear transformation. It is then natural to expect that the role that was played by $f''(z)/f'(z)$ in the Schwarz–Christoffel transformation will now be generalized to an operator that is invariant under bilinear transformations. This quantity is the so-called **Schwarzian derivative**, defined by

$$\{f, z\} \equiv \left(\frac{f''(z)}{f'(z)}\right)' - \frac{1}{2}\left(\frac{f''(z)}{f'(z)}\right)^2 \tag{5.8.1}$$

Indeed, let

$$F = \frac{af + b}{cf + d}, \qquad ad - bc \neq 0 \tag{5.8.2}$$

Then

$$F' = \frac{(ad - bc)f'}{(cf + d)^2}, \qquad \text{or} \qquad (\log F')' = (\log f')' - 2(\log(cf + d))'$$

Hence

$$\frac{F''}{F'} = \frac{f''}{f'} - \frac{2cf'}{cf + d}$$

Using this equation to compute $(F''/F')'$ and $(F''/F')^2$, it follows from Eq. (5.8.1) that

$$\{f, z\} = \{F, z\} \tag{5.8.3}$$

Single-valuedness of $\{f, z\}$ follows in much the same way as the derivation of the single-valuedness of $f''(z)/f'(z)$ in the Schwarz–Christoffel derivation. In the present case, any even number of inversions across circles is a bilinear transformation (see Example 5.7.7). Because the Schwarzian derivative is invariant under a bilinear transformation, it follows that the function $\{f, z\}$ corresponding to any point in the upper half z plane is uniquely obtained. Similar arguments hold for an odd number of inversions and points in the lower half plane. Moreover, the function $\{f, z\}$ takes on real values for real values of z. Hence we can analytically continue $\{f, z\}$ from the upper half to lower half z plane by Schwarz reflection. Consequently, there can be no branches whatsoever and the function $\{f, z\}$ is single valued. Thus the Schwarzian derivative is analytic in the entire z plane except possibly at the points a_ℓ, $\ell = 1, \ldots, n$. The behavior of $f(z)$ at a_ℓ can be found by noting that (after a bilinear transformation) $f(z)$ maps a piece of the real z axis containing $z = a_\ell$ onto two linear segments forming an angle $\pi\alpha_\ell$. Therefore in the neighborhood of $z = a_\ell$

$$f(z) = (z - a_\ell)^{\alpha_\ell} g(z) \tag{5.8.4}$$

where $g(z)$ is analytic at $z = a_\ell$, $g(a_\ell) \neq 0$, and $g(z)$ is real when z is real. This implies that the behavior of $\{f, z\}$ near a_ℓ is given by the following (the reader can verify the intermediate step: $\frac{f''}{f'}(z) = \frac{\alpha_\ell - 1}{z - a_\ell} + \frac{1 + \alpha_\ell}{\alpha_\ell} \frac{g'(a_\ell)}{g(a_\ell)} + \cdots$):

$$\{f, z\} = \frac{1}{2} \frac{1 - \alpha_\ell^2}{(z - a_\ell)^2} + \frac{\beta_\ell}{z - a_\ell} + h(z), \qquad \beta_\ell \equiv \frac{1 - \alpha_\ell^2}{\alpha_\ell} \frac{g'(a_\ell)}{g(a_\ell)} \tag{5.8.5}$$

where $h(z)$ is analytic at $z = a_\ell$. Using these properties of $\{f, z\}$ and Liouville's Theorem, it follows that

$$\{f, z\} = \frac{1}{2} \sum_{\ell=1}^{n} \frac{(1 - \alpha_\ell^2)}{(z - a_\ell)^2} + \sum_{\ell=1}^{n} \frac{\beta_\ell}{z - a_\ell} + c \tag{5.8.6}$$

where $\alpha_1, \ldots, \alpha_n, \beta_1, \ldots, \beta_n, a_1, \ldots, a_n$, are real numbers and c is a constant. We recall that in the case of the Schwarz–Christoffel transformation the analogous constant c was determined by analyzing $z = \infty$. We now use the same idea. If we assume that none of the points $a_1, \ldots, a_n$ coincide with ∞, then $f(z)$ is analytic at $z = \infty$; that is, $f(z) = f(\infty) + c_1/z + c_2/z^2 + \cdots$ near $z = \infty$. Using this expansion in Eq. (5.8.1) it follows that $\{f, z\} = k_4/z^4 + k_5/z^5 + \cdots$ near $z = \infty$. This implies that by expanding the right-hand side of (5.8.6) in a power series in $1/z$, and equating to zero the coefficients of z^0, $1/z$, $1/z^2$, and $1/z^3$, we find that $c = 0$ (the coefficient of z^0) and for the coefficients of $1/z$,

$1/z^2$ and $1/z^3$

$$\sum_{\ell=1}^{n} \beta_\ell = 0, \qquad \sum_{\ell=1}^{n} \left(2a_\ell \beta_\ell + 1 - \alpha_\ell^2 \right) = 0, \qquad \sum_{\ell=1}^{n} \left[\beta_l a_\ell^2 + a_\ell \left(1 - \alpha_\ell^2 \right) \right] = 0$$

$$(5.8.7)$$

In summary, let $f(z)$ be a solution of the third-order differential equation (5.8.6) with $c = 0$, where $\{f, z\}$ is defined by Eq. (5.8.1) and where the real numbers appearing in the right-hand side of Eq. (5.8.6) satisfy the relations given by Eq. (5.8.7). Then the transformation $w = f(z)$ maps the domain of the w plane, bounded by circular arcs forming vertices with angles $\pi \alpha_1, \ldots \pi \alpha_n$, onto the upper half of the z plane. The vertices are mapped to the points $a_1, \ldots, a_n$ of the real z axis.

It is significant that the third-order nonlinear differential Eq. (5.8.6) can be reduced to a second-order linear differential equation. Indeed, if $y_1(z)$ and $y_2(z)$ are two linearly independent solutions of the equation

$$y''(z) + \frac{1}{2} P(z) y(z) = 0 \qquad (5.8.8)$$

then

$$f(z) \equiv \frac{y_1(z)}{y_2(z)} \qquad (5.8.9a)$$

solves

$$\{f, z\} = P(z) \qquad (5.8.9b)$$

The proof of this fact is straightforward. Substituting $y_1 = y_2 f$ into Eq. (5.8.8), demanding that both y_1 and y_2 solve Eq. (5.8.8) and noting that the Wronskian $W = y_2 y_1' - y_1 y_2'$ is a constant for Eq. (5.8.8), it follows that

$$\frac{f''}{f'} = -2 \frac{y_2'}{y_2}$$

which implies Eq. (5.8.9b). This concludes the derivation of the main results of this section, which we express as a theorem.

Theorem 5.8.1 (Mapping of Circular Arcs) If $w = f(z)$ maps the upper half of the z plane onto a domain of the w plane bounded by n circular arcs, and if the points $z = a_\ell$, $\ell = 1, \ldots, n$, on the real z axis are mapped to the vertices of angle $\pi \alpha_\ell$, $\ell = 1, \ldots, n$, then

$$w = f(z) = \frac{y_1(z)}{y_2(z)} \tag{5.8.10}$$

where $y_1(z)$ and $y_2(z)$ are two linearly independent solutions of the linear differential equation

$$y''(z) + \sum_{\ell=1}^{n} \left[\frac{(1 - \alpha_\ell^2)}{4(z - a_\ell)^2} + \frac{1}{2} \sum_{\ell=1}^{n} \frac{\beta_\ell}{z - a_\ell} \right] y(z) = 0 \tag{5.8.11}$$

and the real constants β_ℓ, $\ell = 1, \ldots, n$ satisfy the relations (5.8.7).

Remarks (1) The three identities (5.8.7) are the only general relations that exist between the constants entering Eq. (5.8.11). Indeed, the relevant domain is specified by n circular arcs, that is, $3n$ real parameters (each circle is prescribed by the radius and the two coordinates of the center). However, as mentioned in Section 5.7, three arbitrary points on the real z axis can be mapped to any three vertices (i.e., six real parameters) in the w plane. This reduces the number of parameters describing the w domain to $3n - 6$. On the other hand, the transformation $f(z)$ involves $3n - 3$ independent parameters: $3n$ real quantities $\{\alpha_\ell, \beta_\ell, a_\ell\}_{\ell=1}^{n}$, minus the three constraints (5.8.7). Because three of the values a_ℓ can be arbitrarily prescribed, we see that the $f(z)$ also depends on $3n - 6$ parameters.

  (2) The procedure of actually constructing a mapping function $f(z)$ in terms of a given curvilinear polygon is further complicated by the determination of the constants in Eq. (5.8.11) in terms of the given geometrical configuration. In Eq. (5.8.11) we know the angles $\{\alpha_\ell\}_{\ell=1}^{n}$. We require that the points a_ℓ on the real z axis correspond to the vertices A_ℓ of the polygon. Characterizing the remaining $n - 3$ constants, that is, the n values β_ℓ (the so-called accessory parameters) minus three constraints, by geometrical conditions is in general unknown. The cases of $n = 2$ (a crescent) and $n = 3$ (a curvilinear triangle; see Figure 5.8.2) are the only cases in which the mapping is free of the determination of accessory parameters.

Example 5.8.1 Consider a domain of the w plane bounded by three circular arcs with interior angles $\pi \alpha$, $\pi \beta$, and $\pi \gamma$. Find the transformation that maps this domain to the upper half of the z plane. Specifically, map the vertices with angles $\pi \alpha$, $\pi \beta$, and $\pi \gamma$ to the points ∞, 0, and 1.

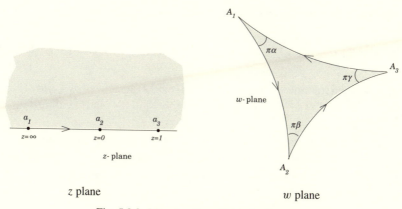

Fig. 5.8.2. Mapping from three circular arcs

We associate with the vertices A_1, A_2, and A_3 the points $a_1(\infty)$, $a_2(0)$, and $a_3(1)$. Calling $\alpha_2 = \beta$, $\alpha_3 = \gamma$, $a_2 = 0$, and $a_3 = 1$, Eq. (5.8.6) with $c = 0$ becomes

$$\{f, z\} = \frac{1 - \beta^2}{2z^2} + \frac{1 - \gamma^2}{2(z - 1)^2} + \frac{\beta_2}{z} + \frac{\beta_3}{z - 1} \tag{5.8.12}$$

When one point, in this case $w = A_1$, is mapped to $z = \infty$, then the terms involving a_1 drop out of the right-hand side of Eq. (5.8.6), and from Eq. (5.8.4), recalling the transformation $z - a_1 \to 1/z$, we find that $f(z) = \gamma z^{-\alpha}$ $[1 + c_1/z + \cdots]$ for z near ∞. Similarly, owing to the identification $a_1 = \infty$, one must reconsider the derivation of the relations (5.8.7). These equations were derived under the assumption that $f(z)$ is analytic at ∞. However, in this example the above behavior of $f(z)$ implies that $\{f, z\} = ((1 - \alpha^2)/2z^2)[1 + D_1/z + \cdots]$ as $z \to \infty$. Expanding the right-hand side of Eq. (5.8.12) in powers of $1/z$ and equating the coefficients of $1/z$ and $1/z^2$ to 0 and $(1 - \alpha^2)/2$, respectively, we find $\beta_2 + \beta_3 = 0$ and $\beta_3 \equiv (\beta^2 + \gamma^2 - \alpha^2 - 1)/2$. Using these values for β_2 and β_3 in Eq. (5.8.12), we deduce that $w = f(z) = y_1/y_2$, where y_1 and y_2 are two linearly independent solutions of Eq. (5.8.11):

$$y''(z) + \frac{1}{4}\left[\frac{1 - \beta^2}{z^2} + \frac{1 - \gamma^2}{(z - 1)^2} + \frac{\beta^2 + \gamma^2 - \alpha^2 - 1}{z(z - 1)}\right] y(z) = 0 \tag{5.8.13}$$

Equation (5.8.13) is related to an important differential equation known as the hypergeometric equation, which is defined as in Eq. (3.7.35c)

$$z(1 - z)\chi''(z) + [c - (a + b + 1)z]\chi'(z) - ab\chi(z) = 0 \tag{5.8.14}$$

where a, b, and c are, in general, complex constants. It is easy to verify that if

$$a = \frac{1}{2}(1 + \alpha - \beta - \gamma), \qquad b = \frac{1}{2}(1 - \alpha - \beta - \gamma), \qquad c = 1 - \beta \quad (5.8.15)$$

(all real), then solutions of Eqs. (5.8.13) and (5.8.14) are related by $\chi = u(z)y(z)$, where $u(z) = z^A / (1 - z)^B$, $A = -c/2$, and $B = \frac{a+b-c+1}{2}$, and therefore $f(z) = y_1/y_2 = \chi_1/\chi_2$.

In summary, the transformation $w = f(z)$ that maps the upper half of the z plane onto a curvilinear triangle with angles $\pi\alpha$, $\pi\beta$, and $\pi\gamma$, in such a way that the associated vertices are mapped to ∞, 0, and 1, is given by $f(z) = \chi_1/\chi_2$, where χ_1 and χ_2 are two linearly independent solutions of the hypergeometric equation (5.8.14) with a, b, and c given by Eqs. (5.8.15).

The hypergeometric equation (5.8.14) has a series solution (see also Section 3.7 and Nehari(1952)) that can be written

$$\chi_1(z; a, b, c) = 1 + \frac{ab}{c}z + \frac{a(a+1)b(b+1)}{c(c+1)2!}z^2 + \cdots \qquad (5.8.16a)$$

as can be directly verified. This function can also be expressed as an integral:

$$\chi_1(z; a, b, c) = \int_0^1 t^{b-1}(1-t)^{c-b-1}(1-zt)^{-a}\, dt \qquad (5.8.16b)$$

where the conditions $b > 0$ and $c > b$ (a, b, c assumed real) are necessary for the existence of the integral.

We shall assume that $\alpha + \beta + \gamma < 1$, $\alpha, \beta, \gamma > 0$; then we see that Eq. (5.8.15) ensures that these conditions hold. Moreover, expanding $(1 - tz)^{-a}$ in a power series in z leads to Eq. (5.8.16b), apart from a multiplicative constant. (To verify this, one can use $\int_0^1 t^{b-1}(1-t)^{c-b-1}\, dt = \Gamma(b)\Gamma(c-b)/\Gamma(c)$.) To obtain $w = f(z)$, we need a second linearly independent solution of Eq. (5.8.14). We note that the transformation $z' = 1 - z$ transforms Eq. (5.8.14) to

$$z'(1 - z')\chi'' + [a + b - c + 1 - (a + b + 1)z']\chi' - ab\chi = 0$$

and we see that the parameters of this hypergeometric equation are $a' = a$, $b' = b$, $c' = a + b - c + 1$, whereupon a second linearly independent solution can be written in the form

$$\chi_2(z; a, b, c) = \chi_1(1 - z, a, b, a + b - c + 1)$$

$$= \int_0^1 t^{b-1}(1 - t)^{a-c}(1 - (1 - z)t)^{-a}\, dt \qquad (5.8.16c)$$

Once again, the condition $\alpha + \beta + \gamma < 1$ ensures the existence of the integral (5.8.16c) because we find that $b > 0$ and $a > c - 1$.

Consequently, the mapping

$$w = f(z) = \frac{\chi_1(z; a, b, c)}{\chi_2(z; a, b, c)} \tag{5.8.16d}$$

taking the upper half z plane to the w plane is now fixed with χ_1 and χ_2 specified as above. The real z axis maps to the circular triangle as depicted in Figure 5.8.2. So, for example, the straight line on the real axis from $z = 0$ to $z = 1$ maps to a circular arc between A_2 and A_3 in the w plane.

We note that the case of $\alpha + \beta + \gamma = 1$ can be transformed into a triangle with straight sides (note that the sum of the angles is π) and therefore can be considered by the methods of Section 5.6. In the case of $\alpha + \beta + \gamma > 1$, one needs to employ different integral representations of the hypergeometric function (cf. Whittaker and Watson (1927)).

In the next example we discuss the properties of $f(z)$ and the analytic continuation of the inverse of $w = f(z)$, or alternatively, the properties of the map and its inverse as we continue from the upper half z plane to the lower half z plane and repeat this process over and over again. This is analogous to the discussion of the elliptic function in Example 5.6.8.

Example 5.8.2 (The Schwarzian Triangle Functions) In Example 5.8.1 we derived the function $w = f(z)$ that maps the upper half of the z plane onto a curvilinear triangle in the w plane. Such functions are known as Schwarzian s functions, $w = s(z)$, or as Schwarzian triangle functions. Now we shall further study this function and the inverse of this function, which is important in applications such as the solution to certain differential equations (e.g. Chazy's Eq. (3.7.52) the Darboux-Halphen system (3.7.53)) which arise in relativity and integrable systems. These inverse functions $z = S(w)$ are also frequently called Schwarzian S functions (capital S) or Schwarzian triangle functions.

We recall from Example 5.6.8 that although the function $w = f(z) = F(z, k)$, which maps the upper half of the z plane onto a rectangle in the w plane, is multivalued; nevertheless, its inverse $z = \text{sn}(w, k)$ is single valued. Similarly, the Schwarzian function $f(z) = s(z)$, which maps the upper half of the z plane onto a curvilinear triangle in the w plane, is also multivalued. While the inverse of this function is not in general single valued, we shall show that in the particular case that the angles of the curvilinear triangle satisfy $\alpha + \beta + \gamma < 1$

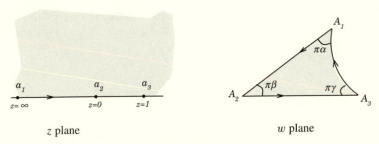

z plane   w plane

Fig. 5.8.3. Two straight segments, one circular arc

and

$$\alpha = \frac{1}{l}, \qquad \beta = \frac{1}{m}, \qquad \gamma = \frac{1}{n}, \qquad l, m, n \in \mathbf{Z}^+, \qquad \alpha, \beta, \gamma \neq 0$$

(5.8.17)

($\mathbf{Z}^+$ is the set of positive integers) the inverse function is single valued.

For convenience we shall assume that two of the sides of the triangle are formed by straight line segments (a special case of a circle is a straight line) meeting at the origin, and that one of these segments coincides with part of the positive real axis.

This is without loss of generality. Indeed, let C_1 and C_2 be two circles that meet at $z = A_2$ at an angle $\pi\beta$. Because $\beta \neq 0$, these circles intersect also at another point, say, A. The transformation $\tilde{w} = (w - A_2)/(w - A)$ maps all the circles through A into straight lines. (Recall from Section 5.7 that bilinear transformations map circles into either circles or lines, but because $w = A$ maps to $\tilde{w} = \infty$, it must be the latter.) In particular, the transformation maps C_1 and C_2 into two straight lines through A_2. By an additional rotation, it is possible to make one of these lines to coincide with the real axis (see Figure 5.8.3, w plane).

It turns out that if

$$\alpha + \beta + \gamma < 1 \tag{5.8.18}$$

then there exists a circle that intersects at right angles the three circles that make up a curvilinear triangle. Indeed, as discussed above, we may without loss of generality consider a triangle with two straight sides (see Figure 5.8.3). Any circle centered at the origin, which we will call C_0, is obviously orthogonal to the two straight sides of the triangle. Let C denote the circle whose part forms the third side of the triangle (see Figure 5.8.4, and note specifically the arc between A_1 and A_3 that extends to the circle C). Equation (5.8.18) implies that the origin of C_0 lies exterior to C because the arc (A_1, A_3) is convex. Hence it

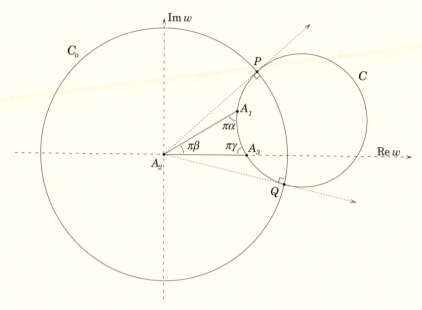

Fig. 5.8.4. Orthogonal circle C_0

is possible to draw tangents from the origin of C_0 to C. If P and Q denote the points of contact of the tangents with C, then the circle C_0 is orthogonal to C. This circle, C_0, is called the orthogonal circle of the triangle. Given the angles $\alpha\pi$, $\beta\pi$, and $\gamma\pi$ and the points A_1 and A_3 (point A_2 is the origin) that are determined by the properties of the equation (i.e., the hypergeometric equation; we discuss this issue later in this example), the circle C and the orthogonal circle C_0 are then fixed.

If either of the angles α or γ are zero, then the lines A_2A_1 or A_2A_3, respectively, correspond to a tangent to the circle C, in which case the vertex A_1 or A_3, respectively lies on the orthogonal circle C_0.

Let $w = s(z; \alpha, \beta, \gamma)$ denote the transformation that maps the upper half of the z plane onto the triangle depicted in Figure 5.8.3. If the angles of this triangle satisfy the conditions (5.8.17) and (5.8.18), then it turns out that the inverse of this transformation, denoted by S

$$w = s(z; \alpha, \beta, \gamma), \qquad z = S(w; \alpha, \beta, \gamma) \qquad (5.8.19)$$

is single valued in the interior of the orthogonal circle associated with this triangle. We next outline the main ideas needed to establish this result.

The derivation is based on two facts. First, the function S has no singularities in the entire domain of its existence except poles of order α^{-1}. Second, the domain of existence of S is simply connected. Using these facts (see the Monodromy Theorem, Section 3.5), it follows that S is single valued. The singularity structure of S follows from the fact that for the original triangle, $s(z) = z^{-\alpha}g_1(z), s(z) = z^{\beta}g_2(z)$, and $s(z) = (z-1)^{\gamma}g_3(z)$ near $z = \infty, z = 0$, and $z = 1$, respectively, where the $g_i(z), i = 1, 2, 3$ are analytic. This follows from the properties of the mapping of two line segments meeting at an angle to the real z axis (see Section 5.6). Hence, $S(w)$ behaves like $(w - A_1)^{-1/\alpha}, w^{1/\beta}$, and $1 + (w - A_3)^{1/\gamma}$, respectively, which shows that if the reciprocals of α, β, γ are positive integers then the only singularity of $S(w)$ is a pole of order $1/\alpha$ at $w = A_1$ corresponding to $z = \infty$ (recall that vertex A_1 corresponds to $z = \infty$). Because the transformation is conformal, there can be no other singularities inside the triangle. All possible analytic continuations of $z = S(w; \alpha, \beta, \gamma)$ to points outside the original triangle can be obtained by reflections about any of the sides of the triangles, by using the Schwarz reflection principle (Theorem 5.7.2).

By the properties of bilinear transformations, we know that an inversion with respect to a circular arc transforms circles into circles, preserves angles, and maps the orthogonal circle onto itself. It follows that any number of inversions of a circular triangle will again lead to a circular triangle situated in the interior of the orthogonal circle. This shows that S cannot be continued to points outside the orthogonal circle and that the vertices are the locations of the only possible singularities. Any point within the orthogonal circle can be reached by a sufficient number of inversions. Indeed, at the boundary of the domain covered by these triangles, there can be no circular arcs of positive radius, because otherwise it would be possible to extend this domain by another inversion. Hence, the boundary of this domain, which is the orthogonal circle, is made up of limit points of circular arcs whose radii tend to zero. This discussion implies that the domain of existence of $S(w)$ is the interior of the orthogonal circle, which is a simply connected domain. The function $S(w)$ cannot be continued beyond the circumference of this circle, so the circumference of the orthogonal circle is a **natural boundary** of $S(w; \alpha, \beta, \gamma)$. The boundary is a dense set of singularities, in this case a dense set of poles of order $1/\alpha$.

Next let us find the analytic expression of the function $w = s(z; \alpha, \beta, \gamma)$. Recall that because $f(z) = s(z)$ satisfies Eq. (5.8.12) then $s(z) = \hat{\chi}_1/\hat{\chi}_2$, where $\hat{\chi}_1$ and $\hat{\chi}_2$ are any two linearly independent solutions of the hypergeometric Eq. (5.8.14) and the numbers a, b, c are related to α, β, γ by Eq. (5.8.15). A solution $\hat{\chi}_2$ of the hypergeometric equation that we will now use is given by Eqs. (5.8.16b): $\hat{\chi}_2 = \chi_1(z; a, b, c)$. A second solution can be obtained by the observation that if χ is a solution of the hypergeometric Eq. (5.8.4), then

$z^{1-c}\chi(z; a', b', c')$ is also a solution of the same equation, where $a' = a - c + 1$, $b' = b - c + 1$, and $c' = 2 - c$. Because the value of $\chi_1(z; a, b, c)$ at $z = 0$ is a nonzero constant, while the value of $z^{1-c}\chi_1(z; a', b', c')$ at $z = 0$ is zero (for $0 < c < 1$), it follows that the Wronskian is nonzero and that these two solutions are linearly independent. Hence

$$w = s(z; \alpha, \beta, \gamma) = \frac{z^{1-c}\chi_1(z; a - c + 1, b - c + 1, 2 - c)}{\chi_1(z; a, b, c)} \qquad (5.8.20)$$

where a, b, and c are given by Eq. (5.8.15). (This is a different representation than that discussed in Example 5.8.1. It is more convenient for this case, two sides being straight lines.) The vertices of the triangle with angles β, γ, α correspond to the points $z = 0, 1, \infty$, respectively. Because $c < 1$, $s(0; \alpha, \beta, \gamma) = 0$; that is, the origin of the z plane corresponds to the origin (vertex A_2 in Figure 5.8.3) in the w plane. We choose the branch of z^{1-c} is such a way that z^{1-c} is real for positive z. Thus $s(z)$ is real if z varies along the real axis from $z = 0$ to $z = 1$; hence one side of the curvilinear triangle is part of the positive real axis of the w plane. For negative z, we find (using $c = 1 - \beta$)

$$z^{1-c} = (e^{i\pi}|z|)^{1-c} = |z|^\beta e^{i\pi\beta}$$

which shows that $-\infty < z < 0$ is mapped by the transformation (5.8.20) onto another side of the triangle, the linear segment that makes the angle $\pi\beta$ with the real axis at the origin.

The remaining portions of the circular triangle in the w plane are fixed by knowledge of the hypergeometric equation and formula (5.8.20) (details of which we do not go into here; the interested reader can consult one of the many references to properties of the hypergeometric equation such as Whittaker and Watson (1927) or Nehari (1952). So, for example, the vertices $S(1; \alpha, \beta, \gamma)$ corresponding to point A_3 and $S(\infty; \alpha, \beta, \gamma)$ corresponding to point A_1, may be calculated from Eq. (5.8.20) and use of the properties of the hypergeometric functions. This yields

$$S(1; \alpha, \beta, \gamma) = \frac{\Gamma(2 - c)\Gamma(c - a)\Gamma(c - b)}{\Gamma(c)\Gamma(1 - a)\Gamma(1 - b)}$$

which is real, and

$$S(\infty; \alpha, \beta, \gamma) = \frac{e^{i\pi(1-c)}\Gamma(b)\Gamma(c - a)\Gamma(2 - c)}{\Gamma(c)\Gamma(b - c + 1)\Gamma(1 - a)}$$

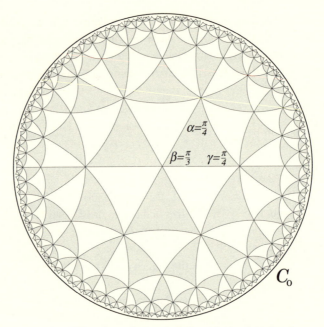

Fig. 5.8.5. Tiling the orthogonal circle with circular triangles

where, given any positive α, β, γ satisfying $\alpha + \beta + \gamma < 1$, we determine a, b, c from Eq. (5.8.15). The fundamental triangle obtained by the map $w = S(z; \alpha, \beta, \gamma)$ is the one depicted in Figures 5.8.3 and 5.8.4 (also see Figure 5.8.5).

The single-valuedness of the inverse function $z = S(w; \alpha, \beta, \gamma)$, $\alpha = 1/\ell$, $\beta = 1/m$, $\gamma = 1/n$, in the case when ℓ, m, n are integers, makes their study particularly important. Successive continuations of the fundamental triangle in the w plane across their sides correspond to reflections from the upper to lower half z plane, and this corresponds to analytically continuing the solution in terms of the hypergeometric functions given by Eq. (5.8.20). Inverting an infinite number of times allows us to eventually "tile" the orthogonal circle C_0. A typical situation is illustrated in Figure 5.8.5, with $m = 3$ and $\ell = n = 4$, that is, $\beta = \pi/3$ and $\alpha = \gamma = \pi/4$. The shaded and white triangles correspond to the upper and lower half planes, respectively.

Finally, we note that when $\beta \to 0$ (A_2 remaining at the origin) and thus $c \to 1$, the triangle degenerates to a line segment. We note that when $c \to 1$, the representation (5.8.20) breaks down because $\hat{\chi}_1$ and $\hat{\chi}_2$ are not linearly independent. The important special case $\alpha = \beta = \gamma = 0$, corresponding to three points lying on the orthogonal circle, is discussed below in Example 5.8.3.

Remark An even number of reflections with respect to circular arcs is a bilinear transformation (see Example 5.7.7). An inversion (reflection) with respect to a circular arc maps circles into circles, preserves the magnitude of an angle, and inverts its orientation. Hence an even number of inversions preserves the magnitude of an angle and its orientation and is a conformal transformation mapping circles into circles; that is, it is equivalent to a bilinear transformation (see Example 5.7.7). Because, these inversions are symmetric with respect to the real axis in the z plane, it follows that an even number of inversions in the w plane will return us to the original position in the z plane. Hence $z = S(w)$ satisfies the functional equation

$$S\left(\frac{aw+b}{cw+d}\right) = S(w) \tag{5.8.21}$$

Functions that satisfy this equation are usually referred to as **automorphic functions**. Such functions can be viewed as generalizations of periodic functions, for example, elliptic functions. Equation (5.8.21) can also be ascertained by studying the Schwarzian equation (5.8.12) for $w = s(z)$ and its inverse $z = S(w)$.

In order to determine the precise form of the bilinear transformation associated with the curvilinear triangles, we first note that these transformations must leave the orthogonal circle invariant. Suppose we normalize this circle to have radius 1. We recall that the most general bilinear transformation taking the unit circle onto itself is (see Example 5.7.6)

$$w = B\left(\frac{z-A}{\bar{A}z-1}\right), \qquad |B| = 1, \qquad |A| < 1 \tag{5.8.22}$$

This shows that under the conditions (5.8.17) and (5.8.18), $S(w; \alpha, \beta, \gamma)$ is a single-valued automorphic function in $|w| < 1$ satisfying the functional equation

$$S\left(B\left(\frac{w-A}{\bar{A}w-1}\right); \alpha, \beta, \gamma\right) = S(w; \alpha, \beta, \gamma) \tag{5.8.23}$$

The bilinear transformations associated with an automorphic function form a group. Indeed, if T and $\tilde{T}$ denote two such bilinear transformations, that is, if S is an automorphic function satisfying Eq. (5.8.21), $S(Tw) = S(w)$ and $S(\tilde{T}w) = S(w)$, then

$$S(T\tilde{T}w) = S[T(\tilde{T}w)] = S[\tilde{T}w] = S(w)$$

Furthermore, if T^{-1} denotes the inverse of T, then

$$S(w) = S(TT^{-1}w) = S(T^{-1}Tw)$$

Example 5.8.3 Find the transformation that maps a curvilinear triangle with zero angles onto the upper half of the z plane.

In this case, $\alpha = \beta = \gamma = 0$, so that $a = b = \frac{1}{2}$, $c = 1$, and the hypergeometric equation (5.8.14) reduces to

$$z(1 - z)\chi''(z) + [(1 - z) - z]\chi'(z) - \frac{1}{4}\chi(z) = 0$$

From this equation we see that if $\chi(z)$ is a solution, then $\chi(1 - z)$ is also a solution (see also Example 5.8.1). We use $w = \hat{\chi}_1/\hat{\chi}_2$, where $\hat{\chi}_1$ and $\hat{\chi}_2$ are two linearly independent solutions of the above hypergeometric equation, specifically

$$\hat{\chi}_2 = \chi(1 - z) \qquad \hat{\chi}_1 = \chi(z)$$

where χ is given by Eq. (5.8.16b) with $\alpha = \beta = \gamma = 0$, $a = b = \frac{1}{2}$, $c = 1$, that is,

$$\chi(z) = \int_0^1 \frac{dt}{\sqrt{t(1 - t)(1 - zt)}}$$

Using the change of variables $t = \tau^2$, as well as $z = k^2$, we find

$$w = f(z) = \frac{\chi(1 - z)}{\chi(z)} = \frac{K'(k)}{K(k)} \qquad z = k^2$$

where the functions K' and K are defined as in Example 5.6.8. The variable k is usually referred to as the modulus of the elliptic function $z = F^{-1}(w, k) = \text{sn}(w, k)$. Because of this connection, the *inverse* of the function $w = K'(k)/K(k)$ is often referred to as the **elliptic modular function**. This function is usually denoted by J. It is actually customary to give this name to the inverse of iK'/K (note the extra factor of i, which is a standard normalization):

$$w = s(z; 0, 0, 0) = \frac{iK'(k)}{K(k)} \quad \Rightarrow \quad z = k^2 = J(w) = S(w; 0, 0, 0)$$
$$(5.8.24)$$

It is useful to determine the location of the "zero angle triangle" in the w plane consistent with Eq. (5.8.24), which we will see has degenerated into a

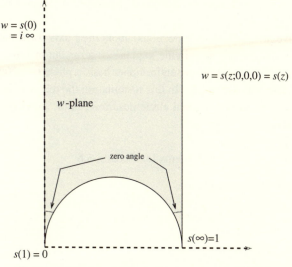

Fig. 5.8.6. Fundamental domain of a zero angle triangle

strip. Note that when $z = 0$, $K(0) = \pi/2$ and $K'(0) = \infty$, and when $z = 1$, $K(1) = \infty$ and $K'(1) = 0$. As z increases from $z = 0$ to $z = 1$, w decreases along the imaginary axis from $i\infty$ to 0. We know that the angles of the "triangle" are zero so at the vertex corresponding to $z = 0$ we have a half circle, cutting out a piece of the upper half w plane, which begins at the origin and intersects the real w axis somewhere to the right of the origin. The fundamental triangle must lie in the first quadrant to be consistent with an orientation that has the upper half z plane to the left as we proceed from $z = 0$ to $z = 1$ to $z = \infty$. Finally, the last critical point has value $w(z = \infty) = s(\infty; 0, 0, 0) = 1$ (see Figure 5.8.6). We could determine this from the properties of the hypergeometric equation or from the following. From Eq. (5.8.24) we have the relationship

$$s(z; 0, 0, 0) \, s(1 - z; 0, 0, 0) = -1 \qquad (5.8.25)$$

Calling $s(z; 0, 0, 0) = s(z)$, for simplicity of notation, and letting $z = \frac{1}{2} + iy$, the Schwarz reflection principle about $y = 0$ implies $s(\frac{1}{2} + iy) = -\bar{s}(\frac{1}{2} - iy)$ because $s(z)$ is pure imaginary for $0 < z < 1$; hence Eq. (5.8.25) yields

$$\left| s\left(\frac{1}{2} + iy\right) \right|^2 = 1$$

and for $y \to \infty$, $|s(\infty)| = 1$. But $s(\infty)$ must lie on the positive real axis, whereupon $s(\infty) = 1$.

Successive inversions (i.e., reflections) about the z axis correspond to inversions of the fundamental strip in the w plane. It turns out that after an infinite number of inversions we fill the entire upper half w plane (see Nehari (1952) for a more detailed discussion). In this formulation the real w axis is a natural boundary; that is, the orthogonal circle described in the previous example is now the real w axis.

Problems for Section 5.8

1. In this problem the equation

$$\{w, z\} = \frac{1 - \beta^2}{2z^2} + \frac{1 - \gamma^2}{2(z - 1)^2} + \frac{\beta^2 + \gamma^2 - \alpha^2 - 1}{2z(z - 1)}$$

   (c.f. Eq. (5.8.12) where $w = f(z)$) is derived. Frequently, it is useful to consider $z = z(w)$ instead of $w = w(z)$.

   (a) Show that $\dfrac{d}{dz} = \dfrac{1}{z'} \dfrac{d}{dw}$ and $\dfrac{d^2 w}{dz^2} = -\dfrac{z''}{z'^3}$, where $z' \equiv \dfrac{dz}{dw}$

   (b) Use these results to establish $\{w, z\} = -\dfrac{1}{(z')^2}\{z, w\}$, where $\{z, w\} = $
   $$\left(\frac{z''}{z'}\right)' - \frac{1}{2}\left(\frac{z''}{z'}\right)^2$$ and hence derive the equation

   $$\{z, w\} + \frac{(z')^2}{2}\left(\frac{1 - \beta^2}{z^2} + \frac{1 - \gamma^2}{(z - 1)^2} + \frac{\beta^2 + \gamma^2 - \alpha^2 - 1}{z(z - 1)}\right) = 0$$

2. In the previous problem, consider the special case $\alpha = \beta = \gamma = 1$ so that $\{z, w\} = 0$. Show that the general solution of this equation (and hence of the mapping) is given by

   $$z = \frac{A}{w - w_0} + B$$

3. Using Eq. (5.8.16d), where χ_1 and χ_2 given by Eqs. (5.8.16b) and (5.8.16c), respectively, show that the vertices with the angles $\pi\alpha$ and $\pi\beta$ (corresponding to $z = 0$ and $z = 1$) are mapped to

   $$w_0 = \frac{\sin \pi\alpha}{\cos\left[\frac{\pi}{2}(\alpha - \beta - \gamma)\right]} \quad \text{and} \quad w_1 = \frac{\cos\left[\frac{\pi}{2}(\alpha + \beta - \gamma)\right]}{\sin \pi\gamma}$$

Hint: Use the identities

$$\int_0^1 t^{r-1}(1-t)^{s-1}\,dt = \frac{\Gamma(r)\Gamma(s)}{\Gamma(r+s)}, \qquad r > 0, \quad s > 0$$

and

$$\Gamma(r)\Gamma(1-r) = \frac{\pi}{\sin \pi r}$$

4. If $\alpha = \gamma$, show that the function $f(z) = \chi_1/\chi_2$, where χ_1 and χ_2 are given by Eqs. (5.8.16b) and (5.8.16c), respectively, satisfy the functional equation

$$f(z)f(1-z) = 1$$

5. Consider the crescent-shaped region shown in the figure below.

    (a) Show that in this case Eq. (5.8.6) reduces to

    $$\{f, z\} = \frac{(1-\alpha)^2(a-b)^2}{2(z-a)^2(z-b)^2}$$

    where a and b are the points on the real axis associated with the vertices.

    (b) Show that the associated linear differential equation (see Eq. 5.8.13) is

    $$y'' + \frac{(1-\alpha^2)(a-b)^2}{4(z-a)^2(z-b)^2}y = 0$$

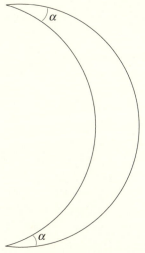

Fig. 5.8.7. Crescent region–Problem 5.8.5

(c) Show that the above equation is equivalent to the differential equation

$$g'' + (1 - \alpha) \left[\frac{1}{z-a} + \frac{1}{z-b} \right] g' - \frac{\alpha(1-\alpha)}{(z-a)(z-b)} g = 0$$

which admits $(z-a)^\alpha$ and $(z-b)^\alpha$ as particular solutions.

(d) Deduce that

$$f(z) = \frac{c_1(z-a)^\alpha + c_2(z-b)^\alpha}{c_3(z-a)^\alpha + c_4(z-b)^\alpha}$$

where $c_1, \ldots, c_4$ are constants, for which $C_1 C_4 \neq C_2 C_3$.

5.9 Other Considerations

5.9.1 Rational Functions of the Second Degree

The most general rational function of the second degree is of the form

$$f(z) = \frac{az^2 + bz + c}{a'z^2 + b'z + c'} \tag{5.9.1}$$

where a, b, c, a', b', c' are complex numbers. This function remains invariant if both the numerator and the denominator are multiplied by a nonzero constant; therefore $f(z)$ depends only on five arbitrary constants. The equation $f(z) - w_0 = 0$ is of second degree in z, which shows that under the transformation $w = f(z)$, every value w_0 is taken twice. This means that this transformation maps the complex z plane onto the doubly covered w plane, or equivalently that it maps the z plane onto a two-sheeted Riemann surface whose two sheets cover the entire w plane. The branch points of this Riemann surface are those points w that are common to both sheets. These points correspond to points z such that either $f'(z) = 0$ or $f(z)$ has a double pole. From Eq. (5.9.1) we can see that there exist precisely two such branch points. We distinguish two cases:
(a) $f(z)$ has a double pole, that is, $w = \infty$ is one of the two branch points.
(b) $f(z)$ has two finite branch points. It will turn out that in case (a), $f(z)$ can be decomposed into two successive transformations: a bilinear one, and one of the type $z^2 + \text{const}$. In case (b), $f(z)$ can be decomposed into three successive transformations: a linear one, a bilinear one, and one of the type $z + 1/z$.

We first consider case (a). Let $w = \infty$ and $w = \lambda$ be the two branch points of $w = f(z)$, and let $z = z_1$ and $z = z_2$ be the corresponding points in the z

plane. The expansions of $f(z)$ near these points are of the form

$$f(z) = \frac{\alpha_{-2}}{(z - z_1)^2} + \frac{\alpha_{-1}}{(z - z_1)} + \alpha_0 + \alpha_1(z - z_1) + \cdots, \qquad \alpha_{-2} \neq 0$$

and

$$f(z) - \lambda = \beta_2(z - z_2)^2 + \beta_3(z - z_2)^3 + \cdots, \qquad \beta_2 \neq 0$$

respectively. The function $\sqrt{(f(z) - \lambda)}$, takes no value more than once (because $f(z)$ takes no value more than twice), and its only singularity in the *entire z* plane is a simple pole at $z = z_1$. Hence from Liouville's Theorem this function must be of the bilinear form (5.7.1). Therefore

$$f(z) = \lambda + \left(\frac{Az + B}{Cz + D} \right)^2 \tag{5.9.2}$$

that is

$$w = \lambda + z_1^2, \qquad z_1 \equiv \frac{Az + B}{Cz + D} \tag{5.9.3}$$

We now consider case (b). Call $w = \lambda$ and $w = \mu$ the two finite branch points. Using a change of variables from $f(z)$ to $g(z)$, these points can be normalized to be at $g(z) = \pm 1$, hence

$$f(z) = \frac{\lambda - \mu}{2} g(z) + \frac{\lambda + \mu}{2} \tag{5.9.4}$$

Let $z = z_1$ and $z = z_2$ be the points in the z plane corresponding to the branch points λ and μ, respectively. Series expansions of $g(z)$ near these points are of the form

$$g(z) - 1 = \alpha_2(z - z_1)^2 + \alpha_3(z - z_1)^3 + \cdots, \qquad \alpha_2 \neq 0$$

and

$$g(z) + 1 = \beta_2(z - z_2)^2 + \beta_3(z - z_1)^3 + \cdots, \qquad \beta_2 \neq 0$$

respectively. The function $f(z)$ has two simple poles: therefore the function $g(z)$ also has two simple poles, which we shall denote by $z = \zeta_1$ and $z = \zeta_2$. Using a change of variables from $g(z)$ to $h(z)$, it is possible to construct a function that has only one simple pole

$$g(z) = \frac{1}{2} \left(h(z) + \frac{1}{h(z)} \right) \tag{5.9.5}$$

Indeed, the two poles of $g(z)$ correspond to $h(z) = \gamma(z-\zeta_1)[1+c_1(z-\zeta_1)+\cdots]$ and to $h(z) = \delta(z - \zeta_2)^{-1}[1 + d_1(z - \zeta_2) + \cdots]$; that is, they correspond to one zero and one pole of $h(z)$. Furthermore, the expansions of $g(z)$ near ± 1 together with Eq. (5.9.5) imply that $h(z)$ is regular at the points $z = z_1$ and $z = z_2$. The only singularity of $h(z)$ in the entire z plane is a pole, hence $h(z)$ must be of the bilinear form (5.7.1). Renaming functions and constants, Eqs. (5.9.4) and (5.9.5) imply

$$w = A'\zeta_2 + B', \qquad \zeta_2 = \frac{1}{2}\left(\zeta_1 + \frac{1}{\zeta_1}\right), \qquad \zeta_1 = \frac{Az + B}{Cz + D} \qquad (5.9.6)$$

The important consequence of the above discussion is that the study of the transformation (5.9.1) reduces to the study of the bilinear transformation (which was discussed in Section 5.7) of the transformation $w = z^2$ and of the transformation $w = (z + z^{-1})/2$.

Let us consider the transformation

$$w = z^2; \qquad w = u + iv, \qquad z = x + iy; \qquad u = x^2 - y^2, \qquad v = 2xy \qquad (5.9.7)$$

Example 5.9.1 Find the curves in the z plane that, under the transformation $w = z^2$, give rise to horizontal lines in the w plane.

Because horizontal lines in the w plane are $v = \text{const}$, it follows that the relevant curves in the z plane are the hyperbolae $xy = \text{const}$. We note that because the lines $u = \text{const}$ are orthogonal to the lines $v = \text{const}$, it follows that the family of the curves $x^2 - y^2 = \text{const}$ is orthogonal to the family of the curves $xy = \text{const}$. (Indeed, the vectors obtained by taking the gradient of the functions $F_1(x, y) = (x^2 - y^2)/2$ and $F_2(x, y) = xy$, $(x, -y)$, and (y, x) are perpendicular to these curves, and clearly these two vectors are orthogonal).

Example 5.9.2 Find the curves in the z plane that, under the transformation $w = z^2$, give rise to circles in the w plane.

Let $c \neq 0$ be the center and R be the radius of the circle. Then $|w - c| = R$, or if we call $c = C^2$, the $w = z^2$ implies

$$|z - C||z + C| = R \qquad (5.9.8)$$

Hence, the images of circles are the loci of points whose distances from two fixed points have a constant product. These curves are called **Cassinians**. The cases of $R > |C|^2$, $R = |C|^2$, and $R < |C|^2$ correspond to one closed curve, to the lemniscate, and to two separate closed curves, respectively. These three

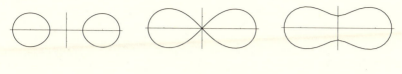

$R < |C|^2$ $\qquad\qquad\qquad\qquad R = |C|^2 \qquad\qquad\qquad\qquad R > |C|^2$

Fig. 5.9.1. Cassinians associated with Eq. (5.9.8)

cases are depicted in Figure 5.9.1, when C is real. Otherwise, we obtain a rotation of angle θ when $C = |C|e^{i\theta}$.

We now consider the transformation

$$
w = \frac{1}{2}\left(z + \frac{1}{z}\right), \qquad u = \frac{1}{2}\left(r + \frac{1}{r}\right)\cos\theta, \qquad v = \frac{1}{2}\left(r - \frac{1}{r}\right)\sin\theta
$$
$$
(5.9.9)
$$

where $z = r\exp(i\theta)$.

Example 5.9.3 Find the image of a circle in the z plane under the transformation (5.9.9).

Let $r = \rho$ be a circle in the z plane. Equation (5.9.9) implies

$$
\frac{u^2}{[\frac{1}{2}(\rho + \rho^{-1})]^2} + \frac{v^2}{[\frac{1}{2}(\rho - \rho^{-1})]^2} = 1, \qquad \rho = \text{const}
$$

This shows that the transformation (5.9.9) maps the circle $r = \rho$ onto the ellipse of semiaxes $(\rho + \rho^{-1})/2$ and $(\rho - \rho^{-1})/2$ as depicted in Figure 5.9.2. Because

$$
\frac{1}{4}(\rho + \rho^{-1})^2 - \frac{1}{4}(\rho - \rho^{-1})^2 = 1
$$

all such ellipses have the same foci located on the u axis at ± 1.

The circles $r = \rho$ and $r = \rho^{-1}$ yield the same ellipse; if $\rho = 1$, the ellipse degenerates into the linear segment connecting $w = 1$ and $w = -1$. We note that because the ray $\theta = \varphi$ is orthogonal to the circle $r = \rho$, the above ellipses are orthogonal to the family of hyperbolae

$$
\frac{u^2}{\cos^2\varphi} - \frac{v^2}{\sin^2\varphi} = 1, \qquad \varphi = \text{const}
$$

which are obtained from Eq. (5.9.9) by eliminating r.

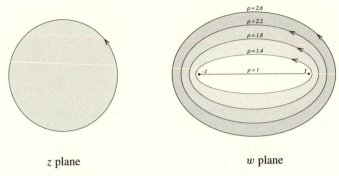

z plane   w plane

Fig. 5.9.2. Transformation of a circle onto an ellipse

Example 5.9.4 (*Joukowski Profiles*) The transformation

$$w = \frac{1}{2}\left(z + \frac{1}{z}\right) \tag{5.9.10}$$

arises in certain aerodynamic applications. This is because it maps the exterior of circles onto the exterior of curves that have the general character of airfoils.

Consider, for example, a circle having its center on the real axis, passing through $z = 1$, and having $z = -1$ as an interior point. Because the derivative of w vanishes at $z = 1$, this point is a critical point of the transformation, and the angles whose vertices are at $z = 1$ are doubled. (Note from Eq. (5.9.10) that the series in the neighborhood of $z = 1$ is $2(w - 1) = (z - 1)^2 - (z - 1)^3 + \cdots$.) In particular, because the exterior angle at point A on C is π (see Figure 5.9.3), the exterior angle at point A' on C' is 2π. Hence C' has a sharp tail at $w = 1$. Note that the exterior of the circle maps to the exterior of the closed curve in the w plane; $|z| \to \infty$ implies $|w| \to \infty$. Because we saw from Example 5.9.3 that the transformation (5.9.10) maps the circle $|z| = 1$ onto the slit $|w| \le 1$, and because C encloses the circle $|z| = 1$, the curve C' encloses the slit $|w| \le 1$.

Suppose that the circle C is translated vertically so that it still passes through $z = 1$ and encloses $z = -1$, but its center is in the upper half plane. Using the same argument as above, the curve C' still has a sharp tail at A' (see Figure 5.9.4). But because C is not symmetric about the x axis, we can see from Eq. (5.9.10) that C' is not symmetric about the u axis. Furthermore, because C does not entirely enclose the circle $|z| = 1$, the curve C' does not entirely enclose the slit $|w| \le 1$. A typical shape of C' is shown in Figure 5.9.4.

By changing C appropriately, other shapes similar to C' can be obtained. We note that C' resembles the cross section of the wing of an airplane, usually referred to as an airfoil.

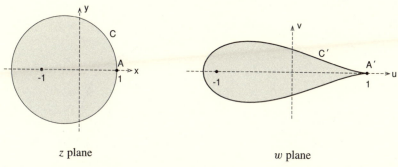

z plane   *w* plane

Fig. 5.9.3. Image of circle centered on real axis under the transformation $w = \frac{1}{2}(z+1/z)$

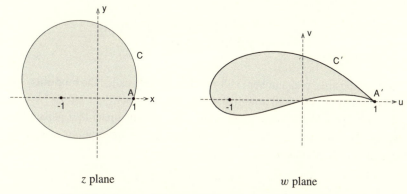

z plane   *w* plane

Fig. 5.9.4. Image of circle whose center is above the real axis

5.9.2 *The Modulus of a Quadrilateral*

Let Γ be a positively oriented Jordan curve (i.e., a simple closed curve), with four distinct points a_1, a_2, a_3, and a_4 being given on Γ, arranged in the direction of increasing parameters. Let the interior of Γ be called Q. The system $(Q; a_1, a_2, a_3, a_4)$ is called a quadrilateral (see Figure 5.9.5).

Two quadrilaterals $(Q; a_1, \ldots, a_4)$ and $(\tilde{Q}; \tilde{a}_1, \ldots, \tilde{a}_4)$ are called conformally equivalent if there exists a conformal map, f, from Q to $\tilde{Q}$ such that $f(a_i) = \tilde{a}_i, i = 1, \ldots, 4$.

If one considers trilaterals instead of quadrilaterals, that is, if one fixes three instead of four points, then one finds that all trilaterals are conformally equivalent. Indeed, it follows from the proof of the Riemann Mapping Theorem that in a conformal mapping any three points on the boundary can be chosen arbitrarily (this fact was used in Sections 5.6–5.8).

Not all quadrilaterals are conformally equivalent. It turns out that the equivalence class of conformally equivalent quadrilaterals can be described in terms

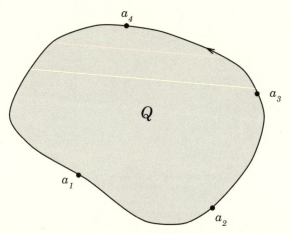

Fig. 5.9.5. Quadrilateral

of a single positive real number. This number, usually called the **modulus**, will be denoted by μ. We shall now characterize this number.

Let h be a conformal map of $(Q; a, b, c, d)$ onto the upper half plane. This map can be fixed uniquely by the conditions that the three points a, b, d are mapped to $0, 1, \infty$, that is,

$$h(a) = 0, \qquad h(b) = 1, \qquad h(d) = \infty$$

Then $h(c)$ is some number, which we shall denote by ξ. Because of the orientation of the boundary, $1 < \xi < \infty$. By letting $\tilde{z} = (az + b)/(cz + d)$ for $ad - bc \neq 0$, we can directly establish that there is a bilinear transformation, that we will call g, that maps the upper half plane onto itself such that the points $z = \{0, 1, \xi, \infty\}$ are mapped to $\tilde{z} = \{1, \eta, -\eta, -1\}$. We find after some calculations that η is uniquely determined from ξ by the equation

$$\frac{\eta + 1}{\eta - 1} = \sqrt{\xi}, \qquad \eta > 1$$

Recall the Example 5.6.8. We can follow the same method to show that for any given $\eta > 1$, there exists a unique real number $\mu > 0$ such that the upper half z plane can be mapped onto a rectangular region R and the image of the points $z = \{0, 1, \xi, \infty\}$, which correspond to $\tilde{z} = \{1, \eta, -\eta, -1\}$, can be mapped to the rectangle with the corners $\{\mu, \mu + i, i, 0\}$. (The number μ can be expressed in terms of η by means of elliptic integrals.) Combining the conformal maps h and g, it follows that $(Q; a, b, c, d)$ is conformally equivalent

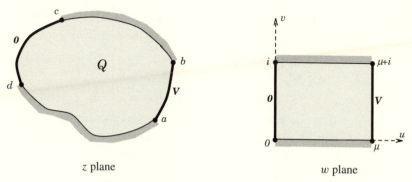

Fig. 5.9.6. Electric current through sheet Q

to the rectangular quadrilateral $(R; \mu, \mu + i, i, 0)$. Thus two quadrilaterals are said to be conformally equivalent if and only if they have the same value μ, which we call the modulus.

Example 5.9.5 (Physical Interpretation of μ) Let Q denote a sheet of metal of specific resistance equal to one. Let the segments (a, b) and (c, d) of the boundary be kept at the potentials V and 0, respectively, and let the segments (b, c) and (d, a) be insulated. Establish a physical meaning for μ.

The charge q on the portion of the sheet a, b is given by (see Example 5.4.3; $\epsilon_o = 1$ here)

$$q = \int_a^b \frac{\partial \Phi}{\partial n} \, ds$$

where $\partial/\partial n$ denotes differentiation in the direction of the exterior normal. The potential Φ is obtained from the solution of (see Figure 5.9.6)

$$\nabla^2 \Phi = 0 \text{ in } Q, \qquad \Phi = V \quad \text{on } (a, b), \qquad \Phi = 0 \quad \text{on } (c, d),$$

$$\frac{\partial \Phi}{\partial n} = 0 \text{ on } (b, c) \text{ and } (d, a).$$

In the w plane we can verify the following solution for the complex potential: $\Omega(w) = \frac{V}{\mu} w$. From the definition of the potential (5.4.11), $\Omega = \Phi + i\Psi$, so $\Phi = \text{Re } \Omega = \frac{V}{\mu} \text{Re } w = \frac{V}{\mu} u$. At the top and bottom we have $\frac{\partial \Phi}{\partial v} = 0$; on $u = 0$, $\Phi = 0$; and for $u = \mu$, $\Phi = V$. Hence we have verified that the solution of this problem in the w plane is given by $\Phi = V \mu^{-1} \text{Re } w$. Furthermore, we know that (see Eq. (5.4.14)) the integral q is invariant under a conformal transformation. Computing this integral in the w plane we find

$$q = \mu^{-1} V, \quad \text{or} \quad V = \mu q$$

Therefore μ^{-1} has the physical meaning of the capacitance of the sheet Q between the "electrodes" (a, b) and (c, d) when the remaining parts of the boundary are insulated.

5.9.3 Computational Issues

Even though Riemann's Mapping Theorem guarantees that there exists an analytic function that maps a simply connected domain onto a circle, the proof is not constructive and doesn't give insight into the determination of the mapping function. We have seen that conformal mappings have wide physical application, and, in practice, the ability to map a complicated domain onto a circle, the upper half plane, or indeed another simple region is desirable. Toward this end, various computational methods have been proposed and this is a field of current research interest. It is outside the scope of this book to survey the various methods or even all of the research directions. Many of the well-known methods are discussed in the books of Henrici, and we also note the collection of papers in *Numerical Conformal Mapping* edited by L. N. Trefethen (1986) where other reviews can be found and specific methods, such as the numerical evaluation of Schwarz–Christoffel transformations, are discussed.

Here we will only describe one of the well-known methods used in numerical conformal mapping. Let us consider the mapping from a unit circle in the z plane to a suitable (as described below) simply connected region in the w plane. We wish to find the mapping function $w = f(z)$ that will describe the conformal mapping. In practice, we really are interested in the *inverse* function, $z = f^{-1}(w)$. Numerically, we determine a set of points for which the correspondence between points on the circle in the z plane and points on the boundary in the w plane is deduced.

We assume that the boundary C in the w plane is a Jordan curve that can be represented in terms of polar coordinates, $w = f(z) = \rho e^{i\theta}$, where $\rho = \rho(\theta)$, and we impose the conditions $f(0) = 0$ and $f'(0) > 0$ (Riemann's Mapping Theorem allows us this freedom) on the unit circle $z = e^{i\varphi}$. The mapping fixes, in principle, $\theta = \theta(\varphi)$ and $\rho = \rho(\theta(\varphi))$. The aim is to determine the **boundary correspondence points**, that is, how points in the z domain, $\varphi = \{\varphi_1, \varphi_2, \ldots, \varphi_N\}$, transform to points in the w domain that is parametrized by $\theta = \{\theta_1, \theta_2, \ldots, \theta_N\}$ (see Figure 5.9.7).

The method we describe involves the numerical solution of a nonlinear integral equation. This equation is a modification of a well-known formula (derived in the homework exercises – see Section 2.6, Exercise 10, and Section 4.3, Exercise 15) that relates the boundary values (on $|z| = 1$) of the real and imaginary parts of a function analytic inside the circle. Specifically, consider $F(z) = u(x, y) + iv(x, y)$, which is analytic inside the circle $z = re^{i\varphi}$, for

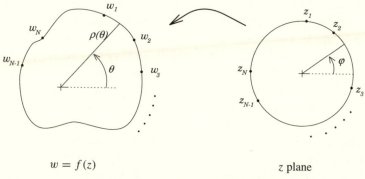

$w = f(z)$ z plane

Fig. 5.9.7. Boundary correspondence points

$r < 1$. Then on the circle $r = 1$ the following equation relating u and v holds:

$$v(\varphi) = v(r = 0) + \frac{1}{2\pi} \int_0^{2\pi} u(t) \cot\left(\frac{\varphi - t}{2}\right) dt \qquad (5.9.11)$$

where the integral is taken as a Cauchy principal value (we reiterate that both t and φ correspond to points on the unit circle). The integral equation we shall consider is derived from Eq. (5.9.11) by considering $F(z) = \log(f(z)/z)$, recalling that $f(0) = 0$, $f'(0) > 0$. Then, using the polar coordinate representation $f(z) = \rho e^{i\theta}$, we see that $F(z) = \log \rho + i(\theta - \varphi)$; hence in Eq. (5.9.11) we take $u = \log \rho(\theta)$. We require that $f'(0)$ be real, thus $v(r = 0) = 0$. On the circle, $v = \theta - \varphi$; this yields

$$\theta(\varphi) = \varphi + \frac{1}{2\pi} \int_0^{2\pi} \log \rho(\theta(t)) \cot\left(\frac{\varphi - t}{2}\right) dt \qquad (5.9.12)$$

Equation (5.9.12) is called **Theodorsen's integral equation.** The goal is to solve Eq. (5.9.12) for $\theta(\varphi)$. Unfortunately, it is nonlinear and cannot be solved in closed form, though a unique solution can be proven to exist. Consequently an approximation (i.e., numerical) procedure is used. The methods are effective when $\rho(\theta)$ is smooth and $|\rho'(\theta)/\rho(\theta)|$ is sufficiently small.

Equation (5.9.12) is solved by functional iteration:

$$\theta^{(n+1)}(\varphi) = \varphi + \frac{1}{2\pi} \int_0^{2\pi} \log \rho\left(\theta^{(n)}(t)\right) \cot\left(\frac{\varphi - t}{2}\right) dt \qquad (5.9.13)$$

where the function $\theta^{(0)}$ is a starting "guess."

Numerically speaking, Eq. (5.9.13) is transformed to a matrix equation; the integral is replaced by a sum, and $\log \rho(\theta)$ is approximated by a finite Fourier series, that is, a trigonometric polynomial, because $\rho(\theta)$ is periodic. Then, corresponding to $2N$ equally spaced points (roots of unity) on the unit circle in the z plane (t: $\{t_1, \ldots, t_{2N}\}$ and φ: $\{\varphi_1, \ldots, \varphi_{2N}\}$), one solves, by iteration, the matrix equation associated with Eq. (5.9.13) to find the set $\theta^{(n+1)}(\varphi_j)$, $j = 1, 2, \ldots, 2N$, which corresponds to an an initial guess $\theta^{(0)}(\varphi_j) = \varphi_j$, $j = 1, 2, \ldots, 2N$, which are equally spaced points on the unit circle. As n is increased enough, the iteration converges to a solution that we call $\hat{\theta}$: $\theta^{(n)}(\varphi_j) \rightarrow \hat{\theta}(\varphi_j)$. These points are the boundary correspondence points. Even though the governing matrix is $2N \times 2N$ and ordinarily the "cost" of calculation is $O(N^2)$ operations, it turns out that special properties of the functions involved are such that fast Fourier algorithms are applicable, and the number of operations can be reduced to $O(N \log N)$.

Further details on this and related methods can be found in Henrici (1977), and articles by Gaier (1983), Fornberg (1980), and Wegmann (1988).

6

Asymptotic Evaluation of Integrals

6.1 Introduction

The solution of a large class of physically important problems can be represented in terms of definite integrals. Frequently, the solution can be expressed in terms of special functions (e.g. Bessel functions, hypergeometric functions, etc.; such functions were briefly discussed in Section 3.7), and these functions admit integral representations (see, e.g., Section 4.6). We have also seen in Section 4.6 that by using integral transforms, such as Laplace transforms or Fourier transforms, the solution of initial and/or boundary value problems for linear PDEs reduces to definite integrals. For example, the solution by Fourier transforms of the Cauchy problem for the Schrödinger equation of a free particle

$$i\Psi_t + \Psi_{xx} = 0 \qquad (6.1.1a)$$

is given by

$$\Psi(x, t) = \frac{1}{2\pi} \int_{-\infty}^{\infty} \hat{\Psi}_0(k) e^{ikx - ik^2 t} \, dk \qquad (6.1.1b)$$

where $\hat{\Psi}_0(k)$ is the Fourier transform of the initial data $\Psi(x, 0)$. Although such integrals provide exact solutions, their true content is not obvious. In order to decipher the main mathematical and physical features of these solutions, it is useful to study their behavior for large x and t. Frequently, such as for wave motion, the interesting limit is $t \to \infty$ with $c = x/t$ held fixed. Accordingly, for the particular case of Eq. (6.1.1b) one needs to study

$$\Psi(x, t) = \int_{-\infty}^{\infty} \hat{\Psi}_0(k) e^{it\phi(k)} \, dk, \qquad t \to \infty \qquad (6.1.2)$$

where $\phi(k) = kc - k^2$.

411

In this chapter we will develop appropriate mathematical techniques for evaluating the behavior of certain integrals, such as Eq. (6.1.2), containing a large parameter (such as $t \to \infty$). Historically speaking, the development of these techniques was motivated by concrete physical problems. However, once these techniques were properly understood it became clear that they are applicable to a wide class of mathematical problems dissociated from any physical meaning. Hence, these techniques were recognized as independent entities and became mathematical methods. The most well-known such methods for studying the behavior of integrals containing a large parameter are: Laplace's method, the method of stationary phase, and the steepest descent method. These methods will be considered in Sections 6.2, 6.3, and 6.4, respectively.

In recent years the solution of several physically important nonlinear PDEs has also been expressed in terms of definite integrals. This enhances further the applicability of the methods discussed in this chapter. Some interesting examples are discussed in Section 6.5.

There are a number of books dedicated to asymptotic expansions to which we refer the reader for futher details. For example: Bleistein and Handelsman, Dingle, Erdelyi and Olver. Many of the methods are discussed in Bender and Orszag, and in the context of an applied complex analysis text see Carrier *et al.*

In order to develop the methods mentioned above, we need to introduce some appropriate fundamental notions and results.

6.1.1 Fundamental Concepts

Suppose we want to find the value of the integral

$$I(\varepsilon) = \int_0^\infty \frac{e^{-t}}{1 + \varepsilon t} \, dt, \qquad \varepsilon > 0$$

for a sufficiently small real positive value of ε. We can develop an approximation to $I(\varepsilon)$ using integrating by parts repeatedly. One integration by parts yields

$$I(\varepsilon) = 1 - \varepsilon \int_0^\infty \frac{e^{-t}}{(1 + \varepsilon t)^2} \, dt$$

Repeating this process N more times yields

$$I(\varepsilon) = 1 - \varepsilon + 2!\varepsilon^2 - 3!\varepsilon^3 + \cdots + (-1)^N N!\varepsilon^N$$

$$+ (-1)^{N+1}(N+1)!\varepsilon^{N+1} \int_0^\infty \frac{e^{-t}}{(1 + \varepsilon t)^{N+2}} \, dt \qquad (6.1.3)$$

Equation (6.1.3) motivates the introduction of several important notions. We assume that ε is sufficiently small, and we use the following terminology:

(a) $-\varepsilon$ *is of order of magnitude* (or simply *is of order*) ε, while $2!\varepsilon^2$ *is of order* ε^2. We denote these statements by $-\varepsilon = O(\varepsilon)$ and $2!\varepsilon^2 = O(\varepsilon^2)$;

(b) $2!\varepsilon^2$ *is of smaller order than* ε, which we denote by $2!\varepsilon^2 \ll \varepsilon$; and

(c) if we approximate $I(\varepsilon)$ by $1 - \varepsilon + 2!\varepsilon^2$, this is an approximation *correct to order* ε^2.

We now make these intuitive notions precise. First we discuss the situation when the parameter, such as ε in Eq. (6.1.3), is real. The following definitions will be satisfactory for our purposes.

Definition 6.1.1 (a) The notation

$$f(k) = O(g(k)), \qquad k \to k_0 \tag{6.1.4}$$

which is read "$f(k)$ is of order $g(k)$ as $k \to k_0$," means that there is a finite constant M and a neighborhood of k_0 where $|f| \leq M|g|$.

(b) The notation

$$f(k) \ll g(k), \qquad k \to k_0 \tag{6.1.5}$$

which is read "$f(k)$ is much smaller than $g(k)$ as $k \to k_0$," means

$$\lim_{k \to k_0} \left| \frac{f(k)}{g(k)} \right| = 0$$

Alternatively, we write Eq. (6.1.5) as

$$f(k) = o(g(k)), \qquad k \to k_0$$

(c) We shall say that $f(k)$ is an approximation to $I(k)$ valid to order $\delta(k)$, as $k \to k_0$, if

$$\lim_{k \to k_0} \frac{I(k) - f(k)}{\delta(k)} = 0 \tag{6.1.6}$$

For example, return to Eq. (6.1.3), where now k is ε and $k_0 = 0$. Consider the approximation $f(\varepsilon) = 1 - \varepsilon + 2!\varepsilon^2$; then, in fact, $\lim_{\varepsilon \to 0} \frac{I(\varepsilon) - f(\varepsilon)}{\varepsilon^2} = 0$. Thus $f(\varepsilon)$ is said to be an approximation of $I(\varepsilon)$ valid to order ε^2.

Equation (6.1.3) involves the ordered sequence $1, \varepsilon, \varepsilon^2, \varepsilon^3, \ldots$. This sequence is characterized by the property that its $(j+1)st$ member is much smaller than its jth member. This property is the defining property of an **asymptotic sequence**. Equation (6.1.3) actually provides an **asymptotic expansion** of $I(\varepsilon)$ with respect to the asymptotic sequence $\{\varepsilon^j\}_{j=0}^{\infty}$, that is, $1, \varepsilon, \varepsilon^2, \ldots$.

Definition 6.1.2 (a) The ordered sequence of functions $\{\delta_j(k)\}$, $j = 1, 2, \cdots$ is called an **asymptotic sequence** as $k \to k_0$ if

$$\delta_{j+1}(k) \ll \delta_j(k), \qquad k \to k_0$$

for each j.

(b) Let $I(k)$ be continuous and let $\{\delta_j(k)\}$ be an asymptotic sequence as $k \to k_0$. The formal series $\sum_{j=1}^{N} a_j \delta_j(k)$ is called an **asymptotic expansion** of $I(k)$, as $k \to k_0$, valid to order $\delta_N(k)$, if

$$I(k) = \sum_{j=1}^{m} a_j \delta_j(k) + O(\delta_{m+1}(k)), \qquad k \to k_0, \qquad m = 1, 2, \ldots, N$$

(6.1.7)

then we write

$$I(k) \sim \sum_{j=1}^{N} a_j \delta_j(k), \qquad k \to k_0$$

(6.1.8)

The notation $\sim$ will be used extensively in this chapter. The notation $I(k) \sim \eta(k), k \to k_0$, means

$$\lim_{k \to k_0} \left| \frac{I(k)}{\eta(k)} \right| = 1$$

With regard to Eq. (6.1.8), the notation $\sim$ implies that each term can be obtained successively via

$$a_n = \lim_{k \to k_0} \left(\frac{I(k) - \sum_{j=0}^{n-1} a_j \delta_j(k)}{\delta_n(k)} \right)$$

When an arbitrarily large number of terms can be calculated, frequently one uses Eq. (6.1.8) with $N = \infty$, despite the fact (as we discuss below) that asymptotic series are often not convergent.

Let us return to Eq. (6.1.3). The right-hand side of this equation is an asymptotic expansion of $I(k)$ provided that the $(n + 1)st$ term is much smaller than the nth term. It is clear that this is true for all $n = 0, 1, \ldots, N - 1$. For $n = N$

because $\varepsilon > 0$ we have $1 + \varepsilon t \geq 1$; thus

$$\int_0^\infty e^{-t}/(1 + \varepsilon t)^{N+2} \, dt \leq \int_0^\infty e^{-t} \, dt = 1$$

hence

$$\left| (-1)^{N+1}(N+1)! \varepsilon^{N+1} \int_0^\infty \frac{e^{-t}}{(1+\varepsilon t)^{N+2}} \, dt \right|$$
$$\leq \left| (-1)^{N+1}(N+1)! \varepsilon^{N+1} \right| \ll \left| (-1)^N N! \varepsilon^N \right|$$

It is important to realize that the expansion Eq. (6.1.3) is *not* convergent. Indeed, for ε fixed the term $(-1)^N N! \varepsilon^N$ tends to infinity as $N \to \infty$. But for fixed N this term vanishes as $\varepsilon \to 0$, and this is the reason that the above expansion provides a good approximation to $I(\varepsilon)$ as $\varepsilon \to 0$.

Example 6.1.1 Find an asymptotic expansion for $J(k) = \int_0^\infty \frac{e^{-kt}}{1+t} \, dt$ as real $k \to \infty$.

Calling $t' = kt$ and $\varepsilon = 1/k$, we see that

$$J = \varepsilon \int_0^\infty \frac{e^{-t'}}{1 + \varepsilon t'} \, dt'$$

Thus from Eq. (6.1.3), with $\varepsilon = 1/k$, we have

$$J(k) = \frac{1}{k} - \frac{1}{k^2} + \frac{2!}{k^3} - \cdots + (-1)^{N-1} \frac{(N-1)!}{k^N} + R_N(k)$$

$$R_N(k) = \frac{(-1)^N N!}{k^{N+1}} \int_0^\infty \frac{e^{-t} \, dt}{(1 + t/k)^{N+1}}$$

(6.1.9)

and from the discussion above we find that

$$|R_N(k)| \leq N!/k^{N+1} \ll 1/k^N$$

Note that Eq. (6.1.9) is exact. As $k \to \infty$, $\frac{1}{k}, \frac{1}{k^2}, \cdots$ form an asymptotic sequence; thus Eq. (6.1.9) provides an asymptotic expansion of $I(k)$ for large k. Again we remark that the above expansion is not convergent: As $N \to \infty$, k fixed, the series does not converge; but as $k \to \infty$, N fixed, $R_N \to 0$ (in the asymptotic expansion, $|R_N(k)| \ll 1/k^N$).

Example 6.1.2 Find an asymptotic expansion for $I(k) = \int_k^\infty \frac{e^{-t}}{t} \, dt$ as real $k \to \infty$.

Integrating by parts N times we find

$$I(k) = e^{-k} \left[\frac{1}{k} - \frac{1}{k^2} + \frac{2!}{k^3} - \cdots + (-1)^{N-1} \frac{(N-1)!}{k^N} \right] + R_N(k)$$

$$R_N(k) = (-1)^N N! \int_k^\infty \frac{e^{-t}}{t^{N+1}} dt \qquad\qquad (6.1.10)$$

As $k \to \infty$, the terms $\frac{e^{-k}}{k}, \frac{e^{-k}}{k^2}, \cdots$ form an asymptotic sequence. We also see that

$$|R_N(k)| < \frac{N!}{k^{N+1}} \int_k^\infty e^{-t} dt = N! e^{-k}/k^{N+1} \ll e^{-k}/k^N$$

thus Eq. (6.1.10) is an asymptotic expansion as $k \to \infty$. As $N \to \infty$ for fixed k, the series is divergent, and $|R_N| \to \infty$. As $k \to \infty$ for fixed N, $R_N \to 0$ (we have an asymptotic expansion).

Asymptotic series frequently give remarkably good approximations. For example, when $k = 10$ and $N = 2$, the error between the exact answer and the first two terms of the series, $R_2(10)$, satisfies $|R_2(10)| < 0.002e^{-10}$, which is clearly very small. In fact, even when $k = 3$ and $N = 2$, we have $|R_2(3)| < 2/(3e)^3 \doteq 3.7 \times 10^{-3}$. However, we cannot take too many terms in the series, because the remainder, which decreases for a while, eventually increases as N increases. In principle, one can find the "optimal" value of N for fixed k for which the remainder is smallest (best approximation); we will not go into this in more detail at this point. In most applications, obtaining the first few terms of the asymptotic expansion is sufficient.

The following property of an asymptotic expansion can be readily established: Given an asymptotic sequence $\{\delta_j(k)\}_{j=1}^\infty$, then the asymptotic expansion of a function $f(k)$ is unique. Specifically, let $f(k) \sim \sum_{j=1}^N a_j \delta_j(k)$ as $k \to k_0$, where $\{\delta_j(k)\}$ is an asymptotic sequence. Then the coefficients a_n are unique. Indeed, if $f(k)$ had another asymptotic expansion $f(k) \sim \sum_{j=1}^N b_j \delta_j(k)$, then calling $c_j = a_j - b_j$ we would find that $0 \sim c_1 \delta_1(k) + c_2 \delta_2(k) + c_3 \delta_3(k) + \cdots$. Dividing by $\delta_1(k)$ and taking the limit $k \to k_0$ implies $c_1 = 0$. Repeating this argument with $\delta_1(k)$ replaced by $\delta_2(k)$, etc., implies that $c_j = 0$ for $j = 1, 2, \ldots$, and hence $a_j = b_j$ for $j = 1, 2, \ldots$.

Until now we have discussed asymptotic expansions when the parameter k is real. However, the previous definitions can be extended to complex values of the parameter k, which we will now call z. Consider a function $f(z)$ that is analytic everywhere outside a circle $|z| = R$. Then we know that $f(z)$ has a convergent Taylor series at infinity of the form $f(z) = a_0 + a_1/z + a_2/z^2 + \cdots$. In this case the convergent Taylor series is equivalent to a convergent asymptotic series with an asymptotic sequence $\{1/z^j\}_{j=0}^{\infty}$. If $f(z)$ is not analytic at infinity, it *cannot* possess an asymptotic expansion valid for all arg z as $z \to \infty$. Typically asymptotic expansions are found to be valid only within some sector of the complex plane; that is, the expansion is constrained by some bounds on arg z.

Often the asymptotic expansion of a given function $f(z)$ has the form

$$f(z) \sim \Phi(z)\big(a_0 + a_1/z + a_2/z^2 + \cdots\big)$$

for z in a sector of the complex plane. The asymptotic sequence is either $\{\Phi(z)/z^j\}_{j=0}^{\infty}$ when considering $f(z)$, or more simply $\{1/z^j\}_{j=0}^{\infty}$ if we choose to work with the asymptotic expansion of $f(z)/\Phi(z)$.

A function f is said to have an **asymptotic power series** in a sector of the z plane as $z \to \infty$ if

$$f(z) \sim a_0 + a_1/z + a_2/z^2 + \cdots$$

The series is generally not convergent. Let another function $g(z)$ have an asymptotic power series representation in the same sector of the form

$$g(z) \sim b_0 + b_1/z + b_2/z^2 + \cdots$$

Then the arithmetic combinations $f + g$ (sum) and fg (product) also have asymptotic power series representations that are obtained by adding or multiplying the series termwise. An asymptotic power series can be integrated or differentiated termwise to yield an asymptotic expansion:

$$f'(z) \sim -a_1/z^2 - 2a_2/z^3 + \cdots,$$

$$\int_z^{\infty} (f(\zeta) - a_0 - a_1/\zeta)\,d\zeta \sim \frac{a_2}{z} + \frac{a_3}{2z^2} + \cdots.$$

We note that more general asymptotic series (as opposed to asymptotic power series) can be integrated termwise but it is, in general, not permissible to differentiate termwise in order to obtain an asymptotic expansion. (The interested reader may wish to consult Erdelyi (1956).)

A given asymptotic expansion can represent two entirely different functions. Suppose as $z \to \infty$ for Re$z > 0$ (i.e., $\frac{-\pi}{2} < \text{arg } z < \frac{\pi}{2}$), the function $f(z)$ is given by the asymptotic power series expansion

$$f(z) \sim \sum_{n=0}^{\infty} \frac{a_n}{z^n}$$

Then the *same* expansion also represents $f(z) + e^{-z}$ in this sector. The reason is that the asymptotic power series representation of e^{-z} for $\mathrm{Re}\, z > 0$ is zero, that is

$$e^{-z} \sim \sum_{n=0}^{\infty} \frac{b_n}{z^n}$$

has $b_n = 0$ for $n \geq 0$ because $\lim_{z \to \infty} z^n e^{-z} = 0$ for $n \geq 0$ and $\mathrm{Re}\, z > 0$. The term e^{-z} is transcendentally small, or said to be "beyond all orders", with respect to the asymptotic power series $\sum_{n=0}^{\infty} \frac{a_n}{z^n}$. An asymptotic power series contains no information about terms beyond all orders.

When $f(z)$ has an asymptotic representation (not necessarily an asymptotic power series) in a sector of the complex plane, it can happen that an entirely different asymptotic representation is valid in an adjacent sector. In fact, even when $f(z)$ is analytic for large but finite values of z, the asymptotic representation can change discontinuously as the sector is crossed. This is usually referred to as the **Stokes phenomenon**, which we discuss in detail in Section 6.7. The Stokes phenomenon arises frequently in applications. Linear second-order ODEs possess convergent series expansions around regular singular points (see Section 3.7). In general they possess asymptotic series representations around irregular singular points, i.e., singular points which are not regular.

Example 6.1.3 Discuss the asymptotic behavior of $I(z) = \sinh(z^{-1})$, $z \to 0$, z complex.

Let $z = r e^{i\theta}$. Then as $z \to 0$ the dominant term is $\frac{1}{2} e^{z^{-1}}$ or $-\frac{1}{2} e^{-z^{-1}}$ depending on whether $\cos \theta$ is > 0 or < 0, respectively. Thus

$$I(z) \sim \frac{1}{2} e^{z^{-1}} \quad \text{as} \quad z \to 0 \quad \text{in} \quad |\arg z| < \frac{\pi}{2}$$

$$\text{and} \quad I(z) \sim -\frac{1}{2} e^{-z^{-1}} \quad \text{as} \quad z \to 0 \quad \text{in} \quad \frac{\pi}{2} < \arg z < \frac{3\pi}{2}$$

We see that the asymptotic expansion of $\sinh(z^{-1})$ changes discontinuously across the ray $\theta = \pi/2$.

In much of this chapter we concentrate on the situation when the large parameter is real. The Stokes phenomenon is discussed in more detail in Section 6.7.

6.1.2 Elementary Examples

It is sometimes possible to determine the behavior of an integral without using any sophisticated asymptotic methods. Next we present two classes of such integrals. The first class involves the evaluation of the integral $\int_a^b f(k,t)\,dt$ as $k \to k_0$, where $f(k,t) \sim f_0(t)$, as $k \to k_0$ uniformly for t in $[a, b]$. The second class involves the evaluation of the integral $\int_k^b f(t)\,dt$ as $k \to \infty$.

To determine the leading behavior of integrals of the first class, we use the following fact. Assume that $f(k,t) \sim f_0(t)$ as $k \to k_0$ uniformly for t in $[a, b]$, and that $\int_a^b f_0(t)\,dt$ is finite and nonzero. Then the limit as $k \to k_0$ and the integral can be interchanged (this is a special case of the fact that asymptotic sequences can be integrated termwise) hence

$$\int_a^b f(k,t)\,dt \sim \int_a^b f_0(t)\,dt, \qquad k \to k_0 \qquad (6.1.11)$$

To determine the behavior of integrals of the second class we use integration by parts.

Example 6.1.4 Find the first two nonzero terms of the asymptotic expansion of $I(k) = \int_0^1 \frac{\sin tk}{t}\,dt$ as $k \to 0$.

Because $\sin z$ has a uniformly convergent Taylor series $\sin z = z - z^3/3! + \cdots$

$$\frac{\sin tk}{t} \sim k - \frac{t^2 k^3}{3!} + \cdots,$$

so $\qquad I(k) \sim \int_0^1 \left(k - \frac{t^2 k^3}{3!} + \cdots \right) dt = k - \frac{k^3}{3 \cdot 3!} + \cdots$

Sometimes it is possible to modify the above technique when Eq. (6.1.11) is not directly applicable. This is illustrated in Examples 6.1.5 and 6.1.6.

Example 6.1.5 Evaluate $I(k) = \int_k^\infty e^{-t^2}\,dt$ as $k \to 0$.

We would like to expand e^{-t^2}, which has an expansion provided that t is finite. Also intuitively we expect that as $k \to 0$, the main contribution to $I(k)$ comes from $\int_0^\infty e^{-t^2}\,dt$. For both of these reasons we write $I(k)$ as

$$I(k) = \int_0^\infty e^{-t^2}\,dt - \int_0^k e^{-t^2}\,dt \qquad (6.1.12)$$

The first integral of Eq. (6.1.12) can be evaluated exactly (see Example 4.2.8, Eq. (4.2.13), and the definition of the gamma function, Eq. (4.5.31)

$$\int_0^\infty e^{-t^2}\,dt = \frac{1}{2}\int_0^\infty e^{-s}s^{-\frac{1}{2}}\,ds = \frac{1}{2}\Gamma\left(\frac{1}{2}\right) = \frac{\sqrt{\pi}}{2}$$

To evaluate the second integral of Eq. (6.1.12), we use the Taylor series $e^{-t^2} = 1 - t^2 + t^4/2 - \cdots$; hence

$$I(k) = \frac{\sqrt{\pi}}{2} - k + \frac{k^3}{3} - \frac{k^5}{10} + \cdots$$

Example 6.1.6 Evaluate $E_1(k) = \int_k^\infty \frac{e^{-t}}{t}\, dt, k \to 0^+$.

As in Example 6.1.5, we write $\int_k^\infty = \int_0^\infty - \int_0^k$. But now the integrand has a logarithmic singularity at $t = 0$. If we subtract (and add) $1/t$ in order to cancel the singularity at $t = 0$, then we are left with an integral ($\int_k^\infty dt/t$) that does not converge. But if we subtract (and add) $1/[t(t + 1)]$, this difficulty is avoided:

$$E_1(k) = \int_k^\infty \frac{dt}{t(t+1)} + \int_0^\infty \left(e^{-t} - \frac{1}{t+1}\right)\frac{dt}{t} - \int_0^k \left(e^{-t} - \frac{1}{t+1}\right)\frac{dt}{t}$$

$$(6.1.13)$$

The first integral of Eq. (6.1.13) equals $\log\frac{1+k}{k} = -\log k + \log(1 + k) = -\log k + k - \frac{k^2}{2} + \frac{k^3}{3} - \cdots$, while the second integral is a constant that we call $-\gamma$. (It can be shown (Abramowitz and Stegun, 1965) that γ is the so-called **Euler constant**, which is approximately 0.577216.) To compute the third integral of Eq. (6.1.13), we note that Taylor series expansions yield

$$e^{-t} - \frac{1}{t+1} = \left(1 - t + \frac{t^2}{2} - \cdots\right) - \left(1 - t + t^2 - \cdots\right) = -\frac{t^2}{2} + \cdots$$

thus the third integral equals $-\frac{k^2}{4} + \cdots$. Hence

$$E_1(k) = -\gamma - \log k + k - \frac{k^2}{4} + \cdots$$

We now consider integrals of the type $\int_k^b f(t)\, dt, k \to \infty$. Such integrals can often be evaluated using integration by parts. Sometimes the integral must be appropriately rewritten before integration by parts is applied.

Example 6.1.7 Evaluate $I(k) = \int_k^\infty e^{-t^2}\, dt, k \to \infty$.

The largest value of e^{-t^2} occurs at the boundary $t = k$. Because integration by parts computes the value of the integrand on the boundaries, we expect that integration by parts will work:

$$I(k) = \int_k^\infty \left(-\frac{1}{2t}\right)\left(-2te^{-t^2}\right) dt = e^{-t^2}\left(-\frac{1}{2t}\right)\Big|_k^\infty - \int_k^\infty e^{-t^2}\frac{dt}{2t^2}$$

or

$$I(k) = \frac{e^{-k^2}}{2k} + O\left(\frac{e^{-k^2}}{k^3}\right)$$

Example 6.1.8 Evaluate $I(k) = \int_0^k t^{-1/2} e^{-t} \, dt$ as $k \to \infty$.

As $k \to \infty$ we expect that the main contribution to $I(k)$ will come from $\int_0^\infty t^{-1/2} e^{-t} \, dt$. Thus

$$I(k) = \int_0^\infty t^{-1/2} e^{-t} \, dt - \int_k^\infty t^{-1/2} e^{-t} \, dt$$

As before (see Example 6.1.5), the first integral above can be evaluated exactly; it is $\Gamma(\frac{1}{2}) = \sqrt{\pi}$. The second integral can be evaluated using integration by parts

$$I(k) = \sqrt{\pi} - \frac{e^{-k}}{\sqrt{k}} + O\left(\frac{e^{-k}}{k^{3/2}}\right)$$

The rigorous justification that the formulae obtained in Examples 6.1.7 and 6.1.8 are actually the first terms in an asymptotic expansion requires one to find a bound on the remainder. (See Example 6.1.4).

Problems for Section 6.1

1. (a) Consider the function

    $$f(\epsilon) = e^{\epsilon}$$

    Find the asymptotic expansion of $f(\epsilon)$ in powers of ϵ, as $\epsilon \to 0$

    (b) Similarly for the function

    $$f(\epsilon) = e^{-\frac{1}{\epsilon}}$$

    find the asymptotic expansion in powers of $1/\epsilon$, as $\epsilon \to 0$

2. Show that both the functions $(1 + x)^{-1}$ and $(1 + e^{-x})(1 + x)^{-1}$ possess the same asymptotic expansion as $x \to \infty$.

3. Find the asymptotic expansions of the following integrals:

    (a) $\displaystyle\int_0^1 \frac{\sin kt}{t} \, dt, \qquad k \to 0$ \qquad (b) $\displaystyle\int_0^k t^{-\frac{1}{4}} e^{-t} \, dt, \qquad k \to 0^+$

(Hint: let $s = t^{1/4}$.)

$$\text{(c)} \quad \int_k^\infty e^{-t^4} dt, \qquad k \to 0$$

4. (a) Show the bound on the remainder for the asymptotic expansion of the integral in Example 6.1.7 is

$$\frac{e^{-k^2}}{4k^3}$$

Hint: note that the remainder $R_2 = \int_k^\infty \frac{e^{-t^2}}{2t^2} dt$, thus

$$|R_2| = \left| \int_k^\infty \frac{(2t)e^{-t^2}}{4t^3} dt \right|,$$

then use the fact that $t > k$.

  (b) Show that a bound on the remainder: $R_2 = \int_k^\infty \frac{e^{-t}}{2t^{3/2}} dt$ in the asymptotic expansion of the integral in Example 6.1.8 is $\frac{e^{-k}}{2k^{3/2}}$.

5. Find the asymptotic expansion of the integral

$$I(z) = \int_a^b \frac{u(x)}{x - z} dx \quad \text{Im } z \neq 0$$

as $z \to \infty$. $\left(\text{Hint: Use } \frac{1}{x-z} = \frac{1}{(-z)(1-\frac{x}{z})}.\right)$ Show that if all the integrals

$$\int_{-\infty}^\infty |u(x)| x^n \, dx < \infty$$

then the analogous result holds when $-a = b = \infty$.

6. Find the asymptotic expansion of $\int_0^k \frac{e^{-t}}{1+t^2} dt$ as $k \to \infty$.

6.2 Laplace Type Integrals

In this section we shall study the asymptotic behavior as $k \to +\infty$ of integrals of the form

$$I(k) = \int_a^b f(t)e^{-k\phi(t)} dt \tag{6.2.1}$$

where $f(t)$ and $\phi(t)$ are real differentiable functions. A special case of such integrals (when $\phi(t) = t$, $a = 0$, $b = \infty$) is the Laplace transform (Section

4.5), which is why integrals such as Eq. (6.2.1) are referred to as Laplace type integrals.

Our analysis of the above integrals will be developed at an intuitive, a heuristic, and a rigorous level. At the intuitive level we identify the region in the t domain that gives the most important contribution to the large k evaluation of Eq. (6.2.1). At the heuristic level, by assuming that our intuition is correct, we derive appropriate mathematical formulae expressing the large k behavior of Eq. (6.2.1). At the rigorous level we establish the validity of our approach by proving that these formulae indeed express the **asymptotic behavior** of Eq. (6.2.1) as $k \to +\infty$.

In order to gain some intuition, suppose first that $I(k) = \int_0^b f(t)e^{-kt}\,dt$. As $k \to \infty$, the integrand becomes exponentially small for all t except for t near 0, because as $t \to 0$ and $k \to \infty$, kt could remain finite. Hence we expect that the major asymptotic contribution comes from the immediate neighborhood of $t = 0$. Thus the problem of analyzing this integral becomes a problem of studying the neighborhood of $t = 0$. In this way a global problem is replaced by a local one, which is precisely why the asymptotic evaluation of such integrals is successful. To render further support to our intuition, let's consider an integral that can be evaluated exactly. Let $J(k) = \int_0^\infty (1+t)e^{-kt}\,dt$, then separating the integrals and integration by parts yields the exact formula $J(k) = \frac{1}{k} + \frac{1}{k^2}$. We expect that for large k, $I(k) \sim \int_0^R e^{-kt}\,dt$, where R is arbitrarily small but $\lim_{k\to\infty} kR = \infty$. Hence we expect that $I(k) \sim \frac{1}{k}$, which indeed agrees with the large k behavior of the exact formula. The generalization of the above argument to integrals such as Eq. (6.2.1) is not difficult. If $t = c$ is the minimum of the function $\phi(t)$ in the interval $a \leq t \leq b$ and if $f(c) \neq 0$, then it is the neighborhood of $t = c$ that gives the dominant contribution to the asymptotic expansion of $I(k)$ for large k. Furthermore, the minimum can occur either at the boundaries or at an interior point, which in the latter case necessarily means $\phi'(t) = 0$. It follows that one only needs to carefully study such (critical) points. We distinguish two cases depending on whether the minimum occurs at the boundaries or at an interior point. The case that $\phi(t)$ is monotonic (i.e., the major contribution to the asymptotics of $I(k)$ comes from the boundaries) is considered in Sections 6.2.1 and 6.2.2. The case that $\phi(t)$ has a local minimum in $[a, b]$ is considered in Section 6.2.3.

6.2.1 Integration by Parts

Suppose that $\phi(t)$ is monotonic in $[a, b]$; then one needs to analyze the behavior of the integrand near the boundaries. Because integration by parts is based on such an analysis, we expect that it is a useful technique for studying Laplace type integrals when $\phi(t)$ is monotonic.

Example 6.2.1 Evaluate $\int_0^\infty (1 + t^2)^{-2} e^{-kt} dt$ as $k \to +\infty$.

Using integration by parts we find

$$\int_0^\infty (1 + t^2)^{-2} e^{-kt} dt = \left. \frac{(1 + t^2)^{-2} e^{-kt}}{-k} \right|_0^\infty + \frac{1}{k} \int_0^\infty (-4t)(1 + t^2)^{-3} e^{-kt} dt$$

$$= \frac{1}{k} + O\left(\frac{1}{k^3}\right)$$

Example 6.2.2 Evaluate $I(k) = \int_0^\infty (t + 1)^{-1} e^{-k(t+2)^2} dt$ as $k \to +\infty$.

Again we use integration by parts by first rewriting $I(k)$ as

$$I(k) = \int_0^\infty \left(\frac{d}{dt} e^{-k(t+2)^2}\right) \frac{(t + 1)^{-1}}{(-2k)(t + 2)} dt$$

$$= \left. e^{-k(t+2)^2} \frac{(t + 1)^{-1}}{(-2k)(t + 2)} \right|_0^\infty + \frac{1}{2k} \int_0^\infty e^{-k(t+2)^2} \frac{d}{dt} \left(\frac{(t + 1)^{-1}}{t + 2}\right) dt$$

$$= \frac{e^{-4k}}{4k} + O\left(\frac{e^{-4k}}{k^2}\right)$$

Example 6.2.3 Evaluate $I(k) = \int_1^2 \exp(k \cosh t) \, dt$ as $k \to \infty$.

The function $\cosh t$ increases monotonically in the interval $[1, 2]$; thus rewriting the integral in the form

$$I(k) = \int_1^2 \frac{d}{dt} \exp(k \cosh t) \frac{dt}{k \sinh t}$$

and integrating by parts yields

$$I(k) \sim \frac{e^{k \cosh 2}}{k \sinh 2}, \qquad k \to +\infty$$

From the above examples it follows that the asymptotic evaluation of integrals of the form $\int_a^b f(t) e^{-kt} dt$ depends on the behavior of $f(t)$ near $t = a$. In particular, if $f(t)$ is sufficiently smooth, the integration by parts approach provides the full asymptotic expansion. The rigorous justification for the results of Example 6.2.1 is obtained from:

$$\int_0^\infty t(1 + t^2)^{-3} e^{-kt} dt \le \int_0^\infty t e^{-kt} dt = \frac{1}{k}$$

The rigorous justification of the integration by parts in general follows from the following lemma.

Lemma 6.2.1 (*Integration by Parts*) Consider the integral

$$I(k) = \int_a^b f(t)e^{-kt}\, dt \tag{6.2.2}$$

where the interval $[a, b]$ is a finite segment of the real axis. Let $f^{(m)}(t)$ denote the mth derivative of $f(t)$, $f^{(0)}(t) \equiv f(t)$. Suppose that $f(t)$ has $N + 1$ continuous derivatives while $f^{(N+2)}(t)$ is piecewise continuous on $a \le t \le b$. Then

$$I(k) \sim \sum_{n=0}^{N} \frac{e^{-ka}}{k^{n+1}} f^{(n)}(a), k \to +\infty \tag{6.2.3}$$

Proof For $m \le N$, integration by parts yields

$$I(k) = \sum_{n=0}^{m-1} \frac{e^{-ka}}{k^{n+1}} f^{(n)}(a) - \sum_{n=0}^{m-1} \frac{e^{-kb}}{k^{n+1}} f^{(n)}(b)$$

$$+ \frac{1}{k^m} \int_a^b e^{-kt} f^{(m)}(t)\, dt, \qquad m = 1, \ldots, N$$

The contribution from the upper endpoint $t = b$ is asymptotically negligible ("beyond all orders") compared to that from $t = a$; indeed, $\lim_{k \to \infty} k^n e^{k(a-b)} = 0$, $n = 1, 2, \ldots$. Also, another integration by parts yields

$$R_{m-1}(k) \equiv \frac{1}{k^m} \int_a^b e^{-kt} f^{(m)}(t)\, dt$$

$$= -\frac{1}{k^{m+1}} \left[e^{-kb} f^{(m)}(b) - e^{-ka} f^{(m)}(a) \right]$$

$$+ \frac{1}{k^{m+1}} \int_a^b e^{-kt} f^{(m+1)}(t)\, dt$$

Thus as $k \to \infty$, $R_{m-1}(k) = O(k^{-(m+1)} e^{-ka})$. For $m = N$ we can decompose $[a, b]$ into subintervals in each of which $f^{(N+2)}(t)$ is continuous. Then a final integration by parts for R_N completes the proof. ∎

We mention two generalizations of the above result. (a) If $b = \infty$, then the above result is also valid provided that as $t \to \infty$ $f(t) = O(e^{\alpha t})$, α real constant, so that $I(k)$ exists for k sufficiently large. In particular, if $a = 0$ then

$I(k)$ becomes the Laplace transform of f and the asymptotic expansion follows from Eq. (6.2.3) by setting $a = 0$.

(b) If $\phi(t)$ is monotonic in $[a, b]$, then the integral $\int_a^b f(t)e^{-k\phi(t)} dt$ can be transformed to Eq. (6.2.2) by the change of variables $\tau = \phi(t)$. Although in practice, this is usually not the most convenient way to evaluate such integrals, it does provide a rigorous justification for the integration by parts approach to such integrals.

Example 6.2.4 Evaluate $I(\epsilon) = \int_0^\infty (1 + \epsilon t)^{-1} e^{-t} dt$ as $\epsilon \to 0$.

This integral can be transformed to a Laplace type integral by letting $kt = \tau$. Then

$$I(k) = \frac{1}{\epsilon} \int_0^\infty (1 + \tau)^{-1} e^{-\frac{\tau}{k}} dt, \qquad \frac{1}{\epsilon} \to \infty$$

Hence, using $f(\tau) = (1 + \tau)^{-1}$, $f^{(n)}(0) = (-1)^n n!$, Eq. (6.2.3) with $k = 1/\epsilon$ yields

$$I(k) \sim \sum_{n=0}^N (-1)^n n! \epsilon^n, \qquad \epsilon \to 0$$

6.2.2 Watson's Lemma

If $f(t)$ is not sufficiently smooth at $t = a$, then the integration by parts approach for the asymptotic evaluation of Eq. (6.2.2) may not work.

Example 6.2.5 Obtain the first two terms of the asymptotic expansion of $I(k) = \int_0^5 (t^2 + 2t)^{-1/2} e^{-kt} dt$ as $k \to \infty$.

The function $f(t)$ is of $O(t^{-1/2})$ as $t \to 0$; thus a straightforward integration by parts fails. Indeed

$$I(k) = \frac{(t^2 + 2t)^{-1/2}}{-k} e^{-kt} \bigg|_0^5 + \frac{1}{k} \int_0^5 e^{-kt} \left(\frac{d}{dt} (t^2 + 2t)^{-1/2} \right) dt$$

and the first term above is singular at $t = 0$.

Intuitively, we expect that the main contribution to $I(k)$ for large k will come near $t = 0$. Thus it would be desirable to expand $(t^2 + 2t)^{-1/2}$ in the neighborhood of the origin. At first glance this seems prohibited because $0 \le t \le 5$. However, owing to the rapid decay of $\exp(-kt)$, $I(k)$ should be asymptotically equivalent to $\int_0^R (t^2 + 2t)^{-1/2} e^{-kt} dt$, where R is sufficiently small but finite.

If $R < 2$, we can expand $(t^2 + 2t)^{-1/2}$ as $t \to 0$. For example, keeping only two terms in this expansion, that is, replacing $(t^2 + 2t)^{-1/2}$ by

$$(2t+t^2)^{-1/2} = (2t)^{-1/2}\left(1 + \frac{t}{2}\right)^{-1/2} \sim (2t)^{-1/2}\left(1 - \frac{t}{4}\right) = (2t)^{-1/2} - \frac{(2t)^{1/2}}{8}$$

we find

$$I(k) \sim \int_0^R e^{-kt}(2t)^{-1/2}\,dt - \frac{1}{8}\int_0^R e^{-kt}(2t)^{1/2}\,dt$$

To evaluate the above integrals in terms of known functions, we replace R by ∞. Again we expect, and below we will show, that this introduces only an exponentially small error (i.e., terms beyond all orders) as $k \to \infty$. Thus

$$I(k) \sim \int_0^\infty e^{-kt}(2t)^{-1/2}\,dt - \frac{1}{8}\int_0^\infty e^{-kt}(2t)^{1/2}\,dt$$

$$= \frac{1}{(2k)^{1/2}}\int_0^\infty e^{-t}t^{-1/2}\,dt - \frac{1}{2(2k)^{3/2}}\int_0^\infty e^{-t}t^{1/2}\,dt$$

$$= \frac{\Gamma(1/2)}{(2k)^{1/2}} - \frac{\Gamma(3/2)}{2(2k)^{3/2}}$$

where $\Gamma(z)$ is the so-called **gamma function** (see Eq. (4.5.31)), which we remind the reader is defined by

$$\Gamma(z) = \int_0^\infty t^{z-1}e^{-t}\,dt, \qquad \mathrm{Re}\,z > 0$$

In the above example, following our intuition that the main asymptotic contribution comes from $t = 0$, we have used the following formal steps: (a) replaced 5 by R, where $R < 2$; (b) expanded $(t^2 + 2t)^{-1/2}$ in a series using the binomial formula; (c) interchanged orders of integration and summation; and (d) replaced R by ∞.

Watson's lemma provides a rigorous justification for the above heuristic approach.

Lemma 6.2.2 (Watson's Lemma) Consider the integral

$$I(k) = \int_0^b f(t)e^{-kt}\,dt, \qquad b > 0 \tag{6.2.4}$$

Suppose that $f(t)$ is integrable in $(0, b)$ and that it has the asymptotic series expansion

$$f(t) \sim t^\alpha \sum_{n=0}^{\infty} a_n t^{\beta n}, \qquad t \to 0^+; \qquad \alpha > -1, \ \beta > 0 \qquad (6.2.5)$$

Then

$$I(k) \sim \sum_{n=0}^{\infty} a_n \frac{\Gamma(\alpha + \beta n + 1)}{k^{\alpha + \beta n + 1}}, \qquad k \to \infty \qquad (6.2.6)$$

If b is finite, we require that for $t > 0$, $|f(t)| \le A$, where A is a constant; if $b = \infty$, we need only require that $|f(t)| \le M e^{ct}$, where c and M are constants.

Proof We break the integral in two parts, $I = I_1(k) + I_2(k)$, where

$$I_1(k) = \int_0^R f(t) e^{-kt} \, dt, \qquad I_2(k) = \int_R^b f(t) e^{-kt} \, dt$$

and R is a positive constant, $R < b$. The integral $I_2(k)$ is exponentially small as $k \to \infty$. For finite b, because $f(t)$ is bounded for $t > 0$, there exists a positive constant A such that $|f| \le A$ for $t \ge R$; thus

$$|I_2(k)| \le A \int_R^b e^{-kt} \, dt = \frac{A}{k} \left(e^{-kR} - e^{-kb} \right)$$

Thus from Definition 6.1.1

$$I_2(k) = O \left(\frac{e^{-kR}}{k} \right) \qquad \text{as } k \to \infty$$

Equation (6.2.5) implies that for each positive integer N

$$I_1(k) = \int_0^R \left[\sum_{n=0}^{N} a_n t^{\alpha + \beta n} + O \left(t^{\alpha + \beta(N+1)} \right) \right] e^{-kt} \, dt, \qquad k \to \infty$$

However,

$$\int_0^R t^{\alpha + \beta n} e^{-kt} \, dt = \int_0^\infty t^{\alpha + \beta n} e^{-kt} \, dt - \int_R^\infty t^{\alpha + \beta n} e^{-kt} \, dt$$

$$= \frac{\Gamma(\alpha + \beta n + 1)}{k^{\alpha + \beta n + 1}} + O \left(\frac{e^{-kR}}{k} \right), \qquad k \to \infty$$

where we have used the definition of the gamma function in the first integral and integration by parts to establish the second. Moreover, using the definition of big-O (Definition 6.1.1)

$$\int_0^R O\left(t^{\alpha+\beta(N+1)}\right) e^{-kt}\, dt \le A_N \int_0^R t^{\alpha+\beta(N+1)} e^{-kt}\, dt$$

$$\le A_N \frac{\Gamma(\alpha + \beta(N+1)+1)}{k^{\alpha+\beta(N+1)+1}}$$

Thus

$$I(k) = \sum_{n=0}^{N} a_n \frac{\Gamma(\alpha+\beta n+1)}{k^{\alpha+\beta n+1}} + O\left(\frac{1}{k^{\alpha+\beta(N+1)+1}}\right), \qquad k \to \infty$$

We note that the assumptions $\alpha > -1$, $\beta > 0$ are necessary for convergence at $t = 0$. Also if $b = \infty$, then it is only necessary that $|f(t)| \le Me^{ct}$ for some real constants M and c, in order to have convergence at $t \to +\infty$. In this case, the estimate of I_2 gives

$$I_2(k) \le M\frac{e^{-(k-c)R}}{k-c} = O\left(\frac{e^{-kR}}{k}\right) \qquad \text{as } k \to \infty$$

∎

Example 6.2.6 Let us return to Example 6.2.5. Find the complete asymptotic expansion of $I(k) = \int_0^\infty (t^2 + 2t)^{-1/2} e^{-kt}\, dt$ as $k \to \infty$.

In this case the binomial formula gives

$$(t^2 + 2t)^{-\frac{1}{2}} = (2t)^{-\frac{1}{2}}\left(1 + \frac{t}{2}\right)^{-\frac{1}{2}} = (2t)^{-\frac{1}{2}} \sum_{n=0}^{\infty} \left(\frac{t}{2}\right)^n \hat{c}_n$$

where $\hat{c}_n$ are the coefficients in the Taylor expansion of $(1+z)^\alpha$ for $|z| < 1$, where $\alpha = -\frac{1}{2}$; more generally, the binomial coefficients of the Taylor expansion of $(1+z)^\alpha$ are given by

$$c_n(\alpha) = \frac{\alpha(\alpha-1)(\alpha-2)\cdots(\alpha-n+1)}{n!} = \frac{\alpha!}{n!(\alpha-n)!}$$

$$= \frac{\Gamma(\alpha+1)}{\Gamma(n+1)\Gamma(\alpha-n+1)}, \qquad n \ge 1$$

$c_0 = 1$, so that $\hat{c}_n = c_n(-\frac{1}{2})$. Thus, Watson's Lemma (e.g. Eq. (6.2.6) with $\alpha = -1/2, \beta = 1$) implies

$$I(k) \sim \sum_{n=0}^{\infty} \frac{\hat{c}_n}{2^{(n+\frac{1}{2})}} \frac{\Gamma\left(n + \frac{1}{2}\right)}{k^{(n+\frac{1}{2})}}, \qquad k \to \infty \qquad (6.2.7)$$

It turns out that the integral $I(k)$ is related to the modified Bessel function of order zero, $K_0(k)$. Modified Bessel functions satisfy the Bessel equation of order p (Eq. (3.7.35a)) with the transformation (rotation) of coordinates $z = ik$, and for k real, $K_p(k) \sim e^{-k}/\sqrt{2k/\pi}$ (independent of p) as $k \to \infty$. Indeed, it can be shown that $K_0(k) = e^{-k}I(k)$. Thus Eq. (6.2.7) yields the large k behavior of $K_0(k)$.

This example is an illustration of the power of Watson's Lemma. Equation (6.2.6) indicates that the asymptotic behavior of $I(k)$ **to all orders** comes from the neighborhood of $t = 0$. Hence it is possible to find an infinite asymptotic expansion just by analyzing the behavior of the integrand in the neighborhood of a single point!

6.2.3 Laplace's Method

In the previous two sections we considered integrals of the type (6.2.1) when $\phi(t)$ is monotonic in $[a, b]$. Now we consider the case that $\phi(t)$ is not monotonic. We suppose, for simplicity, that the local minimum occurs at an interior point c, $a < c < b$, $\phi'(c) = 0$, $\phi''(c) > 0$. Further, we assume that $\phi'(t) \neq 0$ in $[a, b]$ except at $t = c$ and that f, ϕ are sufficiently smooth for the operations below to be justified.

We first proceed heuristically, following the intuitive argument that the main asymptotic contribution comes from the neighborhood of the minimum of $\phi(t)$, which is at $t = c$. A rigorous derivation of the relevant results will be given later.

From Eq. (6.2.1), by expanding both f and ϕ in the neighborhood of c, we expect that for large k, I is asymptotic to

$$\int_{c-R}^{c+R} f(c) \exp\left\{ -k\left[\phi(c) + \frac{(t-c)^2}{2}\phi''(c)\right]\right\} dt$$

where R is small but finite. To evaluate this integral, we let $\tau = \sqrt{\frac{k}{2}\phi''(c)}(t-c)$; thus

$$I \sim e^{-k\phi(c)} f(c) \int_{c-R}^{c+R} e^{-k\frac{(t-c)^2}{2}\phi''(c)} dt = \frac{e^{-k\phi(c)} f(c)}{\sqrt{\frac{k}{2}\phi''(c)}} \int_{-R\sqrt{\frac{k}{2}\phi''(c)}}^{R\sqrt{\frac{k}{2}\phi''(c)}} e^{-\tau^2} d\tau$$

As $k \to \infty$ the last integral becomes $\sqrt{\pi}$; hence

$$\int_a^b f(t)e^{-k\phi(t)}\, dt \sim e^{-k\phi(c)} f(c)\sqrt{\frac{2\pi}{k\phi''(c)}}, \qquad k \to \infty \qquad (6.2.8)$$

The approximation expressed by Eq. (6.2.8) is often referred to as Laplace's formula, and the application of this approach is called Laplace's method. We note that the use of the asymptotic symbol $\sim$ in Eq. (6.2.8) is premature because no error estimate has been obtained. Indeed, the best we can say at the moment is that it seems plausible that the right-hand side of Eq. (6.2.8) represents the leading term of an asymptotic expansion of Eq. (6.2.1) as $k \to \infty$. The rigorous derivation of this result follows from Lemma 6.2.3. In the following, the notation $f(t) \in C^n[a, b]$ means that $f(t)$ has n derivatives and that $f^{(n)}(t)$ is continuous in the interval $[a, b]$.

Lemma 6.2.3 (Laplace's Method) Consider the integral Eq. (6.2.1) and assume that $\phi'(c) = 0$, $\phi''(c) > 0$ for some point c in the interval $[a, b]$. Further, assume that $\phi'(t) \neq 0$ in $[a, b]$ except at $t = c$, $\phi \in C^4[a, b]$, and $f \in C^2[a, b]$. Then if c is an interior point, Eq. (6.2.8) expresses the leading term of an asymptotic expansion of Eq. (6.2.1) as $k \to \infty$, with an error $O(e^{-k\phi(c)}/k^{3/2})$. If c is an endpoint, then the leading term is half that obtained when c is an interior point, and the error is $O(e^{-k\phi(c)}/k)$.

Proof The main idea of the proof is to split $[a, b]$ into two half-open intervals $[a, c)$, $(c, b]$, in each of which $\phi(t)$ is monotonic so that, using a change of variables, Watson's Lemma can be applied. Let $I(k) = I_a(k) + I_b(k)$, where I_a and I_b are integrals to be evaluated inside $[a, c)$ and $(c, b]$, respectively. Let us consider I_b. Because ϕ is monotonic in $[c, b]$, we let $\phi(t) - \phi(c) = \tau$; then I_b becomes

$$I_b = e^{-k\phi(c)} \int_0^{\phi(b)-\phi(c)} \left.\frac{f(t)}{\phi'(t)}\right|_{t=t(\tau)} e^{-k\tau}\, d\tau \qquad (6.2.9)$$

and because $\phi(t)$ is monotonic in this interval, we can invert the change of variables, which we denote as $t(\tau) = \phi^{-1}(\tau + \phi(c))$. To apply Watson's Lemma, we need to determine the behavior of $\left.\frac{f(t)}{\phi'(t)}\right|_{t=t(\tau)}$ as $\tau \to 0^+$. To achieve this, we use the inversion of $\phi(t) - \phi(c) = \tau$ in the neighborhood of $\tau = 0$ and then expand $\frac{f(t)}{\phi'(t)}$ near $t = c$. Even though the function $t = t(\tau)$ is multivalued in the full neighborhood of $\tau = 0$, we bypassed this problem by considering the half interval and noting that as t increases from c to b, τ increases from 0 to

$\phi(b) - \phi(c)$. To find $t = t(\tau)$ near $\tau = 0$ we expand

$$\phi''(c)(t - c)^2 + \frac{\phi'''(c)}{3}(t - c)^3 + O(t - c)^4 = 2\tau \qquad (6.2.10)$$

and solve Eq. (6.2.10) recursively. To the first order, $t - c = \sqrt{\frac{2}{\phi''(c)}}\tau^{1/2}$;
substituting $t - c = \sqrt{\frac{2}{\phi''(c)}}\tau^{1/2} + A\tau^\alpha + \cdots$ with $\alpha > \frac{1}{2}$ in Eq. (6.2.10), we
find $\alpha = 1$ and $A = -\left(\frac{\phi'''(c)}{3(\phi''(c))^2}\right)$. Thus

$$t - c = \sqrt{\frac{2}{\phi''(c)}}\tau^{1/2}\left[1 - \frac{\phi'''(c)}{3\sqrt{2}(\phi''(c))^{3/2}}\tau^{1/2} + O(\tau)\right] \qquad (6.2.11)$$

Using

$$f(t) = f(c) + (t - c)f'(c) + O((t - c)^2)$$

$$\phi'(t) = \phi''(c)(t - c) + \frac{\phi'''(c)(t - c)^2}{2} + O((t - c)^3),$$

we find

$$\frac{f(t)}{\phi'(t)} = \frac{f(c)}{\phi''(c)(t - c)} - \frac{f(c)\phi'''(c)}{2(\phi''(c))^2} + \frac{f'(c)}{\phi''(c)} + O(t - c)$$

Substituting Eq. (6.2.11) in this equation we obtain

$$\frac{f(t)}{\phi'(t)} = \frac{f(c)}{\sqrt{2\phi''(c)}}\tau^{-1/2} + \left(\frac{f'(c)}{\phi''(c)} - \frac{f(c)\phi'''(c)}{3(\phi''(c))^2}\right) + O(\tau^{1/2})$$

$$\equiv a_0\tau^{-1/2} + a_1 + O(\tau^{1/2}),$$

where

$$a_0 = \frac{f(c)}{\sqrt{2\phi''(c)}}$$

and

$$a_1 = \frac{f'(c)}{\phi''(c)} - \frac{f(c)\phi'''(c)}{3(\phi''(c))^2}$$

Substituting this equation in Eq. (6.2.9) and then using Watson's lemma Eqs. (6.2.5)–
(6.2.6) with $\alpha = -1/2$ and $\beta = 1/2$ it follows that

$$I_b(k) = a_0\frac{\Gamma(1/2)}{k^{1/2}} + a_1\frac{\Gamma(1)}{k} + O\left(\frac{1}{k^{3/2}}\right)$$

or

$$I_b(k) = \sqrt{\frac{\pi}{2k\phi''(c)}} f(c) e^{-k\phi(c)} + \left(\frac{f'(c)}{\phi''(c)} - \frac{f(c)\phi'''(c)}{3(\phi''(c))^2} \right) \frac{e^{-k\phi(c)}}{k}$$

$$+ O\left(\frac{e^{-k\phi(c)}}{k^{3/2}} \right), \qquad k \to \infty \tag{6.2.12}$$

Similarly, for the interval $[a, c)$ (note in this interval $t - c = -\sqrt{\frac{2}{\phi''(c)}} \tau^{1/2} + \cdots$),

$$I_a(k) = \sqrt{\frac{\pi}{2k\phi''(c)}} f(c) e^{-k\phi(c)} - \left(\frac{f'(c)}{\phi''(c)} - \frac{f(c)\phi'''(c)}{3(\phi''(c))^2} \right) \frac{e^{-k\phi(c)}}{k}$$

$$+ O\left(\frac{e^{-k\phi(c)}}{k^{3/2}} \right), \qquad k \to \infty \tag{6.2.13}$$

If c is an interior point, then the desired expansion for $I(k)$ is given by the sum of Eqs. (6.2.12) and (6.2.13), which to leading order yields Eq. (6.2.8) with error

$$O\left(\frac{e^{-k\phi(c)}}{k^{3/2}} \right)$$

If $c = b$ or $c = a$, then Eqs. (6.2.12) or (6.2.13) provides the correct expansion.

∎

We note that further terms in the expansion of $I(k)$ can be obtained by determining further terms in the asymptotic expansion of $\frac{f(t)}{\phi'(t)}|_{t=t(\tau)}$. This would require, of course, additional smoothness assumptions about the behavior of f and ϕ near $t = c$. However, because Laplace's method is based on Watson's Lemma that can, in principle, give infinite asymptotic expansions, it follows that Laplace's method can, in principle, give the asymptotic expansion of an integral to all orders. This fact, will be utilized further in connection with the steepest descent method. In the examples presented below, for simplicity of presentation, we give only the leading asymptotic behavior (which in many applications is sufficient).

Example 6.2.7 Evaluate $I(k) = \int_0^\infty e^{-k \sinh^2 t} dt$ as $k \to \infty$.

The relevant minimum of $\phi(t) = \sinh^2 t$ occurs at the endpoint $t = 0$; hence $\phi''(t) = 2\cosh 2t$, $\phi''(0) = 2$, $f(t) = 1$, and Eq. (6.2.12) yields $I(k) \sim \frac{1}{2}\sqrt{\frac{\pi}{k}}$.

Example 6.2.8 Evaluate $I(k) = \int_a^b f(t) e^{k\phi(t)} dt$, $k \to \infty$, when $\phi(t)$ has a unique maximum at an interior point $t = c$. We use the same ideas as those motivating Laplace's method (beginning of Section 6.2.3).

$$I(k) \sim \int_{c-R}^{c+R} f(c) e^{k\left[\phi(c) + \frac{(t-c)^2}{2} \phi''(c)\right]} dt$$

$$= \frac{f(c) e^{k\phi(c)}}{\sqrt{-k\frac{\phi''(c)}{2}}} \int_{-R\sqrt{-\frac{k}{2}\phi''(c)}}^{R\sqrt{-\frac{k}{2}\phi''(c)}} e^{-\tau^2} d\tau$$

$$\sim \sqrt{\frac{2\pi}{-k\phi''(c)}} f(c) e^{k\phi(c)} = \sqrt{\frac{2\pi}{k|\phi''(c)|}} f(c) e^{k\phi(c)} \quad (6.2.14)$$

where we have used $\tau = (t-c)\sqrt{-\frac{k\phi''(c)}{2}}$, noting that $\phi''(c) < 0$ (because $t = c$ is a maximum).

Example 6.2.9 Use Laplace's method to show that for an appropriate class of functions the L_p norm converges to the "maximum" norm, as $p \to \infty$.

The L_p norm of a function g is given by $\|g\|_p = (I(p))^{1/p}$, where $I(p) = \int_a^b |g(t)|^p dt$. We assume that $|g(t)| \in C^4$ and that it has a unique maximum with $g(c) \neq 0$ at $t = c$ inside $[a, b]$. Using Laplace's method, $\phi(t) = \log|g(t)|$, $\phi'(c) = 0$, $\phi''(c) = g''(c)/g'(c)$, and example 6.2.8,

$$I(p) = \int_a^b e^{p \log|g(t)|} dt = \frac{A}{\sqrt{p}} |g(c)|^p \left\{ 1 + O\left(\frac{1}{p}\right) \right\},$$

$$p \to \infty, \qquad A = \sqrt{\frac{2\pi|g(c)|}{|g''(c)|}}$$

Thus

$$\|g\|_p \sim A^{\frac{1}{p}} p^{-\frac{1}{2p}} |g(c)| \left\{ 1 + O\left(\frac{1}{p}\right) \right\} = |g(c)| \left\{ 1 - \frac{\log p}{2p} + O\left(\frac{1}{p}\right) \right\}$$

where we have used

$$p^{-\frac{1}{2p}} = e^{-\frac{1}{2p}\log p} \sim 1 - \frac{\log p}{2p}, \qquad A^{\frac{1}{p}} = e^{\frac{\log A}{p}} \sim 1 + \frac{\log A}{p}$$

Thus $\|g\|_p$ tends to $|g(c)|$ as $p \to \infty$, which is referred to as the maximum norm.

The main ideas used in deriving Eq. (6.2.8) can also be used in dealing with other similar integrals.

Example 6.2.10 Evaluate $I(k) = \int_a^b f(t) e^{k\phi(t)}\, dt, k \to \infty$, with $\phi'(c) = \phi''(c)$ $= \cdots \phi^{(p-1)}(c) = 0$, and $\phi^{(p)}(c) < 0$, where ϕ is maximum at the interior point $t = c$. Again we expand $f(t)$ and $\phi(t)$ near $t = c$:

$$I(k) \sim \int_{c-R}^{c+R} f(c) e^{k\left[\phi(c) + \frac{(t-c)^p}{p!}\phi^{(p)}(c)\right]}\, dt$$

$$= \frac{f(c) e^{k\phi(c)}}{\left(\frac{-k\phi^p(c)}{p!}\right)} \int_{-R\left(-\frac{k}{p!}\phi^p(c)\right)^{1/p}}^{+R\left(-\frac{k}{p!}\phi^p(c)\right)^{1/p}} e^{-\tau^p}\, d\tau$$

$$\sim \frac{f(c) e^{k\phi(c)}}{\left(-\frac{k\phi^{(p)}(c)}{p!}\right)^{\frac{1}{p}}} \left(\frac{2\Gamma\left(\frac{1}{p}\right)}{p}\right) \tag{6.2.15}$$

where we have used $\tau = (t - c)\left(-k\frac{\phi^{(p)}(c)}{p!}\right)^{\frac{1}{p}}$, and $\tau^p = s$

$$\int_{-\infty}^{\infty} e^{-\tau^p}\, d\tau = \int_{-\infty}^{\infty} e^{-s} s^{\frac{1}{p}-1} \frac{ds}{p} = \frac{2\Gamma\left(\frac{1}{p}\right)}{p}$$

We point out that Laplace's method can also work in some cases that are not directly covered by Lemma 6.2.3. For example, it can be used when $f(t)$ either vanishes algebraically or becomes infinite at an algebraic rate. This is illustrated in the following examples.

First we discuss integrals where $f(t)$ behaves algebraically. We will see that this has a significant effect on the asymptotic result.

Example 6.2.11 Evaluate $I = \int_0^5 \sin s e^{-k \sinh^4 s}\, ds$. Note that the minimum of $\sinh^4 s$ is at $s = 0$, which is the location of the major contribution to the integral.

$$I \sim \int_0^R s e^{-ks^4}\, ds = \frac{1}{2} \int_0^{R^{1/2}} e^{-kt^2}\, dt \sim \frac{1}{4\sqrt{k}} \int_0^{\infty} e^{-\tau} \tau^{-\frac{1}{2}}\, d\tau$$

$$= \frac{1}{4}\sqrt{\frac{1}{k}}\, \Gamma\left(\frac{1}{2}\right) = \frac{1}{4}\sqrt{\frac{\pi}{k}}$$

using, in the latter three expressions, the substitutions $s^2 = t$, $\tau = kt^2$, and $\Gamma\left(\frac{1}{2}\right) = \sqrt{\pi}$.

Example 6.2.12 Evaluate $I = \int_0^\infty \frac{e^{-kt^2}}{\sqrt{\sinh t}} \, dt$. Note that the dominant contribution to the integral is at $t = 0$. Again, using $\tau = kt^2$ and the definition of the gamma function

$$I \sim \int_0^\infty \frac{e^{-kt^2}}{t^{1/2}} \, dt = \frac{1}{2k^{1/4}} \int_0^\infty e^{-\tau} \tau^{\frac{1}{4}-1} d\tau = \frac{\Gamma\left(\frac{1}{4}\right)}{2k^{1/4}}$$

If $f(t)$ vanishes exponentially fast, the direct application of Laplace's method fails. However, an intuitive understanding of the basic ideas used in Laplace's method leads to the appropriate modifications needed for the evaluation of such integrals. This is illustrated in the following examples.

Example 6.2.13 Consider $I(k) = \int_0^\infty e^{-kt-\frac{1}{t}} \, dt$. The maximum of e^{-kt} occurs at $t = 0$, but $f(t) = \exp\left(-\frac{1}{t}\right)$ vanishes exponentially fast at $t = 0$. The first step in the evaluation of an integral of the type Eq. (6.2.1) is the determination of the domain of t that yields the dominant contribution to the asymptotic value of the given integral. For the integral $I(k)$ this domain is not the neighborhood of $t = 0$ but the neighborhood of the minimum of $kt + \frac{1}{t}$, that is, the neighborhood of $t = \frac{1}{\sqrt{k}}$. This point depends on k; thus it is a movable maximum. The change of variables $t = \frac{s}{\sqrt{k}}$ maps this maximum to a fixed one, and so it now allows a direct application of Laplace's method:

$$I(k) = \frac{1}{\sqrt{k}} \int_0^\infty e^{-\sqrt{k}(s+\frac{1}{s})} \, ds = \frac{1}{\sqrt{k}} \int_0^\infty e^{-\sqrt{k}\phi(s)} \, ds$$

where $\phi(s) = s + 1/s$. The function $\phi(s)$ has a minimum at $s = 1$ (an interior point), so letting $s = 1 + t$, and expanding, yields

$$I \sim \frac{1}{\sqrt{k}} e^{-2\sqrt{k}} \int_{-\infty}^\infty e^{-\sqrt{k}t^2} \, dt = \frac{e^{-2\sqrt{k}}}{k^{3/4}} \int_0^\infty u^{-1/2} e^{-u} \, du$$

$$= \frac{\sqrt{\pi} e^{-2\sqrt{k}}}{k^{3/4}}, \qquad k \to \infty$$

where we used the substitution $u = \sqrt{k}t^2$, and $\Gamma(\frac{1}{2}) = \sqrt{\pi}$.

Example 6.2.14 (Asymptotic Expansion of the Gamma Function (Stirling's Formula) Consider $\Gamma(k+1) = \int_0^\infty e^{-t} t^k \, dt$. Because $t^k = e^{k \log t}$, the maximum of $\log t$ occurs as $t \to \infty$, but $f(t) = e^{-t}$ vanishes exponentially as $t \to \infty$. The true maximum of $e^{-t} t^k$ occurs at $\frac{d}{dt}(k \log t - t) = 0$, or $t = k$.

This suggests the change of variables $t = sk$, and we find

$$\Gamma(k+1) = k^{k+1} \int_0^\infty e^{-k(s-\log s)}\, ds = k^{k+1} \int_0^\infty e^{-k\phi(s)}\, ds$$

where $\phi(s) = s - \log s$. The function $\phi(s)$ has a minimum at $s = 1$ (an interior point), so letting $s = 1 + t$ with $\phi'(1) = 0$, $\phi''(1) = 1$, it follows from (6.2.8) that

$$\Gamma(k+1) \sim \sqrt{2\pi k} \left(\frac{k}{e}\right)^k$$

using the substitution $u = kt^2/2$, and $\Gamma(\frac{1}{2}) = \sqrt{\pi}$.

Laplace's method can also be used to find the asymptotic behavior of certain sums. The basic idea is to rewrite the sum in terms of an appropriate integral, as illustrated in the following example.

Example 6.2.15 Evaluate

$$I(n) = \sum_{k=0}^n \binom{n}{k} k! n^{-k}$$

as $n \to \infty$ where $\binom{n}{k}$ are the binomial coefficients: $\binom{n}{k} \equiv \frac{n!}{k!(n-k)!}$. (See also Example 6.2.6, where α and n are replaced by n and k, respectively.) We use $\int_0^\infty e^{-nx} x^k\, dx = n^{-k-1} \int_0^\infty e^{-t} t^k\, dt = \Gamma(k-1)n^{-k-1} = k! n^{-k-1}$, $(nx = t)$, to obtain

$$I(n) = \sum_{k=0}^n \binom{n}{k} n \int_0^\infty e^{-nx} x^k\, dx = \int_0^\infty e^{-nx} n \left(\sum_0^n \binom{n}{k} x^k \right) dx$$

where we have exchanged the order of integration and summation (this is allowed because the sum is finite). Because $\sum_{k=0}^n \binom{n}{k} x^k = (1+x)^n$, we find that

$$I(n) = n \int_0^\infty e^{-nx}(1+x)^n\, dx = n \int_0^\infty e^{n[\log(1+x)-x]}\, dx$$

We now use Laplace's method to evaluate the large n asymptotics of the above integral, noting that the function $\phi(x) = \log(1+x) - x$ has its *maximum* at

$x = 0$. We expand, substitute $u = nx^2$, and use $\Gamma\left(\frac{1}{2}\right) = \sqrt{\pi}$ to find

$$I(n) \sim n \int_0^\infty e^{-nx^2/2}\, dx = \sqrt{\frac{n}{2}} \int_0^\infty e^{-u} u^{-1/2}\, du = \sqrt{\frac{\pi n}{2}}, \qquad n \to \infty$$

Problems for Section 6.2

1. Use Lemma 6.2.1 to obtain the first two terms of the asymptotic expansion of

   (a) $\displaystyle\int_1^4 e^{-kt} \sin t\, dt, \quad$ as $k \to \infty$ (b) $\displaystyle\int_5^9 e^{-kt} t^{-1}\, dt, \quad$ as $k \to \infty$

2. Use integration by parts to obtain the first two terms of the asymptotic expansion of

$$\int_1^\infty e^{-k(t^2+1)}\, dt$$

3. Use Watson's Lemma to obtain an infinite asymptotic expansion of

$$I(k) = \int_0^\pi e^{-kt} t^{-\frac{1}{3}} \cos t\, dt$$

   as $k \to \infty$.
   Note that

$$\cos t = \sum_{n=0}^\infty (-1)^n \frac{t^{2n}}{(2n)!} \quad \text{for } -\infty < t < \infty$$

4. Use Watson's lemma to find an infinite asymptotic expansion of

$$I(k) = \int_0^9 \frac{e^{-kt}}{1+t^4}\, dt$$

   as $k \to \infty$.

5. Use Laplace's method to determine the leading behavior (first term) of

$$I(k) = \int_{-\frac{1}{2}}^{\frac{1}{2}} e^{-k \sin^4 t}\, dt$$

   as $k \to \infty$.

6. Show that

$$\int_0^\infty \log\left(\frac{u}{1-e^{-u}}\right) \frac{e^{-ku}}{u}\, du \sim \frac{1}{2k}, \qquad k \to \infty$$

7. Show that for $0 < \alpha < 1$

$$\int_0^\infty \frac{e^{-kx}}{1+k^{\alpha x}x}\, dx \sim \frac{1}{k}, \qquad k \to \infty$$

8. Show that

$$\int_0^\infty \left(1+\frac{u}{k}\right)^{-k} e^{-u}\, du \sim \frac{1}{2} + \frac{1}{8k} - \frac{1}{8k^2}, \qquad k \to \infty$$

6.3 Fourier Type Integrals

In this section we study the asymptotic behavior as $k \to \infty$ of integrals of the form

$$I(k) = \int_a^b f(t)e^{ik\phi(t)}\, dt \qquad (6.3.1)$$

where $f(t)$ and $\phi(t)$ are real continuous functions. The assumption that $f(t)$ is real is actually without loss of generality because if it were complex, $f(t)$ could be decomposed into the sum of its real and imaginary parts. The more general case of $\phi(t)$ complex will be considered in Section 6.4. A special case of $I(k)$ when $\phi(t) = t$ is the Fourier transform, which is why such integrals are referred to as Fourier type integrals.

Our analysis of the above integrals will be similar to that of Laplace type integrals: After identifying intuitively the region of the t domain that gives the dominant contribution to the large k behavior of $I(k)$, we will use a heuristic analysis to develop the relevant formulas; subsequently, the formal approach will be rigorously justified.

For a Laplace type integral it is clear that, because of the decay of the exponential factor in the integrand, its value tends to zero as $k \to \infty$. It turns out that the value of a Fourier type integral also tends to zero as $k \to \infty$. This is a consequence of the fact that as $k \to \infty$, the exponential factor oscillates rapidly and these oscillations are self-canceling. Indeed, there exists a rather general result called the Riemann–Lebesgue lemma that guarantees that $I(k) \to 0$ as $k \to \infty$, provided that $\int_a^b |f(t)|\, dt$ exists, that $\phi(t)$ is continuously differentiable, and that $\phi(t)$ is not constant on any subinterval of $[a, b]$ (see, e.g.,

Titchmarsh (1948)). A special case of this lemma implies that the Fourier transform (when $\phi(t) = t$) of a function $f(t)$ tends to zero as $k \to \infty$, provided that $\int_{-\infty}^{\infty} |f(t)| \, dt$ exists, that is, $f(t) \in L_1$.

Although the Riemann–Lebesgue lemma is very useful in proving certain estimates, it does not provide an evaluation of how fast $I(k)$ vanishes for large k. Computing the leading term of the asymptotic expansion of $I(k)$ demands better understanding. This usually requires knowledge of the zeroes of $\phi'(t)$. Suppose that $t = c$ is a point in (a, b) for which $\phi'(t)$ does not vanish. If Ω_c is a small neighborhood of c, then we expect that $I(k)$ can be approximated by $I_c(k) = f(c) \int_{\Omega_c} e^{ik\phi(t)} \, dt$. As $k \to \infty$, the rapid oscillation of $\exp(ik\phi)$ produces cancelations that, in turn, tend to decrease the value of $I_c(k)$. But if we assume that $\phi'(t)$ vanishes at $t = c$, then, even for large k, there exists a small neighborhood of c throughout which $k\phi$ does not change so rapidly. In this region, $\exp(ik\phi)$ oscillates less rapidly and less cancelation occurs. Thus we expect that the value of $I(k)$ for large k depends primarily on the behavior of f and ϕ near points for which $\phi'(t) = 0$. Such points are called, in calculus, stationary points. Furthermore, in many applications, ϕ has the physical interpretation of a phase. Thus the asymptotic method based on the above arguments is usually referred to as the method of stationary phase.

In analogy with Laplace type integrals we distinguish two cases. The case that ϕ is monotonic in $[a, b]$ is considered in Sections 6.3.1 and 6.3.2, while the case that $\phi'(t)$ vanishes in $[a, b]$ is considered in Section 6.3.3.

6.3.1 Integration by Parts

If $\phi(t)$ is monotonic and $\phi(t)$ and $f(t)$ are sufficiently smooth, an asymptotic expansion of Eq. (6.3.1) can be obtained by the integration by parts procedure.

Lemma 6.3.1 (Integration by Parts)   Consider the integral

$$I(k) = \int_a^b f(t) e^{ikt} \, dt \qquad (6.3.2)$$

where the interval $[a, b]$ is a finite segment of the real axis. Let $f^{(m)}(t)$ denote the mth derivative of $f(t)$. Assume that $f(t)$ has $N + 1$ continuous derivatives and that $f^{(N+2)}$ is piecewise continuous on $[a, b]$. Then

$$I(k) \sim \sum_{n=0}^{N} \frac{(-1)^n}{(ik)^{n+1}} \left[f^{(n)}(b) e^{ikb} - f^{(n)}(a) e^{ika} \right], \qquad k \to \infty \qquad (6.3.3)$$

The proof of this result is essentially the same as that of Lemma 6.2.1 and it is therefore omitted.

We note that in the case $N = 0$, Eq. (6.3.3) reduces to

$$I(k) \sim \frac{1}{ik}\left[f(b)e^{ikb} - f(a)e^{ika}\right], \qquad k \to \infty$$

This is a weak version of the Riemann–Lebesgue lemma for Fourier transform type integrals.

Equation Eq. (6.3.3) can be generalized for integrals of the type (6.3.1) by transforming to a new coordinate $\tau = \phi(t)$ provided that ϕ is monotonic and differentiable $(\phi'(t) \neq 0)$.

Example 6.3.1 Evaluate $I(k) = \int_0^1 \frac{e^{ikt}}{1+t}\,dt$ as $k \to \infty$.
Using integration by parts we find

$$I(k) \sim e^{ik}\left[\frac{1}{2(ik)} + \frac{1}{2^2(ik)^2} + \cdots + \frac{(n-1)!}{2^n(ik)^n} + \cdots\right]$$

$$-\left[\frac{1}{(ik)} + \frac{1}{(ik)^2} + \cdots \frac{(n-1)!}{(ik)^n} + \cdots\right], \qquad k \to \infty$$

6.3.2 *Analog of Watson's Lemma*

We recall that in Section 6.2 we made use of Watson's Lemma in two important ways: (a) to prove the asymptotic nature of the Laplace's method; and (b) to compute Laplace type integrals when $\phi(t)$ is monotonic in $[a, b]$ but $f(t)$ is not sufficiently smooth at $t = a$. In this section we shall use, in a similar manner, a certain lemma that is an analog of Watson's Lemma for Fourier type integrals. Alternatively, it can be thought of as an extension of the Riemann–Lebesgue Lemma. In deriving this lemma we will need to compute the integral $\int_0^\infty t^\gamma \exp(i\mu t)\,dt$, where γ and μ are real, $\gamma > -1$. This type of integral appears often in connection with the asymptotic analysis of Fourier type integrals, so we first present a method for computing them.

Example 6.3.2 Show that

$$I = \int_0^\infty t^{\gamma-1}e^{it}\,dt = e^{\frac{\gamma\pi i}{2}}\Gamma(\gamma), \quad \gamma \text{ real}, \quad 0 < \gamma < 1 \qquad (6.3.4)$$

The technique of computing integrals like Eq. (6.3.4) is to rotate the contour of integration so that the relevant integral can be related to a gamma function (see also Section 4.2). In Example 6.3.2 we rotate the path of integration from the real axis to the imaginary axis in the upper half plane. Using Cauchy's Residue Theorem, noting that there are no singularities in the first quadrant, the above integral becomes

$$I = \int_0^{e^{i\pi/2}\infty} z^{\gamma-1} e^{iz}\, dz = \int_0^{\infty} (e^{i\pi/2} r)^{\gamma-1} e^{-r} i\, dr$$

$$= (e^{i\pi/2})^{\gamma} \int_0^{\infty} r^{\gamma-1} e^{-r}\, dr = e^{\frac{\gamma\pi i}{2}} \Gamma(\gamma)$$

where we have used $z = e^{i\pi/2} r$.

Example 6.3.3 Show that

$$I = \int_0^{\infty} t^{\gamma} e^{i\nu t^p}\, dt = \left(\frac{1}{|\nu|}\right)^{\frac{\gamma+1}{p}} \frac{\Gamma\left(\frac{\gamma+1}{p}\right)}{p} e^{\frac{i\pi}{2p}(\gamma+1)\mathrm{sgn}\nu} \tag{6.3.5}$$

where γ and ν are real constants, $\gamma > -1$, and p is a positive integer.

We first consider the case $\nu > 0$, and as in Example 6.3.2 we use Cauchy's Residue Theorem to rotate the contour of integration in the upper half plane from the real t axis to the ray along angle $\frac{\pi}{2p}$. Then

$$I = e^{\frac{i\pi(\gamma+1)}{2p}} \int_0^{\infty} r^{\gamma} e^{-\nu r^p}\, dr = \frac{e^{i\pi \frac{(\gamma+1)}{2p}}}{p(\nu)^{\frac{\gamma+1}{p}}} \int_0^{\infty} e^{-u} u^{\frac{\gamma+1}{p}-1}\, du$$

where we have used the change of variables $u = \nu r^p$. Using the definition of the gamma function the integral above becomes that of Eq. (6.3.5) for $\nu > 0$. Similarly, if $\nu < 0$ we rotate by an angle $-\pi/2p$.

Lemma 6.3.2 Consider the integral

$$I(k) = \int_0^b f(t) e^{ik\mu t}\, dt, \quad b > 0, \quad \mu = \pm 1, \quad k > 0 \tag{6.3.6}$$

Suppose that $f(t)$ vanishes infinitely smoothly at $t = b$ (i.e., $f(t)$ and all its derivatives vanish at $t = b$) and that $f(t)$ and all its derivatives exist in $(0, b]$. Furthermore, assume that

$$f(t) \sim t^{\gamma} + o(t^{\gamma}), \quad \text{as } t \to 0^+, \quad \gamma \in \mathbb{R} \quad \text{and} \quad \gamma > -1$$

Then

$$I(k) = \left(\frac{1}{k}\right)^{\gamma+1} \Gamma(\gamma + 1) e^{\frac{i\pi}{2}(\gamma+1)\mu} + o\left(k^{-(\gamma+1)}\right), \quad k \to \infty \tag{6.3.7}$$

Proof We will not go into the details of the estimates leading to Eq. (6.3.7). Here we make the (natural) assumption that the leading term of the asymptotic expansion of $I(k)$ comes from the neighborhood of $t = 0$; a rigorous justification of this fact (which is based upon integration by parts where $u = f(t)/t^\gamma$ and $dv = t^\gamma e^{ik\mu t} dt$) is given in Erdelyi (1956). $I(k)$ is approximated by

$$I(k) \sim \int_0^\infty t^\gamma e^{ik\mu t} dt = \left(\frac{1}{k}\right)^{\gamma+1} \Gamma(\gamma+1) e^{\frac{i\pi}{2}(\gamma+1)\mu}, \qquad k \to \infty$$

where we have used Eq. (6.3.5) with $p = 1, \nu = k\mu$. ∎

6.3.3 The Stationary Phase Method

A heuristic analysis of the leading term of the asymptotic expansion of Fourier type integrals closely follows Laplace's method. Indeed, suppose that f is continuous, ϕ is twice differentiable, ϕ' vanishes in $[a, b]$ only at the point $t = c$, and $\phi''(c) \neq 0$. Then we expect that the large k behavior of Eq. (6.3.1) is given by

$$\int_{c-R}^{c+R} f(c) \exp\left\{ik\left[\phi(c) + \frac{(t-c)^2}{2}\phi''(c)\right]\right\} dt$$

where R is small but finite. To evaluate this integral, we let

$$\mu\tau^2 = (t-c)^2 \frac{\phi''(c)}{2}k, \quad \text{or} \quad \tau = (t-c)\sqrt{\frac{|\phi''(c)|k}{2}}$$

where $\mu = \text{sgn}\,\phi''(c)$. Then the above integral becomes

$$f(c)e^{ik\phi(c)}\sqrt{\frac{2}{|\phi''(c)|k}} \int_{-R\sqrt{k|\phi''(c)|/2}}^{R\sqrt{k|\phi''(c)|/2}} \exp[i\mu\tau^2] d\tau$$

As $k \to \infty$ the last integral reduces to $\int_{-\infty}^\infty \exp(i\mu\tau^2) d\tau$, which can be evaluated exactly (see also Eq. (4.2.14) and above)

$$\int_{-\infty}^\infty e^{i\mu\tau^2} d\tau = 2 \int_0^\infty e^{i\mu\tau^2} d\tau = \sqrt{\pi} e^{\frac{i\pi\mu}{4}}$$

where we have used Eq. (6.3.5) with $\gamma = 0$, $p = 2$, and $\nu = \mu$ (recall that $\Gamma(1/2) = \sqrt{\pi}$). Hence our formal analysis suggests that

$$\int_a^b f(t)e^{ik\phi(t)} dt \sim e^{ik\phi(c)} f(c) \sqrt{\frac{2\pi}{k|\phi''(c)|}} e^{\frac{i\pi\mu}{4}}, \qquad \mu = \text{sgn}\phi''(c) \quad (6.3.8)$$

The asymptotic nature of Eq. (6.3.8) is proven below using Lemma 6.3.2. Actually Eq. (6.3.8) can be generalized because Lemma 6.3.2 allows $f(t)$ to be singular at $t = c$. On the other hand, in order to simplify the rigorous considerations, we assume that f and ϕ are infinitely smooth (i.e., all their derivatives exist) on the half-open interval not containing c.

Lemma 6.3.3 (Stationary Phase Method) Consider the integral Eq. (6.3.1) and assume that $t = c$ is the only point in $[a, b]$ where $\phi'(t)$ vanishes. Also assume that $f(t)$ vanishes infinitely smoothly at the two end points $t = a$ and $t = b$, and that both f and ϕ are infinitely differentiable on the half-open intervals $[a, c)$ and $(c, b]$. Furthermore assume that

$$\phi(t) - \phi(c) \sim \alpha(t - c)^2 + o((t - c)^2), \quad f(t) \sim \beta(t - c)^\gamma + o((t - c)^\gamma)$$

$$t \to c; \qquad \gamma > -1 \tag{6.3.9}$$

Then

$$\int_a^b f(t)e^{ik\phi(t)}\,dt \sim e^{ik\phi(c)}\beta\Gamma\left(\frac{\gamma + 1}{2}\right)e^{i\pi\frac{(\gamma+1)}{4}\mu}\left(\frac{1}{k|\alpha|}\right)^{\frac{\gamma+1}{2}}$$

$$+ o\left(k^{-\frac{(\gamma+1)}{2}}\right), \quad k \to \infty, \tag{6.3.10}$$

where $\mu = \operatorname{sgn}\alpha$.

Proof As in the case of Laplace's method we split $[a, b]$ into two half intervals $[a, c)$ and $(c, b]$; let I_a and I_b be the corresponding integrals. The purpose of assuming that $f(t)$ vanishes infinitely smoothly at the end points is to localize the main contribution of the integral to be at the stationary point. To analyze I_b, we let

$$\mu u = \phi(t) - \phi(c), \qquad \mu = \operatorname{sgn}\alpha \tag{6.3.11}$$

thus we can invert Eq. (6.3.11) to obtain $t = t(u)$; hence

$$I_b(k) = e^{ik\phi(c)}\int_0^{|\phi(b)-\phi(c)|} F(u)e^{ik\mu u}\,du, \qquad F(u) = \frac{\mu f(t)}{\phi'(t)} \tag{6.3.12}$$

To evaluate I_b, we need to determine $F(u)$ as $u \to 0^+$. To achieve this, we use the inversion of Eq. (6.3.11) in the neighborhood of $t = c$ (any ambiguity arising in this inversion is resolved by requiring that u increase as t increases);

that is, as $t \to c$

$$\mu u = \phi(t) - \phi(c) \sim \alpha(t-c)^2 \quad \text{or} \quad t - c \sim \left(\frac{u}{|\alpha|}\right)^{1/2} \qquad (6.3.13)$$

Using Eq. (6.3.13), the definition of $F(u)$, and Eq. (6.3.9), we find

$$F(u) \sim \frac{\mu\beta(t-c)^\gamma}{2\alpha(t-c)} = \frac{\beta}{2|\alpha|}(t-c)^{\gamma-1} \sim \frac{\beta u^{\frac{\gamma-1}{2}}}{2|\alpha|^{\frac{\gamma+1}{2}}}, \qquad u \to 0^+ \qquad (6.3.14a)$$

Hence for the integral $I_b(k)$ we know that $F(u)$ vanishes infinitely smoothly as $u \to |\phi(b) - \phi(c)|$, and that its asymptotic behavior as $u \to 0^+$ is given by Eq. (6.3.14a). Hence Lemma 6.3.2 implies that

$$I_b(k) = \frac{1}{2}e^{ik\phi(c)}\beta\Gamma\left(\frac{\gamma+1}{2}\right)e^{i\pi\frac{(\gamma+1)}{4}\mu}\left(\frac{1}{k|\alpha|}\right)^{\frac{\gamma+1}{2}} + o\left(k^{-\left(\frac{\gamma+1}{2}\right)}\right) \qquad (6.3.14b)$$

The leading term of the expansion of I_a is obtained in a completely analogous manner and is the same as (6.3.14b). Adding I_b and I_a we find Eq. (6.3.10). ∎

We note that in the special case when $\gamma = 0$, $\beta = f(c)$, and $\alpha = \phi''(c)/2$, Eq. (6.3.10) reduces to Eq. (6.3.8).

The result of Lemma 6.3.3 can be generalized substantially. It is possible to allow $\phi(t)$ and $f(t)$ to have different asymptotic behaviors as $t \to c^+$ and $t \to c^-$. Using the same ideas as in the above lemma for the integral I_b, if

$$\phi(t) - \phi(c) \sim \alpha_+(t-c)^\nu, \quad f(t) \sim \beta_+(t-c)^\gamma, \quad \text{as } t \to c^+, \quad \gamma > -1 \qquad (6.3.15a)$$

then

$$I_b(k) \sim \frac{1}{\nu}e^{ik\phi(c)}\beta_+\Gamma\left(\frac{\gamma+1}{\nu}\right)e^{i\pi\frac{(\gamma+1)}{2\nu}\mu}\left(\frac{1}{k|\alpha_+|}\right)^{\frac{\gamma+1}{\nu}}, \qquad \mu = \operatorname{sgn}\alpha_+ \qquad (6.3.15b)$$

The integral I_a follows analogously with α_+ and β_+ replaced by α_- and β_- in the expansions of $\phi(t) - \phi(c)$ and $f(t)$ as $t \to c^-$.

There is an important difference between Fourier and Laplace type integrals. For Fourier type integrals, although the stationary points give the dominant contribution, we must also consider the endpoints if more than the leading term is needed. The endpoint contribution is only algebraically smaller than the stationary point contribution. In contrast, we recall that for the Laplace type integrals we have considered so far, the entire asymptotic expansion depends only on the behavior of the integrand in a small neighborhood of the global minimum of ϕ; the points away from the minimum are exponentially small in comparison. Generalizations of Laplace's method to cases where the phase $\phi(t)$ is complex are discussed in Section 6.4.

Example 6.3.4 Evaluate $I(k) = \int_0^{\pi/2} e^{ik\cos t} \, dt$ as $k \to \infty$.

The function $\phi'(t) = -\sin t$ vanishes at $t = 0$ and $\phi''(0) = -1$. Because $t = 0$ is an end point, $I(k)$ is one-half of the value obtained from Eq. (6.3.8), where $\mu = -1$, $f(t) = 1$ (in particular, $f(0) = 1$), $\phi(0) = 1$, and $\phi''(0) = -1$

$$I(k) \sim \sqrt{\frac{\pi}{2k}} e^{i\left(k - \frac{\pi}{4}\right)}$$

Example 6.3.5 Show that if $\phi'(a) = \cdots = \phi^{(p-1)}(a) = 0$, $\phi^{(p)}(a) \neq 0$ if $\phi'(t) \neq 0$ for all t in $(a, b]$, and if $f(t)$ is sufficiently smooth, then, as $k \to +\infty$

$$\int_a^b f(t)e^{ik\phi(t)} \, dt \sim f(a)e^{ik\phi(a)} \left(\frac{p!}{k|\phi^{(p)}(a)|} \right)^{\frac{1}{p}} \frac{\Gamma(1/p)}{p} e^{\frac{i\pi}{2p}\mu},$$

$$\mu = \operatorname{sgn}\phi^{(p)}(a) \qquad\qquad (6.3.16)$$

Equation (6.3.16) is a special case of Eq. (6.3.15b) where

$$\alpha_+ = \frac{\phi^{(p)}}{p!}, \qquad \nu = p, \qquad \gamma = 0, \qquad \beta_+ = f(a), \qquad c = a$$

Equation (6.3.16) can also be derived formally from first principles; expanding $f(t)$ and $\phi(t)$ near $t = a$

$$I(k) \sim f(a)e^{ik\phi(a)} \int_a^\infty e^{ik\frac{\phi^{(p)}(a)}{p!}(t-a)^p} \, dt \sim f(a)e^{ik\phi(a)} \int_0^\infty e^{ik\frac{\phi^{(p)}(a)}{p!}s^p} \, ds$$

where we have replaced the upper limit of integration by ∞, because using integration by parts, the error introduced is $O\left(\frac{1}{k}\right)$. But Eq. (6.3.5) implies that

$$\int_0^\infty e^{ik\frac{\phi^{(p)}(a)}{p!}s^p} \, ds = \left(\frac{p!}{k|\phi^{(p)}(a)|} \right)^{\frac{1}{p}} \frac{\Gamma\left(\frac{1}{p}\right)}{p} e^{\frac{i\pi}{2p}\mu}, \qquad \mu = \operatorname{sgn}\phi^{(p)}(a)$$

thus Eq. (6.3.16) follows.

Example 6.3.6 Evaluate the leading behavior of $J_n(n)$ as $n \to \infty$, where the Bessel function $J_n(x)$ is given by $J_n(x) = \frac{1}{\pi} \int_0^\pi \cos(x \sin t - nt) \, dt$.
We rewrite $J_n(n)$ in the form

$$J_n(n) = \frac{1}{\pi} \, \mathrm{Re} \int_0^\pi e^{in(\sin t - t)} \, dt \sim \frac{1}{\pi} \, \mathrm{Re} \int_0^\infty e^{-int^3/6} \, dt$$

$$= \frac{1}{3\pi} \left(\cos \frac{\pi}{6} \right) \Gamma \left(\frac{1}{3} \right) \left(\frac{6}{n} \right)^{\frac{1}{3}}, \qquad n \to \infty$$

where we have used Eq. 6.3.5, with $\gamma = 0$, $\nu = -n/6$, and $p = 3$.

Problems for Section 6.3

1. Use integration by parts to obtain the asymptotic expansions as $k \to \infty$ of the following integrals up to order $\frac{1}{k^2}$

$$\text{(a)} \int_0^2 (\sin t + t) e^{ikt} \, dt \qquad \text{(b)} \int_0^\infty \frac{e^{ikt}}{1+t^2} \, dt$$

2. Consider

$$\int_0^1 \sqrt{t} e^{ikt} dt \text{ as } k \to \infty$$

(a) Show that

$$\int_0^1 \sqrt{t} e^{ikt} \, dt = -\frac{i}{k} e^{ik} + \frac{i}{2k} \int_0^1 \frac{e^{ikt}}{\sqrt{t}} \, dt$$

(b) Show the leading order term of

$$\int_0^1 \frac{e^{ikt}}{\sqrt{t}} \, dt = \frac{1}{\sqrt{k}} \int_0^k \frac{e^{is}}{\sqrt{s}} \, ds \equiv I$$

satisfies

$$I \sim \frac{\sqrt{\pi}}{k} e^{i\pi/4}$$

(c) Deduce that

$$\int_0^1 \sqrt{t} e^{ikt} \, dt \sim \frac{-i}{k} + \frac{\sqrt{\pi} e^{3i\pi/4}}{2k^{3/2}}$$

as $k \to \infty$.

3. Use the method of stationary phase to find the leading behavior of the following integrals as $k \to \infty$:

$$\text{(a) } \int_0^1 \tan(t)e^{ikt^4}\, dt \qquad \text{(b) } \int_{\frac{1}{2}}^2 (1+t)e^{ik\left(\frac{t^3}{3}-t\right)}\, dt$$

4. Show that the asymptotic expansion of

$$I(k) = \int_{-\infty}^{\infty} \frac{e^{ikt^2}}{1+t^2}\, dt, \qquad \text{as } k \to \infty$$

is given by

$$I(k) \sim \sum_{n=0}^{\infty} \left(\frac{1}{k}\right)^{n+\frac{1}{2}} (-1)^n \Gamma\left(n+\frac{1}{2}\right) e^{i\frac{\pi}{2}\left(n+\frac{1}{2}\right)}$$

5. Show that the integral

$$I(k) = \int_{-\infty}^{\infty} \frac{e^{ikx}}{1+x^2}\, dx$$

has a vanishing power series, i.e., $I(k) = o\left(\frac{1}{k^n}\right)$, for all n, as $k \to \infty$. Use contour integration to establish that $I(k) = \pi e^{-k}$.

6. (a) Show that the integral $I(k) = \int_{-\infty}^{\infty} \frac{\cos kx}{\cosh x}\, dx$ has a vanishing power series, $I(k) = 0\left(\frac{1}{k^n}\right)$, for all n, as $k \to \infty$. Use contour integration to establish that $I(k) = \frac{\pi}{2}\operatorname{sech}\frac{k\pi}{2} \sim \pi e^{-k\pi/2}$.

   (b) The methods of Section 6.4 can sometimes be used to find exponentially small contributions located at poles of the integrand. The interested reader should establish that

$$\int_{-\infty}^{\infty} \frac{\cos xk}{\cosh x\sqrt{x^2+\pi^2}}\, dx \sim \frac{4}{\sqrt{3}} e^{-\frac{\pi k}{2}}$$

   by reading ahead and using the methods of Section 6.4.

6.4 The Method of Steepest Descent

The method of steepest descent is a powerful approach for studying the large k asymptotics of integrals of the form

$$I(k) = \int_C f(z)e^{k\phi(z)}\, dz \qquad (6.4.1)$$

where C is a contour in the complex z plane and $f(z)$ and $\phi(z)$ are analytic functions of z. The basic idea of the method is to utilize the analyticity of the integrand to justify deforming the contour C to a new contour C' on which $\phi(z)$ has a constant imaginary part. Thus if $\phi(z) = u + iv$, the integral $I(k)$ becomes ($\text{Im}\phi = v = \text{constant}$)

$$I(k) = e^{ikv} \int_{C'} f(z)e^{ku(z)} dz \tag{6.4.2}$$

Although z is complex, $u(z)$ is real and hence the ideas used in connection with Laplace type integrals can be used to study Eq. (6.4.2). In this sense the steepest descent method is an extension of Laplace's method to integrals in the complex plane. We note that, if $f(z)$ and $\phi(z)$ have singularities such as poles, important contributions can arise in the deformation process.

It turns out that paths on which v is constant are also paths for which either the decrease of u is maximal (paths of steepest descent) or the increase of u is maximal (paths of steepest ascent). In evaluating $I(k)$ we will use the former paths, which is why this method is called the method of steepest descent. Also, usually the paths of steepest descent will go through a point z_0 for which $\phi'(z_0) = 0$. Such a point is called a **saddle point** (for reasons that will be made clear below), and the method is alternatively referred to as the saddle point method.

We note that we could consider deforming C into a path for which u rather than v is constant (so that v varies rapidly) and then apply an extension of the method of stationary phase. However, we expect intuitively that the self-canceling of oscillations is a weaker decay mechanism than the exponential decay of the exponential factor in the integrand. This is indeed true, which is why the asymptotic expansion of a generalized Laplace integral can be found locally. Namely, it depends only on a small neighborhood of certain points which are called **critical points**. There are points where $\phi'(z) = 0$, singular points of the integrand, and endpoints. By summing up one contributions from all the critical points it is possible to obtained an infinite asymptotic expansion of the generalized Laplace integral. By contrast, in general, without deformations to a Laplace type integral, only the leading term of the asymptotic expansion of a generalized Fourier integral can be found from purely local considerations.

In order to develop the method of steepest descent, we need to better understand the relationship between steepest paths and v equal constant and to investigate the direction of these paths around saddle points.

Steepest Paths Let $\phi(z) = u(z) + iv(z)$, $z = x + iy$. Consider a point z_0 in the complex z plane. A direction away from z_0 in which u is decreasing is called a direction of descent. The direction on which the decrease is maximal is called direction of steepest descent. Similar considerations apply for directions of ascent. We recall that if $f(x, y)$ is a differentiable function of two variables, then the gradient of f is the vector $\nabla f = (\partial f/\partial x, \partial f/\partial y)$. This vector points in the direction of the most rapid change of f at the point (x, y). Thus for any point $z_0 = x_0 + i y_0$ corresponding to $u(z_0) = u(x_0, y_0)$ at which $\nabla u \neq 0$, the direction of steepest ascent coincides with that of ∇u, with u increasing away from $u(z_0)$, while the direction of steepest descent coincides with that of $-\nabla u$, with u decreasing away from $u(z_0)$.

The curves of steepest descent and of steepest ascent associated with a point $z_0 = x_0 + i y_0$ are given by $v(x, y) = v(x_0, y_0)$. Indeed, to show that the curves defined by $v(x, y) = v(x_0, y_0)$ are curves of steepest descent or steepest ascent, we note that the direction normal to such curves is $\nabla v = (\partial v/\partial x, \partial v/\partial y)$, which, using the Cauchy–Riemann equations, equals $(-\partial u/\partial y, \partial u/\partial x)$; hence the direction tangent to such curves is $\nabla u = (\partial u/\partial x, \partial u/\partial y)$. Hence from the vector argument above, $v(x, y) = v(x_0, y_0)$ correspond to the directions of the steepest curves. To show that the steepest curves are $v(x, y) = v(x_0, y_0)$, we define $\delta\phi = \phi(z) - \phi(z_0) = \delta u + i\delta v$. Then $|\delta u| \leq |\delta\phi|$ and equality is achieved; that is, δu is maximal only if $\delta v = 0$, or $v(x, y) = v(x_0, y_0)$.

In practice, we need to establish that the contour can be deformed onto the steepest descent curve, which passes through the saddle point. This requires some global understanding of the geometry. Usually the contribution near the saddle point gives the dominant contribution. However, sometimes the contour cannot be deformed onto a curve passing through a saddle point. In this case, endpoints or singularities of the integrand yield the dominant contribution. Moreover, sometimes the deformation process introduces poles that can lead to significant (and possibly dominant) contributions to the asymptotic expansion.

Saddle Points We say that the point z_0 is a saddle point of order N if the first N derivatives vanish, or alternatively, letting $n = N + 1$

$$\frac{d^m\phi}{dz^m}\bigg|_{z=z_0} = 0, \quad m = 1, \ldots, n-1, \quad \frac{d^n\phi}{dz^n}\bigg|_{z=z_0} = ae^{i\alpha}, \qquad a > 0 \quad (6.4.3)$$

When only the first derivative vanishes $(N = 1)$, we simply say that z_0 is a saddle point, or a "simple" saddle point, and omit the phrase "of order one."

If the first $n - 1$ derivatives vanish, then we will show that for such points there exist n directions of steepest descent and n directions of steepest ascent. If

$$z - z_0 = \rho e^{i\theta} \tag{6.4.4}$$

these directions are given by

$$\text{steepest descent directions: } \theta = -\frac{\alpha}{n} + (2m+1)\frac{\pi}{n}, \qquad m = 0, 1, \ldots, n-1$$

$$\text{steepest ascent directions: } \theta = -\frac{\alpha}{n} + 2m\frac{\pi}{n}, \qquad m = 0, 1, \ldots, n-1$$

$$(6.4.5)$$

where α is defined in Eq. (6.4.3). We note that in these equations we have arbitrarily chosen the integers to be $0, 1, \ldots, n-1$. However, negative values of m (i.e., $m = -1, -2, \ldots, -(n-1)$) might be necessary in dealing with problems where a branch cut is fixed (see Example 6.4.10, part (b)).

To derive Eqs. (6.4.5) we note that

$$\phi(z) - \phi(z_0) \sim \frac{(z-z_0)^n}{n!} \frac{d^n\phi}{dz^n}\bigg|_{z=z_0}$$

$$= \frac{\rho^n e^{in\theta}}{n!} a e^{i\alpha} = \frac{\rho^n a}{n!}[\cos(\alpha + n\theta) + i\sin(\alpha + n\theta)]$$

$$(6.4.6)$$

Since the directions of steepest descent at $z = z_0$ are defined by $\text{Im}(\phi(z) - \phi(z_0)) = 0$ it follows that $\sin(\alpha + n\theta) = 0$, and for u to decrease away from z_0, $\cos(\alpha + n\theta) < 0$. Similarly, the directions of steepest ascent are given by $\sin(\alpha + n\theta) = 0$, $\cos(\alpha + n\theta) > 0$. These relationships are equivalent to Eqs. (6.4.5). This local information is very important in applications, as we will see below.

For $n = 2$, that is, for a simple saddle point, Eqs. (6.4.5) imply

$$\text{descent: } \theta = -\frac{\alpha}{2} + \frac{\pi}{2}, \qquad \theta = -\frac{\alpha}{2} + \frac{3\pi}{2}$$

$$\text{ascent: } \theta = -\frac{\alpha}{2}, \qquad \theta = -\frac{\alpha}{2} + \pi$$

$$(6.4.7)$$

We shall use an arrow to point to the direction of descent. Equations (6.4.7) are depicted in Figure 6.4.1a. The shaded regions represent valleys where the function decreases from the saddle point.

Figure 6.4.1b depicts a typical surface $u(x, y)$ about the point (x_0, y_0) for the case $n = 2$. Locally, the surface is shaped like saddle, which is why a point z_0 at which $d\phi/dz = 0$ is called a (simple) saddle point. The steepest descent path is distinguished by the fact that $\text{Im }\phi = \text{const}$ along this path. If we pick an adjacent path, the real part of ϕ is larger, but the contribution to the integral is smaller owing to the rapid oscillations in the imaginary part of ϕ. Equations (6.4.5) are further illustrated in Example 6.4.1.

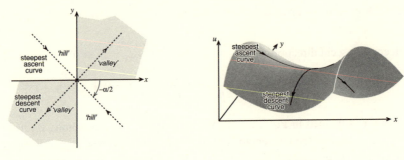

(a) directions of ascent and descent   (b) saddle surface

Fig. 6.4.1. Steepest descent

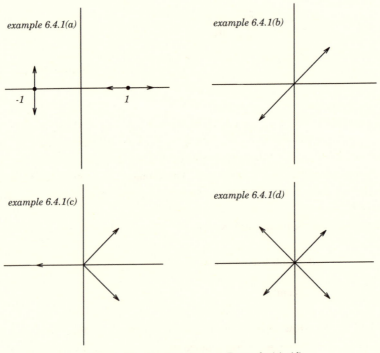

Fig. 6.4.2. Steepest descent, Example (a)–(d)

Example 6.4.1 Find the directions of steepest descent at the saddle point for the functions $\phi(z)$ given below and the points z_0 indicated.

(a) $\phi(z) = z - z^3/3$. Therefore $\phi'(z) = 1 - z^2$, $\phi'' = -2z$. There exist two simple saddle points at $z_0 = 1$ and $z_0 = -1$.

For $z_0 = 1$, $\phi''(1) = -2 = 2e^{i\pi}$, that is, $\alpha = \pi$. Hence Eqs. (6.4.7) imply $\theta = 0, \pi$.

For $z_0 = -1$, $\phi''(-1) = 2$, that is, $\alpha = 0$. Hence Eqs. (6.4.7) imply $\theta = \pi/2, 3\pi/2$.

(b) $\phi(z) = i \cosh z$. Therefore $\phi'(z) = i \sinh z$, $\phi''(z) = i \cosh z$. Consider $z_0 = 0$:

$\phi''(0) = i = e^{\frac{i\pi}{2}}$, that is, $\alpha = \pi/2$. Hence Eq. (6.4.7) imply $\theta = \pi/4$, $5\pi/4$.

(c) $\phi(z) = \sinh z - z$. Therefore $\phi'(z) = \cosh z - 1$, $\phi''(z) = \sinh z$, $\phi'''(z) = \cosh z$.

Consider $z_0 = 0$: $\phi'(0) = \phi''(0) = 0$, $\phi'''(0) = 1$, that is, $\alpha = 0, n = 3$. This is a saddle point of order two; hence Eqs. (6.4.5) imply $\theta = \pi/3, \pi, 5\pi/3$.

(d) $\phi(z) = \cosh z - (z^2/2)$. Therefore $\phi'(z) = \sinh z - z$, $\phi''(z) = \cosh z - 1$, $\phi'''(z) = \sinh z$, $\phi''''(z) = \cosh z$.

Consider $z_0 = 0$: $\phi'(0) = \phi''(0) = \phi'''(0) = 0$, $\phi''''(0) = 1$, that is, $\alpha = 0$, $n = 4$. This is a saddle point of order 3. Hence Eqs. (6.4.5) imply $\theta = \frac{\pi}{4}, \frac{3\pi}{4}, \frac{5\pi}{4}, \frac{7\pi}{4}$.

6.4.1 Laplace's Method for Complex Contours

We consider the integral (6.4.1) and we assume that the contour C can be deformed into a contour C_s through the saddle point of order $n - 1$. Let us consider a portion of the dominant contribution; that is, consider a single path of steepest descent originating from a saddle point z_0 of order $n - 1$. We also will assume that $f(z)$ is of order $(z - z_0)^{\beta-1}$ near z_0, that is, as $z \to z_0$

$$\phi(z) - \phi(z_0) \sim \frac{(z - z_0)^n}{n!} \phi^{(n)}(z_0), \qquad \phi^{(n)}(z_0) = |\phi^{(n)}(z_0)|e^{i\alpha}$$

$$f(z) \sim f_0(z - z_0)^{\beta-1}, \qquad \operatorname{Re} \beta > 0 \qquad (6.4.8)$$

Then we shall show that

$$I(k) \sim \frac{f_0(n!)^{\frac{\beta}{n}} e^{i\beta\theta}}{n} \frac{e^{k\phi(z_0)} \Gamma\left(\frac{\beta}{n}\right)}{(k|\phi^{(n)}(z_0)|)^{\frac{\beta}{n}}} \qquad (6.4.9)$$

To derive Eq. (6.4.9), we note that because C' is a path of steepest descent, we can make the change of variables, $-t = \phi(z) - \phi(z_0)$, where t is real and positive. Recall that along the steepest descent path, $\text{Im}(\phi(z) - \phi(z_0)) = 0$. In the neighborhood of the saddle point, using Eq. (6.4.5),

$$\frac{(z - z_0)^n}{n!}\phi^{(n)}(z_0) = -t, \quad \text{or} \quad |z - z_0| = t^{\frac{1}{n}}\left(\frac{n!}{|\phi^{(n)}(z_0)|}\right)^{\frac{1}{n}} \tag{6.4.10}$$

Then Eq. (6.4.1) yields

$$I(k) \sim e^{k\phi(z_0)}\int_0^\infty \left(-\frac{f(z)}{\phi'(z)}\right)e^{-kt}\,dt \tag{6.4.11}$$

where we used $\phi'(z)\,dz = -dt$ and we replaced the upper limit of integration by ∞ because from Watson's lemma we expect that the dominant contribution comes from the neighborhood of the origin. In order to evaluate Eq. (6.4.11), we need to compute $f(z)/\phi'(z)$ near z_0. Using Eq. (6.4.8)

$$\frac{f(z)}{\phi'(z)} \sim \frac{f_0(z - z_0)^{\beta-1}}{\frac{(z-z_0)^{n-1}}{(n-1)!}\phi^{(n)}(z_0)} = (n-1)!f_0\frac{(z - z_0)^{\beta-n}}{\phi^{(n)}(z_0)} \tag{6.4.12}$$

Also,

$$z - z_0 = |z - z_0|e^{i\theta} \quad \text{and} \quad \phi^{(n)}(z_0) = |\phi^{(n)}(z_0)|e^{i\alpha}$$

Using these equations and Eq. (6.4.10) we find

$$\frac{f(z)}{\phi'(z)} \sim \frac{f_0(n!)^{\frac{\beta}{n}}e^{i\beta\theta-i(\theta n+\alpha)}}{n|\phi^{(n)}(z_0)|^{\frac{\beta}{n}}}t^{\frac{\beta}{n}-1} \tag{6.4.13}$$

Recall that θ is given by Eq. (6.4.5); thus $\theta n + \alpha = (2m + 1)\pi$, that is, $\exp[-i(\theta n + \alpha)] = -1$. Substituting Eq. (6.4.13) in Eq. (6.4.11) and using the definition of the gamma function, we find Eq. (6.4.9).

In order to obtain the higher-order terms in the asymptotic expansion of Eq. (6.4.12), one must solve for $z - z_0$ in terms of a series in t from

$$-t = \phi(z) - \phi(z_0) = -(z - z_0)^n\hat{\phi}(z)$$

where $\hat{\phi}(z)$ is an analytic function of z in a neighborhood of z_0: $\hat{\phi}(z) = \sum_{m=0}^\infty \hat{\phi}_m$ $(z - z_0)^m$. There are n roots of this equation. A series expressing $z - z_0$ in terms of t, $z - z_0 = \sum_{m=1}^\infty c_m t^m$ ($t = 0$ when $z = z_0$) can be obtained by recursively solving $z - z_0 = t^{1/n}[\hat{\phi}(z)]^{1/n}$. This inversion can always be accomplished in principle, by the Implicit Function Theorem. In practice, the necessary devices to accomplish this are often direct and motivated by the particularities of the functions in question (see Example 6.4.8 below). We do not need to go into further details for our purposes.

The asymptotic nature of Eq. (6.4.9) can be rigorously established by solving for z in terms of t as mentioned above, and then using Watson's Lemma. This shows the great advantage of the method of steepest descent: Because it is based on Watson's Lemma, it is possible both to justify it rigorously and to obtain the asymptotic expansion to all orders. A difficulty encountered in practice is the deformation of the original contour of integration onto one or more of the paths of steepest descent. However, the local nature of the method of steepest descent makes even this task relatively simple. This is because quantitative information about the deformed contours is needed only near the critical points; away from these points qualitative information is sufficient (by critical points we mean saddle points, endpoints of integration, and singularities of $f(z)$ and $\phi(z)$). A contour C_1 that coincides with a steepest descent contour C_s for some finite length near the critical point z_0 but that then continues merely as a descent contour is said to be asymptotically equivalent to C_s. It can be shown rigorously (Bleistein and Handelsman, 1986; Olver, 1974) that asymptotic expansions derived from asymptotically equivalent contours differ only by an exponentially small quantity. Therefore, it is really these asymptotically equivalent contours and not the steepest descent curves that are important. One must establish that a given contour can be deformed onto such contours, in which case the problem then becomes local. It is not always easy to see how to deform the contours. Besides the examples provided here, we refer the reader to Bleistein and Handelsman (1986) for a more detailed discussion and further examples.

Steepest descents is one of the most widely used methods of complex variables. The essential idea of deforming a path onto a saddle point dates back to Riemann in his study of hypergeometric equations. Debye (1954) realized that by transforming the entire contour onto a steepest descent path, one could obtain an infinite asymptotic expansion.

Before going into examples let us review the basic steps.

(a) Identify all critical points of the integrand, that is, saddle points, endpoints, and possible singular points.

(b) Determine the paths of steepest descent, C_s. (We note that a detailed determination of C_s is usually not necessary, qualitative features usually are sufficient.)

(c) Deform the original contour C (using Cauchy's Theorem) onto one or more paths of steepest descent C_s or paths that are asymptotically equivalent to C_s. (One needs some global information about the steepest descent paths in order to justify the replacement of C by C_s or by those contours which are asymptotically equivalent to C_s. This step is very important and can be difficult.)

(d) Evaluate the asymptotic expansion by using Eq. (6.4.9), Watson's Lemma, integration by parts, etc. The integrals are of Laplace type.

Having presented the basic elements of the steepest descent method, we now consider several examples that illustrate the main features.

Example 6.4.2 Find the complete asymptotic expansion of $I(k) = \int_0^1 \log t \, e^{ikt} dt$, as $k \to \infty$.

A direct application of the method of stationary phase fails because there is no stationary point. Also, integration by parts fails because $\log t$ diverges at $t = 0$. We solve this problem by the method of steepest descent, which also shows how, by deforming the contour, a Fourier type integral can be mapped onto a Laplace type integral. Let us replace the real variable t by the complex variable z. Note that

$$\phi(z) = iz = i(x + iy) = -y + ix \qquad (6.4.14)$$

Clearly there does not exist any saddle point ($\phi'(z) = i$); we will see that the dominant contribution comes from the endpoints. The steepest paths Im $\phi =$ const are given by $x = $ constant; if $y > 0$, these paths are paths of steepest descent. Thus $x = 0$, $y > 0$ and $x = 1$, $y > 0$ are the paths of steepest descent going through the endpoints. We note that Im $\phi(0) \neq$ Im $\phi(1)$; hence there is no continuous contour joining $t = 0$ and $t = 1$ on which Im ϕ is constant. We connect the two steepest paths by the contour C_2 (see Figure 6.4.3) and use Cauchy's theorem to deform the contour $[0, 1]$. (Since $t = 0$ is an integrable singularity we shall allow the contour to pass through the origin.) Note that here, $i = e^{i\pi/2}$.

$$
\begin{aligned}
I(k) &= \int_{C_1 + C_2 + C_3} \log z \, e^{ikz} \, dz \\
&= i \int_0^R \log(ir) e^{-kr} \, dr + \int_0^1 \log(x + iR) e^{ikx - R} \, dx \\
&\quad - i e^{ik} \int_0^R \log(1 + ir)^{-kr} \, dr
\end{aligned}
$$

Letting $R \to \infty$ we obtain

$$I(k) = i \int_0^\infty \log(ir) e^{-kr} \, dr - i e^{ik} \int_0^\infty \log(1 + ir) e^{-kr} \, dr \qquad (6.4.15)$$

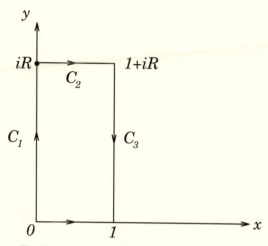

Fig. 6.4.3. Contour of integration for Example 6.4.2

Using $s = kr$, the first integral in Eq. (6.4.15) becomes

$$\frac{i}{k} \int_0^\infty \left(\log\left(\frac{i}{k}\right) + \log s \right) e^{-s} \, ds = -\frac{i \log k}{k} - \frac{\left(i\gamma + \frac{\pi}{2}\right)}{k}$$

where we used the fact that $-\gamma$ is the well-known integral $\int_0^\infty \log s \, e^{-s} ds$ (and the Euler constant, γ, equals $0.577216\ldots$). To compute the second integral in Eq. (6.4.15), we use the Taylor expansion $\log(1 + ir) = -\sum_{n=1}^\infty (-ir)^n/n$, and Watson's Lemma. Thus as $k \to \infty$, the complete asymptotic expansion of the second integral is: $ie^{ik} \sum_{n=1}^\infty (-i)^n (n-1)!/k^{n+1}$. Adding these two contributions we find

$$I(k) \sim -\frac{i \log k}{k} - \frac{i\gamma + \pi/2}{k} + ie^{ik} \sum_{n=1}^\infty \frac{(-i)^n (n-1)!}{k^{n+1}}, \qquad k \to \infty$$

Example 6.4.3 Find the asymptotic behavior of the **Hankel function**

$$H_\nu^{(1)}(k) = \frac{1}{\pi} \int_C e^{ik \cos z} e^{i\nu\left(z - \frac{\pi}{2}\right)} \, dz, \quad \text{as} \quad k \to \infty,$$

where the contour C is illustrated in Figure 6.4.4a. (The Hankel function of the first kind, $H_\nu^{(1)}(z)$, satisfies Bessel's equation (3.7.35a) with the asymptotic behavior

$$H_\nu^{(1)}(x) \sim \sqrt{\frac{2}{\pi x}} e^{i(x - \alpha_\nu)} \qquad \text{as } x \to +\infty$$

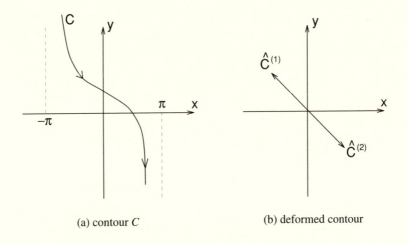

(a) contour C (b) deformed contour

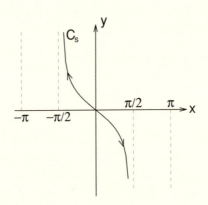

(c) steepest descent

Fig. 6.4.4. Hankel function

where $\alpha_\nu = (2\nu + 1)\frac{\pi}{4}$. The other linearly independent solution of Bessel's equation, $H_\nu^{(2)}(x)$, is the complex conjugate of $H_\nu^{(1)}(x)$, x real.)

In this example $\phi(z) = i \cos z$ and $\phi'(z) = -i \sin z$, thus $z = 0$ is a simple saddle point. In what follows we first obtain the leading-order approximation, then we show how to find higher-order approximations. Calculation of further approximations is usually algebraically tedious, hence in the remaining examples (other than Example 6.4.8) we only find the first approximation.

Because $\phi''(0) = -i = e^{-\frac{i\pi}{2}}$, using Eqs. (6.4.3)–(6.4.5), $\alpha = -\frac{\pi}{2}$ hence Eq. (6.4.7) implies

$$\theta = \frac{3\pi}{4}, \frac{3\pi}{4} + \pi$$

are the directions of steepest descent. We first consider the local deformed contour (depicted in Figure 6.4.4b) passing through the origin, which is as mentioned earlier, asymptotic to C. We use Eq. (6.4.9) with

$$z_0 = 0, \qquad n = 2, \qquad |\phi''(0)| = 1, \qquad \beta = 1,$$

$$f_0 = \frac{e^{-i\nu\frac{\pi}{2}}}{\pi}, \qquad \phi(0) = i.$$

Noting that $\int_C \sim \int_{\hat{C}^{(2)}} - \int_{\hat{C}^{(1)}}$, where the $\hat{C}^{(1)}$ contour has $\theta = \frac{7\pi}{4}$ and $\hat{C}^{(2)}$ has $\theta = \frac{3\pi}{4}$, we need to subtract the formulae obtained with $\theta = \frac{7\pi}{4}$ and $\theta = \frac{3\pi}{4}$, which gives the leading contribution as

$$H_\nu^{(1)}(k) \sim \sqrt{\frac{2}{\pi k}} e^{i\left(k - \frac{\nu\pi}{2} - \frac{\pi}{4}\right)}, \qquad k \to \infty$$

Next we outline the procedure to get higher-order asymptotic corrections to this integral. The steepest descent curves are given by $\mathrm{Im}\,\phi(z) = \mathrm{Im}\,\phi(0)$ or

$$\cos x \cosh y = 1$$

near the saddle point. Recall that for real y

$$\cos iy = \cosh y \qquad \qquad \cosh iy = \cos y$$
$$\sin iy = i \sinh y \qquad \text{and} \qquad \sinh iy = i \sin y.$$

We have

$$\left(1 - \frac{x^2}{2} + \cdots\right)\left(1 + \frac{y^2}{2} + \cdots\right) = 1$$

or

$$\frac{y^2}{2} - \frac{x^2}{2} \approx 0$$

with the steepest descent curve being $y = -x$ as depicted in Figure 6.4.4b. For $|y| \to \infty$ it is clear that $\cos x \sim e^{-|y|}/2$; hence $x \sim +\pi/2$ as $y \to -\infty$, and $x \sim -\pi/2$ as $y \to +\infty$. This qualitative information is sufficient to show us that we can deform our curve C onto a steepest descent curve C_s, as depicted in Figure 6.4.4c. (Note that the integral converges when $\mathrm{Re}\,\phi = \sinh y \sin x < 0$; that is, for $y \to \infty$, $-\pi < x < 0$, and for $y \to -\infty$, $0 < x < \pi$.)

Next we show how to work with the steepest descent transformation: $\phi(z) - \phi(z_0) = -t, t > 0, t$ real. We have

$$i(\cos z - 1) = -t$$

or for z near the saddle point, $z_0 = 0$, $t = 0$

$$\frac{z^2}{2!} - \frac{z^4}{4!} + \cdots = e^{-i\pi/2}t$$

We solve for z as a function of t iteratively:

$$z = \sqrt{2}e^{-i\pi/4}t^{1/2}\left(1 + \frac{1}{24}z^2 + \cdots\right)$$

or

$$z = \sqrt{2}e^{-i\pi/4}t^{1/2} + \frac{\sqrt{2}}{12}e^{-3i\pi/4}t^{3/2} + \cdots$$

Asymptotically the integral is given by

$$H_\nu^{(1)}(k) \sim \frac{1}{\pi}\int_{C_s} e^{ik}e^{-kt}e^{i\nu z(t)}e^{-i\nu\pi/2}\frac{dz}{dt}dt$$

$$\sim \frac{e^{ik}e^{-i\nu\pi/2}}{\pi}\int_{C_s} e^{-kt}\left(1 + i\nu z + \frac{(i\nu z)^2}{2!} + \frac{(i\nu z)^3}{3!} + \cdots\right)\frac{dz}{dt}dt$$

Using the expansion of $z(t)$ above and splitting the steepest descent contour into two pieces, one each from the origin, after some algebra we have

$$H_\nu^{(1)}(k) \sim \frac{2e^{ik}e^{-i\nu\pi/2}}{\pi}\int_0^\infty e^{-kt}\left(c_0 t^{-1/2} + c_1 + c_2 t^{1/2} + c_3 t + \cdots\right)dt$$

$$\sim \frac{2e^{ik}e^{-i\nu\pi/2}}{\pi}\left(\frac{c_0\Gamma\left(\frac{1}{2}\right)}{k^{1/2}} + \frac{c_1}{k} + \frac{c_2\Gamma\left(\frac{3}{2}\right)}{k^{3/2}} + \frac{c_3}{k^2} + \cdots\right)$$

$$\sim \frac{2e^{ik}e^{-i\nu\pi/2}}{\pi}\left(\frac{\sqrt{2\pi}}{2}\frac{e^{-i\pi/4}}{k^{1/2}} + \frac{\nu}{k}\right.$$

$$\left. + \frac{\sqrt{2\pi}}{4}\frac{\left(\frac{1}{4} - \nu^2\right)e^{-3i\pi/4}}{k^{3/2}} + \frac{i\nu(\nu^2 - 1)}{3k^2} + \cdots\right)$$

where we have inserted the requisite values for the coefficients c_0, c_1, c_2, c_3, ..., and used $\Gamma(\frac{1}{2}) = \sqrt{\pi}$ and $\Gamma(\frac{3}{2}) = \frac{1}{2}\Gamma(\frac{1}{2})$. The leading term agrees with that obtained above.

In practice it is advisable to go through the above calculation even for the first term, as opposed to using the formula (6.4.9). We use this formula in order to condense the description.

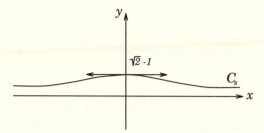

Fig. 6.4.5. Steepest descent contour C_s for example 6.4.4

Example 6.4.4 Evaluate $I(k) = \int_{-\infty}^{\infty} e^{ikt}(1+t^2)^{-k}, k \to \infty$.

We first rewrite this integral in the form Eq. (6.4.1): Using $(1+t^2)^{-k} = e^{-k\log(1+t^2)}$, its integrand becomes $e^{k[it-\log(1+t^2)]}$.

$$\phi(z) = iz - \log(1+z^2), \qquad \phi'(z) = i - \frac{2z}{1+z^2}$$

There exist two simple saddle points (see Figure 6.4.5) given by $z^2 + 2iz + 1 = 0$, or $z_0 = i(\pm\sqrt{2} - 1)$ We consider the saddle point $z_0 = ic, c = \sqrt{2} - 1$

$$\phi''(ic) = -\frac{(1+c^2)}{2c^2} = \frac{(1+c^2)}{2c^2} e^{i\pi}$$

that is, $\alpha = \pi$, (note that $1 - c^2 = 2c$ for $c = \sqrt{2} - 1$); hence Eqs. (6.4.7) imply $\theta = 0, \pi$ are the directions of steepest descent. Also, $\phi(ic) = -c - \log(2c)$. We deform the original contour C ($-\infty < x < \infty$) so that it passes through the point ic, which can be done because the integral converges for $\text{Im}\, t > 0$. We take the branch cut for $\log(1+z^2)$ to be from $\pm i$ to $\pm i\infty$; that is, not through the real z axis. The exact form of the deformed contour with this choice of branch for $\log(1+z^2)$ is given by $\text{Im}\,\phi(z) = \text{Im}\,\phi(ic) = 0$, or $x - \hat{\theta} = 0$, where $\hat{\theta} = \arg(1+z^2)$. The deformed contour C_s approaches the original contour as $|x| \to \infty$. To evaluate the leading order contribution of $I(k)$ on the deformed contour, we use Eq. (6.4.9) with

$$z_0 = ic, n = 2, |\phi''(ic)| = \frac{(1+c^2)}{2c^2}, \qquad f_0 = 1,$$
$$\beta = 1, \qquad \phi(ic) = -c - \log(2c)$$

Thus for the local steepest descent contribution we have, on the right half of C_s, $\theta = 0$ and on the left half of C_s, $\theta = \pi$. Using Eq. (6.4.9) with $\theta = 0$ and $\theta = \pi$ and subtracting the two contributions (which is necessary in order for the steepest descent curve to be consistent with the original contour) we find

$$I(k) \sim \frac{2c}{\sqrt{1+c^2}} \sqrt{\frac{\pi}{k}} e^{-kc} (2c)^{-k}, c = \sqrt{2} - 1, \qquad k \to \infty$$

In what follows we briefly discuss the derivation of the above formula from first principles, rather than simply using Eq. (6.4.9). This should help the reader to do the same in the following problems for which we only use Eq. (6.4.9).

Transforming to the deformed contour C_s and using $\phi(z) - \phi(z_0) = -t$ where t is real and $t > 0$ (recall that $\text{Im}(\phi(z) - \phi(z_0)) = 0$), we have

$$I = \int_{C_s} e^{k\phi(z_0)} e^{-kt} \left(\frac{dz}{dt} \right) dt = e^{k\phi(z_0)} \int_{C_s} e^{-kt} \left(\frac{-dt}{\phi'(z)} \right)$$

$$\sim e^{k\phi(z_0)} \int_{C_s} e^{-kt} \left(\frac{-dt}{(z - z_0)\phi''(z_0)} \right)$$

To find the leading order, we use

$$\frac{\phi''(z_0)}{2} (z - z_0)^2 = -t$$

and noting that $\phi''(z_0) < 0$, we find

$$I \sim 2e^{k\phi(z_0)} \int_0^\infty \frac{e^{-kt} \, dt}{t^{1/2} \sqrt{2|\phi''(z_0)|}}$$

which yields the previous result after substitution and integration. (The factor of 2 corresponded to taking both descent contours, each one emanating from the saddle, into account.)

It is also worth mentioning that with knowledge of the saddle point, the steepest descent contour, and the fact that these calculations are intrinsically local, various modifications of the ideas presented above are possible. For example, we can expand the phase $\phi(z)$ directly in the integrand after deforming the contour, that is

$$I = \int_{C_s} e^{k\phi(z)} \, dz \sim \int_{C_s} e^{k\phi(z_0)} e^{k\phi''(z_0)\frac{(z-z_0)^2}{2}} \, dz$$

Through the saddle point on the steepest descent contour, we can transform to a more convenient variable. Recall that in the neighborhood of the saddle, in general, one needs the correct phase, $z - z_0 = re^{i\theta}$, θ determined by the

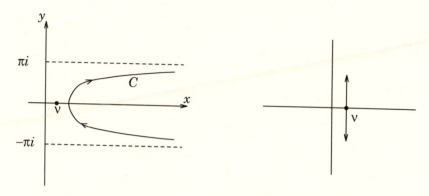

(a) Sommerfeld contour   (b) deformed local steepest descent contour

Fig. 6.4.6. Integrating $e^{k \operatorname{sech} v \sinh z - z}$ as $k \to \infty$

steepest descent contour; that is, calling $\phi''(z_0) = |\phi''(z_0)| e^{i\alpha}$, the steepest descent contour has $2\theta + \alpha = \pi, 3\pi$, so that

$$I \sim 2 e^{k\phi(z_0)} \int_0^\infty e^{-k |\frac{\phi''(z_0)}{2}| r^2} \, dr$$

Directly integrating or transforming to a new variable (e.g. $t = |\frac{\phi''(z_0)}{2}| r^2$) yields the previous results.

Example 6.4.5 Evaluate $I(k) = \int_C e^{k(\operatorname{sech} v \sinh z - z)} \, dz, k \to \infty, v \neq 0$, where C is the so-called Sommerfeld contour illustrated in Figure 6.4.6a and v is a fixed real positive number.

In this example $\phi(z) = \operatorname{sech} v \sinh z - z, \phi'(z) = \operatorname{sech} v \cosh z - 1$, and $z_0 = v$ is a simple saddle point. Also, $\phi''(v) = \tanh v$; hence $\alpha = 0$. From Eq. (6.4.5), $\theta = \pi/2, 3\pi/2$ are directions of steepest descent. We consider a deformed contour passing through the point v; this contour is given by $\operatorname{Im} \phi(z) = 0$ and is asymptotic to C as $z \to \pm i\pi + \infty$. ($\operatorname{Im}\phi(z) = 0$ is given by $\cosh x \sin y = y$; hence $y \to \pm i\pi$ as $x \to \infty$.) We use Eq. (6.4.9) with

$$z_0 = v, n = 2, |\phi''(v)| = \tanh v \qquad \beta = 1, \ f_0 = 1, \ \phi(v) = \tanh v - v$$

(note that $\phi(v) < 0$) and we subtract the formulae obtained with $\theta = \frac{\pi}{2}$ and $\theta = \frac{3\pi}{2}$:

$$I(k) \sim i \sqrt{\frac{2\pi}{k \tanh v}} e^{-k(v - \tanh v)} \qquad k \to \infty$$

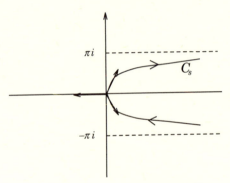

Fig. 6.4.7. Steepest descent contour C_s

It turns out that $I(k) = J_k(k\,\mathrm{sech}\nu)/2i\pi$, where J_k is the Bessel function of order k (see Eq. (3.7.35a)). This example has historical significance. It is the consideration of such an integral that led Debye (1954) to the discovery of the steepest descent technique. As mentioned earlier, Riemann also discovered this technique independently in a problem involving the asymptotic behavior of a hypergeometric function.

Example 6.4.6 Evaluate $I(k) = \int_C e^{k(\sinh z - z)}\, dz, k \to \infty$ where C is the Sommerfeld contour illustrated in Figure 6.4.6a.

This problem corresponds to the case $\nu = 0$ of Example 6.4.5. Here we have $\phi(z) = \sinh z - z$ having a second-order saddle point $z_0 = 0$, which was analyzed in Example 6.4.1(c). We consider the contour C_s defined by $\mathrm{Im}\,\phi(z) = 0$ or $\cosh x \sin y = y$; note that $\cosh x = y/\sin y$, that is, $x \to \infty$ as $y \to \pm\pi$ and C_s is asymptotic to C (see Figure 6.4.7).

We use Eq. (6.4.9) with

$$z_0 = 0, \qquad n = 3, \qquad |\phi'''(0)| = 1,$$
$$f_0 = 1, \qquad \beta = 1, \qquad \phi(0) = 0$$

and subtract the formulae obtained with $\theta = \frac{\pi}{3}$ and $\theta = \frac{5\pi}{3}$

$$I(k) \sim \frac{e^{\frac{i\pi}{3}} - e^{\frac{5i\pi}{3}}}{3}\Gamma\left(\frac{1}{3}\right)\left(\frac{3!}{k}\right)^{\frac{1}{3}}$$

$$= \frac{2i}{3}\sin\left(\frac{\pi}{3}\right)\Gamma\left(\frac{1}{3}\right)\left(\frac{6}{k}\right)^{\frac{1}{3}}, \qquad k \to \infty$$

We remark that $I(k) = J_k(k)/2\pi i$, where J_k denotes the Bessel function of order k (see Eq. (3.7.35a)).

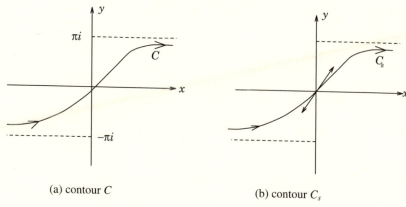

(a) contour C (b) contour C_s

Fig. 6.4.8. Integrating $e^{k(\cosh z - z^2/2)}$ as $k \to \infty$

Example 6.4.7 Evaluate $I(k) = \int_C e^{k(\cosh z - \frac{z^2}{2})}$, $k \to \infty$, where C is the contour illustrated in Figure 6.4.8a.

Note that $\phi(z) = \cosh z - \frac{z^2}{2}$ and the saddle point $z_0 = 0$ was analyzed in Example 6.4.1(d). We consider the contour C_s defined by $\operatorname{Im}\phi(z) = 0$ (i.e., $\sinh x \sin y - xy = 0$), which goes through $z_0 = 0$ and is asymptotic to C (see Figure 6.4.8b). We use Eq. (6.4.9) with

$$z_0 = 0, \quad n = 4, \quad \left|\phi^{(iv)}(0)\right| = 1, \quad f_0 = 1, \quad \beta = 1, \quad \phi(0) = 1$$

and subtract the formula obtained with $\theta = \frac{\pi}{4}$ and $\theta = -\frac{\pi}{4}$

$$I(k) \sim \frac{e^{\frac{i\pi}{4}} - e^{-\frac{i\pi}{4}}}{4} \Gamma\left(\frac{1}{4}\right) \left(\frac{4!}{k}\right)^{\frac{1}{4}} e^k, \quad k \to \infty$$

In Examples 6.4.3–6.4.7 it was possible to replace the contour C by a single contour C_s. However, as was already indicated by Example 6.4.2, sometimes we have to replace the contour C by several contours. This will also be illustrated in Examples 6.4.8 and 6.4.9. These examples differ from Example 6.4.2 in that one of the deformed contours passes through a saddle point (recall that there were no saddle points in Example 6.4.2).

Example 6.4.8 Find the "full" asymptotic expansion of $I(k) = \int_0^1 e^{ikt^2}\, dt$, as $k \to \infty$.

Because $\phi(z) = iz^2$ and $\phi'(z) = 2iz$, $z_0 = 0$ is a simple saddle point. The steepest paths are given by $\operatorname{Im}\phi(z) = $ constant, that is, $x^2 - y^2 = $ constant. Hence the steepest paths going through $z = 0$ and the endpoint $z = 1$ are given

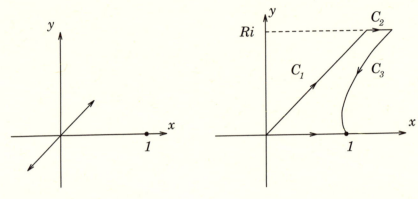

(a) saddle point at 0   (b) steepest descent contours C_1 and C_3

Fig. 6.4.9. Expanding $\int_0^1 e^{ikt^2}\, dt$

by $x = \pm y$ and $x^2 - y^2 = 1$, respectively. The corresponding paths of steepest descent are $x = y$, $y > 0$, and $x = \sqrt{1 + y^2}$. (Note that along the curve $x = -\sqrt{1 + y^2}$ the function e^{iz^2} increases; this is the steepest ascent path.) Because $\operatorname{Im}\phi(0) \neq \operatorname{Im}\phi(1)$ there is no continuous contour joining 0 and 1 on which $\operatorname{Im}\phi$ is constant. We connect the two steepest paths by the contour C_2 (see Figure 6.4.9b) and use Cauchy's theorem to find $I(k) = \int_{C_1} + \int_{C_2} + \int_{C_3}$. Along C_1, $z = e^{i\pi/4}r$; along C_3, $z = (x + iy) = \sqrt{1 + y^2} + iy$, and along C_2, $z = x + iR$, $x > 0$; the function $e^{iz^2} = e^{i(x^2 - R^2) - 2xR}$.

Hence letting $R \to \infty$, we find $\int_{C_2} \to 0$ and

$$I(k) = e^{i\pi/4} \int_0^\infty e^{-kr^2}\, dr - e^{ik} \int_0^\infty e^{-2ky\sqrt{1+y^2}} \left(\frac{y}{\sqrt{1 + y^2}} + i \right) dy.$$
(6.4.16)

The first integral in Eq. (6.4.16) can be evaluated exactly

$$e^{i\pi/4} \int_0^\infty e^{-kr^2}\, dr = \frac{1}{2}\sqrt{\frac{\pi}{k}} e^{\frac{i\pi}{4}}$$
(6.4.17)

To obtain the leading behavior of the second integral in Eq. (6.4.16) as $k \to \infty$, we note that the dominant contribution is near $y = 0$. In this location the integrand can be approximated by $e^{-2ky}(y + i)$, hence

$$\int_0^\infty e^{-2ky\sqrt{1+y^2}} \left(\frac{y}{\sqrt{1 + y^2}} + i \right) dy$$

$$\sim \int_0^\infty e^{-2ky}(y + i)\, dy \sim \frac{i}{2k} + \frac{1}{4k^2}, \qquad k \to \infty$$

(Alternatively we can use integration by parts.) Adding these two contributions we find

$$I(k) \sim \frac{1}{2}\sqrt{\frac{\pi}{k}}e^{\frac{i\pi}{4}} - e^{ik}\left(\frac{i}{2k} + \frac{1}{4k^2}\right), \qquad k \to \infty$$

To obtain the "complete" asymptotic expansion of the second integral in Eq. (6.4.16), we use the change of variables $s = 2y\sqrt{1+y^2}$ so that this integral can be transformed to one for which Watson's Lemma is applicable. Actually, if we go back to the original coordinates, we find for the steepest descent contour

$$z = x + iy = \sqrt{1+y^2} + iy, \quad \text{or} \quad z^2 = 1 + 2iy\sqrt{1+y^2} = 1 + is$$

or more simply $z = (1 + is)^{\frac{1}{2}}$. Hence

$$\int_{C_3} e^{ikz^2}dz = -\frac{i}{2}e^{ik}\int_0^\infty \frac{e^{-ks}}{\sqrt{1+is}}ds \qquad (6.4.18)$$

Using the Taylor expansion (see also Example 6.2.6)

$$(1+is)^{-\frac{1}{2}} = \sum_{n=0}^{\infty} (i)^n c_n\left(-\frac{1}{2}\right)s^n$$

where $c_0(-\frac{1}{2}) = 1$, and for $n \geq 1$

$$c_n(\alpha) = \frac{\alpha(\alpha-1)(\alpha-2)\cdots(\alpha-n+1)}{n!} = \frac{\Gamma(\alpha+1)}{\Gamma(n+1)\Gamma(\alpha-n+1)}$$

and applying Watson's Lemma to evaluate Eq. (6.4.18) and adding the contribution from Eq. (6.4.17), we find

$$I(k) \sim \frac{1}{2}\sqrt{\frac{\pi}{k}}e^{\frac{i\pi}{4}} - \frac{1}{2}ie^{ik}\sum_{n=0}^{\infty}(i)^n\frac{c_n(-\frac{1}{2})n!}{k^{n+1}}, \qquad k \to \infty$$

In this problem we see that the "complete" asymptotic expansion could also be found using the transformation $z = \sqrt{1+y^2} + iy$ and the expansion of the Taylor series $(1+is)^{-1/2}$. In fact, if one is only interested in the first few terms of the expansion of

$$J = \int_0^\infty e^{-2ky\sqrt{1+y^2}}\left(\frac{y}{\sqrt{1+y^2}} + i\right)dy$$

we can let $s = 2y\sqrt{1 + y^2}$, expand for y small, that is, $s = 2y + y^3 + \cdots$, and successively invert to find y in terms of s: $y = \frac{s}{2} - \frac{s^3}{16} + \cdots$. Then the integral for J is given by

$$J = \int_0^\infty e^{-ks} \left(y - \frac{y^3}{2} + \cdots + i \right) \frac{dy}{ds} \, ds$$

After replacing y by the first few terms of its expansion, this yields

$$J = \int_0^\infty e^{-ks} \left(\frac{i}{2} + \frac{s}{4} - \frac{3is^2}{16} + \cdots \right) ds$$

$$= \frac{i}{2k} + \frac{1}{4k^2} - \frac{3i}{8k^2} + \cdots$$

This example provides yet another illustration of the power of the steepest descent method. It gives the full asymptotic expansion of a Fourier type integral. However, it would be false to think that the method of stationary phase is a special case of the steepest descent method. Indeed, the stationary phase formula does not require the strong analyticity assumptions made in the present subsection. Obviously, the two methods are applicable to overlapping classes of integrals, but neither is a special case of the other. If Fourier type integrals involve analytic integrands, then the method of steepest descent is usually the method of choice.

Example 6.4.9 Evaluate $I(k) = \int_{-\infty}^a e^{-kz^2} f(z) \, dz, k \to \infty$, where $0 < \arg a < \frac{\pi}{4}$ and $f(z)$ is analytic and bounded for $0 \le \arg z \le \pi$.

In this example, $\phi(z) = -z^2$, $\phi'(z) = -2z$; hence $z_0 = 0$ is a simple saddle point. Also, $\phi''(0) = -2 = 2e^{i\pi}$, that is, $\alpha = \pi$ and from Eq. (6.4.5) we see that $\theta = 0, \pi$ are the directions of steepest descent for the point $z_0 = 0$. The steepest path $(\text{Im}\phi(z) = \text{Im}\phi(z_0))$ going through $a = x_1 + iy_1$ is given by $xy = x_1 y_1$ (see Figure 6.4.10).

We use Cauchy's Theorem in order to deform the contour C onto steepest descent paths, that is, $I(k) = \int_{C_1} + \int_{C_2} + \int_{C_3}$. Letting $R \to \infty$, we see that $\int_{C_3} \to 0$ and

$$I(k) = \int_{-\infty}^\infty e^{-kx^2} f(x) \, dx - \int_{C_3} e^{-kz^2} f(z) \, dz \qquad (6.4.19)$$

To evaluate the first integral in Eq. (6.4.19), we use Eq. (6.4.9) with

$$z_0 = 0, \qquad n = 2, \qquad |\phi''(0)| = 2,$$

$$f_0 = f(0), \qquad \beta = 1, \qquad \phi(0) = 0$$

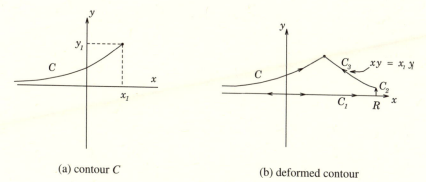

(a) contour C (b) deformed contour

Fig. 6.4.10. Deforming C onto steepest descent curves

and subtract the formulae obtained with $\theta = 0$ and $\theta = \pi$

$$\int_{-\infty}^{\infty} e^{-kr^2} f(r)\, dr \sim f(0)\sqrt{\frac{\pi}{k}}, \qquad k \to \infty$$

To evaluate the second integral in Eq. (6.4.19), we note that on C_3, $z = x + i x_1 y_1 / x$, that is

$$z^2 - a^2 = x^2 - \frac{x_1^2 y_1^2}{x^2} - (x_1^2 - y_1^2) = x^2 - x_1^2 + y_1^2 - y_1^2 \left(\frac{x_1}{x}\right)^2$$

is real and positive ($x \geq x_1$ and $y_1 \leq x_1$). Thus it is convenient to let $z^2 - a^2 = t$, $t > 0$ and use Laplace's method. Expanding $f((a^2 + t)^{1/2})/(a^2 + t)^{1/2}$ near $t = 0$, we find that as $k \to \infty$

$$\int_{C_3} e^{-kz^2} f(z)\, dz \sim -\frac{1}{2} \int_0^{\infty} \frac{e^{-k(a^2+t)} f((a^2 + t)^{1/2})}{(a^2 + t)^{1/2}}\, dt \sim -\frac{e^{-ka^2}}{2ak} f(a)$$

Thus the term due to $\int_{C_3}$ is exponentially smaller than the steepest path through $z = 0$; hence

$$I(k) \sim f(0)\sqrt{\frac{\pi}{k}}, \qquad k \to \infty$$

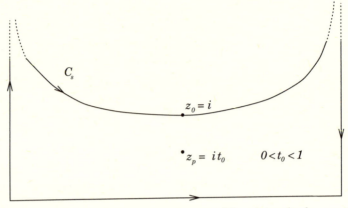

Fig. 6.4.11. steepest descent contour, saddle point, and pole

Example 6.4.10 Consider the integrals

$$\text{(a)} \quad I_1(k) = \int_{-\infty}^{\infty} \frac{e^{ik(t+t^3/3)}}{t^2 + t_0^2} \, dt, \qquad k > 0$$

$$\text{(b)} \quad I_\alpha(k) = \int_{-\infty}^{\infty} \frac{e^{ik(t+t^3/3)}}{(t^2 + 1)^\alpha} \, dt, \qquad k > 0, \ 0 < \alpha < 1$$

where t_0 is a real constant, $t_0 > 0$. Find the leading term of the expansions of these integrals, as $k \to \infty$. (The reader may wish to also see Example 6.5.3, where similar integrals are considered.)

We begin with part (a). The phase is given by $\phi(z) = i(z + z^3/3)$. We have $\phi'(z) = i(1 + z^2)$; hence there are saddle points at $z_0 = \pm i$. Because $\phi''(z) = 2iz$, we have $\phi''(\pm i) = \mp 2$. When $z_0 = i$, Eq. (6.4.5) shows that $\alpha = \pi$; hence the steepest descent directions are $\theta = 0, \pi$. When $z = -i$, $\alpha = 0$ and $\theta = \frac{\pi}{2}, \frac{3\pi}{2}$. The steepest descent curves are given by $\text{Im}\phi(z) = \text{Im}\phi(z_0)$: $-x(3y^2 - x^2 - 3) = 0$. Thus, corresponding to $z_0 = i$, the steepest descent curve is $3y^2 - x^2 - 3 = 0$; for $z_0 = -i$ it is $x = 0$. Because of the given integrand, we can only deform onto the hyperbola $3y^2 - x^2 - 3 = 0$. Indeed, at infinity the exponential $\exp(ik(z + z^3/3))$ decays in the upper half plane when $z = re^{i\theta}$: $0 < \theta < \pi/3$ and $2\pi/3 < \theta < \pi$ and this contains the hyperbola (see Figure 6.4.11). We cannot deform onto the steepest descent contour $x = 0$ (note the integral does not converge at $y = \pm i\infty$). The fact that the value of the integrand at the saddle point at $z = -i$ is exponentially large, $e^{k\phi(-i)} = e^{2k/3}$, immediately indicates that we cannot deform the contour onto a steepest descent curve through $z = -i$. (Since the integral is highly oscillatory, it must vanish as $k \to \infty$). In completing the contour we note that if $0 < t_0 < 1$, by Cauchy's

Theorem we have to include the contribution from the pole at $z_p = it_0$. If $t_0 > 1$ there is no pole within the deformed contour.

We calculate the leading-order steepest descent contributions from Eq. (6.4.9) with $n = 2$, $z_0 = i$, $f_0 = 1/(t_0^2 - 1)$, $\beta = 1$, $\phi(i) = -2/3$, $\phi''(i) = -2$, and $\theta = 0, \pi$. Subtracting the contribution of $\theta = \pi$ from the contribution with $\theta = 0$, and calculating the pole contribution we find,

$$I_1(k) \sim \frac{\sqrt{\frac{\pi}{k}} e^{-\frac{2}{3}k}}{t_0^2 - 1} + \frac{\pi}{t_0} e^{-k(t_0 - t_0^3/3)} \mathcal{H}(1 - t_0) \qquad (6.4.20)$$

where $\mathcal{H}(x)$ is the Heaviside function; $\mathcal{H}(x) = 1$ if $x > 0$ and $\mathcal{H}(x) = 0$ if $x < 0$. Note that for $0 < t_0 < 1$, the pole contribution dominates that from the saddle point.

We also remark that if $t_0 = 1$, Eq. (6.4.20) breaks down. The situation when $t_0 = 1$ is more subtle. When $t_0 = 1$ the steepest descent contour can take the local form

which has the effect of halving the pole contribution and treating the saddle point as a principle value integral. In this case, letting $z = i + s$ in the neighborhood of the saddle yields the following:

$$I_1(k) \sim \fint_{-\infty}^{\infty} \frac{e^{-2k/3} e^{-ks^2} e^{iks^3/3}}{s(2i + s)} \, ds + \frac{\pi}{2t_0} e^{-2k/3}$$

Letting $s = u/\sqrt{k}$ reduces the principle value integral to the form

$$\fint_{-\infty}^{\infty} \frac{e^{-u^2} e^{iu^3/3\sqrt{k}}}{u(2i + u/\sqrt{k})} \, du \sim \fint_{-\infty}^{\infty} \frac{e^{-u^2}}{2iu} \, du = 0$$

(noting that the second integral above is odd); hence when $t_0 = 1$

$$I_1(k) \sim \frac{\pi}{2} e^{-2k/3}$$

Next we consider part (b). Here the important point to note is that the branch cuts are taken from $z = i$ to $z = i\infty$ and $z = -i$ to $z = -i\infty$ with local coordinates $(z^2 + 1) = (z + i)(z - i) = r_1 e^{i\varphi_1} r_2 e^{i\varphi_2}$, and $-\pi/2 < \varphi_1 < 3\pi/2$, and $-3\pi/2 < \varphi_2 < \pi/2$ so that there is no jump across the $\text{Im } z$ axis when $z = 0$. With these choices for φ_1 and φ_2 our saddle point directions at $z = i$ are $\theta = 0, -\pi$, not $\theta = 0, \pi$. So after deforming the contour onto the steepest descent path we can now either use formula (6.4.9) or work out the asymptotic contribution from first principles (always a better idea in practice). For brevity we use Eq. (6.4.9) where $z_0 = i$, $f_0 = \frac{1}{(2i)^\alpha} = \frac{1}{2^\alpha} e^{-i\pi\alpha/2}$ (note that f_0 comes from the contribution of the local coordinate $z + i = r_1 e^{i\varphi_1}$ when $z = i$), $\beta = 1 - \alpha$, $n = 2$, $\phi(i) = -2/3$, and $\phi''(i) = 2e^{i\pi}$. Substituting the contributions corresponding to $\theta = 0$, $\theta = -\pi$ we find

$$I_\alpha(k) \sim \frac{e^{\frac{-i\pi\alpha}{2}} 2^{\frac{1-\alpha}{2}} e^{\frac{-2k}{3}} \Gamma\left(\frac{1-\alpha}{2}\right)}{2^{\alpha+1} (2k)^{\frac{1-\alpha}{2}}} - \frac{e^{\frac{-i\pi\alpha}{2}} 2^{\frac{1-\alpha}{2}} e^{i(\alpha-1)\pi} e^{\frac{-2k}{3}} \Gamma\left(\frac{1-\alpha}{2}\right)}{2^{\alpha+1} (2k)^{\frac{1-\alpha}{2}}}$$

$$= \frac{e^{\frac{-2k}{3}} \Gamma\left(\frac{1-\alpha}{2}\right)}{2^\alpha \; k^{\frac{1-\alpha}{2}}} \cos\frac{\pi\alpha}{2}$$

Problems for Section 6.4

1. Use the method of steepest descent to find the leading asymptotic behavior as $k \to \infty$ of

   (a) $\displaystyle\int_{-\infty}^\infty \frac{t e^{ik\left(\frac{t^3}{3} + t\right)}}{1 + t^4}\, dt$ (b) $\displaystyle\int_{-\infty}^\infty \frac{e^{ik\left(\frac{t^5}{5} + t\right)}}{1 + t^2}\, dt$

2. Show that for $0 < \theta \le \pi/4$

   $$\int_0^\theta e^{-k \sec x}\, dx \sim \sqrt{\frac{\pi}{2k}} e^{-k}, \qquad k \to \infty$$

3. Find the leading asymptotic behavior as $k \to \infty$ of

   $$\int_0^\infty e^{ik\left(\frac{t^4}{4} + \frac{t^3}{3}\right)} e^{-t}\, dt$$

4. In this problem we will find the "complete" asymptotic behavior of

   $$I(k) = \int_0^{\frac{\pi}{4}} e^{ikt^2} \tan t\, dt \qquad \text{as } k \to \infty$$

   (a) Show that the steepest descent paths are given by

   $$x^2 - y^2 = C; \qquad C = \text{constant}$$

   (b) Show that the steepest descent paths that go through $z = 0$ are given by

   $$x = \pm y$$

   and the steepest paths that go through $z = \frac{\pi}{4}$ are given by

   $$x = \pm\sqrt{\left(\frac{\pi}{4}\right)^2 + y^2}$$

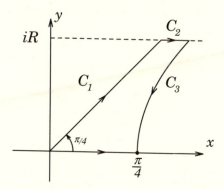

Fig. 6.4.12. Contour for Problem 6.4.4

(c) Note that the steepest descent paths in the first quadrant are $C_1 : x = y$ and $C_3 : x = \sqrt{\left(\frac{\pi}{4}\right)^2 + y^2}$. Construct a path C_2 as shown in the Figure 6.4.12, and therefore

$I(k)$ can be written as $I_1 + I_2 + I_3$ where I_i refers to the integral along contour C_i, for $i = 1, 2, 3$. As $k \to \infty$: show that

$$I_1 \sim \frac{i}{2k} - \frac{e^{i\pi/4}}{4k^{3/2}} \Gamma\left(\frac{1}{2}\right)$$

$$I_2 \sim 0,$$

$$I_3 \sim \frac{-2i}{k\pi} e^{-ik(\pi/4)^2}$$

5. Consider the integral

$$I(k) = \int_0^1 e^{ikt^3}\, dt \quad \text{as } k \to \infty$$

Show that $I(k) = I_1 + I_2 + I_3$ where as $k \to \infty$:

$$I_1 \sim \frac{-i}{3k^{1/3}} \Gamma\left(\frac{1}{3}\right)$$

$$I_2 \sim 0,$$

$$I_3 \sim -\frac{ie^{ik}}{3k} - \frac{2e^{ik}}{9k^2}$$

and the contour associated with I_1 is the steepest descent contour $x = 0$ passing through $z = 0$; the contour I_3 is the steepest descent contour $x^3 - 3xy^2 = 1$ passing through $z = 1$; and the contour associated with I_2

is parallel to the x-axis and is at a large distance from the origin (see also Problem 6.4.4).

6. Consider

$$I(k) = \int_0^1 \frac{e^{ik(t^2+t)}}{\sqrt{t}}\, dt \quad \text{as } k \to \infty$$

Show that $I(k) \sim I_1 + I_2 + I_3$ where

$$I_1 \sim \sqrt{\frac{\pi}{k}}e^{i\pi/4}\left(1 - \frac{3i}{4k} - \frac{105}{32k^2}\right)$$

$$I_2 \sim 0,$$

$$I_3 \sim \frac{i}{3}e^{2ik}\left(-\frac{1}{k} + \frac{7i}{18k^2} + \frac{111}{324k^3} + \cdots\right)$$

and the contour of I_1 passes through the origin (with steepest descent contour: $C_1 : x^2 - y^2 + x = 0$), the contour of I_3 passes through $z = 1$ (with steepest descent contour: $C_3 : x^2 - y^2 + x = 2$) and the contour of I_2 is parallel to the x axis and is at a large distance from the origin (see also Problems 6.4.4 and 6.4.5).

6.5 Applications

We now illustrate how the methods developed in Sections 6.2–6.4 can be used to evaluate integrals that arise from solutions of physically interesting partial differential equations (PDEs).

Example 6.5.1 Consider the "free"Schrödinger equation

$$i\psi_t + \psi_{xx} = 0, \qquad -\infty < x < \infty, \qquad t > 0, \qquad \psi \to 0 \quad \text{as } |x| \to \infty$$

$$\psi(x,0) = \psi_0(x) \qquad \text{with} \int_{-\infty}^{\infty} |\psi_0|^2\, dx < \infty$$

This PDE is the time-dependent Schrödinger equation with a zero potential. It arises in many physical problems, for example, quantum mechanics and also in the study of linear wave communications in fiber optics.

Using a Fourier transform in x, that is

$$\psi(x,t) = \frac{1}{2\pi}\int_{-\infty}^{\infty} b(k,t)e^{ikx}\, dk$$

and substituting this integral into the PDE (assuming the interchange of derivatives and integrals is valid) yields the ordinary differential equation $b_t = -ik^2 b$ or $b(k, t) = b(k, 0)e^{-ik^2 t}$. Thus we find that ψ is given by Eq. (6.1.1b), where $b(k, 0)$ replaces $\Psi_0(k)$. Let us consider the large t behavior of $\psi(x, t)$, where x/t is fixed. Equation (6.1.1b) implies $\psi(x, t) = \frac{1}{2\pi} \int b_0(k)e^{it\phi(k)}\, dk$, where we write $b_0(k) = b(k, 0)$ and

$$\phi(k) = k\frac{x}{t} - k^2, \quad \phi'(k) = \frac{x}{t} - 2k, \quad \phi''(k) = -2 \qquad (6.5.1)$$

Hence, using the method of stationary phase, (there exists a stationary point at $k_0 = x/2t$), the leading order behavior of the integral as $t \to \infty$ is (see Eq. (6.3.8))

$$\psi(x, t) \sim e^{it(\frac{x}{2t})^2} b_0\left(\frac{x}{2t}\right)\sqrt{\frac{\pi}{t}}e^{-\frac{i\pi}{4}} \qquad (6.5.2)$$

From $\psi(x, t) = \frac{1}{2\pi}\int_{-\infty}^{\infty} b_0(k)e^{ik(x-kt)}\, dk$ we see that the solution can be thought of as a superposition of infinitely many wave trains each having a **phase velocity** k. Different waves have different phase speeds; this leads to destructive interference, which in turn implies that the solution **disperses** away as $t \to \infty$. Indeed Eq. (6.5.2) shows that the amplitude decays as $t^{-1/2}$, which is a typical decay rate of such one-dimensional linear dispersive problems.

Equation (6.5.1) indicates that the important propagation velocity is *not* the phase velocity k, but is really the velocity that yields the dominant asymptotic result; that is for this problem $\phi'(k) = 0$ yields $x = 2kt$. This velocity is $2k$. In general the quantity $c_g = x/t$ is called the **group velocity**. The significance of this velocity is that after a sufficiently long time, each wave number k dominates the solution in a region defined by $x \sim c_g(k)t$, where c_g is the group velocity (see also Section 4.6). In other words, different wave numbers (after a sufficiently long time) propagate with the group velocity.

It is straightforward to generalize the above notions to any dispersive linear equation: Equation (6.1.2) is replaced by

$$\psi(x, t) = \int_{-\infty}^{\infty} \hat{\psi}_0(k)e^{i(kx-w(k)t)}\, dk \qquad (6.5.3)$$

where $w(k)$ is a real function of k, called the dispersion relationship. The phase velocity is given by $c_p(k) = w/k$. Note that if $c_p(k)$ is constant, call it c_0, then the solution at any time $t > 0$ is simply the initial function translated by $c_0 t$:

$$\psi(x, t) = \int_{-\infty}^{\infty} \hat{\psi}_0(k) e^{ik(x - c_0 t)} \, dk = \psi_0(x - c_0 t)$$

A linear equation in one space, one time dimension, is called **dispersive** if $w(k)$ is real and $d^2 w / dk^2$ is not identically zero. The group velocity is given by $c_g(k) = dw/dk$, which corresponds to the stationary point

$$\psi(x, t) = \int_{-\infty}^{\infty} \hat{\psi}_0(k) e^{it\phi(k)} \, dk, \qquad \phi(k) = k\frac{x}{t} - w,$$

$$\phi'(k) = \frac{x}{t} - \frac{dw}{dk} = \frac{x}{t} - c_g(k) = 0$$

Hence we expect the dominant contribution to the solution for large t to be in the neighborhood of $\frac{x}{t} = w'(k) = c_g(k)$. In Example 6.5.1, $w(k) = k^2$; thus the phase and group velocities are given by $c_p = k, c_g = 2k$, respectively. On the other hand in the case where $w(k) = c_0 k$, constant, then the phase velocity $c_p = w/k = c_0$ also equals the group velocity $c_g = w'(k) = c_0$. This is an example of a nondispersive wave; note $w''(k) = 0$.

Example 6.5.2 (Burgers equation) The simplest model combining the effects of weak nonlinearity and diffusion is the Burgers equation

$$u_t + u u_x = \varepsilon u_{xx} \tag{6.5.4}$$

This equation appears in various physical applications. For example, it models weak shock waves in compressible fluid dynamics. It is distinguished among other nonlinear equations in that it can be linearized via an explicit transformation. Indeed, Hopf and Cole noted that the Burgers equation can be mapped to the heat equation via the transformation $u = -2\varepsilon v_x / v = -2\varepsilon (\log v)_x$ (see, e.g., Whitham (1974)):

$$u_t = \left(\varepsilon u_x - \frac{1}{2} u^2 \right)_x \quad \text{or} \quad (\log v)_{tx} = \left(\frac{\varepsilon v_{xx}}{v} - \frac{\varepsilon v_x^2}{v^2} + \frac{\varepsilon v_x^2}{v^2} \right)_x$$

or after integrating (assuming for example that $v \to \text{const}$ as $x \to \infty$)

$$v_t = \varepsilon v_{xx} \tag{6.5.5}$$

Let us consider an initial value problem for the Burgers equation, with

$$u(x, 0) = u_0(x)$$

Using a Laplace transform in t to solve Eq. (6.5.5), we can obtain the explicit transformation between u and v. The method is illustrated in Example 4.6.2. The solution of Eq. (6.5.5) is given by Eq. (4.6.13) with t replaced by εt (this change of variables is needed to transform the heat Eq. (4.6.9) without ε to Eq. (6.5.5)). Thus

$$v(x,t) = \frac{1}{2\sqrt{\pi \varepsilon t}} \int_{-\infty}^{\infty} h(\eta)e^{-\frac{(x-\eta)^2}{4\varepsilon t}} \, d\eta$$

where

$$h(x) = v(x,0) = v_0(x)$$

But from $u_0(x) = -2\varepsilon v_{0x}/v_0$ we have

$$v_0(x) = Ae^{-\int_0^x \frac{u_0(\eta')}{2\varepsilon} \, d\eta'} = h(x)$$

and we see that $u(x,t) = -2\varepsilon v_x(x,t)/v(x,t)$ is given by

$$u(x,t) = \frac{\int_{-\infty}^{\infty} \frac{x-\eta}{t} e^{-\frac{G}{2\varepsilon}} \, d\eta}{\int_{-\infty}^{\infty} e^{-\frac{G}{2\varepsilon}} \, d\eta}, \qquad G(\eta;x,t) \equiv \int_0^{\eta} u_0(\eta') \, d\eta' + \frac{(x-\eta)^2}{2t}$$

$$(6.5.6)$$

In the physical application of shock waves in fluids, ε has the meaning of viscosity. Here we are interested in the inviscid limit, that is, $\varepsilon \to 0$, of the Burgers equation (see also Whitham (1974)).

As $\varepsilon \to 0$ we use Laplace's method to evaluate the dominant contributions to the integrals appearing in Eq. (6.5.6). To achieve this, we need to find the points for which $\partial G/\partial \eta = 0$,

$$\frac{\partial G}{\partial \eta} = u_0(\eta) - \frac{x-\eta}{t} \tag{6.5.7}$$

Let $\eta = \xi(x,t)$ be such a point; that is, $\xi(x,t)$ is a solution of

$$x = \xi + u_0(\xi)t \tag{6.5.8}$$

Then Laplace's method implies, using 6.2.8 (or the sum of Eqs. (6.2.12) and (6.2.13)), with $k = 1/2\varepsilon$

$$\int_{-\infty}^{\infty} \frac{x-\eta}{t} e^{-\frac{G}{2\varepsilon}} \, d\eta \sim \frac{x-\xi}{t} \sqrt{\frac{4\pi\varepsilon}{|G''(\xi)|}} e^{-\frac{G(\xi)}{2\varepsilon}},$$

$$\int_{-\infty}^{\infty} e^{-\frac{G}{2\varepsilon}} \, d\eta \sim \sqrt{\frac{4\pi\varepsilon}{|G''(\xi)|}} e^{-\frac{G(\xi)}{2\varepsilon}}$$

6 Asymptotic Evaluation of Integrals

Hence, if Eq. (6.5.8) for a given u_0 has only one solution for ξ, then

$$u(x, t) \sim \frac{x - \xi}{t} = u_0(\xi) \qquad (6.5.9)$$

where ξ is defined by Eq. (6.5.8). Equation (6.5.9) has a simple interpretation: Consider the problem

$$\rho_t + \rho\rho_x = 0, \qquad \rho(x, 0) = u_0(x) \qquad (6.5.10)$$

Equation (6.5.10) is a first-order hyperbolic equation and can be solved by the method of characteristics

$$\frac{d\rho}{dt} = 0$$

on the characteristic $\xi(x, t)$:

$$\frac{dx}{dt} = \rho$$

Integrating these relations, we see that the solution of Eq. (6.5.10) is given by

$$\rho(x, t) = u_0(\xi) \qquad (6.5.11)$$

where $\xi(x, t)$ is defined by $x = \xi + u_0(\xi)t$, provided that "breaking" does not occur, that is, the characteristics do not cross or, equivalently, provided that Eq. (6.5.8) has a single solution. (The reader can verify that if $u_0(\xi)$ is a "hump," for example, $u_0(\xi) = e^{-\xi^2}$, then characteristics must cross because parts on the "top" of the hump will "move faster" than parts on the lower part of the hump.)

The above analysis shows that for appropriate u_0, the $\varepsilon \to 0$ limit of the solution of the Burgers equation is given by the solution of the limit Eq. (6.5.10). However, the relationship between the Burgers equation and Eq. (6.5.10) must be further clarified. Indeed, for some $u_0(x)$, Eq. (6.5.10) gives multivalued solutions (after the characteristics cross), while the solution (6.5.6) is always single valued. This means that, out of all the possible ("weak") solutions that Eq. (6.5.10) can support, there exists a unique solution that is the correct limit of the Burgers equation as $\varepsilon \to 0$. It is interesting that Laplace's method provides us with a way of picking this correct solution: When the characteristics of Eq. (6.5.10) cross, Eq. (6.5.8) admits two solutions, which we denote by ξ_1 and ξ_2 with $\xi_1 > \xi_2$ (both ξ_1 and ξ_2 yield the same values of x and t from

Eq. (6.5.8)). Laplace's method shows that for the sum of these contributions, using Eq. (6.5.8)

$$u(x,t) \sim \frac{u_0(\xi_1)|G''(\xi_1)|^{-\frac{1}{2}}e^{-\frac{G(\xi_1)}{2\varepsilon}} + u_0(\xi_2)|G''(\xi_2)|^{-\frac{1}{2}}e^{-\frac{G(\xi_2)}{2\varepsilon}}}{|G''(\xi_1)|^{-\frac{1}{2}}e^{-\frac{G(\xi_1)}{2\varepsilon}} + |G''(\xi_2)|^{-\frac{1}{2}}e^{-\frac{G(\xi_2)}{2\varepsilon}}} \tag{6.5.12}$$

Hence, owing to the dominance of exponentials

$$u(x,t) \sim u_0(\xi_1) \quad \text{for} \quad G(\xi_1) < G(\xi_2);$$

$$u(x,t) \sim u_0(\xi_2) \quad \text{for} \quad G(\xi_1) > G(\xi_2) \tag{6.5.13}$$

The changeover will occur at those (x,t) for which $G(\xi_1) = G(\xi_2)$, or using the definition of G and the fact that both ξ_1, ξ_2 satisfy Eq. (6.5.8), these values of (x,t) satisfy (by integrating (6.5.8))

$$\frac{(x-\xi_1)^2}{2t} - \frac{(x-\xi_2)^2}{2t} = -\int_{\xi_2}^{\xi_1} u_0(\eta)\,d\eta, \quad \text{or}$$

$$(\xi_1 - \xi_2)\left(-\frac{x}{t} + \frac{(\xi_1+\xi_2)}{2t}\right) = -\int_{\xi_2}^{\xi_1} u_0(\eta)\,d\eta$$

Using Eq. (6.5.8) for ξ_1 and ξ_2, and summing, we find that

$$-\frac{x}{t} + \frac{\xi_1+\xi_2}{2t} = -\frac{1}{2}(u_0(\xi_1) + u_0(\xi_2))$$

hence

$$\frac{1}{2}(u_0(\xi_1) + u_0(\xi_2))(\xi_1 - \xi_2) = \int_{\xi_2}^{\xi_1} u_0(\eta)\,d\eta \tag{6.5.14}$$

Equations (6.5.13) with (6.5.8) show that at $\varepsilon \to 0$ the changeover in the behavior of $u(x,t)$ leads to a discontinuity. In this way the solution of the Burgers equation tends to a **shock wave** as $\varepsilon \to 0$. This solution is that particular solution of the limiting equation (6.5.10) that satisfies the **shock condition** (6.5.14). This condition has a simple geometrical interpretation: given $u_0(\xi)$, find ξ_1, ξ_2 using the requirement that the part between the chord $\xi_1 - \xi_2$ and the curve $u_0(\xi)$ be divided by two equal areas. (Eq. (6.5.14) indicates that the area under $u_0(\xi)$ equals the area enclosed by the trapezoid with base $\xi_1 - \xi_2$.) Then insert a shock (see Figure (6.5.1c) in the multivalued solution ρ ($\rho = u_0(\xi)$) of Eq. (6.5.10) at the position $x = s(t)$, where

$$s(t) = \xi_1 + u_0(\xi_1)t = \xi_2 + u_0(\xi_2)t \tag{6.5.15}$$

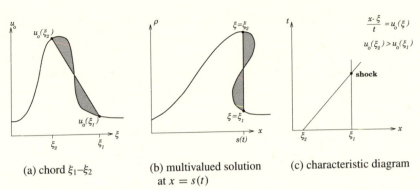

(a) chord ξ_1–ξ_2 (b) multivalued solution   (c) characteristic diagram
                                at $x = s(t)$

Fig. 6.5.1. Shock wave

Note that because Eq. (6.5.10) preserves the area (from Eq. (6.5.11) we see that all points $x = \xi$ evolve linearly according to their amplitude), the equal area property still holds under the curve describing $\rho(x, t)$ (Whitham, 1974). This is sometimes referred to as the "equal area" rule. The mathematical determination of $\xi_1(t)$, $\xi_2(t)$, $s(t)$ involves the simultaneous solution of Eqs. (6.5.14) and both of the equations in (6.5.15).

Example 6.5.3 The Korteweg–deVries (KdV) equation (1895) governs small, but finite, waves in shallow water. The KdV equation, in nondimensional form, is written as

$$V_t + 6VV_x + V_{xxx} = 0 \qquad (6.5.16)$$

In this example, we will consider the linearized KdV equation (see also Eq. (4.6.35))

$$u_t + u_{xxx} = 0, \qquad -\infty < x < \infty, \qquad t > 0,$$
$$u \to 0 \text{ sufficiently rapidly as } |x| \to \infty \qquad (6.5.17)$$

with the initial values

$$u(x, 0) = u_0(x), \qquad u_0 \text{ real}$$

This equation is the small amplitude limit of both the KdV Eq. (6.5.16) and of the modified KdV (mKdV) equation, which arises in plasma physics and lattice dynamics

$$v_t + v_{xxx} \pm 6v^2 v_x = 0 \qquad (6.5.18)$$

Indeed, letting $V = \varepsilon u$ and $v = \varepsilon u$, these equations reduce to Eq. (6.5.17) as $\varepsilon \to 0$. Equations (6.5.16) and (6.5.18), although nonlinear, have the remarkable property that they can be solved exactly. The solution of Eq. (6.6.18) will be described at the end of this section. Using a Fourier transform in x in Eq. (6.5.17), we find that

$$u(x, t) = \frac{1}{2\pi} \int_{-\infty}^{\infty} \hat{u}_0(k) e^{ikx + ik^3 t} \, dk \qquad (6.5.19)$$

where $\hat{u}_0(k)$ is the Fourier transform of $u_0(x)$. Here we are interested in the long-time behavior of $u(x, t)$ for a fixed value of x/t. Equation (6.6.19) can be written as

$$u(x, t) = \frac{1}{2\pi} \int_{-\infty}^{\infty} \hat{u}_0(k) e^{t\phi(k)} \, dk, \qquad \phi(k) = i\left(k^3 + \frac{kx}{t}\right) \qquad (6.5.20)$$

(Also see Example 6.4.10, where a similar problem is discussed.) In this case, the dispersion relationship is $w(k) = k^3$ and the group velocity is $c_g = w'(k) = 3k^2$. We assume that $\hat{u}_0(k)$ can be continued analytically off the real k axis so that we can apply Cauchy's Theorem when needed. We distinguish three cases: $\frac{x}{t} < 0$, $\frac{x}{t} > 0$ (in these two cases we assume $\frac{x}{t} = O(1)$), and $\frac{x}{t} \to 0$.

(a) In the case where $\frac{x}{t} < 0$

$$\phi'(z) = i\left(3z^2 + \frac{x}{t}\right), \qquad \phi''(z) = 6iz$$

there exist two simple real stationary points, $z_\pm = \pm|\frac{x}{3t}|^{\frac{1}{2}}$, and

$$\phi''(z_\pm) = \pm 6i \left|\frac{x}{3t}\right|^{\frac{1}{2}}$$

so that in the notation of Eq. (6.4.8)

$$\alpha_+ = \frac{\pi}{2}, \qquad \alpha_- = \frac{3\pi}{2}$$

Hence using Eqs. (6.4.7), it follows that z_+ and z_- have the steepest descent directions θ given by $\frac{\pi}{4}, \frac{5\pi}{4}$ and $-\frac{\pi}{4}, \frac{3\pi}{4}$, respectively (see Figure 6.5.2a). Calling $z = z_R + iz_I$, the steepest contours are given by $\text{Im}\,\phi(z) = \text{Im}\,\phi(z_\pm)$ or $z_R(z_R^2 - 3z_I^2 - |\frac{x}{t}|) = 0$. The steepest descent contour is the hyperbola; $z_R = 0$ is a steepest ascent contour. Because the exponential decays rapidly in the upper half plane at $z = |z|e^{i\varphi}$, $\varphi = \pi/4, 3\pi/4$ for large $|z|$, we see that

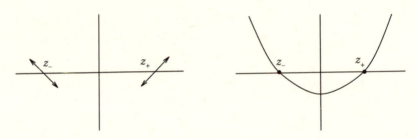

(a) Simple saddle points z_-, z_+ (b) deformed contour

Fig. 6.5.2. Linearized KdV equation

we can deform our original contour ($-\infty < z < \infty$) onto the steepest descent contour going through both saddle points (see Figure 6.5.2b).

Using Eq. (6.4.9) we find that the contributions of z_+ and z_- to the evaluation of u are given (we use $\phi(z_\pm) = \mp 2i(\frac{x}{3t})^{3/2}$) by

$$
\frac{\hat{u}_0\left(\left|\frac{x}{3t}\right|^{\frac{1}{2}}\right) e^{-2it\left|\frac{x}{3t}\right|^{\frac{3}{2}} + \frac{i\pi}{4}}}{2\sqrt{\pi t}\left|\frac{x}{3t}\right|^{\frac{1}{4}}} \quad \text{and} \quad \frac{\hat{u}_0\left(-\left|\frac{x}{3t}\right|^{\frac{1}{2}}\right) e^{2it\left|\frac{x}{3t}\right|^{\frac{3}{2}} - \frac{i\pi}{4}}}{2\sqrt{\pi t}\left|\frac{x}{3t}\right|^{\frac{1}{4}}} \tag{6.5.21}
$$

Also, because $u_0(x)$ is real, we see by taking the complex conjugate of the Fourier transform that $\hat{u}_0(-k) = \overline{\hat{u}_0}(k)$; thus if we write $\hat{u}_0(|\frac{x}{3t}|^{\frac{1}{2}}) = \rho(|\frac{x}{t}|)e^{i\psi(\frac{x}{t})}$, adding the contributions from Eq. (6.5.21) we find

$$
u(x, t) \sim \frac{\rho\left(\frac{x}{t}\right)}{\sqrt{\pi t}\left|\frac{3x}{t}\right|^{\frac{1}{4}}} \cos\left(2t\left|\frac{x}{3t}\right|^{\frac{3}{2}} - \frac{\pi}{4} - \psi\left(\frac{x}{t}\right)\right), \qquad t \to \infty, \ \frac{x}{t} < 0 \tag{6.5.22}
$$

(b) In the case where $\frac{x}{t} > 0$, the saddle points become complex, $z_\pm = \pm i(\frac{x}{3t})^{\frac{1}{2}}$, and $\phi(z_\pm) = \mp 2(\frac{x}{3t})^{1/2}$. The steepest descent paths are found as above. But we can only deform the original contour onto the steepest descent contour through the saddle point z_+ because the exponential $e^{t\phi}$ decays in the upper half plane, but grows in the lower half plane. At the saddle point $z = z_+$ we have

$$
\phi''(z_+) = -6\left(\frac{x}{3t}\right)^{\frac{1}{2}}, \qquad \text{i.e. } \alpha_+ = \pi
$$

thus, from Eq. (6.4.7), $\theta = 0, \pi$ are the directions of steepest descent. Eq.

(6.4.9) implies

$$u(x,t) \sim \frac{\hat{u}_0\left(i\left(\frac{x}{3t}\right)^{\frac{1}{2}}\right)e^{-2\left(\frac{x}{3t}\right)^{\frac{3}{2}}t}}{2\sqrt{\pi t}\left(\frac{3x}{t}\right)^{\frac{1}{4}}}, \qquad t \to 0, \quad \frac{x}{t} > 0 \qquad (6.5.23)$$

(We note again the assumption that $\hat{u}_0(k)$ can be continued off the real axis onto the upper half plane, so that $\hat{u}_0(i(\frac{x}{3t})^{1/2})$ is defined.)

(c) In the case where $\frac{x}{t} \to 0$, both Eqs. (6.5.22) and (6.5.23) clearly break down. To find the asymptotics of $u(x,t)$ in this region, we introduce new "similarity" variables in Eq. (6.5.20):

$$\xi = k(3t)^{\frac{1}{3}}, \qquad \eta = \frac{x}{(3t)^{\frac{1}{3}}}$$

thus

$$u(x,t) = \frac{1}{2\pi(3t)^{\frac{1}{3}}} \int_{-\infty}^{\infty} \hat{u}_0\left(\frac{\xi}{(3t)^{\frac{1}{3}}}\right) e^{i\xi\eta + \frac{i\xi^3}{3}} d\xi \qquad (6.5.24)$$

To find the asymptotic expansion of Eq. (6.5.24) for large t, we can expand $\hat{u}_0$ in a Taylor series near $\xi = 0$. Then Eq. (6.5.24) becomes

$$u(x,t) \sim \frac{1}{2\pi(3t)^{\frac{1}{3}}} \int_{-\infty}^{\infty} e^{i\xi\eta + \frac{i\xi^3}{3}} \left(\hat{u}_0(0) + \frac{\xi}{(3t)^{\frac{1}{3}}}\hat{u}_0'(0) + \cdots\right) d\xi$$

or

$$u(x,t) \sim (3t)^{-\frac{1}{3}}\hat{u}_0(0)A_i(\eta) - i(3t)^{-\frac{2}{3}}\hat{u}_0'(0)A_i'(\eta) \qquad \text{as } t \to \infty \quad (6.5.25)$$

where $A_i(\eta)$ is an integral representation of the Airy function (see Eqs. (4.6.40b) and (4.6.57) in Chapter 4)

$$A_i(\eta) = \frac{1}{2\pi} \int_{-\infty}^{\infty} e^{i\xi\eta + \frac{i\xi^3}{3}} d\xi \qquad (6.5.26)$$

and prime denotes derivatives. The Airy function satisfies the differential equation $A_i'' - \eta A_i = 0$. By analyzing the asymptotic expansion of $A_i(\eta)$ for large η from Eq. (6.5.26) it can be shown that Eq. (6.5.25) in fact matches smoothly with the asymptotic solutions: Eq. (6.5.23) as $\eta \to +\infty$, and Eq. (6.5.22) as $\eta \to -\infty$. (The relevant asymptotic formulae are given in Eqs. (6.7.10), (6.7.11)).

Remark　The modified Korteweg-deVries (mKdV) equation

$$u_t + u_{xxx} - 6u^2 u_x = 0, \qquad -\infty < x < \infty, \qquad t > 0 \qquad (6.5.27)$$

with initial conditions $u(x, 0) = u_0(x)$ decaying sufficiently rapidly as $|x| \to \infty$ can be solved by a new method of mathematical physics called the inverse scattering transform (IST). IST shows that mKdV can be solved as follows (see, for example, Ablowitz and Segur, 1981)

$$u(x, t) = -2K(x, x; t) \qquad (6.5.28a)$$

where $K(x, y; t)$ satisfies the linear integral equation

$$K(x, y; t) + F(x + y; t) - \int_x^\infty \int_x^\infty K(x, z; t)$$
$$\times F(z + s; t) F(s + y; t) \, dz \, ds = 0 \qquad (6.5.28b)$$

for $y \geq x$, where

$$F(x; t) = \frac{1}{2\pi} \int_{-\infty}^\infty r_0(k) e^{i(kx + 8ik^3 t)} \, dk \qquad (6.5.28c)$$

The quantity $r_0(k)$ is called the reflection coefficient, and it can be determined from $u_0(x)$. The integral equation (6.5.28b) is a Fredholm equation (in Section 7.4 we discuss integral equations in more detail), and it can be shown that it has a unique solution. This solution can be obtained by iteration:

$$K(x, y, t) = -F(x + y; t) - \int_x^\infty \int_x^\infty F(x + z; t) F(z + s; t)$$
$$\times F(z + y; t) \, dz \, ds + \cdots \qquad (6.5.29)$$

This is called the Neumann series of the integral equation.

It is important to note that $F(x; t)$ given by Eq. (6.5.28c) is the solution of the linearized equation (6.5.17) where t is replaced by $8t$ and where $\hat{u}_0(k)$ is replaced by $r_0(k)$. As before, there are three cases to consider in the limit $t \to \infty$: $x/t > 0$, $x/t < 0$, and $x/t \to 0$. We only mention the case $x/t > 0$. (The interested reader can consult Ablowitz and Bagor (1981) and Deift and Zhou (1993) for further details, especially with regard to $\frac{x}{t} < 0$.) As before, we assume that $r_0(k)$ is analytically extendible in the upper half plane. Then from Eq. (6.5.23), replacing $t \to 8t$, we have

$$F(x, t) \sim \frac{r_0\left(i\left(\frac{x}{24t}\right)^{1/2}\right)}{\sqrt{8\pi t}\left(\frac{6x}{t}\right)^{1/4}} e^{-16\left(\frac{x}{24t}\right)^{3/2} t} \qquad (6.5.30)$$

Using the fact that the double integral (6.5.29) is exponentially small compared to $F(x + y; t)$, it follows that $K(x, y; t) \sim -F(x + y; t)$. Therefore the asymptotic expansion of the solution of the mKdV equation in this region, from Eqs. (6.5.28a) and (6.5.29), is

$$u(x, t) \sim \frac{r_0 \left(i \left(x/12t \right)^{1/2} \right)}{\sqrt{2\pi t} (x/12t)^{1/4}} e^{-2\left(\frac{x}{3t}\right)^{3/2} t} \tag{6.5.31}$$

Problems for Section 6.5

1. Consider the following initial value problem for the heat equation

$$\psi_t = \psi_{xx}, \qquad -\infty < x < \infty, \qquad t > 0; \qquad \psi \to 0 \text{ as } |x| \to \infty$$

$$\psi(x, 0) = \psi_0(x), \qquad \psi_0(x) \text{ real}, \qquad \int_{-\infty}^{\infty} |\psi_0(x)| \, dx < \infty$$

(a) Use Fourier transforms to show that

$$\psi(x, t) = \frac{1}{2\pi} \int_{-\infty}^{\infty} b_0(k) e^{ikx - k^2 t} \, dk$$

where

$$b_0(k) = \int_{-\infty}^{\infty} e^{-ikx} \psi_0(x) \, dx$$

(b) Show that the large time asymptotic behavior of $\psi(x, t)$ is

$$\psi(x, t) \sim \frac{b_0(k_0)}{\sqrt{4\pi t}} e^{-\frac{x^2}{4t}}$$

where

$$k_0 = \frac{ix}{2t}$$

provided that $b_0(k)$ has an analytic continuation onto the upper half plane.

(c) Establish that the energy $\int_{-\infty}^{\infty} |\psi|^2 \, dx$ always decays, which is consistent with (b) above.

2. The so-called Klein–Gordon equation arises in many areas of physics; for example, relativistic quantum mechanics. This equation is given by

$$u_{tt} - u_{xx} + u = 0$$

Let $-\infty < x < \infty, t > 0$, and consider the following initial value problem: $u(x, 0), u_t(x, 0)$ are both given, they are real, and we assume that $u \to 0$ sufficiently rapidly as $|x| \to \infty$.

(a) Establish conservation of energy:

$$\int_{-\infty}^{\infty} \left(u_t^2 + u_x^2 + u^2 \right) dx = \text{constant}$$

(b) Show that the dispersion relation satisfies $\omega^2 = 1 + k^2$, where $u = e^{i(kx - \omega t)}$ is a special wave solution. Show that the phase speed $c(k)$, $|c(k)| = |\frac{\omega}{k}| \geq 1$ and the group velocity $c_g(k)$, $|c_g(k)| = |\frac{d\omega}{dk}| \leq 1$.

(c) Establish that the Fourier transform solution is

$$u(x, t) = \frac{1}{2\pi} \int_{-\infty}^{\infty} \left(A(k) e^{i(kx + \sqrt{k^2 + 1}\,)t} + B(k) e^{i(kx - \sqrt{k^2 + 1}\,)t} \right) dk$$

Furthermore show that because $u(x, 0)$ and $u_t(x, 0)$ are real

$$u(x, t) = \frac{1}{2\pi} \int_{-\infty}^{\infty} \left(A(k) e^{i(kx + \sqrt{k^2 + 1}\,)t} \right) dx + (*)$$

where $(*)$ denotes the complex conjugate of the first integral.

(d) Show that as $t \to \infty$ for x/t fixed, $|x/t| < 1$,

$$u(x, t) \sim \frac{1}{\sqrt{2\pi}} \frac{t}{(t^2 - x^2)^{3/4}} A\left(-\frac{x}{\sqrt{t^2 - x^2}} \right) e^{i\sqrt{t^2 - x^2} + i\frac{\pi}{4}} + (*)$$

3. (Semidiscrete *free* Schrödinger equation) Consider the following initial value problem for the semidiscrete version of the *free* Schrödinger equation discussed in Example 6.5.1

$$i\frac{d\psi_n}{dt} + \frac{\psi_{n+1} + \psi_{n-1} - 2\psi_n}{h^2} = 0, \quad h \text{ a real constant}$$

$n \in \mathbf{Z}, t > 0$, we assume that $\psi_n \to 0$ sufficiently rapidly as $|n| \to \infty$, and $\psi_n(0) = \psi_{n,0}$ is given. Here $\psi_n(t)$ is the *nth* function of time in a denumberably infinite sequence of such functions.

(a) The discrete analog of the Fourier transform is given by

$$\psi_n(t) = \frac{1}{2\pi i} \oint_C \hat{\psi}(z, t) z^{n-1} \, dz \tag{1}$$

$$\hat{\psi}(z, t) = \sum_{m=-\infty}^{\infty} \psi_m(t) z^{-m} \tag{2}$$

where the contour integral is taken around the unit circle $|z| = 1$. Substitute (2) into (1) and show that you obtain an identity.

(b) The dispersion relation of a semidiscrete problem is obtained by looking for a particular solution in the form $\psi_n(t) = z^n e^{-i\omega t}$. Show that for this semidiscrete problem

$$i\omega(z) = -\frac{(z-1)^2}{zh^2}$$

Note if $z = e^{ikh}$ then

$$i\omega(k) = -\frac{(e^{ikh} - 1)^2}{h^2 e^{ikh}}$$

Explain why this is consistent with the analogous result obtained in Example 6.5.1 of the text.

(c) Use (a) and (b) to establish

$$\psi_n(t) = \frac{1}{2\pi i} \oint_C \hat{\psi}_0(z) z^{n-1} e^{i \frac{(z-1)^2}{zh^2} t} \, dz$$

where C is the unit circle or if $z = e^{i\theta h}$

$$\psi_n(t) = \frac{h}{2\pi} \int_0^{2\pi} \hat{\psi}_0(\theta) e^{inh\theta + 2it \frac{\cos h - 1}{h^2}} \, d\theta$$

(d) Show that there are stationary points when $\frac{n}{t} = 2\frac{\sin\theta}{h^2}$. If $|\frac{nh^2}{2t}| \ll 1$, obtain the leading term in the asymptotic solution as $t \to \infty$; and show that the amplitude of the solution decays as $t^{-\frac{1}{2}}$ as $t \to \infty$.

4. (Fully discrete heat equation) Consider the following explicit fully discrete evolution equation.

$$\frac{\psi_n^{m+1} - \psi_n^m}{\Delta t} = \frac{\psi_{n+1}^m + \psi_{n-1}^m - 2\psi_n^m}{(\Delta x)^2}$$

(a) The dispersion relation of a discrete equation is obtained by looking for a solution of the form $\psi_n^m = z^n \Omega^m$ or $\psi_n^m = e^{ikn\Delta x}e^{i\omega \Delta tm}$ (see also Problem 3 above). In this case, show that

$$\Omega = 1 + \sigma \left(z + \frac{1}{z} - 2 \right)$$

or

$$e^{i\omega \Delta t} = 1 + 2\sigma (\cos k\Delta x - 1), \quad \sigma = \frac{\Delta t}{(\Delta x)^2}$$

(b) For *stability* we need $|\Omega| < 1$. Show that this leads to the condition $\sigma < \frac{1}{2}$.

(c) Use the results of Problem 3, part (a) above, to obtain the Fourier transform solution in the form

$$\psi_n^m = \frac{1}{2\pi i} \int_C \hat{\psi}_0(z) z^{n-1} \left(1 + \sigma \left(z + \frac{1}{z} - 2 \right) \right)^m dz$$

$$= \frac{\Delta x}{2\pi} \int_0^{2\pi} \hat{\psi}_0(k) e^{ink\Delta x} (1 + 2\sigma (\cos k\Delta x - 1))^m dk$$

(d) When $|\Omega| < 1$, $|\frac{n\Delta x^2}{2m\Delta t}| \ll 1$, obtain the *long-time* solution as $m \to \infty$, and show that the solution approaches zero as $m \to \infty$.

6.6 The Stokes Phenomenon

In Section 6.1 we discussed the fact that even if a function $f(z)$ is analytic and single valued in the neighborhood of infinity but not *at* infinity, its asymptotic expansion can change discontinuously (see Example 6.1.3). If $f(z)$ is analytic *at* infinity, its asymptotic expansion coincides with its power series, and the series converges. In this case, the asymptotic expansion of $f(z)$ changes *continuously* at infinity. If the point at infinity is a branch point, the asymptotic expansion changes *discontinuously* across the branch cut. It is interesting that, if the point at infinity is a local essential singular point, and even though $f(z)$ is single valued in the neighborhood of infinity, the asymptotic expansion has lines (rays) across which it changes discontinuously. This manifestation is usually referred to as the **Stokes phenomenon**, after George Stokes (Stokes, 1864), who first discovered this.

We use a concrete example to illustrate a nontrivial case of the Stokes phenomenon. This example also shows how the methods developed in this chapter can be employed when the large parameter is complex.

Example 6.6.1 Evaluate $f(z) = \int_0^\infty \frac{e^{-zt}}{1+t^4}\, dt$, $z \to \infty$, z complex.

Let $z = re^{i\theta}$, then $\exp(-zt) = \exp[-rt(\cos\theta + i\sin\theta)]$. Thus as $t \to \infty$ the integral converges provided that $\cos\theta > 0$, or $|\theta| < \frac{\pi}{2}$. Then the asymptotic analysis of $f(z)$ can be treated as if z were real. The dominant contribution comes from the neighborhood of t near zero. For t near zero, $\frac{1}{1+t^4} \sim 1 - t^4$ thus

$$f(z) \sim \int_0^\infty e^{-zt}\, dt - \int_0^\infty e^{-zt} t^4\, dt = \frac{1}{z} - \frac{4!}{z^5} \tag{6.6.1}$$

where we have used $s = zt$ and the definition of gamma function (see Eq. 4.5.31–4.5.32) to compute the second integral in Eq. (6.6.1). Hence

$$f(z) \sim \frac{1}{z} - \frac{4!}{z^5}, \qquad z \to \infty \qquad \text{in } |\arg z| < \frac{\pi}{2} \tag{6.6.2}$$

Equation (6.6.2) gives the correct asymptotic expansion of $f(z)$ for $|\arg z| < \frac{\pi}{2}$. In order to find the behavior of $f(z)$ outside this region, we must understand the origin of the constraint on the argument of z.

This constraint comes from the requirement that $\text{Re}(zt) > 0$. Therefore, if t is also complex, then $\arg z$ will satisfy a different constraint. Let us, for example, use contour integration to change the path of integration from the real axis to the negative imaginary axis, assuming the contribution owing to the large contour at infinity vanishes. Taking into consideration the pole at $e^{-\frac{i\pi}{4}}$ we find

$$f(z) = \int_0^{\infty e^{-\frac{i\pi}{2}}} \frac{e^{-z\zeta}}{1+\zeta^4}\, d\zeta + \frac{\pi i}{2} \frac{e^{-ze^{-\frac{i\pi}{4}}}}{e^{-\frac{3i\pi}{4}}} \tag{6.6.3}$$

The integral in Eq. (6.6.3) converges provided that $\text{Re}(z\zeta) > 0$ or $\cos(\theta - \frac{\pi}{2}) > 0$, or $0 < \theta < \pi$. The original integral representing $f(z)$ is well defined for $-\frac{\pi}{2} < \arg z < \frac{\pi}{2}$, while the integral in the right-hand side of Eq. (6.6.3) is well defined for $0 < \arg z < \pi$. Because there is an overlap between these two domains, it follows that Eq. (6.6.3) provides the concrete analytic continuation of $f(z)$. Therefore, we now know how to compute $f(z)$ for $-\frac{\pi}{2} < \arg z < \pi$. To evaluate the asymptotic behavior of $f(z)$ in this extended region, we note that in the pole contribution in Eq. (6.6.3)

$$e^{-ze^{-\frac{i\pi}{4}}} = e^{-|z|(\cos(\arg z - \frac{\pi}{4}) + i\sin(\arg z - \frac{\pi}{4}))}$$

hence $\exp(-ze^{-\frac{i\pi}{4}})$ is exponentially small in $-\frac{\pi}{4} < \arg z < \frac{3\pi}{4}$ but exponentially large in $\frac{3\pi}{4} < \arg z < \pi$. Equations (6.6.1) and (6.6.3) imply that

$$f(z) \sim \frac{1}{z} - \frac{4!}{z^5}, \qquad z \to \infty \quad \text{in} \quad -\frac{\pi}{4} < \arg z < \frac{3\pi}{4}$$

$$f(z) \sim \frac{\pi i}{2} e^{\frac{3i\pi}{4}} e^{-ze^{-\frac{i\pi}{4}}}, \qquad z \to \infty \quad \text{in} \quad \frac{3\pi}{4} < \arg z < \pi$$

$$\text{and} \quad -\frac{\pi}{2} < \arg z < -\frac{\pi}{4}$$

We see therefore that the asymptotic expansion of $f(z)$ changes discontinuously across the rays $\arg z = -\frac{\pi}{4}$ and $\frac{3\pi}{4}$. In fact the asymptotic sequence is entirely different in the two regions. In the first region there is an asymptotic power series representation of $f(z)$. In the second region there is an exponentially large contribution. On the rays $\arg z = -\frac{\pi}{4}, \frac{3\pi}{4}$ one needs both contributions in the asymptotic expansion.

The approach used in Example 6.6.1 provides an extension of Watson's Lemma for integrals where the large parameter is complex:

Lemma 6.6.1 (Watson's Lemma in the complex plane.) Consider the integral

$$I(z) = \int_0^\infty f(t)e^{-zt}\, dt, \qquad z \in \mathbb{C} \tag{6.6.4}$$

Assume that $f(t)$ has the asymptotic series expansion

$$f(t) \sim \sum_{n=0}^\infty a_n t^{\alpha n}, \qquad t \to 0^+; \qquad \alpha_0 > -1 \tag{6.6.5}$$

(a) If $f(t) = O(e^{\beta t})$ as $t \to \infty$ and $f(t)$ is continuous in $(0, \infty)$, then

$$I(z) \sim \sum_{n=0}^\infty a_n \frac{\Gamma(\alpha n + 1)}{z^{\alpha n+1}} \quad \text{as } z \to \infty \qquad \text{for } |\arg(z-\beta)| < \frac{\pi}{2} \tag{6.6.6}$$

(b) If $f(t) = O(e^{\beta t})$ as $t \to \infty$ and $f(t)$ is analytic in $\phi_1 \le \arg t \le \phi_2$, then $I(z)$ has the asymptotic behavior given by Eq. (6.6.6) but for z in

$$-\frac{\pi}{2} - \phi_2 < \arg(z-\beta) < \frac{\pi}{2} - \phi_1 \tag{6.6.7}$$

Proof The proof is similar to that of Lemma 6.2.2. We concentrate only on the differences between the two proofs:

(a) For convergence of the integral $I(z)$ as $t \to \infty$ one needs $\mathrm{Re}(z - \beta) > 0$, which implies that $|\arg(z - \beta)| < \pi/2$. Then the asymptotic expansion in Eq. (6.6.6) follows from the approach used in Lemma 6.2.2.

(b) We use the analyticity of $f(t)$ and rotate the contour by an angle ϕ such that $\phi_1 \le \phi \le \phi_2$. Then from Cauchy's Theorem, $I(z)$ is given by

$$I_\phi(z) = \int_0^{\infty e^{i\phi}} f(t)e^{-zt}\, dt \quad \text{for} \quad -\frac{\pi}{2} - \phi < \arg(z - \beta) < \frac{\pi}{2} - \phi \quad (6.6.8)$$

the constraint on arg z follows from the fact that for convergence of the integral one needs $-\frac{\pi}{2} < \arg(z - \beta) + \phi < \frac{\pi}{2}$. Since $-\phi_2 < \phi < \phi_1$, adding these constraints implies that for z in the sector given by Eq. (6.6.7) the large z asymptotics follow as usual.

Therefore $I(z)$ is valid for $-\frac{\pi}{2} < \arg(z - \beta) < \frac{\pi}{2}$ and $I_\phi(z)$ is valid for $-\frac{\pi}{2} - \phi < \arg(z - \beta) < \frac{\pi}{2} - \phi$. Because there exists an overlap between the two sectors

$$I(z) = I_\phi(z) \quad \text{for} \quad \frac{-\pi}{2} < \arg(z - \beta) < \frac{\pi}{2} - \phi \quad (6.6.9)$$

and $I_\phi(z)$ provides the analytic continuation of $I(z)$. It is clear that the largest possible sector in the z plane where the above procedure of analytic continuation holds, is given by Eq. (6.6.7). ∎

Example 6.6.2 Find the sector in the complex z plane in which the large z asymptotics of $I(z) = \int_0^\infty (1 + t)^{-1/2} e^{-zt}\, dt$ can be computed via Watson's Lemma.

Taking a branch cut along the negative real axis in the complex t plane, it follows that $(1 + t)^{-1/2}$ is analytic and single valued in $-\pi < \arg t < \pi$, that is, $\phi_1 = -\pi$, $\phi_2 = \pi$. Note that convergence of the integral is obtained for $-\pi/2 < \arg z < \pi/2$ and $\beta = 0$. Therefore Eq. (6.6.7) implies that Watson's lemma holds in

$$\frac{-3\pi}{2} < \arg z < \frac{3\pi}{2} \quad (6.6.10)$$

Example 6.6.3 Discuss the asymptotic expansion as $z \to \infty$ of $\mathrm{erf}(z) = \int_z^\infty e^{-t'^2}\, dt'$, z complex.

Using the change of variables $t' = z(1 + t)^{1/2}$ it follows that

$$\text{erf}(z) = \frac{z}{2} e^{-z^2} \int_0^\infty e^{-z^2 t} (1 + t)^{-1/2} \, dt \qquad (6.6.11)$$

Expanding $(1 + t)^{-1/2}$ in a Taylor series for small t, Eqs. (6.6.6) and (6.6.10) imply (note $\beta = 0$ and z is replaced by z^2, therefore arg z is replaced by 2 arg z in these equations)

$$\text{erf}(z) \sim e^{-z^2} \sum_{n=0}^\infty \frac{(-1)^n 1 \cdot 3 \cdots (2n - 1)}{2^{n+1} z^{2n+1}}, \qquad z \to \infty$$

$$\text{in} \quad -\frac{3\pi}{4} < \arg z < \frac{3\pi}{4} \qquad (6.6.12)$$

It is possible to extend even further the validity of $\text{erf}(z)$. Indeed writing $\int_{-\infty}^{-z} = \int_{-\infty}^{\infty} - \int_{-z}^{\infty}$ and computing the integral $\int_{-\infty}^{\infty}$ exactly, it follows that

$$\text{erf}(z) = \sqrt{\pi} - \text{erf}(-z) \qquad (6.6.13)$$

where we note that $\int_{-\infty}^{-z} e^{-t'^2} \, dt' = \int_z^\infty e^{-t'^2} \, dt'$ by replacing $t' \to -t'$. Let $\sum$ denote the sum appearing in Eq. (6.6.12). Using the transformation $t' = -z(1 + t)^{1/2}$, we find that the asymptotic expansion for large z of $-\text{erf}(-z)$ is also given by the right-hand side of Eq. (6.6.11). Thus the asymptotic expansion of $-\text{erf}(-z)$ is also given by $e^{-z^2} \sum$, provided that $-3\pi/4 < \arg (-z) < 3\pi/4$, or $\pi/4 < \arg z < 7\pi/4$ (where we have used $-z = |z| e^{-i\pi}$, so that arg $(-1) = -\pi$). Hence, using Eqs. (6.6.12) and (6.6.13), it follows that

$$\text{erf}(z) \sim \begin{cases} e^{-z^2} \sum & \text{for} \quad -\dfrac{3\pi}{4} < \arg z < \dfrac{3\pi}{4} \\ \sqrt{\pi} + e^{-z^2} \sum & \text{for} \quad \dfrac{\pi}{4} < \arg z < \dfrac{7\pi}{4} \end{cases} \quad \text{as } z \to \infty$$

$$(6.6.14)$$

From Eq. (6.6.14) we see that despite the fact that $\text{erf}(z)$ is analytic and single valued for any finite z, its asymptotic expansion for large z suffers a discontinuity as we vary $\theta = \arg z$ through $3\pi/4$. The function $\text{erf}(z)$ does have an essential singular point at $z = \infty$, hence the Stokes phenomenon is expected.

Furthermore, note that e^{-z^2} is exponentially large in the region $\pi/4 < \theta < 3\pi/4$. We say that e^{-z^2} is dominant and the other contribution $\sqrt{\pi}$ is recessive in this region. Indeed $\sqrt{\pi}$ is negligible in comparison with e^{-z^2} in this region. When $\theta = 3\pi/4$, e^{-z^2} is purely oscillatory, and $\sqrt{\pi}$ is no longer small by comparison. Such rays are called **antistokes lines** (the rays $\theta = \pi/4, 7\pi/4, \ldots$

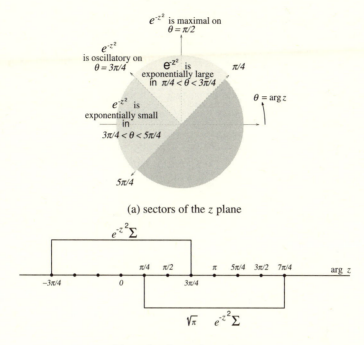

(a) sectors of the z plane

(b) overlapping regions of arg z

Fig. 6.6.1. Large z behavior of erf(z) as a function of $\theta = \arg z$

are also called antistokes lines). When $\theta = \pi/2$, Im $z^2 = 0$ and e^{-z^2} is maximal. Such rays are called the **stokes lines**. So when we traverse in $\theta = \arg z$ through $\theta = \pi/4$ to $\theta = 3\pi/4$ then we "pick up" an exponentially small term in comparison with $e^{-z^2} \Sigma$. This situation is graphically depicted in Figure 6.6.1.

The situation illustrated by Example 6.6.3 for erf(z) is actually typical of solutions to linear second-order differential equations such as those discussed in Section 3.7. Calling the solution of such a differential equation $y(z)$, its asymptotic expansion for $|z| \to \infty$ frequently takes the form

$$y(z) \sim A_+(z)e^{\phi_+(z)} + A_-(z)e^{\phi_-(z)} \tag{6.6.15}$$

where Re $\phi_+(z) \gg$ Re $\phi_-(z)$ in some sector $\alpha < \theta < \beta$, $\theta = \arg z$; the terms $+$ and $-$ refer to the dominant and recessive parts of the solution, respectively, and $A_\pm(z)$ are represented by asymptotic power series. In fact, the function $y = $ erf(z) satisfies the simple second-order differential equation

$$\frac{d^2y}{dz^2} + 2z\frac{dy}{dz} = 0 \tag{6.6.16a}$$

which can be readily solved via $\frac{d}{dz}\left(\frac{dy}{dz}e^{z^2}\right) = 0$ to find

$$y(x) = A \int_z^\infty e^{-t^2}\, dt + B \qquad (6.6.16b)$$

We can see that Eq. (6.6.16b) is consistent with the asymptotic representation (6.6.14) (in the region, $\frac{\pi}{4} < \arg z < \frac{3\pi}{4}$, $\phi_+ = 0$ and $\phi_- = -z^2$) and therefore with Eq. (6.6.15).

The rays for which Re ϕ_+ = Re ϕ_-, are called the antistokes lines, and in this case both terms of Eq. (6.6.15) are of the same order. The rays for which Im ϕ_+ = Im ϕ_-, are called the stokes lines, and in this case one exponential dominates over the other.

The situation is complicated because we have two asymptotic series with differing exponential factors. Apart from the regions near the antistokes lines where both exponentials are of the same asymptotic order, the usual asymptotic techniques do not work with both exponentials and their power series $A_\pm(z)$ present. This inadequacy manifests itself by the Stokes phenomenon, that is, discontinuities in asymptotic expansions.

*6.6.1 Smoothing of Stokes Discontinuities

The Stokes phenomenon has been a topic that has interested mathematicians for over a century. (A review can be found in Dingle (1973).) In this section we present a simple example that shows how, within the framework of conventional asymptotic analysis, one can analyze the fundamental issues involving Stokes discontinuities and develop a mathematical formulation in which there are no discontinuities, that is, the Stokes discontinuities are "smoothed." This point of view and analysis was recently developed by M. Berry (1989). In the example that follows we use an asymptotic method which is somewhat different from Berry's. (We appreciate important conversations with M. Kruskal on this issue.)

We consider the exponential integral

$$E(z) = \int_a^z \frac{e^t}{t}\, dt \qquad (6.6.17)$$

where a is an arbitrary constant to be fixed below. It is worth mentioning that $E(z)$ satisfies a simple second-order ODE:

$$y'' + \left(\frac{1}{z} - 1\right)y' = 0$$

as can be verified directly. Consequently, Eq. (6.6.17) falls within the framework of the linear ODEs mentioned at the end of the previous subsection, that is, Eqs. (6.6.15)–(6.6.16a,b). If we take $a = -\infty$, then it is easily seen that the value of $E(z)$ depends sensitively on the contour from $-\infty$ to z. For example, if we let $z = x + i\epsilon, x > 0$, that is, take the indented contour above the singularity at $t = 0$,

we obtain a different answer than if we let $z = x - i\epsilon$, $x > 0$, and use an indented contour *below* $t = 0$:

Clearly, as $\epsilon \to 0$ we have

$$\lim_{\epsilon \to 0}(E(x + i\epsilon) - E(x - i\epsilon)) = 2\pi i$$

Note that along a large circular contour from $z = -R$ to $z = iR$ as $R \to \infty$, the integral $\int_{-R}^{iR} \frac{e^t}{t} dt$ is small. Hence we can avoid any ambiguity by taking $a = i\infty$.

Let $z = |z|e^{i\theta}$, and for now restrict θ: $0 < \theta \le \pi/2$, $|z|$ large. Repeated integration by parts yields the following asymptotic series as $|z| \to \infty$:

$$E(z) = \int_{i\infty}^{z} \frac{e^t}{t} dt \sim \frac{e^z}{z}\left(1 + \frac{1}{z} + \frac{2!}{z^2} + \frac{3!}{z^3} + \cdots \frac{(N-1)!}{z^{N-1}} + \frac{N!}{z^N} + \cdots\right)$$

(6.6.18)

The series (6.6.18) is clearly asymptotic but not convergent. It is typical of such expansions that their terms decrease and then increase. It turns out that the approximate minimum term is obtained when $z \sim N$ for large N. This follows by using Stirling's formula (see Example 6.2.14), that is,

$$\Gamma(N + 1) = N! \sim \sqrt{2\pi N}\left(\frac{N}{e}\right)^N$$

and finding the value of z for which $N!/z^N \sim \sqrt{2\pi N}(\frac{N}{ez})^N$ is smallest. We find that $z \sim N$ and the value of the smallest term in the asymptotic expansion (6.6.18) is approximately $\frac{e^z N!}{z^{N+1}} \sim \sqrt{2\pi/N}$.

We argue that it is the difference between $E(z)$ and the first N terms of the asymptotic expansion that should have a smooth asymptotic structure for large N. So we consider

$$J(z; N) = \int_{i\infty}^{z} \frac{e^t}{t} \, dt - \frac{e^z}{z} \sum_{k=0}^{N-1} \frac{k!}{z^k}.$$

Integrating by parts yields

$$J(z; N) = N! \int_{i\infty}^{z} \frac{e^t}{t^{N+1}} \, dt$$

$$= N! \int_{i\infty}^{z} \frac{e^t - N \log t}{t} \, dt \tag{6.6.19}$$

Changing variables $t = N\tau$ and calling $z = N\hat{z}$ in Eq. (6.6.19) yields

$$J(z; N) = \frac{N!}{N^N} \int_{i\infty}^{\hat{z}} \frac{e^{N(\tau - \log \tau)}}{\tau} \, d\tau \tag{6.6.20}$$

At this point we can use the method of steepest descents to evaluate the asymptotic contribution as $N \to \infty$. Call the phase $\phi(\tau) = \tau - \log \tau$, that is, $e^{N(\tau - \log \tau)} = e^{N\phi(\tau)}$. We have $\phi'(\tau) = 1 - \frac{1}{\tau}$, hence there is a saddle point at $\tau = 1$. The steepest descent curves are $\text{Im}\phi(\tau) = 0$. Calling $\tau = u + iv$, they are given by $v - \tan(v/u) = 0$, and near the saddle point $u = 1$, $v = 0$ we have $v - (\frac{v}{u} - \frac{1}{3}(\frac{v}{u})^3 + \cdots) = 0$ or $v(1 - u) - \frac{v^3}{3u^2} + \cdots = 0$. It is clear that we can deform our contour onto the steepest descent path (see Figure 6.6.2).

Because the endpoint z will play an important role in the asymptotic calculation, we will derive the necessary approximation, rather than using standard formulae. In the neighborhood of the saddle point, $\tau = 1 + i\zeta$, $\phi(\zeta) = 1 + i\zeta - \log(1 + i\zeta)$, hence using the Taylor series for $\log(1 + i\zeta)$

$$\phi(\zeta) \sim 1 + i\zeta - \left(i\zeta - \frac{(i\zeta)^2}{2} + \cdots \right) \sim 1 - \frac{\zeta^2}{2}$$

The deformation of the contour in the neighborhood of the saddle therefore yields

$$J(z, N) \sim \frac{N! e^N}{N^N} \int_{\infty}^{(\hat{z}-1)e^{-i\pi/2}} e^{-N\zeta^2/2} i \, d\zeta \tag{6.6.21}$$

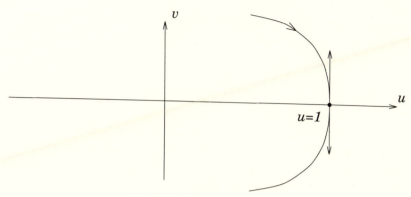

Fig. 6.6.2. Steepest descent path

or calling $\zeta = \sqrt{2/N}\,\eta$, and using Stirling's approximation, from Example (6.2.14) $N! \sim \sqrt{2\pi N}\left(\frac{N}{e}\right)^N$, we have

$$J(z, N) \sim 2\sqrt{\pi}i \int_{\infty}^{(\hat{z}-1)e^{-i\pi/2}\sqrt{N/2}} e^{-\eta^2}\, d\eta \qquad (6.6.22)$$

In fact, we now see explicitly how the smoothing of Stokes discontinuity occurs. Recall that $\hat{z} = z/N$. The smoothed variable is

$$\tilde{z} = (\hat{z} - 1)e^{-i\pi/2}\sqrt{\frac{N}{2}} = \left(\frac{z}{N} - 1\right)e^{-i\pi/2}\sqrt{\frac{N}{2}} \qquad (6.6.23)$$

(Note that the quantity $(\hat{z} - 1)$ is purely imaginary on the steepest descent path near the saddle.) So if (a) $(\hat{z} - 1)e^{-i\pi/2} > 0$, then $J(z, N) \sim 0$; if (b) $\hat{z} = 1$, then $J(z, N) \sim i\pi$; and if (c) $(\hat{z} - 1)e^{-i\pi/2} < 0$, then $J(z, N) \sim 2\pi i$. The smoothing takes place on the scale $(\hat{z} - 1)e^{-i\pi/2} \sim C/\sqrt{N}$, where C is a constant – that is, in a "layer of width" $O(1/\sqrt{N})$ around the point $\hat{z} = 1$, (along the imaginary direction), or in terms of z, in a width of $O(\sqrt{N})$ around the point $z = N$. Hence the asymptotic expansion of $E(z)$ can be expressed as follows:

$$E(z) \sim \frac{e^z}{z}\left(1 + \frac{1}{z} + \frac{2!}{z^2} + \cdots + \frac{(N-1)!}{z^{N-1}}\right) + 2\pi i S(z) \qquad (6.6.24)$$

where $S(z) = \frac{1}{2\pi i} J(z, N)$. $J(z, N)$ varies smoothly and is given asymptotically by Eq. (6.6.22). Note that in an appropriate scale, that is, $z \sim O(N)$, or

$\tilde{z} = O(1)$, $S(\theta)$ is a rapidly changing function

$$S(z) = \begin{cases} 0 & \text{if } \tilde{z} > 0 \\ \frac{1}{2} & \text{if } \tilde{z} = 0 \\ 1 & \text{if } \tilde{z} < 0 \end{cases} \qquad (6.6.25)$$

where $\tilde{z}$ is defined in Eq. (6.6.23).

We see that the rapid variation is in the vicinity of the Stokes ray $\text{Im} z = 0$, where the term $\frac{e^z}{z}(1 + 1/z + \cdots)$ is maximally dominant as compared to a constant. When we cross the Stokes ray we see that the asymptotic expansion of $E(z)$, Eq. (6.6.24), is such that the term $(e^z/z)(1 + \cdots)$ "picks up" a constant.

In fact, the same analysis can be developed for solutions of second-order ODEs whose asymptotic form is given by Eq. (6.6.15). The coefficient $A_-(z)$ in Eq. (6.6.15) changes rapidly in the neighborhood of the Stokes line $\text{Im} \, \phi_+ = \text{Im} \, \phi_-$, but in suitable coordinates the variation is smooth[6].

Problems for Section 6.6

1. Investigate Stokes phenomena for the integral

    $$I(k) = \int_0^1 e^{kt^3} \, dt \quad \text{as } k \to \infty; \qquad k = re^{i\theta}$$

    (a) Show that there are Stokes lines at $|\arg k| = \pi/2$.
    (b) If $0 \le \theta < \frac{\pi}{2}$, show that

    $$\int_0^1 e^{(r\cos\theta + ir\sin\theta)t^3} \, dt \to \infty$$

    (c) If $\theta = \frac{\pi}{2}$, show that

    $$I(k) \sim \frac{e^{i\pi/6}}{3r^{1/3}} \Gamma\left(\frac{1}{3}\right)$$

    (d) If $\frac{\pi}{2} < \theta < \frac{3\pi}{2}$, show that

    $$I(k) \sim \frac{1}{3(|r|\cos\theta)^{\frac{1}{3}}} \Gamma\left(\frac{1}{3}\right)$$

2. Show that as $k \to \infty$

    $$\int_0^\infty \frac{e^{kx}}{\Gamma(x+1)} \, dx \sim \frac{-1}{k} + \exp(\exp k), \quad \text{Re} \, k > 0, \, |\text{Im} \, k| < \pi$$

3. Show that as $k \to \infty$

$$\int_1^\infty e^{-kx}(x^2-1)^{\nu-\frac{1}{2}}\, dx$$

$$\sim e^{-k}\frac{\Gamma(\nu+\frac{1}{2})}{\sqrt{2k}}\left(\frac{2}{k}\right)^\nu \sum_{n=0}^\infty \frac{\Gamma\left(n+\nu+\frac{1}{2}\right)}{\Gamma\left(\nu+\frac{1}{2}-n\right)n!}(2k)^{-n}$$

for $|\arg k| < \frac{\pi}{2}$.

4. Show that as $k \to \infty$

$$\int_0^{\frac{\pi}{4}} e^{-k\sec x}\, dx \sim \sqrt{\frac{\pi}{2k}}e^{-k}, \quad |\arg k| < \frac{\pi}{2}$$

5. In this problem we investigate the asymptotic expansion of

$$I(k) = \int_0^1 \frac{1}{\sqrt{t}}e^{ik(t^2+t)}\, dt \quad \text{as } k \to \infty$$

Note if we call $\phi(t) = t^2 + t$, then $\operatorname{Im}\phi(t) = y(1+2x)$, where $t = x + iy$.

(a) Show that the path of integration can be deformed as indicated in Figure 6.6.3. Where the steepest descent contours C_1 and C_2 are given by

$$C_1 : x^2 - y^2 + x = 0$$
$$C_2 : x^2 - y^2 + x = 1$$

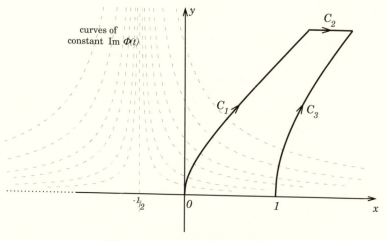

Fig. 6.6.3. Figure for Problem 6.6.5

(b) Show that by changing variable $si = t^2 + t$ and $2 + i\eta = t^2 + t$ in the integral associated with the contours C_1 and C_2, respectively, one obtains

$$I(k) = \int_0^\infty \frac{e^{-ks} i \, ds}{\left(-\frac{1}{2} + \frac{1}{2}\sqrt{1+4is}\right)^{\frac{1}{2}} \sqrt{1+4is}}$$

$$- \int_0^\infty \frac{e^{2ik} e^{-k\eta} i \, d\eta}{\left(-\frac{1}{2} + \frac{1}{2}\sqrt{9+4i\eta}\right)^{\frac{1}{2}} \sqrt{9+i4\eta}}$$

(c) Show that by using Watson's Lemma

$$I(k) = \sqrt{\frac{\pi}{k}} e^{i\frac{\pi}{4}} \left(1 - \frac{3i}{4k} - \frac{105}{32k^2} + \cdots\right)$$

$$+ \frac{i}{3} e^{2ik} \left(-\frac{1}{k} + \frac{7i}{18k^2} + \frac{111}{324k^3} + \cdots\right)$$

*6.7 Related Techniques

*6.7.1 WKB Method

Frequently, differential equations arise in which a small or large parameter appears. In this section we briefly introduce the reader to the most basic of these problems. The equation we consider is

$$y'' + \lambda^2 u(x) y = 0 \tag{6.7.1}$$

where λ is a large parameter and we take x to be real. This equation is related to the time-independent Schrödinger equation, where λ is inversely proportional to Planck's constant. It was analyzed by Wentzel, Kramers, and Brillouin (hence the name WKB) in the early part of this century; especially important was their contribution regarding turning points, which is discussed later in this section. Actually there is a much larger history owing to the work of Liouville, Green, Jeffreys, etc. (see also the discussion in Carrier et al. 1966). The asymptotic solution of Eq. (6.7.1) can be obtained by looking for a solution of the form

$$y(x) = e^{\lambda \phi(x)} \left(z_0(x) + \frac{1}{\lambda} z_1(x) + \cdots\right) \tag{6.7.2}$$

Substituting Eq. (6.7.2) into Eq. (6.7.1), and factoring out $e^{\lambda\phi}$, one finds

$$\lambda^2(\phi'^2 + u(x))\left(z_0(x) + \frac{1}{\lambda}z_1(x) + \frac{1}{\lambda^2}z_2(x) + \cdots\right)$$

$$+ \lambda(2\phi'z_0' + \phi''z_0) + (2\phi'z_1 + \phi''z_1 + z_0'') + \cdots = 0 \qquad (6.7.3)$$

Solving Eq. (6.7.3) asymptotically yields the formal WKB solutions (for a rigorous discussion see Olver (1974)). We obtain only the leading-order contribution here. At $O(\lambda^2)$ we have $\phi'^2 = -u(x)$, which has two solutions denoted as $\phi_\pm(x)$:

$$\phi_\pm(x) = \pm i \int u^{1/2}(x)\,dx \qquad (6.7.4)$$

At $O(\lambda)$, $2\phi'z_0' + \phi''z_0 = 0$; hence $z_0'/z_0 = -\phi''/2\phi'$, so that $z_{0\pm} = \text{const}/\sqrt{\phi'_\pm}$ or

$$z_{0\pm} = \frac{C_\pm}{u^{1/4}(x)} \qquad (6.7.5)$$

Eqs. (6.7.4) and (6.7.5) yield the asymptotic solution for large λ

$$y(x) \sim \frac{1}{u^{1/4}(x)}\left(C_+ e^{i\lambda\int u^{1/2}dx} + C_- e^{-i\lambda\int u^{1/2}dx}\right) \qquad (6.7.6)$$

assuming $u(x) \neq 0$.

However, if $u(x)$ vanishes, then Eq. (6.7.6) breaks down. In such a case, we need a transition region. Suppose the only zero of $u(x)$ has the form $u(x) = -kx^m + \cdots$, $k > 0$ (say m is a positive integer), then $y(x)$ near $x = 0$ satisfies

$$y'' - k\lambda^2 x^m y = 0. \qquad (6.7.7)$$

The point near $x = 0$, is referred to as a **turning point** (a change in the behavior of the asymptotic expansion). We can rescale to eliminate the dependence on λ; that is, letting $z = \lambda^q x$. Eq. (6.7.7) yields

$$\frac{d^2y}{dz^2} - k\lambda^{2-qm-2q}z^m y = 0$$

We take $q = 2/(m + 2)$ to find the "universal" differential equation in this region:

$$\frac{d^2y}{dz^2} - kz^m y = 0 \qquad (6.7.8)$$

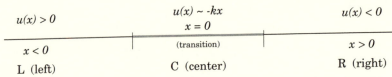

$u(x) > 0$	$u(x) \sim -kx$ $x = 0$	$u(x) < 0$
$x < 0$	(transition)	$x > 0$
L (left)	C (center)	R (right)

Fig. 6.7.1. Regions of solution to $y'' + \lambda^2 u(x)y = 0$ with a turning point $u(x) = -kx + \cdots$, as $x \to 0$

One can use power series to solve the linear equation (6.7.8). In fact Eq. (6.7.8) is related to Bessel functions of an imaginary argument. Namely, $y = x^{1/2} Z_p$ $(i\sqrt{k}\frac{x^s}{s})$, where $s = \frac{m+2}{2}$, $p = \frac{1}{m+2}$, and $w = Z_p(i\sqrt{k}x)$ is either of the two linearly independent solutions that satisfy $x^2 w'' + xw' - (kx^2 + p^2)w = 0$; see also Section 3.7.

Suppose $m = 1$. Then the solution of Eq. (6.7.8) is given by Airy functions:

$$ y = C_1 \text{Ai}\left(k^{1/3}z\right) + C_2 \text{Bi}\left(k^{1/3}z\right) \tag{6.7.9} $$

where $\text{Ai}(x)$ and $\text{Bi}(x)$ are the two standard solutions of Eq. (6.7.8) with $k = 1$ and $m = 1$ (see, also Eqs. (4.6.40a,b)). Asymptotic expansions of $\text{Ai}(x)$ and $\text{Bi}(x)$ for $x \to \pm\infty$ are well known. They can be computed by the methods of this chapter from the integral representations of these functions; see Eq. (4.6.40b). We simply quote them here (see, e.g., Abramowitz and Stegun, 1965). As $x \to +\infty$

$$ \text{Ai}(x) \sim \frac{1}{2\sqrt{\pi}x^{1/4}} e^{-\frac{2}{3}x^{3/2}} $$

$$ \text{Bi}(x) \sim \frac{1}{\sqrt{\pi}x^{1/4}} e^{\frac{2}{3}x^{3/2}} \tag{6.7.10} $$

and as $x \to -\infty$

$$ \text{Ai}(x) \sim \frac{1}{\sqrt{\pi}|x|^{1/4}} \sin\left(\frac{2}{3}|x|^{3/2} + \frac{\pi}{4}\right) $$

$$ \text{Bi}(x) \sim \frac{1}{\sqrt{\pi}|x|^{1/4}} \cos\left(\frac{2}{3}|x|^{3/2} + \frac{\pi}{4}\right). \tag{6.7.11} $$

Recalling that $z = \lambda^{2/3}x$, the asymptotic solution through a transition region takes the following form (see Figure 6.7.1); we also transform Eq. (6.7.9) from

z to x with $z = \lambda^{2/3}x$:

$$y_L(x) \sim \frac{1}{|u(x)|^{1/4}} \left(C_+^L e^{i\lambda \int_0^x u^{1/2}\, dx} + C_-^L e^{-i\lambda \int_0^x u^{1/2}\, dx} \right)$$

$$y_C(x) \sim C_1 \text{Ai}\left(k^{1/3}\lambda^{2/3}x \right) + C_2 \text{Bi}\left(k^{1/3}\lambda^{2/3}x \right)$$

$$y_R(x) \sim \frac{1}{|u(x)|^{1/4}} \left(C_+^R e^{\lambda \int_0^x |u|^{1/2}\, dx} + C_-^R e^{-\lambda \int_0^x |u|^{1/2}\, dx} \right)$$

where $u(x)^{1/2}$ stands for the positive square root.

These solutions match smoothly from one region to the other provided the constants are chosen properly. This can be achieved as follows. Taking the outer limits of the inner solution and the inner limits of the outer solution,

$$y_C(x) \underset{x \to +\infty}{\sim} \frac{1}{\sqrt{\pi}(k^{1/2}\lambda)^{1/6}x^{1/4}} \left(C_1 \frac{e^{-\theta}}{2} + C_2 e^{\theta} \right),$$

$$y_C(x) \underset{x \to -\infty}{\sim} \frac{1}{\sqrt{\pi}(k^{1/2}\lambda)^{1/6}(-x)^{1/4}} \left(C_1 \left(\frac{e^{i\theta+i\pi/4} - e^{i\theta-i\pi/4}}{2i} \right) \right.$$
$$\left. + C_2 \left(\frac{e^{i\theta+i\pi/4} - e^{i\theta-i\pi/4}}{2} \right) \right),$$

$$y_R(x) \underset{x \to 0^+}{\sim} \frac{1}{k^{1/4}x^{1/4}} \left(C_+^R e^{\theta} + C_-^R e^{-\theta} \right),$$

$$y_L(x) \underset{x \to 0^-}{\sim} \frac{1}{k^{1/4}(-x)^{1/4}} \left(C_+^L e^{-\theta} + C_-^L e^{\theta} \right),$$

where $\theta = \frac{2}{3}k^{1/2}\lambda|x|^{3/2}$. These formulae match smoothly so long as

$$C_+^R = \frac{C_2}{\sqrt{\pi}} \left(\frac{k}{\lambda} \right)^{1/6},$$

$$C_-^R = \frac{C_1}{2\sqrt{\pi}} \left(\frac{k}{\lambda} \right)^{1/6},$$

$$C_+^L = \frac{1}{2\sqrt{\pi}} \left(\frac{k}{\lambda} \right)^{1/6} \left(C_1 e^{i\pi/4} + C_2 e^{-i\pi/4} \right),$$

$$C_-^L = \frac{1}{2\sqrt{\pi}} \left(\frac{k}{\lambda} \right)^{1/6} \left(C_1 e^{-i\pi/4} + C_2 e^{i\pi/4} \right). \tag{6.7.12}$$

Higher-order turning points, that is, $m > 1$, and problems where $u(x)$ has zeroes at more than one point can be studied in a similar fashion. The Airy functions $\text{Ai}(x)$ and $\text{Bi}(x)$ are often called *turning point functions* or *connection functions*. We see that they change character from exponential decay $(x \rightarrow +\infty)$ to oscillation $(x \rightarrow -\infty)$.

*6.7.2 The Mellin Transform Method

The methods discussed so far are most useful for evaluating integrals that involve a large parameter appearing in the exponential. Now we mention a method that, when applicable, can be rather versatile because the large parameter need not appear exponentially. This method, although often quick and easy to apply, is not widely known. (We appreciate L. Glasser for bringing this analysis to our attention).

It turns out that given a function $f(x)$, $x\epsilon[0, \infty)$, there exists an integral representation of $f(x)$ that is quite convenient for discussing its behavior for large x. This representation is

$$f(x) = \frac{1}{2\pi i} \int_{c-i\infty}^{c+i\infty} x^{-s} F(s) \, ds, \qquad \alpha < c < \beta \qquad (6.7.13)$$

where $F(s)$ is called the **Mellin transform of** $f(x)$, and which is given by

$$F(s) = \int_0^\infty x^{s-1} f(x) \, dx, \qquad \alpha < \text{Re}(s) < \beta \qquad (6.7.14)$$

(A derivation is sketched in Exercise 6.7.5.) The real constants α and β are determined by the behavior of $f(x)$ as $x \rightarrow \infty$ and $x \rightarrow 0^+$ and are chosen in such a way that the above integrals converge. Extensive tables of these pairs are available. The derivation of Eqs. (6.7.13) and (6.7.14) follows from the Fourier transform formulae. From (6.7.13) we can formally obtain the Fourier transform by using the substitutions $\log x = u$, $ik = s$, and noting that $x^s = e^{s \log x}$.

Because $x^{-s} = e^{-s \log x}$, it follows that as $x \rightarrow \infty$ the integrand in Eq. (6.7.14) decays exponentially fast for s in the right half of the complex s plane. Thus if $F(s)$ is meromorphic in this part of the complex s plane, the asymptotic expansion of $f(x)$ can be found from Eq. (6.7.13) by closing the contour into the right half plane and summing the residues of the integrand.

Example 6.7.1 Find the asymptotic behavior of

$$f(x) = \int_0^\infty \sin(x\tau)e^{-\tau^h}\,d\tau, \quad h > 0 \qquad (6.7.15)$$

for large x.

Our aim is to write $f(x)$ in the form (6.7.13). To achieve this we find the Mellin transform of $f(x)$, that is, compute the integral

$$F(s) = \int_0^\infty \left(\int_0^\infty x^{s-1}\sin(x\tau)\,dx \right)e^{-\tau^h}\,d\tau, \qquad -1 < \mathrm{Re}(s) < 1$$

First, from contour integration, we find that

$$\int_0^\infty x^{s-1}\sin(x\tau)\,dx = \tau^{-s}\Gamma(s)\sin\left(\frac{\pi s}{2}\right), \qquad -1 < \mathrm{Re}(s) < 1 \quad (6.7.16)$$

This integral can be computed by letting $\sin x\tau = \frac{e^{ix\tau}-e^{-ix\tau}}{2i}$ and using equation (6.3.5) with $\gamma = s - 1$, $\nu = \tau$ and $p = 1$. Thus from Eq. (6.7.16) it follows that

$$F(s) = \Gamma(s)\sin\left(\frac{\pi s}{2}\right)\int_0^\infty \tau^{-s}e^{-\tau^h}\,d\tau$$

By changing variables ($u = \tau^h$), this latter integral can also be evaluated in terms of a Gamma function; thus we have

$$F(s) = \frac{1}{h}\Gamma(s)\Gamma\left(\frac{1-s}{h}\right)\sin\left(\frac{\pi s}{2}\right) \qquad (6.7.17a)$$

Substituting $F(s)$ from Eq. (6.7.17) into Eq. (6.7.13), we find

$$f(x) = \frac{1}{2\pi i h}\int_{c-i\infty}^{c+i\infty} x^{-s}\Gamma(s)\Gamma\left(\frac{1-s}{h}\right)\sin\left(\frac{\pi s}{2}\right)ds \qquad (6.7.17b)$$

Because x^{-s} decays exponentially fast as $x \to +\infty$ in the right half s plane (the formula for $f(x)$ was derived when $-1 < \mathrm{Re}(s) < 1$, however it can be established that the above the integral converges for $\mathrm{Re}(s) > 0$), we wish to calculate any possible residues from pole singularities in the integrand. Neither $\Gamma(s)$ nor $\sin(\pi s/2)$ have poles, but in fact $\Gamma(\frac{1-s}{h})$ has poles for $s = 1 + mh$, $m = 0, 1, 2, \ldots$. This follows from the fact that $\Gamma(z)$ has poles for any negative

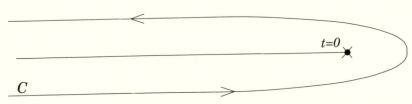

Fig. 6.7.2. Hankel's contour for the Gamma function

integer z. To see this, we note that the integral used to define the Gamma function (see Eq. (4.5.31))

$$\Gamma(z) = \int_0^\infty e^{-t} t^{z-1} \, dt \tag{6.7.18}$$

is useful to us only for Re $z > 0$ due to the singular behavior at $t = 0$. A suitable definition of the gamma function for all z, owing to Hankel, is

$$\Gamma(z) = \frac{1}{2i \sin \pi z} \int_C e^t t^{z-1} \, dt \tag{6.7.19}$$

where C is the contour given in Figure 6.7.2.

If Re $z > 0$, then the integral in Eq. (6.7.19) agrees with that of Eq. (6.7.18). This can be seen by carrying out the calculation with $t = e^{i\pi} r$ on the top part and $t = e^{-i\pi} r$ on the bottom part of the contour and using the definition of $\sin \pi z$. But for Re $z \le 0$, only Eq. (6.7.19) converges. Equation (6.7.19) implies that the only singularities of $\Gamma(z)$ are poles at $z = -n, n = 0, 1, 2, \ldots$. The residue of one of these poles is given by

$$\operatorname{Res}\Gamma(z = -n) = \frac{(-1)^n}{2\pi i} \int_C e^t t^{-(n+1)} \, dt \tag{6.7.20}$$

Expanding e^t in a Taylor series, the integral over C picks up a nontrivial contribution due to the term $\frac{1}{n!t}$, hence

$$\operatorname{Res}\Gamma(z = -n) = \frac{(-1)^n}{n!} \tag{6.7.21}$$

Thus the singularities of the integrand of the integral (6.7.17b) in the right half s plane are the simple poles $s = 1 + mh, m = 0, 1, 2, \ldots$. The corresponding residues for $f(x)$ as $x \to \infty$ are found from Eq. (6.7.21); that is, the residue of $\Gamma(\frac{1-s}{h})$ at $s = 1 + mh$ is $(-1)^m/m!$. Thus the full asymptotic expansion of $f(x)$ is

$$f(x) \sim \frac{1}{x} \sum_{m=0}^{\infty} \frac{(-1)^m}{m!} \frac{\Gamma(1+mh)}{x^{mh}} \cos\left(\frac{\pi mh}{2}\right), \qquad x \to \infty$$

Remark As $x \to 0^+$, x^{-s} decays exponentially fast for s in the left half of the complex s plane. So by closing the contour to the left, we can find in a similar way the behavior of $f(x)$ for small x.

Problems for Section 6.7

1. Consider the boundary value problem

    $$\epsilon y'' + y = 0, \quad y(0) = 0, \quad y(1) = 1, \quad \epsilon \to 0$$

    The above equation is a particular case of Eq. (6.7.1): ($u(x) = 1$, $\lambda = \frac{1}{\sqrt{\epsilon}}$). Show that in this case the WKB approximation yields the exact solution

    $$y(x) = \frac{\sin\left(\frac{x}{\sqrt{\epsilon}}\right)}{\sin\left(\frac{1}{\sqrt{\epsilon}}\right)}$$

2. Consider the equation

    $$\frac{d^4 y}{dx^4} + \lambda^4 u(x) y = 0$$

    (a) Suppose $u(x) > 0$ on $-\infty < x < \infty$. Find the leading term of the solution of the above equation as $\lambda \to \infty$.

    (b) Suppose $u(x)$ has a simple zero at $x = x_0$, that is, $u(x) = u_0(x - x_0) + \cdots$. Show that the equation governing the solution of the above equation in the local neighborhood near $x = x_0$ is

    $$\frac{d^4 y}{dx^4} + \lambda^4 u'(x_0)(x - x_0) y \sim 0$$

3. Consider the equation

    $$\frac{dy}{dx} + \lambda u(x) y + q(x) y^2 = 0$$

    where $u(x) > 0$ on $-\infty < x < \infty$.

(a) Show that the leading term of the asymptotic expansion of the solution as $\lambda \to \infty$ is

$$y \sim A \exp\left(-\lambda \int_{-\infty}^{x} u(x)\,dx\right), \qquad A \text{ constant}$$

(b) Show that the above equation can be transformed to a linear equation for $W(x)$ using: $y(x) = \alpha(x)\frac{W'}{W}$.
Derive the asymptotic expansion for $W(x)$ and then obtain the result in (a) for $y(x, \lambda)$.

4. Consider the differential equation

$$-\epsilon^2 y'' + \left(e^{-\frac{x^2}{4}} - 1\right)y = 0, \qquad \epsilon \to 0$$

This is a prototypical example of a second order turning point because since near $x = 0$ this equation becomes

$$\epsilon^2 y'' + \frac{x^2}{4}y = 0, \qquad \epsilon \to 0$$

(i.e., $u(x)$ in Eq. (6.7.1) has a double zero) Let I and III denote the regions for which $x > 0$ and $x < 0$, respectively; let II denote the neighborhood of $x = 0$.

(a) Show that in II

$$y_{\text{II}}(x) \sim \alpha D_{-\frac{1}{2}}\left(e^{-\frac{i\pi}{4}}\frac{x}{\sqrt{\epsilon}}\right) + \beta D_{-\frac{1}{2}}\left(-e^{-\frac{i\pi}{4}}\frac{x}{\sqrt{\epsilon}}\right)$$

where $D_\nu(t)$ denotes the parabolic cylinder function that solves

$$-y'' + \left(\frac{t^2}{4} - \nu - \frac{1}{2}\right)y = 0$$

and α and β are arbitrary constants.

(b) Using the well-known asymptotic formulae

$$D_\nu(t) \sim \begin{cases} t^\nu e^{-\frac{t^2}{4}} & t \to \infty, \quad |\arg t| < \frac{3\pi}{4} \\ t^\nu e^{-\frac{t^2}{4}} - \frac{\sqrt{2\pi}}{\Gamma(-\nu)}e^{i\pi\nu}t^{-\nu-1}e^{\frac{t^2}{4}}, & t \to \infty, \quad \frac{\pi}{4} < \arg t < \frac{5\pi}{4} \end{cases}$$

where $\Gamma(z)$ is the Gamma function, show that

$$y_{II}(x)$$

$$\sim \begin{cases} \epsilon^{\frac{1}{4}}x^{-\frac{1}{2}}\left[\left(\alpha e^{\frac{i\pi}{8}} + \beta e^{-\frac{3i\pi}{8}}\right)e^{\frac{ix^2}{4\epsilon}} + \beta\sqrt{2}e^{\frac{i\pi}{8}}e^{-\frac{ix^2}{4\epsilon}}\right], & \frac{x}{\sqrt{\epsilon}} \to \infty \\ \epsilon^{\frac{1}{4}}(-x)^{-\frac{1}{2}}\left[\left(\alpha e^{-\frac{3i\pi}{8}} + \beta e^{\frac{i\pi}{8}}\right)e^{\frac{ix^2}{4\epsilon}} + \alpha\sqrt{2}e^{\frac{i\pi}{8}}e^{-\frac{ix^2}{4\epsilon}}\right], & \frac{x}{\sqrt{\epsilon}} \to -\infty \end{cases}$$

(c) Show that in region I

$$y_I \sim A\left(1 - e^{-\frac{x^2}{4}}\right)^{-\frac{1}{4}}\exp\left[-\frac{i}{\epsilon}\int_0^x \sqrt{1 - e^{-\frac{t^2}{4}}}\,dt\right]$$

where A is a constant. Furthermore, show that

$$y_I \sim \begin{cases} Ae^{-\frac{iI}{\epsilon}}e^{-\frac{ix}{\epsilon}}, & x \to \infty \\ A\left(\frac{2}{x}\right)^{\frac{1}{2}}e^{-\frac{ix^2}{4\epsilon}}, & x \to 0^+ \end{cases}$$

where

$$I \equiv \int_0^\infty \left(\sqrt{1 - e^{-\frac{t^2}{4}}} - 1\right)dt$$

(d) By matching y_{II} and y_I show that

$$A = \epsilon^{\frac{1}{4}}\beta e^{\frac{i\pi}{8}}, \qquad \alpha e^{\frac{i\pi}{8}} + \beta e^{-\frac{3i\pi}{8}} = 0$$

(e) Show that in III

$$y_{III} \sim \left(1 - e^{-\frac{x^2}{4}}\right)^{-\frac{1}{4}}\left\{B\exp\left[\frac{i}{\epsilon}\int_x^0 \sqrt{1 - e^{-\frac{t^2}{4}}}\,dt\right]\right.$$

$$\left. + C\exp\left[-\frac{i}{\epsilon}\int_x^0 \sqrt{1 - e^{-\frac{t^2}{4}}}\,dt\right]\right\}$$

where B and C are constants. Furthurmore, show that

$$y_{III} \sim \begin{cases} Be^{\frac{iI}{\epsilon}}e^{-\frac{ix}{\epsilon}} + Ce^{-\frac{iI}{\epsilon}}e^{\frac{ix}{\epsilon}}, & x \to -\infty \\ \left(-\frac{2}{x}\right)^{\frac{1}{2}}\left(Be^{\frac{ix^2}{4\epsilon}} + Ce^{-\frac{ix^2}{4\epsilon}}\right), & x \to 0^- \end{cases}$$

(f) By matching y_{III} and y_{II}, show that

$$B\sqrt{2} = \epsilon^{\frac{1}{4}}\left(\alpha e^{-\frac{3i\pi}{8}} + \beta e^{\frac{i\pi}{8}}\right), \qquad C = \alpha e^{\frac{i\pi}{8}}\epsilon^{\frac{1}{4}}$$

Note that the steps (d) and (f) provide global matching across region II.

5. In Chapter 4 we derived the direct and inverse Laplace transforms. Namely, if $g(\tau)$ is such that $\int_0^\tau |g(\xi)| \, d\xi < \infty$ for all $\tau > 0$ and $g(\tau) = O(e^{\alpha\tau})$ as $\tau \to \infty$ for some real constant α, then the *right one-sided* Laplace transform pair is given by

$$G(s) = \int_0^\infty e^{-s\tau} g(\tau) \, d\tau, \qquad g(\tau) = \frac{1}{2\pi i} \int_{c-i\infty}^{c+i\infty} e^{s\tau} G(s) \, ds$$

for Re $c > \alpha$.

(a) Show that the following analogous formulae hold for the *left one-sided* Laplace transform pair if (β real), $\int_0^\tau |g(-\xi)| \, d\xi < \infty$, and $g(\tau) = O(e^{\beta\tau})$ as $\tau \to -\infty$:

$$G(s) = \int_{-\infty}^0 e^{-s\tau} g(\tau) \, d\tau, \qquad g(\tau) = \frac{1}{2\pi i} \int_{c-i\infty}^{c+i\infty} e^{s\tau} G(s) \, ds$$

for Re $c < \beta$.

(b) Conclude that for $\alpha <$ Re $c < \beta$ the *two-sided* Laplace transform pair holds:

$$G(s) = \int_{-\infty}^\infty e^{-s\tau} g(\tau) \, d\tau, \qquad g(\tau) = \frac{1}{2\pi i} \int_{c-i\infty}^{c+i\infty} e^{s\tau} G(s) \, ds$$

(c) Transform the pair in (b) using the variable $\tau = -\log t$, and define $f(t) = g(-\log t)$ to find the Mellin transform:

$$M(s) = \int_0^\infty t^{s-1} f(t) \, dt, \qquad f(t) = \frac{1}{2\pi i} \int_{c-i\infty}^{c+i\infty} t^{-s} M(s) \, ds$$

for $\alpha <$ Re $c < \beta$.

6. Mellin (Parseval's formula for Mellin transforms): Consider the integral

$$I = \frac{1}{2\pi i} \int_{c-i\infty}^{c+i\infty} H(s) F(1-s) \, ds$$

where $H(s)$, $F(s)$ are the Mellin transforms associated with $h(t)$, $f(t)$, respectively, and assume $H(s)$, $F(1-s)$ are analytic in some common vertical strip: $\alpha < s < \beta$ and $\alpha <$ Re $c < \beta$, α, β real.

(a) Show that

$$I = \frac{1}{2\pi i} \int_{c-i\infty}^{c+i\infty} F(1-s) \left(\int_0^{\infty} h(t) t^{s-1} \, dt \right) ds$$

(b) Assuming the integrals can be interchanged, derive the *Parseval* relation for Mellin transforms:

$$I = \int_0^{\infty} h(t) f(t) \, dt = \frac{1}{2\pi i} \int_{c-i\infty}^{c+i\infty} H(s) F(1-s) \, ds$$

(Note: if $\int_{-\infty}^{\infty} |F(1-c-iy)| \, dy < \infty$, $\int_{-\infty}^{\infty} t^{c-1} |h(t)| \, dt < \infty$, then the integrals in part (a) can be interchanged.)

7. Establish the following Mellin transforms:

(a) $f(t) = e^{-t}$, $\quad F(s) = \Gamma(s) \quad \mathrm{Re}\, s > 0$

(b) $f(t) = e^{it}$, $\quad F(s) = e^{\frac{i\pi s}{2}} \Gamma(s)$, $\quad 0 < \mathrm{Re}\, s < 1$

(c) $f(t) = \frac{1}{1+t}$, $\quad F(s) = \frac{\pi}{\sin \pi s}$, $\quad 0 < \mathrm{Re}\, s < 1$

(d) $f(t) = \frac{1}{t} e^{-\frac{1}{t}}$, $\quad F(s) = \Gamma(1-s)$, $\quad \mathrm{Re}\, s < 1$

These particular transforms are useful in applications.

8. Consider the integral

$$I = \int_0^{\infty} h(\lambda t) f(t) \, dt$$

(a) Use the Parseval formula for Mellin transforms derived in Example 6.7.6 to establish

$$I = \int_0^{\infty} h(\lambda t) f(t) \, dt = \frac{1}{2\pi i} \int_{c-i\infty}^{c+i\infty} \lambda^{-s} H(s) F(1-s) \, ds$$

where c lies in a (common) strip of analyticity of $G(s) = H(s) F(1-s)$.

(b) Assume that

$$\lim_{|y| \to \infty} G(x+iy) = 0, \qquad \int_{-\infty}^{\infty} |G(R+iy)| \, dy < \infty$$

Show that

$$I = -\sum \text{res}\{\lambda^{-s}G(s)\} + \frac{1}{2\pi i} \int_{R-i\infty}^{R+i\infty} \lambda^{-s}G(s)\,ds$$

and that as $\lambda \to \infty$

$$\frac{1}{2\pi i} \int_{R-i\infty}^{R+i\infty} \lambda^{-s}G(s)\,ds$$

$$= \frac{1}{2\pi} \int_{-\infty}^{\infty} \lambda^{-R-iy}G(R+iy)\,dy = O(\lambda^{-R}).$$

The above formulae are useful for asymptotic evaluation of certain classes of integrals.

9. Consider

$$I = \int_0^\infty \frac{e^{-\frac{1}{t}}}{t(1+\lambda t)}\,dt; \qquad \lambda \to \infty$$

(a) Show that Example 6.7.8 applies where $f(t) = \frac{e^{-\frac{1}{t}}}{t}$, $h(t) = \frac{1}{1+t}$. Use the results of Example 6.7.8 to establish that

$$I = \frac{1}{2\pi i} \int_{c-i\infty}^{c+i\infty} \lambda^{-s} \frac{\pi \Gamma(s)}{\sin \pi s}\,ds, \quad 1 > \text{Re } c > 0$$

(b) Show that for $s = x + iy$, $y \to \infty$

$$\frac{\pi}{\sin \pi s} \sim O(e^{-\pi|y|})$$

$$\Gamma(s) = O\left(|y|^{x-\frac{1}{2}}e^{-\frac{\pi}{2}|y|}\right)$$

(for $\Gamma(s)$ use Stirling's formula of the gamma function derived in Section 6.2.3)) and therefore the asymptotic expansion of I is obtained from the poles of $\frac{\pi}{\sin \pi s}$ for Re $s > 0$; that is,

$$I \sim \sum_{m=0}^{\infty} \frac{(-1)^m m!}{\lambda^{m+1}}$$

10. Consider

$$I(\lambda) = \int_0^\infty \frac{f(t)}{1+\lambda t}\,dt, \qquad \lambda \to \infty,$$

where

$$f(t) \sim_{t \to 0^+} \sum_{m=0}^{\infty} \alpha_m t^{a_m}, \quad a_m \neq \text{integer}, \quad \int_0^{\infty} |f(t)| \, dt < \infty$$

(a) Using the results of Examples 6.7.7 and 6.7.8, show that $H(s) = \frac{\pi}{\sin \pi s}$, and hence

$$I(\lambda) = \frac{1}{2\pi i} \int_{c-i\infty}^{c+i\infty} \lambda^{-s} \left(\frac{\pi}{\sin \pi s} \right) F(1-s) \, ds$$

(b) Establish the asymptotic formula,

$$I(\lambda) \sim \sum_{m=0}^{\infty} (-1)^m \lambda^{-1-m} F(1-m) - \sum_{m=0}^{\infty} \lambda^{-(1+a_m)} \left(\frac{\pi \alpha_m}{\sin \pi a_m} \right)$$

that is, that the asymptotic form of $I(\lambda)$ is composed of the pole contributions from $H(s)$ and $F(1-s)$, respectively, and that the poles of $F(1-s)$ are determined by the $t \to 0^+$ behavior of $f(t)$.

7

Riemann–Hilbert Problems

7.1 Introduction

It is remarkable that a large number of diverse problems of physical and mathematical significance involve the solution of the so-called Riemann–Hilbert (RH) problem. Let us mention a few such problems.

(1) Find a function $w(z) = u(x, y) + iv(x, y)$, u, v real, analytic inside a region enclosed by a contour C, such that

$$\alpha(t)u(t) + \beta(t)v(t) = \gamma(t), \qquad t \text{ on } C \qquad (7.1.1)$$

where α, β, and γ are given, real functions. In the special case of $\alpha = 1$, $\beta = 0$, C a circle this problem reduces to deriving the well-known Poisson formula (see Problem 10 of Section 2.6 for the case of a circle, and the discussion in Section 4.6 for when C is the real axis.)

(2) Solve the linear singular integral equation,

$$f(t) + \fint_a^b \frac{\alpha(t')}{t' - t} f(t') \, dt' = \beta(t) \qquad (7.1.2)$$

where $\alpha(t)$ and $\beta(t)$ are given functions and $\fint$ denotes a principal value integral (see Sections 4.3 and 7.2). Such equations arise in many applications. For example, the equation $\fint_0^b \frac{f(t') \, dt'}{t' - t} = \beta$, plays an important role in airfoil theory.

(3) Solve the linear integral equation

$$f(t) + \int_0^\infty \alpha(t - t') f(t') \, dt' = \beta(t), \qquad t > 0 \qquad (7.1.3)$$

where α and β are given, integrable functions. (In Section 7.3 some of the basic notions associated with integral equations will be introduced).

514

(4) Solve the time independent wave equation (Helmholtz equation)

$$\varphi_{xx} + \varphi_{yy} + k^2\varphi = 0, \quad -\infty < x < \infty, \quad y \geq 0, \quad k \text{ constant, real,}$$
(7.1.4)

where $\varphi(x, 0) = f(x)$ for $-\infty < x \leq 0$, $(\partial\varphi/\partial y)(x, 0) = g(x)$ for $0 < x < \infty$, and φ satisfies an appropriate boundary ("radiation") condition at infinity.

(5) Derive the inverse of the Radon transform. The Radon transform is a generalized Fourier transform and plays a fundamental role in the mathematical foundation of computerized tomography.

(6) Solve an inverse scattering problem associated with the time-independent Schrödinger equation

$$\psi_{xx} + (q(x) + k^2)\psi = 0, \quad -\infty < x < \infty \quad (7.1.5)$$

that is, reconstruct the potential $q(x)$ from appropriate inverse scattering data. Inverse problems arise in many areas of application, for example, geophysics, image reconstruction, quantum mechanics, etc. In many cases, they can be solved using Riemann–Hilbert problems.

(7) Solve the following initial value problem for the Korteweg–deVries (KdV) equation

$$u_t + u_{xxx} + uu_x = 0, \quad -\infty < x < \infty, \quad t > 0$$
$$u(x, 0) = u_0(x); \quad u \to 0 \text{ as } |x| \to \infty$$
(7.1.6)

Many other nonlinear PDEs in as well as many nonlinear ODEs can also be related to RH problems.

The class of functions in which the above problems are solved will be stated when these problems are considered in detail in this chapter.

The above list is by no means exhaustive. Several aspects of RH theory were motivated and developed owing to the relation of RH problems with problems arising in physical application, for example, elasticity and hydrodynamics (Freund, 1990; Gakhov, 1966; Muskhelishvili, 1977).

Problems 1–5 above are associated with **scalar** RH problems. The simplest such problem involves finding two analytic functions $\Phi^+(z)$ and $\Phi^-(z)$, defined inside and outside a closed contour C of the complex z plane such that

$$\Phi^+(t) - g(t)\Phi^-(t) = f(t), \quad t \text{ on } C \quad (7.1.7)$$

for given functions $g(t)$ and $f(t)$. This problem can be solved in closed form; its solution is intimately related to the Cauchy type integral

$$\Phi(z) = \frac{1}{2\pi i} \int_C \frac{\varphi(t)\, dt}{z - t} \quad (7.1.8)$$

where φ is a certain function related to f and g. A generalization of the above problem allows C to be an open contour; this problem can also be solved in closed form. In this chapter we will not use the notation $\oint_C$; the specification of the RH problem will suffice to denote whether the contour is closed or open. Cauchy type integrals will be discussed in Section 7.2, and scalar RH problems for both open and closed contours will be discussed in Section 7.3. Several applications, including the solution of Problems 1–5 will be given in Section 7.4.

Problem 1 above was first formulated by Riemann in 1851. In 1904, Hilbert reduced this problem to a RH problem of the form (7.1.7), which he also expressed in terms of a singular integral equation of the form (7.1.2). In 1908, Plemelj gave the first closed form solution of a simple RH problem (an RH problem of "zero index," see Section 7.3). The closed form solution of a general scalar RH problem was given by Gakhov (1938). Integral equations of the form (7.1.3) were studied by Carleman, who solved such an equation in 1932 using a method similar to the so-called Wiener–Hopf method. This method, introduced originally in 1931, was also in connection with the solution of a particular integral equation of the type (7.1.3). The Wiener–Hopf method, which also can be used for the solution of Problem 4, actually reduces to solving a certain RH problem. (The interested reader can find relevant references in the books of Gakhov (1966) and Muskhelishvili (1977).) The derivation of transforms, such as the Radon transform, via RH techniques appears to be rather recent (Fokas and Novikov, 1991).

Problems 6 and 7 are associated with **vector** RH problems. The formulation of such problems is similar to that for scalar ones, where Φ^+ and Φ^- are now vectors instead of scalars. Unfortunately, in general, vector RH problems cannot be solved in closed form; their solution can be given in terms of linear integral equations of Fredholm type. In most of this chapter we concentrate on scalar RH problems. Vector RH problems will be introduced briefly in Section 7.5.

There exists a significant generalization of the RH problem that is called a $\bar{\partial}$ (DBAR) problem. This problem involves solving the equation

$$\frac{\partial \Phi(x, y)}{\partial \bar{z}} = g(x, y), \qquad z \in D, \quad z = x + iy \tag{7.1.9}$$

for $\Phi(x, y)$, where g is given, D is some domain in the complex z plane, and $\bar{z}$ is the complex conjugate of z. To appreciate the relationship between $\bar{\partial}$ and RH problems, it is convenient to consider the particular RH problem

$$\Phi^+(x) - \Phi^-(x) = f(x), \qquad -\infty < x < \infty \tag{7.1.10}$$

where $\Phi^+(z)$ and $\Phi^-(z)$ are analytic in the upper and lower half complex z plane. Let $\Phi(z) = \Phi^+(z)$ for $y > 0$ and $\Phi(z) = \Phi^-(z)$ for $y < 0$. Solving the RH problem (7.1.10) means finding a function $\Phi(z)$ that is analytic in the entire z complex plane except on the real axis, where it has a prescribed "jump." The quantity $\partial\Phi/\partial\bar{z}$ measures the departure of Φ from analyticity; if $\Phi(z)$ is analytic everywhere in D, then $\partial\Phi/\partial\bar{z} = 0$ in D. But for Eq. (7.1.10), $\partial\Phi/\partial\bar{z}$ vanishes everywhere in the z complex plane *except* on the real axis, where it is given by $f(x)\delta(y)$ and where $\delta(y)$ denotes the Dirac delta function. Thus the RH problem (7.1.10) can be viewed as a special case of a DBAR problem where $g(x, y) = f(x)\delta(y)$.

DBAR problems have recently appeared in applications in connection with the solution of multidimensional inverse problems and with the solution of certain nonlinear PDEs in x, y, t. A brief introduction to DBAR problems will be given in Section 7.6. Applications of vector RH problems and of DBAR problems will be discussed in Section 7.7.

7.2 Cauchy Type Integrals

Consider the integral

$$\Phi(z) = \frac{1}{2\pi i} \int_L \frac{\varphi(\tau)}{\tau - z}\, d\tau \tag{7.2.1}$$

where L is a smooth curve (L may be an arc or a closed contour) and $\varphi(\tau)$ is a function satisfying the **Hölder condition** on L, that is for any two points τ and τ_1 on L

$$|\varphi(\tau) - \varphi(\tau_1)| \leq \Lambda|\tau - \tau_1|^\lambda, \qquad \Lambda > 0,\ 0 < \lambda \leq 1 \tag{7.2.2}$$

If $\lambda = 1$, the Hölder condition becomes the so-called Lipschitz condition. For example, a differentiable function $\varphi(\tau)$ satisfies the Hölder (Lipschitz) condition with $\lambda = 1$. (This follows from the definition of a derivative.) If $\lambda > 1$ on L, it follows, from the definition of the derivative, that $d\varphi/d\tau = 0$ and hence $\varphi = \text{const}$ on L, which is a trivial case. We have encountered integrals similar to Eq. (7.2.1) before, for example, Cauchy's Integral Theorem in Section 2.6. The integral (7.2.1) is well defined and $\Phi(z)$ is analytic provided that z is not on L (see Section 3.2, Eqs. (3.2.28) and (3.2.29)). We also note that from the series expansion, as $|z| \to \infty$ off L, we have $\Phi(z) \sim c/z$ where $c = -\frac{1}{2\pi i} \int_L \varphi(\tau)\, d\tau$. However, if z is on L, this integral becomes ambiguous; to give it a unique meaning we must know *how* z approaches L. We denote by $+$ the region that is on the left of the positive direction of L and by $-$ the region

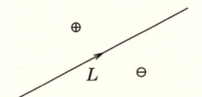

Fig. 7.2.1. Regions on either side of L

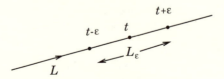

Fig. 7.2.2. Definition of L_ε

on the right (see Figure 7.2.1). It turns out that $\Phi(z)$ has a limit $\Phi^+(t)$, t on L, when z approaches L along a curve entirely in the $+$ region. Similarly, $\Phi(z)$ has a limit $\Phi^-(t)$, when z approaches L along a curve entirely in the $-$ region. These limits, which play a fundamental role in the theory of RH problems, are given by the so-called Plemelj formulae.

Lemma 7.2.1 (Plemelj Formulae) Let L be a smooth contour (closed or open) and let $\varphi(\tau)$ satisfy a Hölder condition on L. Then the Cauchy type integral $\Phi(z)$, defined in Eq. (7.2.1), has the limiting values $\Phi^+(t)$ and $\Phi^-(t)$ as z approaches L from the left and the right, respectively, and t is not an endpoint of L. These limits are given by

$$\Phi^\pm(t) = \pm\frac{1}{2}\varphi(t) + \frac{1}{2\pi i}\fint_L \frac{\varphi(\tau)}{\tau - t}\,d\tau \qquad (7.2.3^\pm)$$

In Eq. $(7.2.3^\pm)$, $\fint$ denotes the principal value integral defined by

$$\fint_L \frac{\varphi(\tau)\,d\tau}{\tau - t} = \lim_{\varepsilon \to 0} \int_{L - L_\varepsilon} \frac{\varphi(\tau)\,d\tau}{\tau - t} \qquad (7.2.4)$$

where L_ε is the part of L that has length 2ε and is centered around t, as depicted in Figure 7.2.2.

Proof The derivation of Eq. (7.2.3) is straightforward if $\varphi(\tau)$ is analytic at t. In this case, we can use the Cauchy Theorem to deform the contour L to two contours, $L - L_\varepsilon$ and C_ε, where C_ε is a semicircle of radius ε centered at t (see

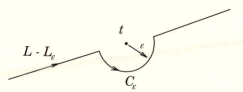

Fig. 7.2.3. Contours $L - L_\varepsilon$ and C_ε

Figure 7.2.3).

$$\Phi^+(t) = \lim_{\varepsilon \to 0} \frac{1}{2\pi i} \int_{L-L_\varepsilon} \frac{\varphi(\tau)}{\tau - t} \, d\tau + \lim_{\varepsilon \to 0} \frac{1}{2\pi i} \int_{C_\varepsilon} \frac{\varphi(\tau)}{\tau - t} \, d\tau \qquad (7.2.5)$$

Using $\tau = t + \varepsilon e^{i\theta}$, the second integral in Eq. (7.2.5) becomes $\frac{1}{2\pi i} \int_{-\pi}^{0} \varphi(t) i \, d\theta$
$= \frac{1}{2}\varphi(t)$, and Eq. (7.2.5) reduces to Eq. (7.2.3$^+$); similarly for Eq. (7.2.3$^-$). If
$\varphi(\tau)$ satisfies a Hölder condition at t, then the proof is more complicated; we
give only the essential idea. In this case it is convenient to consider

$$\Psi(z) = \frac{1}{2\pi i} \int_L \frac{\varphi(\tau) - \varphi(t)}{\tau - z} \, d\tau \qquad (7.2.6)$$

and to assume that L is closed. (If L is open, we can supplement it by an
arbitrary curve so that it becomes a closed one, provided we set $\varphi(\tau) = 0$ on
the additional curve.) Let D^+, D^- denote the areas inside and outside L. Then
the Cauchy Theorem implies

$$\frac{1}{2\pi i} \int_L \frac{d\tau}{\tau - z} = \begin{cases} 1, & z \in D^+ \\ 0, & z \in D^- \\ \dfrac{1}{2}, & z \in L \end{cases}$$

Thus, Eq. (7.2.6) yields

$$\Psi^+(t) = \Phi^+(t) - \varphi(t), \qquad \Psi^-(t) = \Phi^-(t) \qquad (7.2.7)$$

It can be shown (Gakhov, 1966; Muskhelishvili, 1977) that if $\varphi(t)$ satisfies a
Hölder condition, then the function $\Psi(z)$ behaves as a continuous function as
z passes through t. Taking z on L in Eq. (7.2.6), we have

$$\Psi^+(t) = \Psi^-(t) = \frac{1}{2\pi i} \oint \frac{\varphi(\tau)}{\tau - t} \, d\tau - \frac{\varphi(t)}{2\pi i} \oint \frac{d\tau}{\tau - t}$$

$$= \frac{1}{2\pi i} \oint \frac{\varphi(\tau) \, d\tau}{\tau - t} - \frac{\varphi(t)}{2} \qquad (7.2.8)$$

Equations (7.2.7) and (7.2.8) imply Eq. (7.2.3$^\pm$). ∎

In the above formulation we have assumed that L is a finite contour, otherwise $\varphi(\tau)$ must satisfy an additional (uniformity) condition. Suppose, for example, that L is the real axis; then we assume that $\varphi(\tau)$ satisfies a Hölder condition for all finite τ, and that as $t \to \pm\infty$, $\varphi(\tau) \to \varphi(\infty)$, where

$$|\varphi(\tau) - \varphi(\infty)| < \frac{M}{|\tau|^\mu}, \qquad M > 0, \ \mu > 0 \qquad (7.2.9)$$

Equations $(7.2.3^\pm)$ are equivalent to

$$\Phi^+(t) - \Phi^-(t) = \varphi(t), \qquad \Phi^+(t) + \Phi^-(t) = \frac{1}{\pi i}\oint \frac{\varphi(\tau)}{\tau - t}\, d\tau \quad (7.2.10)$$

Equations (7.2.10) will be extensively used hereafter. The function $\Phi(z)$ is said to be **sectionally analytic**; see Section 7.3. Functions that are the boundary values of $\Phi(z)$ as $z \to L$ from the left and the right will sometimes be referred to as $\oplus$ and $\ominus$ functions.

Example 7.2.1 Find $\Phi^\pm(t)$ corresponding to $\Phi(z) = \frac{1}{2\pi i}\oint_C (\tau + \frac{1}{\tau})\frac{d\tau}{\tau - z}$, where C is the unit circle (see Figure 7.2.4).

To compute $\Phi^+(z)$, we consider $\Phi(z)$ with z inside the circle, thus using contour integration $\Phi^+(z) = z + \frac{1}{z} - \frac{1}{z} = z$. Similarly, to compute $\Phi^-(z)$ we consider $\Phi(z)$ with z outside the circle and use contour integration to find $\Phi^-(z) = -\frac{1}{z}$. Therefore on the contour $z = t$

$$\Phi^+(t) = t, \qquad \Phi^-(t) = -\frac{1}{t}$$

Also, using contour integration, it follows that $\frac{1}{i\pi}\!\!\oint(\tau + \frac{1}{\tau})\frac{d\tau}{\tau - t} = (t - \frac{1}{t})$; therefore Eqs. (7.2.10) are verified. We note that $\Phi^+(z)$ is indeed analytic

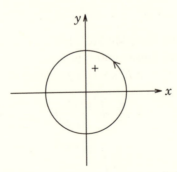

Fig. 7.2.4. Inside and outside unit circle C

inside the unit circle, while $\Phi^-(z)$ is analytic outside the unit circle. Taking this into consideration, as well as that $\Phi^+(t) - \Phi^-(t) = \varphi(t)$, it follows that in this simple example, $\Phi^+(z) = z$ and $\Phi^-(z) = -1/z$ could have been found by inspection.

Example 7.2.2 Find $\Phi^\pm(t)$ corresponding to $\Phi(z) = \frac{1}{2\pi i} \int_{-\infty}^{\infty} \frac{2}{\tau^2+1} \frac{d\tau}{\tau-z}$. We split $2/(t^2+1)$ as follows:

$$\frac{2}{t^2+1} = \frac{i}{t+i} - \frac{i}{t-i}$$

Furthermore, $\frac{i}{(z-i)}$ is analytic in the lower half plane, while $\frac{i}{(z+i)}$ is analytic in the upper half plane. Hence this suggests that

$$\Phi^+(t) = \frac{i}{t+i} \qquad \Phi^-(t) = \frac{i}{t-i}$$

These formulae can be verified by contour integration. For example, computing $\Phi(z)$ with z in the upper half plane (we can consider a large semicircular contour in the lower half plane), we find $\Phi^+(z) = \frac{i}{(z+i)}$.

The above analysis is sufficient for studying RH problems on closed contours. However, in order to study RH problems for open contours, we need to analyze the behavior of $\Phi(z)$ near the end points. Also, as is necessary in some applications we allow $\varphi(t)$ to have integrable singularities at the endpoints.

Lemma 7.2.2 Consider the Cauchy type integral

$$\Phi(z) = \frac{1}{2\pi i} \int_a^b \frac{\varphi(\tau)}{\tau - z} d\tau \tag{7.2.11}$$

where $\varphi(t)$ satisfies a Hölder condition on any closed interval $a'b'$ of the arc ab, except possibly at the ends where it satisfies

$$\varphi(t) = \frac{\tilde{\varphi}(t)}{(t-c)^\gamma}, \qquad \gamma = \alpha + i\beta, \qquad 0 \le \alpha < 1 \tag{7.2.12}$$

where c is either a or b, α and β are real constants, and $\tilde{\varphi}(t)$ satisfies a Hölder condition. Then the following limits are valid:

(1) $\gamma = 0$.

(a) As $z \to c$, with z not on the contour

$$\Phi(z) = \pm \frac{\varphi(c)}{2\pi i} \log \frac{1}{z-c} + \Phi_0(z) \qquad (7.2.13)$$

(b) $t \to c$,

$$\Phi(t) = \pm \frac{\varphi(c)}{2\pi i} \log \frac{1}{t-c} + \Psi_0(t) \qquad (7.2.14)$$

In these formulae the upper sign is taken for $c = a$ and the lower for $c = b$. The function $\Phi_0(z)$ is bounded and tends to a definite limit as $z \to c$ along any path. The function $\Psi_0(t)$ satisfies a Hölder condition near c. The function $\log \frac{1}{z-c}$ denotes any branch that is single valued near c with the branch cut taken to go through the contour.

(2) $\gamma \neq 0$.

(a) As $z \to c$, with z not on the contour

$$\Phi(z) = \pm \frac{e^{\pm \gamma \pi i}}{2i \sin \gamma \pi} \frac{\tilde{\varphi}(c)}{(z-c)^\gamma} + \Phi_0(z) \qquad (7.2.15)$$

(b) as $t \to c$

$$\Phi(t) = \pm \frac{\cot \gamma \pi}{2i} \frac{\tilde{\varphi}(c)}{(t-c)^\gamma} + \Psi_0(t) \qquad (7.2.16)$$

In these formulae the signs are chosen as in part **1**. If $\alpha = 0$, then the behavior of $\Phi_0(z)$ and $\Psi_0(t)$ are as in part **1**, while if $\alpha > 0$

$$|\Phi_0(z)| < \frac{A_0}{|z-c|^{\alpha_0}}, \qquad \Psi_0(t) = \frac{\tilde{\Psi}_0(t)}{|t-c|^{\alpha_0}}, \qquad \alpha_0 < \alpha$$

where A_0 and α_0 are real constants and $\tilde{\Psi}_0(t)$ satisfies a Hölder condition near c. The function $(z-c)^\gamma$ is any branch that is single valued near c with the branch cut taken to go through the contour and with the value $(t-c)^\gamma$ on the left side of the contour.

Proof The proof is rather involved; it can be found in Muskhelishvili (1977) or Gakhov (1966). However, the leading order asymptotic form of these formulae can be intuitively understood by an application of Plemelj formulae. Let us first consider the case of $\gamma \neq 0$ with $c = a$.

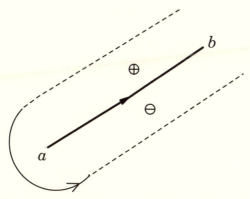

Fig. 7.2.5. Contour around the branch cut from a to ∞ through b

We take the branch cut of $(z - a)^\gamma$ from the endpoint a to ∞ going through b, and we select the branch that tends to $(t - a)^\gamma$ on the left of the cut, that is

$$(t - a)^\gamma = [(t - a)^\gamma]^+ \qquad (7.2.17a)$$

To find the value of $(t - a)^\gamma$ on the right of the cut, we follow the contour indicated in Figure 7.2.5, thus

$$[(t - a)^\gamma]^- = e^{2i\pi\gamma}[(t - a)^\gamma]^+ \qquad (7.2.17b)$$

Equations (7.2.17a,b) can be written as

$$[(t - a)^{-\gamma}]^+ - [(t - a)^{-\gamma}]^- = (1 - e^{-2i\pi\gamma})(t - a)^{-\gamma}$$

or

$$\frac{e^{i\pi\gamma}}{2i \sin \pi\gamma}[(t - a)^{-\gamma}]^+ - \frac{e^{i\pi\gamma}}{2i \sin \pi\gamma}[(t - a)^{-\gamma}]^- = (t - a)^{-\gamma} \qquad (7.2.18)$$

Equation (7.2.18) shows that the function $(t - a)^{-\gamma}$ can be written as the difference of $\oplus$ and $\ominus$ functions. Furthermore, we expect that the largest contribution to the integral will come from the locations where $\varphi(t)$ is singular, that is, the endpoints. Hence as $z \to a$

$$\Phi(z) \sim \frac{\tilde{\varphi}(a)}{2\pi i} \int_a^b \frac{(\tau - a)^{-\gamma}}{\tau - z} \, d\tau$$

(The reader can find the basic definitions of asymptotic symbols "$\sim$" and order relations "big-O" and "small-O" in Section 6.1.) Thus using the Plemelj formulae and Eq. (7.2.18) we find, for z not on the contour

$$\Phi(z) \sim \frac{e^{\gamma \pi i}}{2i \sin \gamma \pi} \frac{\tilde{\varphi}(a)}{(z-a)^\gamma}$$

Also, for $z = t$ on the contour

$$\Phi(t) = \frac{1}{2}(\Phi^+(t) + \Phi^-(t)) \sim \frac{e^{\gamma \pi i}}{2i \sin \gamma \pi} \frac{\tilde{\varphi}(a)}{2} [[(t-a)^{-\gamma}]^+ + [(t-a)^{-\gamma}]^-]$$

$$= \frac{\cot \gamma \pi}{2i} \frac{\tilde{\varphi}(a)}{(t-a)^\gamma}$$

where we have used $[(t-a)^{-\gamma}]^+ = (t-a)^{-\gamma}, [(t-a)^{-\gamma}]^- = e^{-2i\pi\gamma}(t-a)^{-\gamma}$. ∎

In the case $\gamma = 0$ near $z = a$, we rewrite the integral (7.2.11) as

$$\Phi(z) = \frac{1}{2\pi i} \int_a^b \frac{\varphi(\tau) - \varphi(a)}{\tau - z} d\tau + \frac{\varphi(a)}{2\pi i} \int_a^b \frac{d\tau}{\tau - z}$$

Hence by integration

$$\Phi(z) = \frac{\varphi(a)}{2\pi i} \int \left(\frac{z-b}{z-a}\right) + \frac{1}{2\pi i} \int_a^b \frac{\varphi(\tau) - \varphi(a)}{\tau - z} d\tau.$$

The second term is not singular, as $z \to a$, thus the leading asymptotic contribution is given by

$$\Phi(z) \sim \frac{\varphi(a)}{2\pi i} \log\left(\frac{1}{z-a}\right)$$

As we approach the contour to leading order, we obtain the same dominant contribution on either side of the branch cut, so that on the contour in the neighborhood of $t = a$

$$\Phi(t) \sim \frac{\varphi(a)}{2\pi i} \log\left(\frac{1}{t-a}\right)$$

Problems for Section 7.2

1. Consider the integral

$$\Phi(z) = \frac{1}{2\pi i} \int_C \frac{d\tau}{\tau(\tau - 4)(\tau - z)}$$

where C denotes the unit circle.

(a) Use Cauchy's Theorem to show that $\Phi(z) = \frac{1}{4(z-4)} \equiv \Phi^+(z)$ for z inside the unit circle, and $\Phi(z) = \frac{1}{4z} \equiv \Phi^-(z)$ outside the unit circle.

(b) Use Cauchy's Theorem to show that if t is on the unit circle, then the principal value integral is given by

$$\Phi(t) = \frac{1}{2\pi i} \oint_C \frac{d\tau}{\tau(\tau - 4)(\tau - t)} = \frac{t - 2}{2t(t - 4)}$$

(c) Use the results of (a) and (b) to verify the Plemelj formulae for the above integral.

2. Consider the integral

$$\Phi(z) = \frac{1}{2\pi i} \int_{-\infty}^{\infty} \frac{1}{\tau^2 + 4} \frac{d\tau}{\tau - z}$$

Show that $\Phi(z) = -\frac{1}{4i(z+2i)} \equiv \Phi^+(z)$ if z is in the upper half plane and $\Phi(z) = \frac{-1}{4i(z-2i)} \equiv \Phi^-(z)$ if z is in the lower half plane.
Hint: Use

$$\frac{1}{\tau^2 + 4} = \frac{1}{4i} \left(\frac{1}{\tau - 2i} - \frac{1}{\tau + 2i} \right)$$

3. Consider the integral

$$\Phi(z) = \frac{1}{2\pi i} \int_L \frac{\varphi(\tau)}{\tau - z} d\tau$$

where L is a finite, closed contour. Assume that the mth derivative of $\varphi(t)$ satisfies the Hölder condition.

(a) Show that

$$\Phi^{(m)}(z) = \frac{m!}{2\pi i} \int_L \frac{\varphi(\tau)}{(\tau - z)^{m+1}} d\tau$$

where $\Phi^{(m)}(z)$ denotes the m th derivative.

(b) Use integration by parts to obtain

$$\Phi^{(m)}(z) = \frac{1}{2\pi i} \int_L \frac{\varphi^{(m)}(\tau)}{\tau - z} d\tau$$

(c) Use the Plemelj formulae to establish

$$\Phi^{(m)+}(t) = \frac{1}{2}\varphi^{(m)}(t) + \frac{1}{2\pi i}\oint_L \frac{\varphi^{(m)}(\tau)}{\tau - t}\, d\tau$$

$$\Phi^{(m)-}(t) = -\frac{1}{2}\varphi^{(m)}(t) + \frac{1}{2\pi i}\oint_L \frac{\varphi^{(m)}(\tau)}{\tau - t}\, d\tau$$

4. Show that the change of variables

$$\zeta = \frac{z - i}{z + i}, \qquad \sigma = \frac{\tau - i}{\tau + i}$$

maps a Cauchy type integral over the real axis in the z plane to a Cauchy type integral over the unit circle in the ζ plane.

5. Consider the integral

$$U(z) = \frac{1}{2\pi}\int_0^{2\pi} u(\theta)\frac{e^{i\theta} + z}{e^{i\theta} - z}\, d\theta$$

where u is a real function. This integral is usually referred to as a Schwarz type integral. Establish the following relationship between Schwarz type and Cauchy type integrals

$$U(z) = \frac{1}{2\pi i}\int_C \frac{2u(-i\log\tau)}{\tau - z}\, d\tau - \frac{1}{2\pi}\int_0^{2\pi} u(\theta)\, d\theta$$

where C denotes the unit circle.
   Hint: Use the transformation $\tau = e^{i\theta}$ and note that

$$\frac{\tau + z}{\tau - z} = -1 + \frac{2\tau}{\tau - z}$$

6. Let $u(x, y) + iv(x, y)$ be a function that is analytic inside the unit circle. It can be shown (see also Section 7.4) that

$$u(x, y) + iv(x, y) = \frac{1}{2\pi}\int_0^{2\pi} u(\theta)\frac{e^{i\theta} + z}{e^{i\theta} - z}\, d\theta + iv_0$$

where $u(\theta)$ is the limiting value of $u(x, y)$ as $x + iy$ approaches the unit circle, and v_0 is a real constant.

(a) Using the result of Problem 7.2.5, show that the limit of the above expression as $x + iy$ approaches the unit circle is

$$u(\theta) + iv(\theta) = u(\theta) + \frac{1}{2\pi i}\oint_C \frac{2u(-i\log\tau)}{\tau - t}\,d\tau$$
$$-\frac{1}{2\pi}\int_0^{2\pi} u(\theta)\,d\theta + iv_0$$

(b) Combining the two integrals above, establish that

$$v(\theta) = \frac{1}{2\pi i}\oint_0^{2\pi} u(\varphi)\frac{e^{i\varphi} + e^{i\theta}}{e^{i\varphi} - e^{i\theta}}\,d\varphi + v_0$$

(c) Show that this equation can also be written as

$$v(\theta) = -\frac{1}{2\pi}\int_0^{2\pi} u(\varphi)\cot\left(\frac{\varphi - \theta}{2}\right)d\varphi + v_0$$

This equation, which expresses the boundary value of the imaginary part of an analytic function in terms of the real part, is called the Hilbert inversion formula.

(d) Consider the function $-i(u + iv)$. For this function the real part is v and the imaginary part is $-u$. Deduce that

$$u(\theta) = \frac{1}{2\pi}\int_0^{2\pi} v(\varphi)\cot\left(\frac{\varphi - \theta}{2}\right)d\varphi + u_0$$

where u_0 is a real constant.

See Section 5.9.3 where an application of these inversion formulae is discussed.

7.3 Scalar Riemann–Hilbert Problems

The machinery introduced in Section 7.2, namely, the formulae that express the behavior of a Cauchy integral as z approaches any point on the contour, will now be used to solve scalar RH problems. We first introduce some definitions.

(1) Let C be a simple, smooth, closed contour dividing the complex z plane into two regions D^+ and D^-, where the positive direction of C will be taken as that for which D^+ is on the left (see Figure 7.3.1). A scalar function $\Phi(z)$ defined in the entire plane, except for points on C, will be called **sectionally**

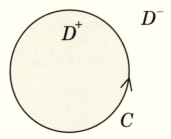

Fig. 7.3.1. Simple closed contour C and the "+", "−" regions

analytic if (a) the function $\Phi(z)$ is analytic in each of the regions D^+ and D^- except, perhaps, at $z = \infty$, and (b) as z approaches any point t on C along any path that lies wholly in either D^+ or D^-, the function $\Phi(z)$ approaches a definite limiting value, $\Phi^+(t)$ or $\Phi^-(t)$, respectively. The values $\Phi^{\pm}(t)$ are called the boundary values of the function $\Phi(z)$.

It then follows that $\Phi(z)$ is continuous in the closed region $D^+ + C$ if it is assigned the value $\Phi^+(t)$ on C. A similar statement applies for the region $D^- + C$.

(2) The sectionally analytic function $\Phi(z)$ has a finite degree at infinity if for some finite integer m, $\lim_{z \to \infty} \Phi(z)/|z|^m = 0$. The function $\Phi(z)$ is said to have degree κ at infinity if

$$\Phi(z) \sim c_\kappa z^\kappa + O(z^{\kappa-1})q \quad \text{as } z \to \infty, \quad c_\kappa \text{ a nonzero constant} \quad (7.3.1)$$

(3) The index of a function $\varphi(t)$ with respect to C is the increment of its argument in traversing a curve C in the positive direction, divided by 2π, that is

$$\operatorname{ind} \varphi(t) := \frac{1}{2\pi}[\arg \varphi(t)]_C = \frac{1}{2\pi i}[\log \varphi(t)]_C = \frac{1}{2\pi i}\int_C d(\log \varphi(t)) \quad (7.3.2)$$

Example 7.3.1 Show that if the function $\varphi(t)$ is analytic inside a closed contour C except at a finite number of points where it may have poles, then its index equals to the difference between the number of zeroes and the number of poles inside the contour.

If $\varphi(t)$ is differentiable, then Eq. (7.3.2) yields

$$\operatorname{ind} \varphi(t) = \frac{1}{2\pi i}\int_C \frac{\varphi'(t)}{\varphi(t)} dt = N - P \quad (7.3.3)$$

where N and P are the number of zeroes and poles of $\varphi(t)$, respectively, and where a multiple zero or pole is counted according to its multiplicity. This is really the argument principle (see Theorem 4.4.1) and can be derived from contour integration.

To illustrate this result, let us compute the index of $\varphi(t) = t^n$ where C is an arbitrary finite contour enclosing the origin. Because t^n has a zero of order n inside the contour and no poles, it follows that $\mathrm{ind}(t^n) = n$.

7.3.1 Closed Contours

The scalar **homogeneous** RH problem for closed contours is formulated as follows: Given a closed contour C and a function $g(t)$ that satisfies a Hölder condition on C with $g(t) \neq 0$ on C, find a sectionally analytic function $\Phi(z)$, with finite degree at infinity, such that

$$\Phi^+(t) = g(t)\Phi^-(t) \qquad \text{on } C \qquad (7.3.4)$$

where $\Phi^\pm(t)$ are the boundary values of $\Phi(z)$ on C. We assume that the index of $g(t)$ is κ. Here we only consider RH problems in a simply connected region. The extension to a multiply connected region can be dealt with by similar methods, cf. Gakhov (1966).

The solution of this RH problem with degree m at infinity is given by

$$\Phi(z) = X(z)P_{m+\kappa}(z) \qquad (7.3.5)$$

where $P_{m+\kappa}(z)$ is an arbitrary polynomial of degree $m + \kappa$, and $X(z)$, called the fundamental solution of Eq. (7.3.4), is given by

$$X(z) \equiv \begin{cases} e^{\Gamma(z)}, & z \text{ in } D^+ \\ z^{-\kappa}e^{\Gamma(z)}, & z \text{ in } D^- \end{cases} \qquad (7.3.6a)$$

$$\Gamma(z) \equiv \frac{1}{2\pi i} \int_C \frac{d\tau \log(\tau^{-\kappa}g(\tau))}{\tau - z} \qquad (7.3.6b)$$

with $\kappa = \mathrm{ind}\, g(t)$ on C. We assume C encloses the origin so that $\mathrm{ind}(\tau^{-\kappa}) = -\kappa$. Note that this definition of $X(z)$ implies the normalization $X^-(z) \sim z^{-\kappa}$ as $z \to \infty$. One can modify this normalization as needed in a given problem; for example, $X(z)$ can be multiplied by a constant.

These results follow from an application of the Plemelj formulae. However, care must be taken to ensure that the function appearing in the integrand of the Cauchy integral used in the Plemelj formulae satisfies a Hölder condition. The index of $g(t)$ is κ, and as such, $\log g(t)$ does not satisfy the Hölder condition, and the arguments leading to the Plemelj formulae fail. To remedy this, we note that the index of $t^{-\kappa}g(t)$ is zero, and hence $\log(t^{-\kappa}g(t))$ satisfies a Hölder condition. This suggests rewriting Eq. (7.3.4) in the form

$$\Phi^+(t) = (t^{-\kappa}g(t))t^{\kappa}\Phi^-(t)$$

or

$$\log\Phi^+(t) - \log(t^{\kappa}\Phi^-(t)) = \log(t^{-\kappa}g(t))$$

A special solution of this equation (the fundamental solution) $\Phi = X(z)$ is obtained by letting $\log X^+(t) = \Gamma^+(t)$ and $\log(t^{\kappa}X^-(t)) = \Gamma^-(t)$, which implies

$$\Gamma^+(t) - \Gamma^-(t) = \log(t^{-\kappa}g(t))$$

The representation (7.3.6b) for $\Gamma(z)$ follows from the Plemelj formulae. Equation (7.3.5) follows from the fact that $X(z) \sim z^{-\kappa}$ as $z \to \infty$ (recall that $\Gamma(z) \sim O(1/z)$ as $z \to \infty$). Note that $X^+(z)$ is nonvanishing in D^+ and $X^-(z)$ is nonvanishing in D^- except perhaps as $z \to \infty$.

Thus the solution of Eq. (7.3.4) that satisfies the requisite condition at infinity is therefore given by $\Phi(z) = X(z)P_{m+\kappa}(z)$. Note that multiplying the fundamental solution $X(z)$ by a polynomial that is analytic for all z has no effect on Eq. (7.3.4).

RH problems are closely related to singular integral equations. For this and other applications one is often interested in finding all solutions of Eq. (7.3.4) that vanish at infinity. The solution (7.3.5) implies the following.

(a) if $\kappa > 0$, then there exist κ linearly independent solutions of Eq. (7.3.4) vanishing at infinity; this follows from the fact that as $z \to \infty$, $z^{-\kappa}P_{m+\kappa}(z)$, and for decaying solutions we require $m = -1$. The polynomial $P_{\kappa-1}(z) = A_0 + A_1 z + A_2 z^2 + \cdots + A_{\kappa-1}z^{\kappa-1}$ has κ arbitrary constants.

(b) If $\kappa \leq 0$, then there exists no nontrivial solution of Eq. (7.3.4) vanishing at infinity. Stated differently, the fundamental solution $X(z)$ grows algebraically at infinity for $\kappa < 0$ or is bounded at infinity for $\kappa = 0$. Consequently, to have a vanishing solution we must take $P_{m+\kappa}(z) = 0$ and we only have the trivial solution.

The so-called **inhomogeneous** RH problem differs from the homogeneous one in that Eq. (7.3.4) is replaced by

$$\Phi^+(t) = g(t)\Phi^-(t) + f(t), \qquad t \text{ on } C \qquad (7.3.7)$$

where $f(t)$ is also a Hölder condition on C. The solution of this problem (derived below) is given by

$$\Phi(z) = X(z)[P_{m+\kappa}(z) + \Psi(z)], \quad \Psi(z) \equiv \frac{1}{2\pi i} \int_C \frac{f(\tau)\, d\tau}{X^+(\tau)(\tau - z)} \quad (7.3.8)$$

where $X(z)$ is given by Eq. (7.3.6a).

To derive Eq. (7.3.8), we rewrite $g(t)$ as $X^+(t)/X^-(t)$ using Eq. (7.3.6a), hence Eq. (7.3.7) becomes

$$\frac{\Phi^+(t)}{X^+(t)} - \frac{\Phi^-(t)}{X^-(t)} = \frac{f(t)}{X^+(t)}$$

Then a special solution $\frac{\Phi}{X}(z) = \Psi(z)$ is obtained from the Plemelj formulae, and the equation for $\Phi(z)$ given by Eq. (7.3.8) follows.

Again (thinking ahead to applications) it is useful to find solutions vanishing at infinity. Because for large z, $X(z) = O(z^{-\kappa})$ and $\Psi(z) = O(z^{-1})$, it follows that:

(a) If $\kappa > 0$, then there exist κ linearly independent solutions given by Eq. (7.3.8) with $m = -1$ (This follows by the same argument as discussed in part (a) of the homogeneous problem, above.)

(b) If $\kappa = 0$, then there exists a unique solution $X(z)\Psi(z)$; here we need to take $P_{m+\kappa}(z) = 0$.

(c) If $\kappa < 0$, then there exists a unique solution $X(z)\Psi(z)$ if and only if the orthogonality conditions

$$\int_C \frac{f(\tau)\tau^{n-1}}{X^+(\tau)}\, d\tau = 0, \quad n = 1, 2, \ldots, -\kappa \quad (7.3.9)$$

hold . As in (b), we need to take $P_{m+\kappa}(z) = 0$. These orthogonality conditions follow from the asymptotic expansion of $\Psi(z)$ for large z and the requirement that $\Psi(z) \sim O(z^{-|\kappa|-1})$ as $z \to \infty$, because $X(z) \sim z^{|\kappa|}$.

Note that

$$\frac{1}{2\pi i} \int_C \frac{f(\tau)}{X^+(\tau)} \frac{1}{\tau - z}\, d\tau$$

$$= \frac{1}{2\pi i} \frac{(-1)}{z} \int_C \frac{f(\tau)}{X^+(\tau)} \frac{d\tau}{(1 - \tau/z)}$$

$$\sim \frac{-1}{2\pi i} \int_C \frac{f(\tau)}{X^+(\tau)} \left(\frac{1}{z} + \frac{\tau}{z^2} + \frac{\tau^2}{z^3} + \cdots + \frac{\tau^{n-1}}{z^n} + \cdots \right) d\tau$$

The behavior of $\Psi(z)$ and $X(z)$ is such that all coefficients of z^{-n}, $n = 1, 2, \ldots, |\kappa|$ must vanish in order for $\Phi(z) \to 0$ as $z \to \infty$.

Example 7.3.2 Solve the RH problem (7.3.7) with $g(t) = t/(t^2 - 1)$, $f(t) = (t^3 - t^2 + 1)/(t^2 - t)$; C encloses the points $0, 1, -1$, and $\Phi(z)$ vanishes at infinity.

Because $g(t)$ is analytic inside C and it has one zero and two poles, it follows from Example 7.3.1 that $\kappa = \operatorname{ind} g = -1$. The fundamental solution $X(z)$ satisfies, $X^+(t)/X^-(t) = g(t)$ (recall $X^-(z) \to z^{-\kappa}$ as $z \to \infty$ hence

$$X^+(t) = \frac{t}{(t-1)(t+1)} X^-(t), \qquad X^-(z) \to z \quad \text{as } z \to \infty \qquad (7.3.10)$$

The solution of Eq. (7.3.10) can be found by inspection: Because $t/(t-1)(t+1)$ is analytic outside C (C encloses both $t = \pm 1$), it follows that the right-hand side Eq. (7.3.10) is a $\ominus$ function; thus it follows that

$$X^+(t) = \frac{t}{(t-1)(t+1)} X^-(t) = \hat{X}^-(t)$$

and a solution follows: $X^+(t) = \hat{X}^-(t) = c = \text{constant}$. To satisfy the boundary condition $X^-(z) \to z$ as $z \to \infty$, we take $c = 1$; thus $X^+(z) = 1$, $X^-(z) = (z^2 - 1)/z$, and

$$X^+(t) = 1, \qquad X^-(t) = \frac{t^2 - 1}{t}$$

To compute $\Psi(z)$ via Eq. (7.3.8), we need to evaluate the Cauchy integral associated with $f(t)/X^+(t) = f(t) = t + 1/[t(t - 1)]$. This can also be done by inspection namely $\Psi(z)$ satisfies the RH problem: $\Psi^+(t) - \Psi^-(t) = t + 1/t(t - 1)$, which has the solution

$$\Psi^+(z) = z, \qquad \Psi^-(z) = -\frac{1}{z(z-1)}$$

where $\Psi^-(z) \to 0$ as $z \to \infty$. Of course these formulae can be verified by contour integration; for example, from Eq. (7.3.8), if $z \in D^+$

$$\Psi^+(z) = \frac{1}{2\pi i} \int_C \left(\tau + \frac{1}{\tau(\tau - 1)} \right) \frac{d\tau}{\tau - z} = z$$

Similarly, we have seen from general considerations that a solution of this RH problem could exist only if the orthogonality condition (7.3.9) is satisfied for $n = 1$ (because $\kappa = -1$). Let us verify this. Evaluating Eq. (7.3.9) with $n = 1$, $X^+ = 1$, and $f(t) = t + \frac{1}{t(t-1)}$, we find

$$\int_C \left(\tau + \frac{1}{\tau(\tau - 1)} \right) d\tau = 0$$

Thus the above inhomogeneous RH problem has the unique solution

$$\Phi^+(z) = z, \qquad \Phi^-(z) = -\frac{z + 1}{z^2}$$

7.3.2 Open Contours

Let a and b be the endpoints of L, as in Figure 7.3.2. The scalar homogeneous RH problem for open contours is defined as in Eq. (7.3.4) buts now the closed contour C is replaced by an open contour L. (Sometimes this is called a discontinuous RH problem because in the RH problem (7.3.4), $g(t)$ could be thought of as a discontinuous function taking the value zero on a portion of a closed contour. Similarly, the closed contour problem, discussed in Section 7.3.1, is sometimes referred to as a continuous RH problem.) We again seek a sectionally analytic solution $\Phi(z)$ with finite degree at infinity. In some applications it is useful to find solutions that are not bounded at the endpoints but which have integrable singularities at these points. Therefore we shall allow $\Phi(z)$ to have such behavior at the endpoints of the contour.

In analogy with Section 7.3.1 we first look for a fundamental solution $X(z)$ to Eq. (7.3.4). We add the condition that as z approaches c, where c is either of the two endpoints of L, that is, $c = a$ or $c = b$, then

$$X(z) \sim O(z - c)^\nu, \qquad z \to c, \qquad |\mathrm{Re}\, \nu| < 1 \qquad (7.3.11)$$

The condition that $-1 < \mathrm{Re}\, \nu$ follows from the requirement that $X(z)$ be integrable at c; requiring $\mathrm{Re}\, \nu < 1$ is without loss of generality because the general solution will be obtained from $X(z)$ by multiplication with an arbitrary polynomial.

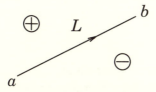

Fig. 7.3.2. Open contour L

The function $e^{\Gamma(z)}$, where

$$\Gamma(z) = \frac{1}{2\pi i} \int_L \frac{\log g(\tau)\, d\tau}{\tau - z} \qquad (7.3.12)$$

provides a particular solution of the RH problem under consideration. Because L is open, the function $\log g(\tau)$ will satisfy the Hölder condition so long as $g(\tau)$ does; hence we see that the notion of index is different from that where L is closed. Later we will define the index on the basis of the properties of the fundamental solution at infinity. Equation (7.2.13) implies that for $g(\tau)$ satisfying a Hölder condition

$$\Gamma(z) \to \begin{cases} -\dfrac{1}{2\pi i} \log g(a) \log(z - a) + \Gamma_a(z), & z \to a \\[2mm] \dfrac{1}{2\pi i} \log g(b) \log(z - b) + \Gamma_b(z), & z \to b \end{cases} \qquad (7.3.13)$$

where $\Gamma_a(z)$ and $\Gamma_b(z)$ are bounded functions tending to definite limits at $z = a$ and $z = b$, respectively. Thus

$$e^{\Gamma(z)} = \begin{cases} O(z - a)^{\alpha + iA}, & z \to a \\[2mm] O(z - b)^{\beta + iB}, & z \to b \end{cases} \qquad (7.3.14)$$

where

$$\alpha + iA \equiv -\frac{1}{2\pi i} \log g(a), \qquad \beta + iB \equiv \frac{1}{2\pi i} \log g(b) \qquad (7.3.15)$$

and α, A, β, B are real numbers. The fundamental solution $X(z)$ of the RH problem (7.3.4) on an open contour is defined by

$$X(z) = (z - a)^{\lambda}(z - b)^{\mu} e^{\Gamma(z)} \qquad (7.3.16)$$

where λ and μ are integers chosen to satisfy

$$-1 < \lambda + \alpha < 1, \qquad -1 < \mu + \beta < 1 \qquad (7.3.17)$$

Let us summarize the above result: Consider the RH problem

$$\Phi^+(t) = g(t)\Phi^-(t), \qquad t \text{ on } L \qquad (7.3.18)$$

where $g(t)$ satisfies a Hölder condition and is nonzero on L and Φ^+, Φ^- are the limits of a sectionally analytic function $\Phi(z)$ as z approaches L from the left and right, respectively. The fundamental solution of this problem $X(z)$ is given by Eq. (7.3.16) where the integers λ and μ are defined by Eq. (7.3.17) and $\Gamma(z)$ is given by Eq. (7.3.12). This solution has the following properties:

(i) The boundary values of $X(z)$ are given by

$$X^+(t) = \sqrt{g(t)}X(t), \qquad X^-(t) = \frac{X(t)}{\sqrt{g(t)}} \qquad (7.3.19)$$

where

$$X(t) \equiv (t-a)^\lambda (t-b)^\mu e^{\Gamma_p(t)}, \qquad \Gamma_p(t) = \frac{1}{2\pi i}\oint_L \frac{\log g(\tau)\,d\tau}{\tau - t} \qquad (7.3.20)$$

and we have used

$$\Gamma_\pm(t) = \pm\frac{1}{2}\log g(t) + \frac{1}{2\pi i}\oint_L \frac{\log g(\tau)}{\tau - t}\,d\tau$$

The functions $X^\pm$ satisfy Eq. (7.3.18); the square root appearing in $X^+(t)$ is defined by $\sqrt{g(t)} = e^{\frac{1}{2}\log g(t)}$.

(ii) Because $\Gamma(z) \to 0$ as $z \to \infty$

$$X(z) = O(z^{\lambda+\mu}) \qquad \text{as } z \to \infty \qquad (7.3.21)$$

(iii) From Eqs. (7.3.13)–(7.3.15),

$$X(z) = \begin{cases} O((z-a)^{\lambda+\alpha+iA}), & z \to a \\ O((z-b)^{\mu+\beta+iB}), & z \to b \end{cases} \qquad (7.3.22)$$

We note that Eqs. (7.3.17) uniquely determine λ and μ only if α and β are integers; in this case $\lambda = -\alpha$, $\mu = -\beta$. We call these ends **special ends**. In fact, if α and β are not special, then there are two solutions; for example, we can write $\alpha = M + \hat\alpha$ where M is an integer and $0 < \hat\alpha < 1$, in which case Eqs. (7.3.17) imply that $\lambda + M = 0$ or $\lambda + M = -1$. Similar statements apply for the case when β is not an integer. If the end b is nonspecial, then $X(z)$ will be bounded near b if $\mu + \beta > 0$, and $X(z)$ will be unbounded near b if $\mu + \beta < 0$; furthermore, if $X(z)$ is bounded near b, it will necessarily be zero there. Similar statements apply if the end a is nonspecial.

Obviously

$$\Phi(z) = X(z)P(z) \qquad (7.3.23)$$

where $P(z)$ is an arbitrary polynomial, is also a solution of the RH problem (7.3.18). From Eq. (7.3.22) we see that $\Phi(z)$ is bounded near any special end and if $\Phi(z)$ is bounded near any nonspecial end, then it is zero at this end. If the degree of $P(z)$ is m, then beacuse $X(z) = O(z^{\mu+\lambda})$ as $z \to \infty$, we find that $\Phi(z) = O(z^{\mu+\lambda+m})$ as $z \to \infty$.

Having obtained a fundamental solution for the homogeneous RH problem, the solutions of the inhomogeneous RH problem can be constructed in the same manner as in Section 7.3.1 (see Eqs. (7.3.7) and (7.3.8)), and is given by

$$\Phi(z) = X(z)[P(z) + \Psi(z)], \qquad \Psi(z) \equiv \frac{1}{2\pi i} \int_L \frac{f(\tau)\,d\tau}{X^+(\tau)(\tau - z)} \qquad (7.3.24)$$

where again we assume that $f(t)$ satisfies a Hölder condition. Equation (7.2.15) implies that if $X(z)$ is bounded near a nonspecial end, then $\Phi(z)$ is also bounded near this end (but it is not, in general, zero at this end. Note that if $X(z)$ vanishes, from Lemma 7.2.2, $\Psi(z)$ will have an integrable singularity that, in general, is canceled when we multiply by $X(z)$). The function $\Phi(z)$ is also bounded near a special end c, unless $g(c) = 1$, that is, $\lambda = 0$ or $\mu = 0$, in which case beacuse $X(z)$ is bounded, $\Psi(z)$ and therefore $\Phi(z)$ have logarithmic singularities (see Eq. (7.2.13)).

In some applications it is important to find those solutions of the inhomogeneous RH problem that vanish at infinity. Noting that $X(z) = O(z^{\lambda+\mu})$ at infinity, and comparing with the case of the closed contour RH problem where $X(z) = O(z^{-\kappa})$ at infinity, the index of the open contour problem is naturally defined by

$$\kappa = -(\lambda + \mu) \qquad (7.3.25)$$

The same arguments leading to κ linearly independent solutions for $\kappa \geq 0$, or the necessity for the additional orthogonality conditions (7.3.9) follow in the same way as in the case of a closed contour.

Example 7.3.3 Solve the RH problem

$$\Phi^+(t) + \Phi^-(t) = f(t), \qquad t \text{ on } L \qquad (7.3.26)$$

where L is an open contour such as that depicted in Figure 7.3.2. In this case, $g(t) = -1$, so that $(1/2\pi i) \log g(t) = n + 1/2$, where the integer n depends on the choice of the branch of the logarithm. It does not matter which branch we choose. (We will see that $n = 0$ is without loss of generality.) From Eq. (7.3.12), $\Gamma(z) = \left(n + \frac{1}{2}\right) \log \frac{z-b}{z-a}$; hence $e^{\Gamma(z)} = \left(\frac{z-b}{z-a}\right)^{n+1/2}$. Equations (7.3.15)–(7.3.17) imply that $\alpha = -n - 1/2$, $\beta = n + 1/2$, and consequently $\lambda = n$ or $\lambda = n + 1$; $\mu = -n$ or $\mu = -n - 1$. For $e^{\Gamma(t)}$ we take some definite branch of this function; a cut is chosen along the contour from a to b and we define $e^{\Gamma(z)} = \left(\frac{z-b}{z-a}\right)^n R(z)$, where

$$R(z) \equiv \left(\frac{z-b}{z-a}\right)^{\frac{1}{2}}, \qquad R^-(t) = -R^+(t) = -R(t) \qquad (7.3.27)$$

The particular form of the fundamental solution $X(z)$ depends on the requirement of boundedness:

(a) Solution unbounded at both ends: $\lambda + \mu = -1 = -\kappa$

$$X(z) = \frac{1}{[(z-a)(z-b)]^{\frac{1}{2}}} \qquad (7.3.28a)$$

(b) Solution bounded at a (or b), but unbounded at b (or a): $\lambda + \mu = 0 = -\kappa$

$$X(z) = \left(\frac{z-a}{z-b}\right)^{\frac{1}{2}} \quad \left(\text{or } X(z) = \left(\frac{z-b}{z-a}\right)^{\frac{1}{2}}\right) \qquad (7.3.28b)$$

(c) Solution bounded at both ends: $\lambda + \mu = 1 = -\kappa$

$$X(z) = [(z-a)(z-b)]^{\frac{1}{2}} \qquad (7.3.28c)$$

It is clear that the choice of integer n is immaterial, so $n = 0$ without loss of generality.

In what follows we consider solutions $\Phi(z)$ to the inhomogeneous problem (7.3.26) such that $\Phi(z) \to 0$ as $z \to \infty$. For case (a) the index $\kappa = 1$, the solution $\Phi(z)$ will depend on one arbitrary constant, and $\Phi(z)$ is given by Eq. (7.3.24) with $P(z) = A_0 = \text{constant}$. Because $X = \frac{R(z)}{z-b}$ (note that $R(z) \to 1$ as $z \to \infty$), we have

$$\Phi(z) = \frac{R(z)}{z-b}\left[\frac{1}{2\pi i}\int_L \frac{\tau-b}{R(\tau)}\frac{f(\tau)\,d\tau}{\tau-z} + A_0\right], \qquad \text{and}$$
$$\Phi(z) \to 0 \text{ as } z \to \infty \qquad (7.3.29a)$$

For case (b) the index $\kappa = 0$ and the solution of Eq. (7.3.26) is unique; here $X(z) = (R(z))^{-1}$

$$\Phi(z) = \frac{1}{2\pi i R(z)}\int_L R(\tau)\frac{f(\tau)}{\tau-z}\,d\tau \qquad (7.3.29b)$$

and $\Phi(z) \to 0$ as $z \to \infty$.

For case (c) the index $\kappa = -1$, $X = (z - a)R(z)$, the solution is given by

$$\Phi(z) = \frac{(z-a)R(z)}{2\pi i} \int_L \frac{1}{R(\tau)(\tau - a)} \frac{f(\tau)}{\tau - z} d\tau \qquad (7.3.29c)$$

and $\Phi(z) \to 0$ as $z \to \infty$ if and only if the following orthogonality condition is satisfied,

$$\int_L \frac{f(\tau)}{R(\tau)(\tau - a)} d\tau = 0 \qquad (7.3.29d)$$

To compute $\Phi^{\pm}(t)$ we use Eqs. (7.2.3), (7.3.29a–d), and (7.3.27). For example, in case (a)

$$\Phi^{\pm}(t) = \pm \frac{R(t)}{t-b} \left[\pm \frac{1}{2} \frac{(t-b)}{R(t)} f(t) + \frac{1}{2\pi i} \oint_L \frac{\tau - b}{R(\tau)} \frac{f(\tau)}{t - \tau} d\tau + A_0 \right] \qquad (7.3.30)$$

7.3.3 Singular Integral Equations

In this section we discuss how solutions of certain "singular integral equations" can be reduced to and solved via Riemann–Hilbert problems. First we introduce some basic notions. It is outside the scope of this book to discuss, in any detail, the general theory of integral equations. There are numerous texts on this subject that the reader can consult.

An equation of the form

$$g(t)\phi(t) = f(t) + \lambda \int_a^b K(t, \tau)\phi(\tau) d\tau \qquad (7.3.31)$$

where t lies on a given contour from a to b (usually this is the real axis from a to b) in which $g(t)$, $f(t)$, and $K(t, \tau)$ are given integrable functions, is called a **linear integral equation** for the function $\phi(t)$. The function $K(t, \tau)$ is called the **kernel**. If $g(t) = 0$ everywhere on the contour from a to b, the integral equation is said to be of the **first kind**. If $g(t) \neq 0$ anywhere on the contour, then we can divide by $g(t)$ and the equation is said to be of the **second kind**.

If $K(t, \tau) = 0$ for $t < \tau$, then the equation is called a **Volterra integral equation**

$$g(t)\phi(t) = f(t) + \lambda \int_a^t K(t, \tau)\phi(\tau) d\tau \qquad (7.3.32)$$

Otherwise the equation is said to be a **Fredholm integral equation**. For convenience in the rest of the discussion we assume that $g(t) = 1$.

A Volterra equation of the second kind can always be solved by iteration so long as

$$\int_a^b \int_a^t |K(t, \tau)|^2 \, dt \, d\tau < \infty \qquad \text{and} \qquad \int_a^b |f(t)|^2 \, dt < \infty$$

Namely, the following series, called the Neumann series, converges for all λ:

$$\phi(t) = f(t) + \sum_{n=1}^{\infty} \lambda^n \int_a^t K_n(t, \tau) f(\tau) \, d\tau \qquad (7.3.33a)$$

where K_n is defined iteratively:

$$K_n(t, \tau) = \int_a^t K_{n-1}(t, t') K(t', \tau) \, dt', \qquad n \geq 2 \qquad (7.3.33b)$$

and $K_1(t, \tau) = K(t, \tau)$.

Even in the case of a Fredholm integral equation of the second kind, the Neumann series, that is Eqs. (7.3.33a,b) with the upper integration limit t replaced by b, converges for sufficiently small λ, provided that

$$\int_a^b \int_a^b |K(t, \tau)|^2 \, dt \, d\tau < \infty, \qquad \int_a^b |f(t)|^2 \, dt < \infty,$$

In general, this Fredholm equation has a solution so long as λ is not one of a set of special values, called **eigenvalues**. This solution is of the form

$$\phi(t) = f(t) + \int_a^b \frac{\hat{D}(t, \tau; \lambda)}{D(\lambda)} \phi(\tau) \, d\tau \qquad (7.3.34)$$

where $\hat{D}(t, \tau; \lambda)$ and $D(\lambda)$ are entire functions of the complex variable λ. The eigenvalues satisfy $D(\lambda) = 0$. They are isolated and since $D(\lambda)$ is entire only a finite number of them can lie in a bounded region. The function $\phi(t)$ that corresponds to an eigenvalue is called an **eigenfunction**. In the special case when $K(t, \tau) = K(\tau, t)$, the function $K(t, \tau)$ is said to be **symmetric**. There are many further results known when the kernel is symmetric; this is called Hilbert–Schmidt Theory. When $K(t, \tau) = \sum_{j=1}^{n} M_j(t) N_j(\tau)$, $K(t, \tau)$ is called a **degenerate kernel**, and the integral equation can be reduced to a matrix eigenvalue problem.

An integral equation is usually said to be singular if (a) the contour extends to infinity, for example, $a = -\infty$ or $b = \infty$ or both, and/or (b) the kernel is unbounded. A "weakly singular" kernel is of the form

$$K(t, \tau) = \frac{\hat{K}(t, \tau)}{(t - \tau)^\alpha} \tag{7.3.35}$$

where $0 < \operatorname{Re}\alpha < 1$ and $|\hat{K}(t, \tau)| < \infty$. The case when $(t - \tau)^\alpha$ is replaced by a logarithm, (e.g. $\log(t - \tau)$, $(\log(t - \tau))^{-1}$, etc.) is also weakly singular. A Cauchy singular kernel has $\alpha = 1$, and in this case the contour integral is usually taken to be a principal value integral. For weakly singular kernels, standard methods will work; perhaps after redefining the kernel of the integral equation by doing a finite number of iterations (recall in the standard case we only need $\int_a^b \int_a^b |K(t, \tau)|^2 \, dt \, d\tau < \infty$). However, in the case of a Cauchy singular kernel, the situation is more complicated. Here we only discuss some of the prototypical cases. An extensive discussion of Cauchy singular integral equations can be found in Muskhelishvili (1977).

We will now show that the solution of scalar RH problems provides an effective way of solving certain singular integral equations. We note the following equivalence: Solving the singular integral equation

$$a(t)\varphi(t) + \frac{b(t)}{\pi i} \fint_L \frac{\varphi(\tau)}{\tau - t} \, d\tau = c(t) \tag{7.3.36}$$

where $a(t), b(t), c(t)$ satisfy the Hölder condition on L with $a \pm b \neq 0$ on L, is equivalent to finding the sectionally analytic function

$$\Phi(z) = \frac{1}{2\pi i} \int_L \frac{\varphi(\tau)}{\tau - z} \, d\tau \tag{7.3.37}$$

associated with the RH problem

$$\Phi^+(t) = g(t)\Phi^-(t) + f(t) \quad \text{for } t \text{ on } L; \quad \Phi^-(\infty) = 0,$$

$$g(t) \equiv \frac{a(t) - b(t)}{a(t) + b(t)}, \qquad f(t) \equiv \frac{c(t)}{a(t) + b(t)} \tag{7.3.38}$$

To show that Eq. (7.3.36) reduces to the RH problem (7.3.38), we use the Plemelj formula for $\Phi(z)$, that is,

$$\varphi(t) = \Phi^+(t) - \Phi^-(t), \qquad \frac{1}{\pi i} \fint_L \frac{\varphi(\tau)}{\tau - t} = \Phi^+(t) + \Phi^-(t)$$

Substituting these equations in Eq. (7.3.36), we immediately find Eq. (7.3.38). The converse is also true: If $\Phi(z)$ solves the RH problem (7.3.38) with the boundary condition $\Phi^-(\infty) = 0$, then $\varphi = \Phi^+ - \Phi^-$ solves Eq. (7.3.36) (see Muskhelishvili, 1977).

A singular integral equation of the form (7.3.36) is usually referred to as a dominant singular equation. It plays an important role in studying the solvability of the more general equation

$$a(t)\varphi(t) + \frac{1}{i\pi} \int_L \left(\frac{K(t,\tau)\phi(\tau)}{t - \tau} \right) d\tau = c(t). \tag{7.3.39}$$

Writing $\hat{K}(t,\tau) = \hat{K}(t,t) + (\hat{K}(t,\tau) - \hat{K}(t,t))$, and calling $b(t) = \hat{K}(t,t)$ and $F(t,\tau) = \frac{1}{i\pi}(\hat{K}(t,\tau) - \hat{K}(t,t))/(t - \tau)$, we obtain

$$a(t)\varphi(t) + \frac{b(t)}{\pi i} \int_L \frac{\varphi(\tau)}{\tau - t} d\tau + \int_L F(t,\tau)\varphi(\tau)\, d\tau = c(t) \tag{7.3.40}$$

Under mild assumptions on $\hat{K}(t,\tau)$ the kernel $F(t,\tau)$ is a Fredholm kernel. The study of such equations with $F(t,\tau) \neq 0$ is more complicated than Eq. (7.3.36) and is outside the scope of this book. Here we only note that if $F(t,\tau)$ is degenerate, that is, if $F(t,\tau) = \sum_1^n M_j(t)N_j(\tau)$, then Eq. (7.3.40) can also be solved in closed form. An example of this follows.

Example 7.3.4 Solve the singular integral equation

$$(t + t^{-1})\varphi(t) + \frac{t - t^{-1}}{\pi i} \int_C \frac{\varphi(\tau)}{\tau - t} d\tau$$
$$- \frac{1}{2\pi i} \int_C (t + t^{-1})(\tau + \tau^{-1})\varphi(\tau)\, d\tau = 2t^2 \tag{7.3.41}$$

where C is the unit circle.

The Fredholm kernel $F(t,\tau) = (t + t^{-1})(\tau + \tau^{-1})$ is degenerate. According to the above remark we expect to be able to solve Eq. (7.3.41) in closed form. Let

$$A \equiv \frac{1}{2\pi i} \int_C (\tau + \tau^{-1})\varphi(\tau)\, d\tau$$

Then Eq. (7.3.41) yields the dominant singular integral equation

$$(t + t^{-1})\varphi(t) + \frac{t - t^{-1}}{\pi i} \int_C \frac{\varphi(\tau)}{\tau - t} d\tau = 2t^2 + A(t + t^{-1})$$

Using the Plemelj formula, this equation is equivalent to the scalar RH problem

$$(t + t^{-1})(\Phi^+ - \Phi^-) + (t - t^{-1})(\Phi^+ + \Phi^-) = 2t^2 + A(t + t^{-1})$$

or

$$\Phi^+(t) = t^{-2}\Phi^-(t) + t + \frac{A}{2}(1 + t^{-2}), \qquad \Phi^-(\infty) = 0 \qquad (7.3.42)$$

The analytic function $g(t) = t^{-2}$ has a second-order pole inside C; thus from Eq. (7.3.3) the index $\kappa = \operatorname{ind} g = -2$. From Eq. (7.3.6a,b) the fundamental solution $X(z)$ of the homogeneous RH problem satisfies $X^-(z) = O(z^2)$ as $z \to \infty$; it turns out that $X(z)$ can be found by inspection:

$$t^2 X^+(t) = X^-(t)$$

that is

$$X^+(t) = 1, \qquad X^-(t) = t^2$$

Because the index is -2, the solution of Eq. (7.3.42) exists (see Eq. (7.3.9), where in this case $f(t) = t + (A/2)(1 + t^{-2}))$ if and only if

$$\int_C \left[\tau + \frac{A}{2}(1 + \tau^{-2})\right] d\tau = 0, \qquad \int_C \left[\tau + \frac{A}{2}(1 + \tau^{-2})\right] \tau\, d\tau = 0$$

The first of these functions is automatically satisfied, and the second equation above implies that $A = 0$. Thus from Eq. (7.3.8), using $f(t) = t$, $X^+(z) = 1$, and $X^-(z) = z^2$

$$\Phi(z) = \frac{X(z)}{2\pi i} \int_C \frac{\tau\, d\tau}{\tau - z} = \begin{cases} z, & z \text{ inside the circle} \\ 0, & z \text{ outside the circle} \end{cases}$$

Therefore $\Phi^+(t) = t$, $\Phi^-(t) = 0$, and $\varphi(t) = \Phi^+(t) - \Phi^-(t) = t$. Hence Eq. (7.3.41) has the unique solution $\varphi(t) = t$ provided that $A = 0$, which is indeed the case because from the definition of A

$$\frac{1}{2\pi i} \int_C (\tau + \tau^{-1})\varphi(\tau)\, d\tau = \frac{1}{2\pi i} \int_C (\tau + \tau^{-1})\tau\, d\tau = 0$$

Problems for Section 7.3

1. Consider the RH problem that satisfies the same jump condition as that of Example 7.3.2 discussed in the text, that is

$$\Phi^+(t) = \frac{t}{t^2 - 1}\Phi^-(t) + \frac{t^3 - t^2 + 1}{t^2 - t}$$

but now for the case that the contour C encloses the point 0 and does not enclose the points 1 and -1.

(a) Show that the index is 1.

(b) Establish that the solution of the homogeneous problem for which $X(z) \to z^{-1}$ as $z \to \infty$ is

$$X^+(z) = \frac{1}{z^2 - 1}, \qquad X^-(z) = \frac{1}{z}$$

(c) Show that the general solution of the inhomogeneous problem vanishing at ∞ is

$$\Phi^+(z) = z + \frac{\alpha}{z^2 - 1}, \qquad \Phi^-(z) = -\frac{z+1}{z^2} + \frac{\alpha}{z}$$

where α is an arbitrary constant.

2. Consider the RH problem that satisfies the same jump condition, as in Problem 7.3.1 above, but now for the case that the contour C encloses the points 0, 1, and -1.

(a) Show that the index is -1.

(b) Establish that the solution of the homogeneous problem for which $X(z) \to z$ as $z \to \infty$, is

$$X^+(z) = 1, \qquad X^-(z) = \frac{z^2 - 1}{z}$$

(c) Show that the solution of the inhomogeneous problem vanishing at ∞ exists if and only if the following solvability condition is valid:

$$\int_C \frac{\tau^3 - \tau^2 + 1}{\tau^2 - \tau} \, d\tau = 0$$

(d) Show that the solvability condition is satisfied and that the unique solution of the inhomogeneous problem vanishing at ∞ is

$$\Phi^+(z) = z, \qquad \Phi^-(z) = -\frac{z+1}{z^2}$$

3. Consider the following scalar inhomogeneous RH problem defined on a finite closed contour C

$$\Phi^+(t) = \frac{p(t)}{q(t)} \Phi^-(t) + f(t)$$

where $p(t)$ and $q(t)$ are polynomials, which do not vanish on C. The polynomials $p(t)$ and $q(t)$ can always be written in the form

$$p(z) = p_+(z)p_-(z), \qquad q(z) = q_+(z)q_-(z)$$

where $p_+(z)$ and $q_+(z)$ are polynomials with zeroes inside C, while $p_-(z)$ and $q_-(z)$ are polynomials with zeroes outside C.

(a) Let m_+ and n_+ be the number of zeroes of p_+ and q_+, respectively. Show that the associated index of the RH problem is $m_+ - n_+$.

(b) Show that the canonical solution $X(z)$ of the homogeneous RH problem satisfies

$$\frac{q_-(t)}{p_-(t)} X^+(t) = \frac{p_+(t)}{q_+(t)} X^-(t)$$

$$X^-(z) \to z^{n_+ - m_+}, \quad z \to \infty$$

Note that this RH problem can be solved by inspection because the left-hand side above is a $\oplus$ function, while the right hand side above is a $\ominus$ function.

4. Consider the singular integral equation

$$(t + t^{-1})\varphi(t) + \frac{t - t^{-1}}{\pi i} \fint_C \frac{\varphi(\tau)}{\tau - t} \, d\tau = 2t^2 + \alpha(t + t^{-1})$$

where α is a constant and C is the unit circle.

(a) Show that the associated RH problem is given by

$$\Phi^+ = t^{-2}\Phi^- + t + \frac{\alpha}{2}(1 + t^{-2})$$

(b) Establish that the canonical solution $X(z)$ of the corresponding homogeneous problem is

$$X^+ = 1, \qquad X^- = z^2$$

(c) Show thast the solvability conditions imply

$$\int_C \left[t + \frac{\alpha}{2}(1 + t^{-2}) \right] dt = 0, \qquad \int_C \left[t^2 + \frac{\alpha}{2}(t + t^{-1}) \right] dt = 0$$

and deduce that the above integral equation is solvable if and only if $\alpha = 0$. Its unique solution is $\varphi(t) = t$.

5. Consider the singular integral equation

$$(t^2 + t - 1)\varphi(t) + \frac{t^2 - t - 1}{\pi i} \oint_C \frac{\varphi(\tau)}{\tau - t} d\tau = 2\left(t^3 - t + 1 + \frac{1}{t}\right)$$

where C is a finite, closed contour that encloses the point 0 but does not enclose the points 1 and -1.

(a) Show that the associated RH problem is the one considered in Problem 7.3.1.

(b) Deduce that the general solution of this integral equation is

$$\varphi(t) = \frac{t^3 + t + 1}{t^2} + \frac{\alpha(1 + t - t^2)}{t^3 - t}$$

where α is an arbitrary constant.

6. Consider the singular integral equation of Problem 7.3.5, but now in the case where C encloses 0, 1, and -1. Use the result of Problem 7.3.2 to deduce that its unique solution is given by

$$\varphi(t) = \frac{1 + t + t^3}{t^2}$$

7. Show that the singular integral equation

$$(t^2 - 2)\varphi(t) + \frac{3t}{\pi} \oint_C \frac{\varphi(\tau)}{\tau - t} d\tau = 2t(t - 2i)(1 + \alpha t)$$

where α is a constant, and C is a finite closed contour that encloses $-i$ but does not enclose $i, 2i, -2i$, is solvable if and only if $\alpha = -i$. In this case show that its unique solution is

$$\varphi(t) = \frac{-2it(t - 2i)}{t + 2i}$$

8. Consider the following scalar RH problem on the real axis:

$$\Phi^- = \Phi^+ \qquad \text{for } |x| > 1$$

$$\Phi^- = (1 + \alpha^2)\Phi^+ \qquad \text{for} |x| < 1$$

with $\Phi \to 1$ as $z \to \infty$, where α is a real constant. Show that its solution is given by

$$\Phi(z) = \left(\frac{z-1}{z+1}\right)^{\nu}, \qquad \nu = \frac{1}{2\pi i}\log(1+\alpha^2)$$

7.4 Applications of Scalar Riemann–Hilbert Problems

We shall see that the methods and techniques developed in Sections 7.2 and 7.3 provide effective ways to solve several problems of mathematical and physical interest.

Example 7.4.1 (The Hilbert Problem) Let D^+ denote the interior of the unit circle C. Solving Laplace's equation with Dirichlet boundary conditions is equivalent to solving what is sometimes called the Riemann problem of Dirichlet form. This consists of finding a function harmonic in D^+, continuous in $D^+ + C$, and satisfying the boundary condition

$$u(t) = f(t), \qquad t \text{ on } C \tag{7.4.1}$$

where $f(t)$ is a real, continuous function on C.

We shall map this problem to an appropriate RH problem. In order to apply the RH theory developed earlier, we assume that $f(t)$ satisfies the Hölder condition; however, it is clear that the final formula is valid even if $f(t)$ is just continuous. Let

$$\Phi^+(z) = u(x, y) + iv(x, y)$$

be a function analytic in D^+ such that $u(x, y) \to u(t)$ as $z \to C$ from D^+ (u and v are real functions). The basic idea needed in this problem is to construct a suitable function analytic in D^-, that is, analytic outside the circle. By the Schwarz reflection principle (Sections 5.6, 5.7), because the function $\overline{\Phi^+(\frac{1}{\bar{z}})}$ is analytic outside the unit circle, we can write

$$\Phi^-(z) = \overline{\Phi^+(1/\bar{z})} \tag{7.4.2}$$

where the bar denotes the complex conjugate. The function $\Phi^-(z)$ is indeed analytic in D^-, and on C $\Phi^-(t) = \overline{\Phi^+(t)} = u(t) - iv(t)$, that is

$$\Phi^+(t) = u(t) + iv(t), \qquad \Phi^-(t) = u(t) - iv(t)$$

Using these equations, the boundary condition (7.4.1) becomes

$$\Phi^+(t) + \Phi^-(t) = 2f(t), \qquad t \text{ on } C \tag{7.4.3}$$

Thus the Dirichlet problem on the circle formulated above is mapped onto the RH problem of finding a sectionally analytic function $\Phi(z)$ satisfying the boundary condition (7.4.3). The function $\Phi(z)$ must also satisfy the constraint (7.4.2); in particular, applying Eq. (7.4.2) at infinity, it follows that $\Phi^-(\infty)$ is bounded, because $\Phi^+(0)$ is bounded.

The method to find the solution of the RH problem (Eq. (7.4.3)) was given in Section 7.3. There exist several approaches to satisfying the constraint (7.4.2). Our approach is direct; that is, we first find the most general solution of Eq. (7.4.3) that is bounded at infinity, and then we choose the relevant arbitrary constants by imposing Eq. (7.4.2).

The fundamental solution of Eq. (7.4.3) satisfies

$$X^+(z) = -X^-(z), \qquad X^-(z) \to 1 \text{ as } z \to \infty$$

We have used the fact that $\kappa = \text{ind}(-1) = 0$ and the Eqs. (7.3.6a,b). Hence $g(t) = -1 = e^{i\pi + 2n\pi i}$, so that

$$
\begin{aligned}
\Gamma(z) &= \frac{i\pi + 2n\pi i}{2\pi i} \int_C \frac{d\tau}{\tau - z} \\
&= \begin{cases} i\pi + 2n\pi i & z \in D^+ \\ 0 & z \in D^- \end{cases}
\end{aligned}
$$

Taking $e^{\Gamma(z)}$, then from Eq. (7.3.6a) we have $X^+(z) = -X^-(z) = -1$. The most general solution of Eq. (7.4.3) *bounded* at infinity is $\Phi(z) = X(z)(A_0 + \Psi(z))$, where $A_0 = \text{constant}$ (a "zeroth-order" polynomial) and from Eq. (7.3.8), $\Psi(z) = \frac{1}{2\pi i} \int_C \frac{2 f(\tau) d\tau}{X^+(\tau)(\tau - z)} = -\frac{1}{i\pi} \int_C \frac{f(\tau) d\tau}{\tau - z}$. Thus

$$\Phi^+(z) = \frac{1}{\pi i} \int_C \frac{f(\tau) \, d\tau}{\tau - z} - A_0 \tag{7.4.4}$$

Next we use Eq. (7.4.2). If we call

$$F(z) = \int_C \frac{f(\tau) \, d\tau}{\tau - z} \quad \Rightarrow \quad \overline{F\left(\frac{1}{\bar z}\right)} = z \int_C \frac{f(\tau) \, d\tau}{\tau(\tau - z)} = F(z) - \int_C \frac{f(\tau)}{\tau} \, d\tau$$

where we have used the fact that $f(\tau)$ is real, $\bar\tau = 1/\tau$ (recall that $\tau = e^{i\theta}$ on C) and $\overline{d\tau} = -d\tau/\tau^2$. Thus

$$\overline{\Phi^+\left(\frac{1}{\bar z}\right)} = -\overline{A_0} + \frac{1}{i\pi} \int_C \frac{f(\tau)}{\tau} \, d\tau - \frac{1}{i\pi} F(z)$$

On the other hand, the solution of the RH problem implies that

$$\Phi^-(z) = A_0 - \frac{1}{i\pi} F(z)$$

Thus $\overline{\Phi^+(1/\bar{z})} = \Phi^-(z)$ yields

$$\frac{A_0 + \bar{A}_0}{2} = \frac{1}{i\pi} \int \frac{f(\tau)\,d\tau}{\tau}, \qquad \text{or}$$

$$A_0 + \bar{A}_0 = A_{0R} = \frac{1}{2\pi i} \int_C \frac{f(\tau)}{\tau}\,d\tau = \frac{1}{2\pi} \int_0^{2\pi} f(\theta)\,d\theta$$

and $A_{0I} \equiv -B$ for arbitrary B **constant**. By combining terms

$$\Phi^+(z) = \frac{1}{\pi i} \int_C \frac{f(\tau)\,d\tau}{\tau - z} - \frac{1}{2\pi i} \int_C \frac{f(\tau)\,d\tau}{\tau} + iB \qquad (7.4.5)$$

where B is an arbitrary real constant, we have obtained the solution of the Dirichlet problem, $u(x, y) = \text{Re}\,\Phi^+(z)$.

This example can be generalized in several ways.

(a) The boundary condition (7.4.1) can be replaced by the more general "mixed" boundary condition

$$\alpha(t)u(t) - \beta(t)v(t) = \gamma(t), \qquad t \text{ on } C$$

In this case, using $\Phi^\pm(t) = u(t) \pm iv(t)$, Eq. (7.4.3) is replaced by

$$\Phi^+(t)\left(\frac{\alpha(t) + i\beta(t)}{2}\right) + \Phi^-(t)\left(\frac{\alpha(t) - i\beta(t)}{2}\right) = \gamma(t), \qquad t \text{ on } C$$

or

$$\Phi^+(t) = -\frac{\alpha(t) - i\beta(t)}{\alpha(t) + i\beta(t)}\Phi^-(t) + \frac{2\gamma(t)}{\alpha(t) + i\beta(t)}, \qquad t \text{ on } C$$

Solving this RH problem (via the method described in Section 7.3) for the sectionally analytic functions $\Phi^\pm(z)$, which are bounded at infinity and which satisfy the constraint (7.4.2), yields the solution to the mixed problem on the circle C.

(b) In principle, the circle C can be replaced by a more general contour. However, one must first use a conformal mapping to reduce the problem to one on a circle, whereupon the Schwarz reflection principle can be used, and then proceed as above.

(c) The Hilbert problem on a half plane (C is now the real axis) could be solved by mapping the real axis onto a circle. Alternatively, it can be solved directly (see also Section 7.4.1 where RH problems on the real axis are discussed in more detail) by using the fact that if $\Phi^+(z)$ is analytic in the upper half z plane, then by the Schwarz reflection principle (Sections 5.6, 5.7)

$$\Phi^-(z) = \overline{\Phi^+(\bar{z})} \tag{7.4.6}$$

is analytic in the lower half complex plane. For example, the solution of the Dirichlet problem (see Eqs. (7.4.1) and (7.4.3)) on the real axis follows the same route as that for the circle. As before, the functions $X(z)$ and $\Phi(z)$ satisfy $X^+(z) = -X^-(z) = -1$, $\Phi(z) = X(z)(A_0 + \Psi(z))$, where

$$\Psi(z) = -\frac{1}{i\pi} \int_{-\infty}^{\infty} \frac{f(\tau)}{\tau - z} d\tau$$

(note that $\int_C$ is replaced by $\int_{-\infty}^{\infty}$). Thus

$$\Phi^+(z) = -A_0 + \frac{1}{i\pi} \int_{-\infty}^{\infty} \frac{f(\tau)}{\tau - z} d\tau$$

$$\overline{\Phi^+(\bar{z})} = -\bar{A}_0 - \frac{1}{i\pi} \int_{-\infty}^{\infty} \frac{f(\tau)}{\tau - z} d\tau \qquad (f(\tau) \text{ is real for real } \tau)$$

$$\Phi^-(z) = A_0 - \frac{1}{i\pi} \int_{-\infty}^{\infty} \frac{f(\tau)}{\tau - z} d\tau$$

Using Eq. (7.4.6) we have $A_0 + \bar{A}_0 = 0$ so that $A_{0R} = 0$ and A_{0I} is arbitrary: $A_{0I} = -iB$ for arbitrary B. Hence $\Phi^+(z)$ is given by

$$\Phi^+(z) = \frac{1}{\pi i} \int_{-\infty}^{\infty} \frac{f(\tau) \, d\tau}{\tau - z} + iB \tag{7.4.7}$$

where B is an arbitrary real constant. We can rewrite this, letting $z = x + iy$, as

$$\Phi^+(z) = iB + \frac{1}{i\pi} \int_{-\infty}^{\infty} \frac{f(\tau)((\tau - x) + iy)}{(\tau - x)^2 + y^2} d\tau$$

Therefore the solution of the Dirichlet problem $u(x, y) = \mathrm{Re}\,\Phi^+(z)$ yields the Poisson formula for the half plane

$$u(x, y) = \frac{1}{\pi} \int_{-\infty}^{\infty} \frac{f(\tau) y \, dx}{(\tau - x)^2 + y^2}$$

(see Eq. (4.6.6)).

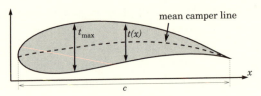

Fig. 7.4.1. Airfoil

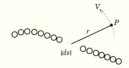

Fig. 7.4.2. Vortex sheet ds

Example 7.4.2 (*Planar Wing Theory*) The mathematical construction of an airfoil assumes that a thin envelope of material is wrapped around a mean camper line (see Figure 7.4.1). The mean line lies halfway between the upper and lower surfaces of the airfoil and intersects the chord line at the leading and trailing edges.

Experiments performed by NACA as early as 1929 established that the lifting characteristics of an airfoil are negligibly influenced by either viscosity or the thickness function, provided that the ratio of the maximum thickness, t_{max}, to the chord length c, that is, t_{max}/c, is small, and the airfoil is operating at a small angle of attack (i.e., the angle between the incident flow and the mean camper line is assumed to be small). This motivated the development of what is known as **thin airfoil theory**. In this approximate theory, viscosity is neglected and the airfoil is replaced by its mean camper line. The flow pattern past the airfoil is found by placing a vortex sheet (an idealized surface in a fluid in which there is a discontinuity in velocity across the surface) on the mean line and by requiring that the mean line be a streamline of this flow.

It can be shown that, according to the law of Biot and Savart, the circumferential velocity induced at a point P by an element of a vortex sheet ds (see Figure 7.4.2) is given by $dV = \gamma \, ds/2\pi r$, where $\gamma(s)$ is the strength of the sheet per unit length. (It was shown in Section 5.4 that the velocity potential associated with a point vortex located at $z = \zeta$ is $\Omega(z) = \frac{\gamma}{2\pi i} \log(z - \zeta)$, where $\gamma/2\pi$ is the strength of the vortex. The incremental velocity potential associated with a varying sheet strength $\gamma(s) \, ds$ is $d\Omega/dz = \frac{\gamma(s)ds}{2\pi i}/(z - \zeta)$. The incremental circumferential velocity, $dV = \gamma \, ds/2\pi r$, follows from this relationship.) This velocity field is consistent with the assumption of zero viscosity because it is continuous and irrotational everywhere except on the sheet

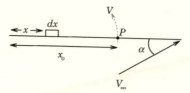

Fig. 7.4.3. Planar wing

where there is a jump in velocity and the vorticity is infinity. Let V_∞ be the velocity at infinity and let α be the angle of attack. Then it can be shown that the above flow is uniquely determined provided that the circulation around the airfoil satisfies the so-called **Kutta condition**, which implies that γ must be zero at the trailing edge.

For simplicity we make a further approximation and assume that the wing is planar, that is, the mean line is a straight line (see Figure 7.4.3).

The velocity due to an element dx of the vortex sheet dx on a point P of the planar airfoil is $dV = \gamma(x)\,dx/2\pi(x_0 - x)$; hence

$$V = \frac{1}{2\pi} \fint_0^c \frac{\gamma(x)\,dx}{x_0 - x}$$

The condition for the mean line to be a streamline is that it must line up with the impinging flow: $V + V_\infty \alpha = 0$ ($\sin\alpha \sim \alpha$ because α is small). Thus the problem of determining the unknown vortex strength reduces to

$$\frac{1}{2\pi} \fint_0^c \frac{\gamma(x)\,dx}{x - x_0} = -\alpha V_\infty \tag{7.4.8}$$

subject to the Kutta condition $\gamma(c) = 0$.

We will see that this equation can be solved by the formulae given in Example 7.3.3 of Section 7.3. Let $\Phi(z) = \frac{1}{2\pi i} \int_0^c \frac{\gamma(x)}{x-z}\,dx$; then by the Plemelj formula

$$\gamma(x) = \Phi^+(x) - \Phi^-(x) \tag{7.4.9a}$$

$$\Phi^+(x) + \Phi^-(x) = 2i\alpha V_\infty \tag{7.4.9b}$$

Equation (7.4.9b) is the same RH problem as that of Eq. (7.3.26), where we identify $b = c$, $a = 0$. Following Example 7.3.3, the fundamental solution bounded at c but not at 0 is $X(z) = (\frac{c-z}{z})^{1/2}$, where we have multiplied $X(z)$ by $(-1)^{1/2}$ for convenience. The index $\kappa = 0$, so we have a unique solution to the forced problem $\Phi(z) \to 0$ as $|z| \to \infty$. From Eq. (7.3.24) we find that the solution of the RH problem (7.4.9b) bounded at $x = c$ but unbounded at $x = 0$ is given by

$$\Phi(z) = \frac{X(z)}{2\pi i} \int_0^c \frac{2i\alpha V_\infty \, dx}{X^+(x)(x - z)}, \qquad X^+(x) = \sqrt{\frac{c - x}{x}}$$

where the branch cut is taken along the real axis. Thus, taking the limits to the real axis, we have

$$\Phi^\pm(x) = \pm \sqrt{\frac{c - x}{x}} \left(\frac{1}{2\pi i} \fint_0^c \frac{2i\alpha V_\infty}{X^+(x')(x' - x)} \, dx' \pm \frac{i\alpha V_\infty}{\sqrt{\frac{c-x}{x}}} \right)$$

hence using Eq. (7.4.9a)

$$\Phi^+(x) - \Phi^-(x) = \gamma(x) = \frac{2\alpha V_\infty \sqrt{\frac{c-x}{x}}}{\pi} \fint_0^c \frac{dx'}{(x' - x)} \sqrt{\frac{x'}{c - x'}} \qquad (7.4.10)$$

We can transform the integral appearing in Eq. (7.4.10) by letting $x = \frac{c}{2}(1 - \cos\theta)$ and $x' = \frac{c}{2}(1 - \cos\theta')$, that is

$$\sqrt{\frac{c - x}{x}} = \sqrt{\frac{1 + \cos\theta}{1 - \cos\theta}} = \sqrt{\frac{(1 + \cos\theta)^2}{1 - \cos^2\theta}} = \frac{1 + \cos\theta}{\sin\theta}$$

and Eq. (7.4.10) becomes

$$\gamma(x) = \frac{2\alpha V_\infty}{\pi} \left(\frac{1 + \cos\theta}{\sin\theta} \right) I, \qquad I = \fint_0^\pi \frac{\sin^2\theta' \, d\theta'}{(1 + \cos\theta')^{1/2}(\cos\theta - \cos\theta')}$$

It can be shown that the integral $I = \pi$; hence

$$\gamma(x) = 2\alpha V_\infty \frac{(1 + \cos\theta)}{\sin\theta} \qquad (7.4.11)$$

The solution $\gamma(x)$ satisfies $\gamma(c) = 0$ ($\theta = \pi$), but γ (and hence the velocity) becomes infinity at the leading edge ($\theta = 0$). This is a consequence of neglecting viscosity.

The lift per unit span is given by $L = \int_0^c \rho V_\infty \gamma \, dx$, and carrying out the integral yields

$$L = 2\alpha\rho V_\infty^2 \int_0^\pi \left(\frac{1 + \cos\theta}{\sin\theta} \right) \left(\frac{c}{2} \sin\theta \right) d\theta = \pi\alpha\rho V_\infty^2 c$$

Example 7.4.3 (A Generalized Abel Integral Equation) We shall show that the integral equation

$$\alpha(x) \int_a^x \frac{\varphi(t)\,dt}{(x-t)^\mu} + \beta(x) \int_x^b \frac{\varphi(t)\,dt}{(t-x)^\mu} = \gamma(x), \qquad 0 < \mu < 1 \quad (7.4.12)$$

can be reduced to the classical Abel equation

$$\int_a^x \frac{\Psi(t)\,dt}{(x-t)^\mu} = g(x), \qquad 0 < \mu < 1 \qquad (7.4.13)$$

which can be solved directly. We assume that the real functions $\alpha(x)$, $\beta(x)$ defined on $[a, b]$ do not vanish simultaneously and that they satisfy a Hölder condition. Furthermore, it is convenient to assume that $\gamma(x) = (x - a)$ $(b - x)^\varepsilon \Gamma(x)$, $\Gamma(x)$ real, and $\varepsilon > 0$, where $\frac{d\Gamma}{dx}(x)$ satisfies the Hölder condition. We seek a solution in the class

$$\varphi(x) = \frac{\eta(x)}{[(x-a)(b-x)]^{1-\mu-\varepsilon}}, \qquad \eta(x) \text{ Hölder} \qquad (7.4.14)$$

(Knowledge of the function classes for $\gamma(x)$ and $\eta(x)$ are not essential in order to understand what follows.)

The form of the integrals appearing in Eq. (7.4.12) suggests the introduction of the analytic function $\int_a^b \frac{\varphi(t)\,dt}{(t-z)^\mu}$; however, this function is $O(z^{-\mu})$ at infinity. It is convenient to have a function of order $O(z^{-1})$ at infinity and to define

$$\Phi(z) = R(z) \int_a^b \frac{\varphi(t)\,dt}{(t-z)^\mu}, \qquad R(z) \equiv [(z-a)(b-z)]^{\frac{1}{2}(\mu-1)} \quad (7.4.15)$$

We introduce a branch cut from a to b and we choose the argument of $t - z$ to be $-\pi$ as $z \to x$ above the cut with $a < t < x$; then $t - z$ will have argument π as $z \to x$ below the cut with $a < t < x$, and $t - z$ has 0 argument when $z = x < t$. (A choice of local angles, see Section 2.3, is as follows: $t - z = (z - t)e^{-i\pi}$, $z - t = re^{i\theta}$, $0 < \theta < 2\pi$, so that on top of the cut $t - x = (x - t)e^{-i\pi}$, and on the bottom of the cut $t - x = (x - t)e^{i\pi}$.) If we choose $R^+(x) = R(x)$, then $R^-(x) = e^{2i\pi \frac{1}{2}(\mu-1)} R(x) = -e^{i\pi\mu} R(x)$. (Similarly, a choice of local angles for $f(z) = (z - a)(b - z) = (z - a)(z - b)e^{-i\pi} = r_1 r_2 e^{i(\theta_1 + \theta_2 - \pi)}$ with $0 \le \theta_1, \theta_2 < 2\pi$, so on top of the cut $R^+(x) = [(x - a)(b - x)]^{(\mu-1)/2}$, and on the bottom $R^-(x) = e^{(\mu-1)\pi i}[(x - a)(b - x)]^{(\mu-1)/2}$.) Taking the limit

$z \to x \pm i\varepsilon$

$$\Phi^+(x) = R(x) \left[e^{i\mu\pi} \int_a^x \frac{\varphi(t)\,dt}{(x-t)^\mu} + \int_x^b \frac{\varphi(t)\,dt}{(t-x)^\mu} \right]$$

$$\Phi^-(x) = -e^{i\mu\pi} R(x) \left[e^{-i\mu\pi} \int_a^x \frac{\varphi(t)\,dt}{(x-t)^\mu} + \int_x^b \frac{\varphi(t)\,dt}{(t-x)^\mu} \right]$$

These equations are the analog of the Plemelj formula for the integral (7.4.15). They can also be written in the form

$$\int_a^x \frac{\varphi(t)\,dt}{(x-t)^\mu} = \frac{e^{i\mu\pi}\Phi^+(x) + \Phi^-(x)}{R(x)(e^{2i\mu\pi} - 1)} \tag{7.4.16a}$$

$$\int_x^b \frac{\varphi(t)\,dt}{(t-x)^\mu} = \frac{e^{i\mu\pi}\Phi^-(x) + \Phi^+(x)}{R(x)(1 - e^{2i\mu\pi})} \tag{7.4.16b}$$

Using Eqs. (7.4.16), the integral Equation (7.4.12) reduces to the RH problem

$$\Phi^+(x) = \frac{\beta(x)e^{i\mu\pi} - \alpha(x)}{\alpha(x)e^{i\mu\pi} - \beta(x)} \Phi^-(x) + \frac{(e^{2i\mu\pi} - 1)\,\gamma(x)R(x)}{\alpha(x)e^{i\mu\pi} - \beta(x)} \tag{7.4.17}$$

This RH problem is also supplemented with the boundary conditions (from Eq. (7.4.15))

$$\Phi(z) = O(z^{-1}), \quad z \to \infty; \qquad \Phi(z) = O\big((z-a)^{\frac{\mu-1}{2}}\big), \quad z \to a; \tag{7.4.18}$$
$$\Phi(z) = O(b-z)^{\frac{\mu-1}{2}}, \quad z \to b$$

Conversely, as in the case of singular integral equations with Cauchy kernels, it can be shown (Carleman, 1922 a,b) that the RH problem satisfying Eqs. (7.4.17) and (7.4.18) is equivalent to the integral equation (7.4.12).

The above RH problem can be solved for Φ^+, Φ^- by the method presented in Section 7.3; after finding $\Phi^+(t)$ and $\Phi^-(t)$, $\varphi(t)$ is found by solving the classical Abel equation (7.4.16a) (or (7.4.16b), which is really the same as (7.4.16a) apart from a negative sign) described below.

We briefly sketch the method of solution to Abel's equation. Write Abel's equation in the form

$$K_\mu(\Psi) = \int_a^x \frac{\Psi(t)}{(x-t)^\mu}\,dt = g(x), \qquad g(a) = 0$$

where the operator K_μ (acting on a function f) is defined as

$$K_\mu(f) = \int_a^x \frac{dt}{(x-t)^\mu} f(t)$$

Consider

$$K_{1-\mu}(K_\mu(\Psi)) = \int_a^x \frac{dx'}{(x-x')^{1-\mu}} \int_a^{x'} \frac{\Psi s(t)\, dt}{(x'-t)^\mu} = \int_a^x g(x')\frac{dx'}{(x-x')^{1-\mu}}$$

$$(7.4.19)$$

We interchange integrals to find on the left-hand side

$$K_{1-\mu}(K_\mu\Psi) = \int_a^x \left(\int_t^x \frac{dx'}{(x-x')^{1-\mu}(x'-t)^\mu} \right) \Psi(t)\, dt$$

The inside integral can be evaluated exactly, by the change of variables, $x' = t + (x-t)u$, whereupon

$$\int_t^x \frac{dx'}{(x-x')^{1-\mu}(x'-t)^\mu} = \int_0^1 \frac{du}{(1-u)^{1-\mu}\, u^\mu} = \frac{\pi}{\sin \mu\pi}$$

the latter being a well-known integral. Therefore Eq. (7.4.19) yields

$$\int_a^x \Psi(t)\, dt = \frac{\sin \mu\pi}{\pi} \int_a^x \frac{g(x')}{(x-x')^{1-\mu}}\, dx'$$

By taking the derivative of this equation, the solution of Abel's equation (7.4.13) is given by

$$\Psi(x) = \frac{\sin \mu\pi}{\pi} \frac{d}{dx} \int_a^x \frac{g(x')\, dx'}{(x-x')^{1-\mu}} \qquad (7.4.20)$$

As an illustration of the above approach we consider an equation solved by Carleman (1922 a):

$$\int_a^b \frac{\varphi(t)\, dt}{|x-t|^\mu} = \int_a^x \frac{\varphi(t)\, dt}{(x-t)^\mu} + \int_x^b \frac{\varphi(t)\, dt}{(t-x)^\mu} = \gamma(x), \qquad 0 < \mu < 1$$

$$(7.4.21)$$

This equation is a special case of Eq. (7.4.12) where $\alpha = \beta = 1$; thus Eq. (7.4.17) becomes

$$\Phi^+(x) = \Phi^-(x) + (1 + e^{\mu\pi i})R(x)\gamma(x)$$

The unique solution of this RH problem, satisfying Eqs. (7.4.18) is found by the method of Section 7.3.2: see Eqs. (7.3.18) where we note that $g(t) = 1 = e^{2\pi i n}$ for integer n, and therefore

$$\Gamma(z) = \frac{1}{2\pi i} \int_a^b \frac{\log(e^{2\pi i n})}{t - z}\, dt = n \log \left(\frac{z - b}{z - a} \right)$$

Thus

$$X(z) = (z - a)^\lambda (z - b)^\mu e^{\Gamma(z)} = (z - a)^{\lambda - n}(z - b)^{\mu + n}$$

and hence, $\lambda = n$, $\mu = -n$, and $\lambda + \mu = 0 = \kappa$ (index). The decaying solution is given by Eq. (7.3.24) with $P(z) = 0$, $X(z) = 1$, $f(x) = (1 + e^{\mu \pi i})R(x)$ $\gamma(x)$:

$$\Phi(z) = \frac{1 + e^{\mu \pi i}}{2\pi i} \int_a^b \frac{R(t)\gamma(t)}{t - z}\, dt \qquad (7.4.22)$$

Using Eq. (7.4.22) to determine $\Phi^+(x)$ and $\Phi^-(x)$ (from Eq. (7.2.3$^\pm$)), and substituting these values in Eq. (7.4.16a) we find

$$\int_a^x \frac{\varphi(t)\, dt}{(x - t)^\mu} = \frac{\gamma(x)}{2} - \frac{\cot\left(\frac{\mu\pi}{2}\right)}{2\pi} \frac{1}{R(x)} \fint_a^b \frac{R(t)\gamma(t)}{t - x}\, dt$$

Solving this Abel equation (by using Eqs. (7.4.13) and (7.4.20)), it follows that the solution of the singular integral equation (7.4.20) is given by

$$\varphi(x) = \frac{\sin \mu\pi}{2\pi} \frac{d}{dx} \int_a^b \frac{\gamma(t)\, dt}{(x - t)^{1 - \mu}}$$

$$- \frac{\cos^2\left(\frac{\mu\pi}{2}\right)}{\pi^2} \frac{d}{dx} \int_a^x \frac{1}{(x - t)^{1 - \mu}} \frac{1}{R(t)} \left(\fint_a^b \frac{R(\tau)\gamma(\tau)}{\tau - t}\, d\tau \right) dt$$

Example 7.4.4 (Integral Equations With Logarithmic Kernels) We shall show that the weakly singular integral equation

$$\alpha(x) \int_a^b \log|t - x|\varphi(t)\, dt - \pi i \beta(x) \int_a^x \varphi(t)\, dt = \gamma(x) \qquad (7.4.23)$$

where $\alpha(x)$, $\beta(x)$, and $\gamma(x)$ satisfy the Hölder condition, can also be reduced to a RH problem (we also assume that $\alpha^2 - \beta^2 \neq 0$).

Associated with Eq. (7.4.23), we consider the function

$$\Phi(z) = \int_a^b \log\left(1 - \frac{a - t}{a - z}\right)\varphi(t)\, dt \qquad (7.4.24a)$$

or equivalently

$$\Phi(z) = \int_a^b \log(t - z)\varphi(t)\, dt - A \log(a - z); \quad A \equiv \int_a^b \varphi(t)\, dt \quad (7.4.24b)$$

We introduce a branch cut from a to b, and we choose the argument of $t - z$ to be $-\pi$ (as in Example 7.4.3); also we choose as $-\pi$ the argument of $a - z$, as $z \to x$ above the cut. (The choice of local angles is $a - z = (z - a)e^{-i\pi}$, $z - a = re^{i\theta}, 0 < \theta < 2\pi$, and $t - z = (z - t)e^{-i\pi}, z - t = \rho e^{i\phi}, 0 < \phi < 2\pi$, so that $\arg(t - z) = 0$ when $z = x < t$.) Then as $z \to x \pm i\epsilon$, splitting the integral $\int_a^b$ to $\int_a^x + \int_x^b$ and noting that

$$\log(t - x) = \log|x - t|e^{\pm i\pi} \quad \text{as } z \to x \pm i\varepsilon \quad \text{for } x > t$$
$$\log(t - x) = \log|x - t| \quad \text{as } z \to x \pm i\varepsilon \quad \text{for } x < t$$

we have

$$\Phi^\pm(x) = \int_a^b \log|x - t|\varphi(t)\, dt \mp \pi i \int_a^x \varphi(t)\, dt - A \log(x - a) \pm \pi i A \tag{$7.4.25^\pm$}$$

These equations are the analog of the Plemelj formulae for the integrals (7.4.24a,b). Subtracting and adding, they are equivalent to

$$-2\pi i \int_a^x \varphi(t)\, dt = \Phi^+(x) - \Phi^-(x) - 2\pi i A \tag{7.4.26a}$$

$$2 \int_a^b \log|t - x|\varphi(t)\, dt = \Phi^+(x) + \Phi^-(x) + 2A \log(x - a) \tag{7.4.26b}$$

Substituting these equations into Eq. (7.4.23), this integral reduces to the RH problem

$$\Phi^+(x) = \frac{\beta(x) - \alpha(x)}{\beta(x) + \alpha(x)} \Phi^-(x) + \frac{2\gamma(x) + 2\pi i A\beta(x) - 2A\alpha(x) \log(x - a)}{\alpha(x) + \beta(x)}. \tag{7.4.27}$$

Equations (7.4.24a,b) also imply that $\Phi(z)$ is of order $O(z^{-1})$ as $z \to \infty$ and it has a certain behavior as z approaches the endpoints. (We usually assume that $\varphi(x)$ has an integrable singularity at the endpoints and that $\Phi(z)$ will be bounded at $x = a$ and $x = b$.) After finding $\Phi(z)$ satisfying these boundary conditions using the method of Section 7.3, we obtain $\varphi(x)$, which is given by (see Eq. (7.4.26a)

$$-2\pi i\varphi(x) = \frac{d}{dx}(\Phi^+(x) - \Phi^-(x)) \tag{7.4.28}$$

The solution $\varphi(x)$ depends linearly on the constant A; this constant can be found by integrating the expression (7.4.28), because $A = \int_a^b \varphi(t)\, dt$.

7.4.1 Riemann–Hilbert Problems on the Real Axis

In many applications one encounters RH problems formulated on the real axis. For such problems it turns out that the space of integrable functions is more convenient to work with than the space of Hölder functions. We shall consider, in more detail, the set (which is, in fact, a ring) Λ_1 of functions of the form

$$c + \int_{-\infty}^{\infty} f(x)e^{-ikx}\,dx, \qquad c \in \mathbb{C}, \quad k \in \mathbb{R} \qquad (7.4.29)$$

where $f(x)$ is continuous and

$$f(x) \in L_1 \qquad (7.4.30)$$

that is, $\int_{-\infty}^{\infty}|f(x)|\,dx$ exists

Let Λ_1^- denote the subset of Λ_1 consisting of functions of the form $c + \int_0^{\infty} f(x)e^{-ikx}\,dx$. Letting $k = k_R + ik_I$, it is clear that functions in Λ_1^- are analytically continuable in the lower half k-complex plane (note that $x > 0$), which we denote by π^-. Similarly, Λ_1^+ denotes the subset of Λ_1 consisting of functions of the form $c + \int_{-\infty}^0 f(x)e^{-ikx}\,dx$, which are analytic in the upper half k-complex plane (note that $x < 0$), denoted by π^+.

Example 7.4.5 (Integral equations with displacement kernels – also see the Wiener–Hopf method discussed in Example 7.4.7) We shall show that the integral equation

$$\varphi(x) - \int_0^{\infty} g(x - t)\varphi(t)\,dt = f(x), \qquad x > 0 \qquad (7.4.31)$$

where $f(x)$ and $g(x)$ are continuous and belong to the space of integrable functions L_1, can be reduced to a RH problem. We assume that the Fourier transform of g (see Section 4.5 for the definition of the Fourier transform), which we denote by $\hat{G}(k) = \int_{-\infty}^{\infty} g(x)e^{-ikx}\,dx$, satisfies the following conditions: (a) $1 - \hat{G}(k) \neq 0$ for $k \in \mathbb{R}$, and (b) the index of $(1 - \hat{G}) \equiv \operatorname{ind}(1 - \hat{G}) = 0$, where

$$\operatorname{ind}(1 - \hat{G}) = \frac{1}{2\pi i}[\log(1 - \hat{G})]_{-\infty}^{\infty} = \frac{1}{2\pi}[\arg(1 - \hat{G})]_{-\infty}^{\infty} \qquad \text{for } k \in \mathbb{R}$$

We seek a solution $\varphi(x)$ whose Fourier transform belongs to Λ_1.

In order to take the Fourier transform of Eq. (7.4.31), we extend this equation so that it is valid for all x. Let $\tilde{\varphi}(x) = \varphi(x)$ for $x > 0$ and $\tilde{\varphi}(x) = 0$ for $x < 0$. Then Eq. (7.4.31) can be rewritten as

$$\tilde{\varphi}(x) - \int_{-\infty}^{\infty} g(x - t)\tilde{\varphi}(t)\,dt = f(x), \qquad x > 0 \qquad (7.4.32a)$$

For $x < 0$

$$\tilde{\varphi}(x) - \int_{-\infty}^{\infty} g(x-t)\tilde{\varphi}(t)\,dt = -\int_0^{\infty} g(x-t)\varphi(t)\,dt = h(x), \qquad x < 0$$

(7.4.32b)

where $h(x)$ is some unknown function. Let

$$\{\tilde{\varphi}(x), \tilde{f}(x), \tilde{h}(x)\} = \begin{cases} \{0, 0, h(x)\} & x < 0 \\ \{\varphi(x), f(x), 0\} & x > 0 \end{cases}$$

Equations (7.4.32a,b) can be written as the single equation

$$\tilde{\varphi}(x) - \int_{-\infty}^{\infty} g(x-t)\tilde{\varphi}(t)\,dt = \tilde{f}(x) + \tilde{h}(x), \qquad -\infty < x < \infty$$

Denoting by $\hat{\Phi}, \hat{F}, \hat{H}, \hat{G}$, the Fourier transforms of $\tilde{\varphi}, \tilde{f}, \tilde{h}, g$, respectively, and taking the Fourier transform of the above equation (using the Fourier transform of a convolution product) we find

$$(1 - \hat{G}(k))\hat{\Phi}(k) = \hat{F}(k) + \hat{H}(k) \qquad (7.4.33a)$$

But because $\tilde{\varphi} = 0$ when $x < 0$, we have

$$\hat{\Phi}(k) = \int_0^{\infty} \tilde{\varphi}(x)e^{-ikx}\,dx, \qquad \text{i.e. } \hat{\Phi}(k) \in \Lambda_1^-$$

Similarly, $\hat{F}(k) \in \Lambda_1^-$ and $\hat{H}(k) \in \Lambda_1^+$. Thus Eq. (7.4.33a) defines the RH problem

$$\hat{H}^+(k) = (1 - \hat{G}(k))\hat{\Phi}^- - \hat{F}^-(k), \qquad -\infty < k < \infty. \qquad (7.4.33b)$$

This is an RH problem where $\hat{F}^-$ is known (because $f(x)$ is given) and the unknowns are $\hat{H}^+$ and $\hat{\Phi}^-$. To connect with the previous notation, we will denote the unknown $\hat{H}^+$ as $\hat{\Phi}^+$. Because we have so far assumed that $\text{ind}(1 - \hat{G}) = 0$, the solution of this RH problem vanishing at infinity is unique. Equation (7.4.34) can be solved by the formulae developed in Section 7.3 with the "closed" contour now being $L = (-\infty, \infty)$. We also assume that $\hat{G}(k)$ and $\hat{F}(k)$ satisfy the Hölder condition (which is not guaranteed in the space Λ_1). In particular, the fundamental solution satisfies

$$\hat{X}^+(k) = (1 - \hat{G}(k))\hat{X}^-(k), \qquad \hat{X}^\pm(\infty) = 1, \qquad -\infty < k < \infty,$$

and it is given (see Eq. (7.3.6a,b)) by

$$\hat{X}(k) = \exp\left[\frac{1}{2\pi i}\int_{-\infty}^{\infty} \frac{\log(1 - \hat{G}(\tau))d\tau}{\tau - k}\right], \qquad k \in \mathbb{C}.$$

The solution to the nonhomogeneous problem is given by Eq. (7.3.8) where $P(z) = 0$ and $f(t) = -\hat{F}^-(t)$, that is,

$$\hat{\Phi}(k) = \frac{-X(z)}{2\pi i} \int_{-\infty}^{\infty} \frac{\hat{F}^-(\tau)\,d\tau}{X^+(\tau)(\tau - k)}, \qquad k \in \mathbb{C}.$$

Thus the solution $\varphi(x)$ to Eq. (7.4.31) is obtained by finding $\hat{\Phi}^\pm(k)$ and then taking the inverse Fourier transform of $\hat{\Phi}^-(k)$.

We also note that an extensive theory for equations of the form (7.4.31) has been developed in Gohberg and Krein (1958), including important results involving systems of equations of the form (7.4.31) where ϕ and f are vectors and g a matrix. (We discuss the matrix case in Section 7.7.) In connection with such scalar equations these authors have considered the following **factorization** problem: Given $1 - \hat{G}(k) \in \Lambda_1$ and $1 - \hat{G} \neq 0$, find $\hat{G}^\pm \in \Lambda_1^\pm$ such that (the analysis of Gohberg and Krein does not require $\hat{G}$ to satisfy a Hölder condition!)

$$(1 - \hat{G})(k) = \hat{\eta}^+(k)\,\hat{\eta}^-(k), \qquad \hat{\eta}^\pm(\infty) = 0, \qquad -\infty < k < \infty. \tag{7.4.34}$$

A factorization is called proper if either $\hat{\eta}^+(k) \neq 0$ in π^+ or if $\hat{\eta}^-(k) \neq 0$ in π^-. A factorization is called canonical if both $\hat{\eta}^\pm(k) \neq 0$ in $\pi^\pm$. They show that $(1 - \hat{G})$ admits a canonical factorization if and only if

$$(1 - \hat{G}(k)) \neq 0 \qquad \text{and} \qquad \text{ind}\,(1 - \hat{G}) = 0, \qquad -\infty < k < \infty.$$

Furthermore, the canonical factorization is the only proper one. These results are consistent with the RH theory presented here. Indeed, because $\hat{\eta}^- \neq 0$ in π^-, it follows that $(\hat{\eta}^-)^{-1}$ is analytic in π^- and Eq. (7.4.34) can be written as

$$X^+(k) = (1 - \hat{G})^{-1}(k)X^-(k) \quad \text{on} \quad -\infty < k < \infty, \quad \text{and} \quad X^\pm(\pm\infty) = 1$$

where

$$X^+(k) = (\eta^+(k))^{-1}, \quad \text{and} \quad X^-(k) = \eta^-(k)$$

Furthermore, because $\text{ind}(1 - \hat{G}) = 0$, the solutions (found by taking the logarithm and using the Plemelj formulae) $X^+(k)$ and $X^-(k)$ are unique.

Actually, one can go further. Suppose $\text{ind}(1 - \hat{G}(k)) = \kappa$; then the unique factorization into nonvanishing functions $\eta^\pm(k)$ analytic for $\text{Im}\,k \neq 0$ is given by

$$\eta^+(k) = m(k)\left(\frac{k+i}{k-i}\right)^\kappa \eta^-(k) \qquad \text{on} \quad -\infty < k < \infty$$

where we define $m(k) = (1 - \hat{G}(k))$. We will show below that the index of the function $m(k) \left(\frac{k+i}{k-i} \right)^{\kappa}$ is zero, and therefore the solution to this RH problem can be readily found by taking the logarithm and using the Plemelj formulae.

It is significant that the case of nonzero index has a nice interpretation in terms of the original integral equation. It can be shown that if $\kappa > 0$ there exists a κ-dimensional family (κ linearly independent solutions) of homogeneous solutions to Eq. (7.4.31):

$$\varphi(x) - \int_0^\infty g(x - t)\phi(t)\,dt = 0$$

If $\kappa < 0$, there exists a $|\kappa|$-dimensional family of solutions to the adjoint problem

$$\Psi(x) - \int_0^\infty g(t - x)\Psi(t)\,dt = 0$$

where, in addition, there needs to be supplemented $|\kappa|$ conditions of the form

$$\int_0^\infty f(t)\Psi(t)\,dt = 0$$

corresponding to each of the $|\kappa|$ linearly independent solutions $\Psi(t)$.

At this point it is convenient to discuss the more general RH problem on the real axis that arises frequently in applications and that is not only associated with integral equations such as Eq. (7.4.31). Consider

$$\Phi^+(k) = m(k)\Phi^-(k) + f(k) \qquad \text{on} \quad -\infty < k < \infty$$

where we wish to find the sectionally analytic function $\Phi(k)$ analytic for $\text{Im}\,k > 0$ and $\text{Im}\,k < 0$, satisfying the above boundary condition on the real k axis. We look for solutions Φ that are bounded at infinity, usually we specify $m(\pm\infty) = 1$, $f(\pm\infty)$ bounded, assume that $m(k)$ and $f(k)$ satisfy Hölder conditions, and that the index

$$\text{ind}(m(k)) = \frac{1}{2\pi i}[\log m(k)]_{-\infty}^\infty = \frac{1}{2\pi}[\arg m(k)]_{-\infty}^\infty = \kappa$$

First we find the fundamental solution of the homogeneous problem

$$X^+(k) = m(k)X^-(k) \qquad \text{on} \quad -\infty < k < \infty$$

where $X^\pm(\infty) = 1$. The process is similar to those used for the closed contour, except that the factor multiplying $m(k)$ must be modified to achieve a zero index. (Previously, we used $k^{-\kappa}$, which is not acceptable in this case because $k = 0, \infty$ now lie on the contour.) We note that $\text{ind}\left(\frac{k-i}{k+i}\right) = 1$. This follows from

$$
\begin{aligned}
\text{ind}\left(\frac{k-i}{k+i}\right) &= \frac{1}{2\pi i}\left[\log\frac{k-i}{k+i}\right]_{-\infty}^{\infty} \\
&= \frac{1}{2\pi i}\left[\log\frac{r_1 e^{i\theta_1}}{r_2 e^{i\theta_2}}\right]_{-\infty}^{\infty} \\
&= \frac{1}{2\pi i}(\theta_d(\infty) - \theta_d(-\infty)) \\
&= 1
\end{aligned}
$$

where $\theta_d = \theta_1 - \theta_2$. It is useful to take local angles to be $-\pi/2 \le \theta_1 < 3\pi/2$ and $-\pi/2 \le \theta_2 < 3\pi/2$ (or any angles for which $\log\left(\frac{k-i}{k+i}\right)$ does not have a branch cut on the real axis). Thus $\text{ind}\left(\left(\frac{k-i}{k+i}\right)^{-\kappa}\right) = -\kappa$ and we rewrite the homogeneous problem as

$$
X^+(k) = \left(\left(\frac{k-i}{k+i}\right)^{-\kappa} m(k)\right)\left(\frac{k-i}{k+i}\right)^{\kappa} X^-(k) \qquad \text{on} \quad -\infty < k < \infty
$$

so that the first factor in the right-hand side of this equation has zero index. Taking the logarithm of this equation

$$
\begin{aligned}
&\log X^+(k) - \log\left(\left(\frac{k-i}{k+i}\right)^{\kappa} X^-(k)\right) \\
&= \log\left(\left(\frac{k-i}{k+i}\right)^{-\kappa} m(k)\right) \qquad \text{on} \quad -\infty < k < \infty
\end{aligned}
$$

Using the Plemelj formulae yields the fundamental homogeneous solution

$$
X^+(k) = e^{\Gamma^+(k)}
$$

$$
X^-(k) = \left(\frac{k+i}{k-i}\right)^{\kappa} e^{\Gamma^-(k)}
$$

where

$$
\Gamma(k) = \frac{1}{2\pi i}\int_{-\infty}^{\infty} \frac{\log\left(\left(\frac{\ell-i}{\ell+i}\right)^{-\kappa} m(\ell)\right)}{\ell - k}\, d\ell
$$

We see that if $\kappa < 0$, then $X^-(k)$ has a pole of order $|\kappa|$ at $k = -i$; hence $X^-(k)$ is not analytic for $\text{Im } k < 0$, and there is no solution to the homogeneous problem satisfying the analyticity requirements.

Next we solve the forced problem, $\Phi^+ = m\Phi^- + f$ by using $m = X^+/X^-$, which leads to

$$\frac{\Phi^+(k)}{X^+(k)} - \frac{\Phi^-(k)}{X^-(k)} = \frac{f(k)}{X^-(k)} \qquad \text{on} \qquad -\infty < k < \infty.$$

The forced solution follows from the Plemelj formulae. For $\kappa \geq 0$

$$\Phi(k) = X(k) \left[\frac{1}{2\pi i} \int_{-\infty}^{\infty} \frac{f(\ell)}{X^-(\ell)(\ell - k)} \, d\ell + \frac{P_\kappa(k)}{(k + i)^\kappa} \right]$$

where $P_\kappa(k)$ is a polynomial of degree κ; $P_\kappa(k) = a_0 + a_1 k + a_2 k^2 + \cdots + a_\kappa k^\kappa$. Note that because $X^+(k) = e^{\Gamma^+(k)}$ and $X^-(k) = \left(\frac{k+i}{k-i} \right)^\kappa e^{\Gamma^-(k)}$, we see that for $\kappa \geq 0$ the above expression for $\Phi(k)$ is analytic for $\mathrm{Im}\, k > 0$ and $\mathrm{Im}\, k < 0$ and bounded at infinity for large k. Thus when $\kappa \geq 0$ we have a solution with $\kappa + 1$ arbitrary constants.

On the other hand, when $\kappa < 0$ there is in general no solution. With suitable extra conditions added, the most general solution to the forced problem is given by

$$\Phi(k) = X(k) \left[\frac{1}{2\pi i} \int_{-\infty}^{\infty} \frac{f(\ell)}{X^-(\ell)\ell - k)} \, d\ell + c \right]$$

where c is an appropriate constant (determined below), and $X^\pm(k)$ are those obtained above. This formula leads to $\Phi^-(k)$ having a pole of order $|\kappa|$ at $k = -i$, which can be removed by choosing c to be

$$c = \frac{1}{2\pi i} \int_{-\infty}^{\infty} \frac{f(k)}{X^-(k)(k + i)} \, dk$$

and requiring the supplementary conditions

$$\frac{1}{2\pi i} \int_{-\infty}^{\infty} \frac{f(\ell)}{X^-(\ell)(\ell + i)^m} \, d\ell = 0, \qquad m = 2, 3, \ldots, |\kappa| - 1$$

These conditions follow from the Taylor expansion of $1/(\ell - k)$ for k near $-i$, that is

$$\frac{1}{\ell + i - k} = \frac{1}{\ell + i} \left(1 + \frac{k}{\ell + i} + \frac{(k)^2}{(\ell + i)^2} + \cdots \right)$$

and requiring boundedness of the solution. Note also that if we require $\Phi(k) \to$ 0 as $k \to \infty$, with $f(\infty) = 0$, then $C = 0$ and we need to have the supplementary conditions satisfied for $m = 1, 2, 3, \ldots, |\kappa|$.

Example 7.4.6 (The Fourier and Radon transforms) The Radon transform is defined by

$$\tilde{q}(k, p) = \int_l q(x_1, x_2)\, d\tau$$

where the integral is taken along a line l with direction determined by the unit vector $\underline{k} = \left(\frac{1}{\sqrt{1+k^2}}, \frac{k}{\sqrt{1+k^2}}\right)$, at a distance p from the origin, and τ is a parameter on this line (see Figure 7.4.4).

This transform plays a fundamental role in the mathematical formulation of computerized tomography. By computerized tomography (CT), we mean the reconstruction of a function from knowledge of its line integrals, irrespective of the particular field of application. However, the most prominent application of CT is in diagnostic radiology. Here a cross section of the human body is scanned by a thin X-ray beam whose intensity loss is recorded by a detector and processed by a computer to produce a two-dimensional image that in turn is displayed on a screen.

A simple physical model is as follows (see Figure 7.4.5). Let $f(x_1, x_2)$ be the X-ray attenuation coefficient of the tissue at the point $\underline{x} = (x_1, x_2)$. This means that X-rays traversing a small distance $\Delta\tau$ along the line l suffer the relative intensity loss

$$\frac{\Delta I}{I} = -f(x_1, x_2)\Delta\tau$$

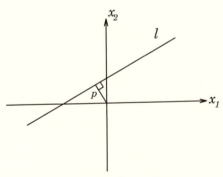

Fig. 7.4.4. Line l, distance p

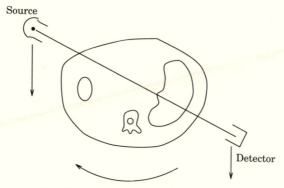

Fig. 7.4.5. Simple physical model of CT

Let I_0 and I_1 be the initial and final intensity of the beam, before and after leaving the body, respectively. In the limit $\Delta \tau \to 0$ it follows from the above equation that

$$\frac{I_1}{I_0} = e^{-\int_l f(x_1, x_2)\, d\tau}$$

that is, the scanning process determines an integral of the function $f(x_1, x_2)$ along each line l. Given all these integrals, one wishes to reconstruct the function f.

Let $\underline{k}$ be a unit vector along l and let $\underline{k}^{\perp}$ be the unit vector orthogonal to $\underline{k}$, that is, $\underline{k}^{\perp} = \left(-\frac{k}{\sqrt{1+k^2}}, \frac{1}{\sqrt{1+k^2}}\right)$. Then any point $\underline{x} = (x_1, x_2)$ can be written as $\underline{x} = p\underline{k}^{\perp} + \tau\underline{k}$, or $x_1 = (\tau - pk)/\sqrt{1+k^2}$ and $x_2 = (\tau k + p)/\sqrt{1+k^2}$. Note that for fixed p, as τ varies, $\underline{x}$ moves along the line l as depicted in Figure 7.4.4. Therefore the Radon transform can also be written as

$$\tilde{q}(k, p) = \int_{-\infty}^{\infty} q\left(\frac{\tau - pk}{\sqrt{1+k^2}}, \frac{\tau k + p}{\sqrt{1+k^2}}\right) d\tau$$

Next we show how the reconstruction of functions via the classical Fourier transform and the Radon transform can be viewed as RH problems. More generally, we mention that in recent years a new method has been discovered (see, e.g., Ablowitz and Clarkson, 1991; Ablowitz and Segur, 1981; Fokas and Zakharov, 1993) for solving certain nonlinear PDEs. This method, which is called the inverse scattering transform (IST), can be thought as a nonlinear version of the transform methods discussed in Chapter 4. For example, the Fourier transform has a certain nonlinear version that can be used to solve several equations of physical interest such as the Korteweg–deVries equation, the nonlinear Schrödinger equation, and others. We will discuss this further in Section 7.7. Similarly, the Radon transform has a nonlinear version that can be used to solve a special case (a 2 + 1 dimensional reduction) of the so-called self-dual Yang–Mills equation. Both of these nonlinear transforms can be derived by using the analytic properties of solutions of certain linear eigenvalue problems and then formulating appropriate RH problems. These eigenvalue problems, appropriately simplified, can also be used to derive the corresponding linear transforms. This is interesting for both conceptual and practical considerations. Conceptually, it unifies the RH theory and the theory of linear transforms by establishing that in some sense the latter is a special case of the former. Practically, it also provides constructive approaches to deriving linear transforms.

First we illustrate the approach for the classical Fourier transform. Then we derive the Radon transform.

7.4.2 The Fourier Transform

Consider the linear differential equation

$$\mu_x - ik\mu = q(x), \qquad q \to 0 \quad \text{as } |x| \to \infty, \quad -\infty < x < \infty, \quad k \in \mathbb{C} \tag{7.4.35}$$

where $\mu_x = \partial\mu/\partial x$, and assume that q and q_x are integrable: $\int_{-\infty}^{\infty} |q|\, dx < \infty$ and $\int_{-\infty}^{\infty} |q_x|\, dx < \infty$, that is, q and q_x belong to L_1. We define a solution $\mu(x, k)$ of Eq. (7.4.35), which is bounded for all $k \in \mathbb{C}$. Let μ be defined by

$$\mu(x, k) = \begin{cases} \mu^+(x, k), & k_I \geq 0, \\ \mu^-(x, k), & k_I \leq 0, \end{cases} \quad k = k_R + ik_I$$

where μ^+ and μ^- are the following particular solutions of Eq. (7.4.35)

$$\mu^+(x, k) = \int_{-\infty}^{x} q(\xi)e^{ik(x-\xi)}\, d\xi \tag{7.4.36a}$$

$$\mu^-(x, k) = -\int_{x}^{\infty} q(\xi)e^{ik(x-\xi)}\, d\xi \tag{7.4.36b}$$

which are obtained by direct integration of Eq. (7.4.35). We see that μ^+ is analytic in the upper half plane ($x - \xi > 0$; $k_I > 0$), while μ^- is analytic in the lower half plane ($x - \xi < 0$; $k_I < 0$). Furthermore, the large x behavior of both μ^+ and μ^- is uniquely determined by $\hat{q}(k)$, which is defined by

$$\hat{q}(k) \equiv \int_{-\infty}^{\infty} q(x)e^{-ikx}\, dx, \qquad k \in \mathbb{R} \tag{7.4.37}$$

Indeed,

$$\lim_{x \to -\infty} \left(e^{-ikx}\mu^-\right) = -\hat{q}(k), \qquad \lim_{x \to \infty} \left(e^{-ikx}\mu^+\right) = \hat{q}(k) \qquad (7.4.38)$$

Equation (7.4.37) defines $\hat{q}$ in terms of q. We now want to invert this relationship. This problem can be solved in an elementary way because $\hat{q}(k)$ is the Fourier transform of $q(x)$. However, we shall use instead a method that can also be used in other more complicated cases. Namely, we shall formulate a (RH) problem. Subtracting Eq. (7.4.36a,b), we find

$$\mu^+(x, k) - \mu^-(x, k) = e^{ikx}\hat{q}(k), \qquad k \in \mathbb{R} \qquad (7.4.39)$$

Equations (7.4.36a,b), using integration by parts, imply

$$\mu = O\left(\frac{1}{k}\right), \qquad \text{as } k \to \infty$$

Equation (7.4.39), with $\mu \to 0$ as $k \to \infty$, defines an elementary RH problem for the sectionally holomorphic function $\mu(x, k)$. Its unique solution is given by

$$\mu(x, k) = \frac{1}{2\pi i} \int_{-\infty}^{\infty} \frac{e^{i\ell x}\hat{q}(\ell)}{\ell - k} \, d\ell, \qquad k \in \mathbb{C} \qquad (7.4.40)$$

Given $\hat{q}(\ell)$, Eq. (7.4.40) yields $\mu(x, k)$, which then implies $q(x)$ through Eq. (7.4.35). An elegant formula for q can be obtained by comparing the large k asymptotics of Eqs. (7.4.35) and (7.4.40). Equation (7.4.35) implies $q = -i \lim_{k\to\infty}(k\mu)$, while Eq. (7.4.40) yields

$$\lim_{k \to \infty} (k\mu) = -\frac{1}{2\pi i} \int_{-\infty}^{\infty} e^{ikx}\hat{q}(k) \, dk$$

Hence

$$q(x) = \frac{1}{2\pi} \int_{-\infty}^{\infty} e^{ikx}\hat{q}(k) \, dk \qquad (7.4.41)$$

Equations (7.4.37) and (7.4.41) are the usual formulae for the direct and inverse Fourier transform.

7.4.3 The Radon Transform

In this subsection we consider the linear differential equation

$$\mu_{x_1} + k\mu_{x_2} = q(x_1, x_2), \qquad -\infty < x_1 < \infty, \quad -\infty < x_2 < \infty, \qquad k \in \mathbb{C}$$

$$(7.4.42)$$

where $\mu_{x_1} = \frac{\partial \mu}{\partial x_1}$, $\mu_{x_2} = \frac{\partial \mu}{\partial x_2}$, and $|q(x_1, x_2)| < c(1 + |x|)^{-(2+\epsilon)}$, where $|x| = \sqrt{x_1^2 + x_2^2}$, c is a constant, and $\epsilon > 0$. As before, we will define a suitable solution $\mu(x_1, x_2, k)$ of Eq. (7.4.42). Let μ^+ and μ^- be the following particular solutions of Eq. (7.4.42)

$$\mu_\pm(x_1, x_2, k) = \pm \frac{1}{2\pi i} \int_{\mathbb{R}^2} \frac{q(y_1, y_2)\, dy_1\, dy_2}{(x_2 - y_2) - k(x_1 - y_1)}, \qquad k \in \mathbb{C}^\pm, \ \mathrm{Im}\, k \neq 0 \tag{7.4.43}$$

where the integral $\int_{\mathbb{R}^2}$ is taken over the entire plane. To derive Eqs. (7.4.43), it is convenient to define the associated Green's function

$$G_{x_1} + k G_{x_2} = \frac{1}{(2\pi)^2} \int_{\mathbb{R}^2} e^{i(x_1 \xi_1 + x_2 \xi_2)}\, d\xi_1\, d\xi_2, \quad \mathrm{Im}\, k \neq 0 \tag{7.4.44}$$

where we have used $\delta(x_1) = \frac{1}{2\pi} \int_R e^{i x_1 \xi_1}\, d\xi_1$. Thus

$$G(x_1, x_2, k) = \frac{1}{i(2\pi)^2} \int_{\mathbb{R}^2} \frac{e^{i(x_1 \xi_1 + x_2 \xi_2)}}{\xi_1 + k \xi_2}\, d\xi_1\, d\xi_2 \tag{7.4.45}$$

Using contour integration to evaluate the above integral, it follows (this integral is evaluated in the exercises) that for $\mathrm{Im}(k) \neq 0$

$$G = \frac{\mathrm{sgn}\, \mathrm{Im}(k)}{2\pi i (x_2 - k x_1)} \tag{7.4.46}$$

Using the Plemelj formulae to evaluate the limit of the integrals (7.4.43) as $k \to k_R \pm i0$, it follows that (see the exercises)

$$\mu_\pm(x_1, x_2, k) = \pm \frac{1}{2\pi i} \int_{-\infty}^{\infty} dy_1 \fint_{-\infty}^{\infty} dy_2 \frac{q(y_1, y_2)}{(x_2 - y_2) - k(x_1 - y_1)}$$
$$+ \frac{1}{2} \left(\int_{-\infty}^{x_1} - \int_{x_1}^{\infty} \right) q(y_1, x_2 - k(x_1 - y_1))\, dy_1, \qquad k \in \mathbb{R}$$

where a principal value integral is assumed for one of the integrals. Subtracting the above equations, we obtain

$$(\mu_+ - \mu_-)(x_1, x_2, k) = \frac{1}{i\pi} \int_{-\infty}^{\infty} dy_1 \fint_{-\infty}^{\infty} dy_2\, \frac{q(y_1, y_2)}{(x_2 - y_2) - k(x_1 - y_1)},$$
$$k \in \mathbb{R} \tag{7.4.47}$$

The right-hand side of this equation can be written in terms of the Radon transform of the function $q(x_1, x_2)$ defined by

$$\tilde{q}(\kappa, p) = \int_{-\infty}^{\infty} q\left(\frac{\tau - pk}{\sqrt{1 + k^2}}, \frac{\tau k + p}{\sqrt{1 + k^2}}\right) d\tau \qquad (7.4.48)$$

Indeed, changing variables from (y_1, y_2) to (p', τ') where

$$y_1 = \frac{\tau' - p'k}{\sqrt{1 + \kappa^2}}, \qquad y_2 = \frac{\tau'k + p'}{\sqrt{1 + \kappa^2}} \qquad (7.4.49)$$

and using Eq. (7.4.47) and the Jacobian of the transformation

$$J = \left| \det \begin{pmatrix} \dfrac{\partial y_1}{\partial \tau'} & \dfrac{\partial y_2}{\partial \tau'} \\ \dfrac{\partial y_1}{\partial p'} & \dfrac{\partial y_2}{\partial p'} \end{pmatrix} \right| = 1$$

it follows that

$$\mu_+(x_1, x_2, k) - \mu_-(x_1, x_2, k) = \frac{1}{i\pi} \fint_{-\infty}^{\infty} \frac{\tilde{q}(k, p')\, dp'}{x_2 - kx_1 - p'\sqrt{1 + k^2}},$$

$$k \in \mathbb{R} \qquad (7.4.50a)$$

Equation (7.4.43) implies that

$$\mu = O\left(\frac{1}{k}\right), \qquad k \to \infty \qquad (7.4.50b)$$

hence Eqs. (7.4.50a,b) define an elementary RH problem for the sectionally analytic function $\mu(x_1, x_2, \kappa)$. Its unique solution is

$$\mu(x_1, x_2, k) = \frac{1}{2\pi i} \int_{-\infty}^{\infty} \frac{dk'}{k' - k} \left(\frac{1}{i\pi} \fint_{-\infty}^{\infty} \frac{\tilde{q}(k', p')\, dp'}{x_2 - k'x_1 - p'\sqrt{1 - k'^2}}\right),$$

$$\text{Im}\, k \neq 0 \qquad (7.4.51)$$

Comparing the large-k asymptotics of Eqs. (7.4.42), (7.4.43), and (7.4.51), it follows that

$$q = \lim_{k \to \infty} \frac{\partial}{\partial x_2} (k\mu)$$

or

$$q(x_1, x_2) = \frac{1}{2\pi^2} \frac{\partial}{\partial x_2} \int_{-\infty}^{\infty} dk \int_{-\infty}^{\infty} \frac{\tilde{q}(k, p)}{x_2 - kx_1 - p\sqrt{1 + k^2}} \, dp, \quad k \in \mathbb{R}$$
(7.4.52)

Equations (7.4.48) and (7.4.52) are the usual direct and inverse Radon transforms.

Example 7.4.7 (The Wiener–Hopf Method) In this example we consider the so-called Sommerfeld half-plane diffraction problem; that is, we are interested in obtaining the reflected and diffracted acoustic wave field generated by a plane wave incident on a semiinfinite plane. This problem is prototypical and appears in several physical applications.

The incident wave is given by $\tilde{\varphi}_I = \varphi_I e^{i\lambda t} = \exp[-i\lambda(x\cos\theta + y\sin\theta)]e^{i\lambda t}$, $-\frac{\pi}{2} < \theta < \frac{\pi}{2}$. The total field, $\tilde{\varphi}_T$, can be thought to be the sum of the potential $\tilde{\varphi}_I$ due to the incident wave (in the absence of the plane), and the potential $\tilde{\varphi}$ due to the disturbance produced by the presence of the plane, that is, $\tilde{\varphi}_T = \tilde{\varphi} + \tilde{\varphi}_I$. Both $\tilde{\varphi}_T$ and $\tilde{\varphi}_I$ satisfy the linear wave equation, so it follows that $\tilde{\varphi} = \varphi e^{i\lambda t}$ and $\tilde{\varphi}_T = \varphi_T e^{i\lambda t}$ also satisfy this equation, which leads to the so-called **reduced wave equation** or **Helmholtz equation** for φ; namely, $\Box^2 \tilde{\varphi}_T = \Box^2 \tilde{\varphi}_I = 0$, where $\Box^2 = \frac{\partial^2}{\partial x^2} + \frac{\partial^2}{\partial y^2} - \frac{\partial^2}{\partial t^2}$, and therefore

$$\nabla^2 \varphi + \lambda^2 \varphi = \varphi_{xx} + \varphi_{yy} + \lambda^2 \varphi = 0$$
(7.4.53)

Next, we formulate the boundary conditions satisfied by φ.

(a) In this physical problem, $\tilde{\varphi}$ represents the velocity potential of an acoustic wave field and $\partial \tilde{\varphi}/\partial y$ represents the vertical velocity. The normal velocity on the plate must vanish; hence $\frac{\partial \varphi_T}{\partial y}(x, 0) = 0$ for $x \leq 0$, thus on the plate

$$\left.\frac{\partial \varphi}{\partial y}\right|_{y=0} = -\left.\frac{\partial \varphi_I}{\partial y}\right|_{y=0}$$

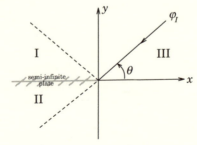

Fig. 7.4.6. Sommerfeld half-plane diffraction problem

hence,

$$\varphi_y(x, 0) = i\lambda \sin\theta e^{-i\lambda x \cos\theta} \qquad \text{for } -\infty < x \le 0 \qquad (7.4.54)$$

(b) We look for a solution in which the normal velocity $\frac{\partial \varphi_T}{\partial y}(x, 0)$ is continuous for all x, but $\varphi_T(x, 0)$ is continuous only for positive x, that is,

$$\begin{array}{ll} \varphi_y(x, 0) & \text{is continuous for } -\infty < x < \infty \\ \varphi(x, 0) & \text{is continuous for } 0 < x < \infty \end{array} \qquad (7.4.55)$$

(c) At infinity, φ satisfies a **radiation condition**, which means that

$$\varphi \sim c e^{-i\lambda r} \qquad \text{as } r \to \infty \qquad (7.4.56)$$

so that $\tilde{\varphi} \sim c e^{-i\lambda(r-t)}$ consists of outgoing waves.

(d) At the edge of the plate, $\tilde{\varphi}_T$ and $\tilde{\varphi}$ are allowed to have integrable singularities, which allows energy in the wave field to remain finite.

We shall show that the solution of the reduced wave equation (7.4.53) satisfying these boundary conditions can be obtained by solving a RH problem. The method of solving this type of problem is usually referred to as the Wiener–Hopf method. It is similar to the method used in Example 7.4.5. Equation (7.4.53) is solved by a Fourier transform (see also Sections 4.5–4.6) and then the boundary conditions (7.4.54)–(7.4.56) are used to determine the relevant regions of analyticity. This leads to a RH problem, which can be solved uniquely after taking into consideration the edge conditions.

Taking the x-Fourier transform of Eq. (7.4.53)

$$\hat{\Phi}(k, y) = \int_{-\infty}^{\infty} \varphi(x, y) e^{ikx} \, dy \qquad (7.4.57)$$

we find (note that we have replaced k by $-k$ in the definition of the Fourier transform)

$$\frac{d^2\hat{\Phi}}{dy^2} - \gamma^2 \hat{\Phi} = 0, \qquad \gamma = (k^2 - \lambda^2)^{\frac{1}{2}} = (k - \lambda)^{\frac{1}{2}}(k + \lambda)^{\frac{1}{2}} \qquad (7.4.58)$$

We define the multivalued function γ in such a way that $\gamma \to |k|$ as $k \to \infty$. In particular, we shall take the branch cut associated with $(k + \lambda)^{1/2}$ to be from $-\infty$ to $-\lambda$, and define $(k+\lambda)^{1/2}$ for real κ to be the limit of the analytic function $(k + \lambda)^{1/2}$ as k approaches the negative real axis from above. Similarly, the branch cut of $(k-\lambda)^{1/2}$ is from λ to ∞, and $(k-\lambda)^{1/2}$ for real k is the limit of the analytic function $(k - \lambda)^{1/2}$ as k approaches the positive real axis from below. In this sense, the functions $(k + \lambda)^{1/2}$ and $(k - \lambda)^{1/2}$ are $\oplus$ and $\ominus$ functions, respectively. (Note: Convenient local angles are given by $(k - \lambda) = r_1 e^{i\theta_1}$ for $-2\pi \leq \theta_1 < 0$, and $(k + \lambda) = r_2 e^{i\theta_2}$ for $-\pi \leq \theta_1 < \pi$.)

The potential φ is discontinuous at $y = 0$; hence $\hat{\Phi}$ changes form at $y = 0$

$$\hat{\Phi}(k, y) = \begin{cases} A_1(k)e^{-\gamma y} + B_1(k)e^{\gamma y}, & y > 0 \\ A_2(k)e^{-\gamma y} + B_2(k)e^{\gamma y}, & y < 0 \end{cases}$$

The requirements of boundedness as $|y| \to \infty$ and of the radiation condition can be used to show that $A_2 = B_1 = 0$. Indeed, for $|k| > |\lambda|$, γ is real and positive, thus boundedness as $|y| \to \infty$ implies $A_2 = B_1 = 0$. For $|k| < |\lambda|$, γ is imaginary and although both exponentials are bounded, only one of them is consistent with the radiation condition (7.4.47). Thus

$$\hat{\Phi}(k, y) = \begin{cases} A_1(k)e^{-\gamma y}, & y > 0 \\ B_2(k)e^{\gamma y}, & y < 0 \end{cases}$$

The continuity of φ_y (i.e., Eq. (7.4.55a)) implies $A_1(k) = -B_2(k) = A(k)$. Hence

$$\hat{\Phi}(k, y) = \begin{cases} A(k)e^{-\gamma y}, & y > 0 \\ -A(k)e^{\gamma y}, & y < 0 \end{cases} \tag{7.4.59}$$

As in Example 7.4.5, we see that the definitions

$$\hat{\Phi}^+(k, y) = \int_0^\infty \varphi(x, y)e^{ikx}dx, \qquad \hat{\Phi}^-(k, y) = \int_{-\infty}^0 \varphi(x, y)e^{ikx}dx \tag{7.4.60}$$

imply that $\hat{\Phi}^+(k, y)$ and $\hat{\Phi}^-(k, y)$ are analytically extendible into the upper and lower half k planes, respectively. Continuity of $\varphi(x, 0)$ for positive x, and continuity of $\varphi_y(x, 0)$ for all x (i.e., Eqs. (7.4.55)) imply

$$\hat{\Phi}^+(k, 0+) = \hat{\Phi}^+(k, 0-) \equiv \hat{\Phi}^+(k, 0)$$

$$\hat{\Phi}^{+'}(k, 0+) = \hat{\Phi}^{+'}(k, 0-) \equiv \hat{\Phi}^{+'}(0)$$

$$\hat{\Phi}^{-'}(k, 0+) = \hat{\Phi}^{-'}(k, 0-) \equiv \hat{\Phi}^{-'}(0)$$

where prime denotes the derivative with respect to y. Since $\varphi(x, y)$ can be discontinuous across $y = 0$ for $-\infty < x < \infty$, we have $\hat{\Phi}^-(k, 0+) \neq \hat{\Phi}^-(k, 0-)$. Note that

$$\hat{\Phi}^+(k, 0) + \hat{\Phi}^-(k, 0+) = \lim_{y \to 0^+} \left[\int_0^\infty \varphi e^{ikx} dx + \int_{-\infty}^0 \varphi e^{ikx} dx \right]$$

$$= \lim_{y \to 0^+} \hat{\Phi}(k, y) = A(k) \tag{7.4.61a}$$

and similarly, we find

$$\hat{\Phi}^{+}(k,0) + \hat{\Phi}^{-}(k,0-) = \lim_{y \to 0^-} \hat{\Phi}(k,y) = -A(k) \qquad (7.4.61b)$$

$$\hat{\Phi}^{+'}(k,0) + \hat{\Phi}^{-'}(k,0) = \lim_{y \to 0} \frac{\partial \hat{\Phi}}{\partial y}(x,y) = -\gamma A(k) \qquad (7.4.61c)$$

The functions $\hat{\Phi}^{\pm'}(k,0)$ are analytic in the upper/lower half planes; the derivative in y does not affect the analyticity in k.

We now show that Eq. (7.4.61c) actually defines a RH problem. Indeed, subtracting Eqs. (7.4.61a,b) it follows that $2A(k) = \hat{\Phi}^{-}(k,0+) - \hat{\Phi}^{-}(k,0-)$, that is, $A(k)$ is analytic in the lower half k-complex plane. Also from Eq. (7.4.54) $\Phi^{-'}(k,0)$ can be computed

$$\hat{\Phi}^{-'}(k,0) = \int_{-\infty}^{0} e^{ikx} i\lambda \sin\theta \, e^{-i\lambda \cos\theta} \, dx = \frac{\lambda \sin\theta}{k - i\epsilon - \lambda \cos\theta}$$

Note for convergence of the integral we assume k is slightly extended into the lower half plane: $k \to k - i\epsilon$ for $\epsilon > 0$. Therefore, calling $\Psi^{-}(k) \equiv A(k)$ and $\Psi^{+}(k) \equiv \hat{\Phi}^{+'}(k,0)$, Eq. (7.4.61c) defines the RH problem on the "closed contour" $L : -\infty < k < \infty$:

$$\Psi^{+}(k) = -\gamma \Psi^{-}(k) - \frac{\lambda \sin\theta}{k - i\epsilon - \lambda \cos\theta} \qquad (7.4.62)$$

Writing γ as $(k + \lambda)^{1/2}(k - \lambda)^{1/2}$, and dividing by $(k + \lambda)^{1/2}$, it follows that

$$\frac{\Psi^{+}(k)}{(k + \lambda)^{1/2}} = -(k - \lambda)^{\frac{1}{2}} \Psi^{-}(k) - \frac{\lambda \sin\theta}{(k + \lambda)^{\frac{1}{2}}(k - i\epsilon - \lambda \cos\theta)} \qquad (7.4.63)$$

The function $\frac{1}{(k+\lambda)^{1/2}}$, as mentioned earlier, is analytic in the upper half plane (a $\oplus$ function), and the function $(k - \lambda)^{1/2}$ is analytic in the lower half plane (a $\ominus$ function). Calling

$$\frac{\Psi^{+}(k)}{(k + \lambda)^{1/2}} = \mu^{+}(k) \quad \text{and} \quad -(k - \lambda)^{1/2} \Psi^{-}(k) = \mu^{-}(k)$$

we see that Eq. (7.4.63) is now a standard RH problem for which the method of Section 7.3 applies. In fact it can be solved "directly." Note that

$[(k + \lambda)^{1/2}(k - i\epsilon - \lambda \cos \theta)]^{-1}$ can be written as the sum of

$$\frac{1}{(k - i\epsilon - \lambda \cos \theta)} \left[\frac{1}{(k + \lambda)^{\frac{1}{2}}} - \frac{1}{(\lambda + 2i\epsilon + \lambda \cos \theta)^{\frac{1}{2}}} \right] \quad \text{and}$$

$$\frac{1}{(k - i\epsilon - \lambda \cos \theta)(\lambda + 2i\epsilon + \lambda \cos \theta)^{\frac{1}{2}}}$$

which are $\oplus$ (the apparent singularity at $k = i\epsilon + \lambda \cos \theta$ cancels) and $\ominus$ functions of k, respectively. Therefore

$$\frac{\Psi^+(k)}{(k + \lambda)^{\frac{1}{2}}} = -\frac{\lambda \sin \theta}{(k - i\epsilon - \lambda \cos \theta)}$$

$$\times \left[\frac{1}{(k + \lambda)^{\frac{1}{2}}} - \frac{1}{(\lambda + 2i\epsilon + \lambda \cos \theta)^{\frac{1}{2}}} \right] \quad (7.4.64a)$$

$$(k - \lambda)^{\frac{1}{2}} \Psi^-(k) = -\frac{\lambda \sin \theta}{(k - i\epsilon - \lambda \cos \theta)(\lambda + 2i\epsilon + \lambda \cos \theta)^{\frac{1}{2}}} \quad (7.4.64b)$$

Having obtained $\Psi^-(k)$, that is, $A(k)$ in Eq. (7.4.59), then the definition of the Fourier transform implies

$$\varphi(x, y) = \frac{\text{sgn}(y)}{2\pi} \int_{-\infty}^{\infty} A(k) e^{-\gamma y} e^{-ikx} \, dk$$

and hence from the definition $A(k) = \Psi^-(k)$

$$\varphi(x, y) = -\frac{\text{sgn}(y)\lambda \sin \theta}{2\pi}$$

$$\times \lim_{\epsilon \to 0^+} \int_{-\infty}^{\infty} \frac{e^{-ikx - \gamma|y|}}{(k - \lambda)^{\frac{1}{2}}(k - i\epsilon - \lambda \cos \theta)(\lambda + 2i\epsilon + \lambda \cos \theta)^{\frac{1}{2}}} \, dk$$

$$= -\frac{\text{sgn}(y)}{2\pi}(\lambda - \lambda \cos \theta)^{\frac{1}{2}} \lim_{\epsilon \to 0^+} \int_{-\infty}^{\infty} \frac{e^{-ikx - \gamma|y|}}{(k - \lambda)^{\frac{1}{2}}(k - i\epsilon - \lambda \cos \theta)} \, dk$$

$$= -\frac{\text{sgn}(y)}{2\pi}(\lambda - \lambda \cos \theta)^{\frac{1}{2}} \int_{C_u} \frac{e^{-ikx - \gamma|y|}}{(k - \lambda)^{\frac{1}{2}}(k - \lambda \cos \theta)} \, dk \quad (7.4.65)$$

where C_u represents the contour from $-\infty$ to ∞ indented *underneath* the pole $k = \lambda \cos \theta$. (Note that we have used $\lambda \sin \theta / (\lambda + \lambda \cos \theta)^{\frac{1}{2}} = (\lambda - \lambda \cos \theta)^{\frac{1}{2}}$ to simplify the form of $A(k)$.) It can be verified that this solution satisfies the edge conditions, that is, it has an integrable singularity. We also note that integral (7.4.65) can actually be evaluated in closed form, but because we do not need to go further, we leave the result as an integral.

Traditionally, the splitting process involved in Wiener–Hopf problems uses strips where the analytic functions have overlapping regions of analyticity. This can be achieved by introducing a small amount of damping: the Helmholtz equation is replaced by

$$\nabla^2 \varphi + \lambda^2 \varphi - i\epsilon\lambda\varphi = 0$$

Alternatively, in the above analysis we can modify λ, that is, $\lambda \to \lambda_1 + i\lambda_2$ where $\lambda_2 > 0$. It can then be shown that $\frac{\Psi^+(k)}{(k+\lambda_1+i\lambda_2)^{1/2}}$ is actually analytic for $\operatorname{Im} k > -\lambda_2$, and $\Psi^-(k)(k - \lambda_1 - i\lambda_2)^{1/2}$ is analytic for $\operatorname{Im} k < \lambda_2 \cos\theta$.

From the above discussion it follows that the RH method is actually more general than the Wiener–Hopf method: (a) The RH problem is formulated on a contour and not in a strip. Hence it can be used to solve problems where the overlapping region of analyticity is just a curve. (b) The function $g(t)$ appearing in the RH problem need be defined only on a contour. Therefore there is no need for $g(t)$ to be derived from a function of a complex variable that is analytic in a strip of the complex plane.

Problems for Section 7.4

1. Let C be a finite, closed contour and $f(t)$ a given function satisfying the Hölder condition.

    (a) Show that the solution of the singular integral equation

    $$\frac{1}{\pi i}\oint_C \frac{\varphi(\tau)}{\tau - t}\, d\tau = f(t)$$

    is given by

    $$\varphi(t) = \frac{1}{\pi i}\oint_C \frac{f(\tau)}{\tau - t}\, d\tau$$

    Hint: Note that the associated RH problem is

    $$\Phi^+(t) + \Phi^-(t) = f(t)$$

    and thus for $z \notin C$

    $$\Phi^\pm(z) = \pm\frac{1}{2i\pi}\int_C \frac{f(\tau)}{\tau - z}\, d\tau$$

(b) Use (a) to establish the so-called Poincaré–Bertrand formula:

$$f(t) = -\frac{1}{\pi^2} \oint_C \frac{d\tau}{\tau - t} \left(\oint_C \frac{f(\tau')}{\tau' - \tau} \, d\tau' \right)$$

2. Define the Hilbert transform of a suitably decaying function $f(x)$ by

$$(Hf)(x) = \frac{1}{\pi} \fint_{-\infty}^{\infty} \frac{f(\xi)}{\xi - x} \, d\xi$$

(a) Use the results of Problem 7.4.1 to show that, in the space of suitably decaying functions, $H(Hf(x)) = -1$ (or in shorthand notation we often write $H^2 = -1$).

(b) Note that the functions

$$\Phi^+(x) = f(x) - i(Hf)(x) \text{ and } \Phi^-(x) = f(x) + i(Hf)(x)$$

are analytic in the upper and lower half complex planes, respectively. Use this fact to establish the following property of the Hilbert transform

$$H[fHg + gHf] = -fg + (Hf)(Hg)$$

Hint: the product of the two $\oplus$ functions $f - iHf$ and $g - iHg$ is also a $\oplus$ function.

3. Verify the following identity

$$\oint_C \frac{d\tau}{\tau - t} \oint_C \frac{f(\tau, \tau')}{\tau' - \tau} \, d\tau' = \oint_C \frac{d\tau}{\tau - t} \int_C \frac{f(\tau, \tau') - f(\tau, \tau)}{\tau' - \tau} \, d\tau'$$

$$+ \int_C \frac{f(\tau, \tau) - f(t, t)}{\tau - t} \, d\tau \oint_C \frac{d\tau'}{\tau' - \tau}$$

$$+ f(t, t) \oint_C \frac{d\tau}{\tau - t} \oint_C \frac{d\tau'}{\tau' - \tau}$$

where C is a closed, finite contour.

Note that the first two terms in the right-hand side involve only single singular integrals. By reversing the order of integration in these two integrals and by using the formulae

$$\oint_C \frac{d\tau}{\tau - t} \oint_C \frac{d\tau'}{\tau' - \tau} = -\pi^2$$

(See also Problem 7.4.1b), establish the so-called Poincaré–Bertrand transposition formula:

$$\oint_C \frac{d\tau}{\tau - t} \oint_C \frac{f(\tau, \tau')}{\tau' - \tau} d\tau' = \int_C d\tau' \oint_C \frac{f(\tau, \tau')}{(\tau - t)(\tau' - \tau)} d\tau - \pi^2 f(t, t)$$

4. Consider the following integral

$$I(t) = \oint_{-1}^{1} \frac{(1 - \tau)^{\alpha - 1}}{(1 + \tau)^\alpha (\tau - t)} d\tau, \qquad -1 < t < 1$$

where $0 < \alpha < 1$.

(a) Let

$$\Phi(z) = \frac{1}{2\pi i} \int_{-1}^{1} \frac{(1 - \tau)^{\alpha - 1}}{(1 + \tau)^\alpha (\tau - z)} d\tau$$

Show that $I(t) = i\pi(\Phi^+(t) + \Phi^-(t))$ and that $\Phi(z)$ satisfies the RH problem

$$\Phi^+(t) - \Phi^-(t) = \frac{(1 - t)^{\alpha - 1}}{(1 + t)^\alpha}$$

(b) Verify that the function,

$$\Phi(z) = -\frac{1}{2i \sin \alpha \pi} \frac{(z - 1)^{\alpha - 1}}{(z + 1)^\alpha}$$

solves the above RH problem.

(c) Deduce that

$$I(t) = \pi(\cot \alpha \pi)(1 - t)^{\alpha - 1}(1 + t)^{-\alpha}$$

5. Consider the following boundary value problem:

$$\varphi_{xx} + \varphi_{yy} + \lambda^2 \varphi = 0, \qquad -\infty < x < \infty, \quad 0 < y < \infty$$
$$\varphi_y(x, 0) = g(x) \qquad \text{for } -\infty < x < 0$$
$$\varphi(x, 0) = f(x) \qquad \text{for } 0 < x < \infty$$

where φ satisfies a radiation condition at infinity, λ is a real constant, and $f(x)$ and $g(x)$ are given, suitably decaying functions.

(a) If $\Phi(k, y)$ is defined by

$$\Phi(k, y) = \int_{-\infty}^{\infty} \phi(x, y)e^{ikx}dx$$

show that

$$\Phi(k, y) = A(k)e^{-\gamma y}, \quad \gamma = (k^2 - \lambda^2)^{\frac{1}{2}}$$

where the function $\gamma(k)$ is defined in Example 7.4.7 of the text.

(b) Show that

$$\Phi_-(k, 0) \equiv \int_{-\infty}^{0} \phi(x, 0)e^{ikx}dx, \text{ and}$$

$$\Phi'_+(k, 0) \equiv \int_{0}^{\infty} \phi_y(x, 0)e^{ikx}dx$$

satisfy the following RH problem

$$\Phi'_+(k, 0) + \Phi'_-(k, 0) = -\gamma[\Phi_+(k, 0) + \Phi_-(k, 0)]$$

where $\Phi_+(k, 0)$ and $\Phi'_-(k, 0)$ are given by

$$\Phi_+(k, 0) = \int_{0}^{\infty} f(x)e^{ikx}dx, \quad \Phi'_-(k, 0) = \int_{-\infty}^{0} g(x)e^{ikx}dx$$

6. Consider the following linear differential equation for $\mu(x, k)$ with $q_x \equiv \frac{\partial q}{\partial x}(x)$ as a forcing function:

$$(x - k)\mu_x + \frac{1}{2}\mu = q_x \quad \alpha < x < \infty, \quad q(\alpha) = 0, \quad k \in \mathbb{C} \quad (1)$$

(a) Show that a solution of this equation, bounded for all complex k, is given by

$$\mu(x, k) = \int_{\alpha}^{x} \frac{q_\xi(\xi)\, d\xi}{(k - \xi)^{\frac{1}{2}}(k - x)^{\frac{1}{2}}}$$

(b) Define the function $(k - \xi)^{\frac{1}{2}}(k - x)^{\frac{1}{2}}$ in such a way that there is a branch cut between ξ and x. By splitting the integral $\int_{\alpha}^{x}$ in the form $\int_{\alpha}^{k} + \int_{k}^{x}$, show that for k real,

$$(\mu^+ - \mu^-)(x, k) = \begin{cases} \dfrac{2\hat{q}(k)}{(k-x)^{\frac{1}{2}}}, & \alpha < k < x \\ 0, & k > x \end{cases}$$

where $\hat{q}(k)$ is the following *Abel transform* of $q(x)$:

$$\hat{q}(k) = \int_\alpha^k \frac{q_\xi(\xi)}{\sqrt{k - \xi}} \, d\xi \tag{2}$$

(c) Deduce that

$$\mu(x, k) = \frac{1}{\pi} \int_\alpha^x \frac{\hat{q}(k')}{\sqrt{x - k'}} \frac{dk'}{k' - k}, \quad \operatorname{Im} k \neq 0 \tag{3}$$

(d) Comparing the large k behavior of equations (1) and (3) show that

$$q(x) = -\lim_{k \to \infty} (k\mu) = \frac{1}{\pi} \int_\alpha^x \frac{\hat{q}(k)}{\sqrt{x - k}} \, dk \tag{4}$$

Equations (2) and (4) are the usual direct and inverse Abel transforms. (An appropriate *nonlinearization* of this transform can be used to solve a certain two-dimensional reduction of Einstein's equation).

*7.5 Matrix Riemann–Hilbert Problems

Matrix or vector RH problems are, in general, far more complicated than scalar RH problems. The solution cannot, in general, be found in closed form (i.e., not in terms of explicit integrals); it is characterized through a system of linear integral equations. We do not intend to present a complete theory of such RH problems here; our aim is simply to introduce important aspects. Emphasis will be given to those results that are constructive. For simplicity, we consider only the case of closed contours.

The vector homogeneous RH problem for a closed contour C is defined as follows. Given a contour C and an $N \times N$ matrix $G(t)$ that satisfies a Hölder condition and is nonsingular on C (i.e., all the matrix elements $\{G\}_{ij}$ satisfy a Hölder condition and $\det G(t) \neq 0$ on C), find a sectionally analytic vector function $\mathbf{\Phi}(z)$,[1] with finite degree at infinity[2] such that

$$\mathbf{\Phi}^+(t) = G(t)\mathbf{\Phi}^-(t), \qquad t \text{ on } C \tag{7.5.1}$$

The meaning of $+$ and $-$ is the same as in Section 7.3 (meant for each component of $\mathbf{\Phi}(z)$).

---

[1] $\mathbf{\Phi}(z)$ is a column vector: $\mathbf{\Phi}(z) = (\phi_1(z), \phi_2(z), \ldots, \phi_n(z))^T$, where T represents the transpose.

[2] If all components of the vector $\mathbf{\Phi}(z)$ have finite degree at infinity, we say that $\mathbf{\Phi}(z)$ has finite degree at infinity. We say that $\mathbf{\Phi}(z)$ has degree k at infinity if k is the highest degree of any of its components.

In contrast to the scalar case, where the existence or nonexistence of solutions is *a priori* determined in terms of the index, the existence and uniqueness of solutions of Eq. (7.5.1) must be investigated by analyzing certain integral equations. This difficulty arises from the fact that the solutions of Eq. (7.5.1) depend on a set of integers, often called **partial indices** or the **individual indices** $\kappa_1, \ldots, \kappa_N$, which cannot *a priori* be calculated from $G(t)$. We will see that only their sum

$$\kappa = \kappa_1 + \cdots + \kappa_N = \text{ind} \det G(t) \qquad (7.5.2)$$

can be explicitly calculated in terms of $G(t)$.

Solving Eq. (7.5.1) means finding a fundamental solution matrix $X(z)$ composed of solution vectors $X_1(z), \cdots, X_N(z)$. The individual indices are defined by the behavior of the functions $X_1, \ldots, X_N$ at infinity, so that the individual index κ_1 is related to the behavior of $X_1(z)$ as $z \to \infty$, etc. Suppose we are looking for solutions of Eq. (7.5.1) with a sufficiently large degree at infinity. Among these solutions there exist some with the lowest possible degree $(-\kappa_1)$. Let X_1 denote a solution with degree $(-\kappa_1)$, that is, $X_1 = \hat{X}_1(z)/z^{\kappa_1}$, where $\hat{X}_1(z)$ is analytic in D. From the remaining solutions, consider all those that cannot be obtained from X_1 by $\phi(z) = P_1(z)X_1(z)$, where $P_1(z)$ is an arbitrary polynomial. Among these solutions, pick the one with the lowest possible degree. Call X_2 and $(-\kappa_2)$ the solution and degree, respectively, of one of these solutions. Denote as $(-\kappa_3)$ the lowest degree of those solutions that are not related to $X_1(z)$ and $X_2(z)$ by any relationship of the form $\phi(z) = P_1(z)X_1(z) + P_2(z)X_2(z)$, where $P_1(z)$ and $P_1(z)$ are arbitrary polynomials. The vector X_3 is one of these solutions. We repeat the process to obtain the matrix $X(z) = (X_1(z), X_2(z), \ldots, X_N(z))^T$. It can be shown (Vekua, 1967) that the solution matrix $X(z)$ constructed this way has the following two properties:

$$\det X(z) \neq 0 \qquad \text{for all finite } z, \qquad \text{and}$$
$$\det(z^{\kappa_1} X^1, \cdots, z^{\kappa_N} X^N) \neq 0 \qquad \text{at infinity} \qquad (7.5.3)$$

Furthermore, any solution of the homogeneous RH problem is given by

$$\Phi(z) = X(z)P(z) \qquad (7.5.4)$$

where $P(z)$ is an arbitrary polynomial vector, that is, each component of $P(z)$ is an arbitrary polynomial. In this way we think of Eq. (7.5.1) as an RH problem for the matrix-valued function Φ (or X).

The relationship (7.5.2) can now be derived from $X^+ = GX^-$ and the properties (7.5.3). We take $\frac{1}{2\pi i}$ of the logarithm of the determinant around the contour C:

$$\frac{1}{2\pi i}[\log \det X^+]_C = \frac{1}{2\pi i}[\log \det G]_C + \frac{1}{2\pi i}[\log \det X^-]_C$$

where $[f]_C$ denotes the change of f over the closed curve C taken in the $+$ direction. The function $\det X^+$ is analytic and nonvanishing in D^+; hence $\frac{1}{2\pi i}[\log \det X^+]_C = 0$, and from

$$X^-(z) = \left(\frac{\hat{X}^{-(1)}(z)}{z^{\kappa_1}}, \frac{\hat{X}^{-(2)}(z)}{z^{\kappa_2}}, \ldots, \frac{\hat{X}^{-(n)}(z)}{z^{\kappa_n}} \right)$$

we find that $\frac{1}{2\pi i}[\log \det X^-]_C = -(\kappa_1 + \kappa_2 + \cdots + \kappa_N)$ because $\hat{X}^-$ is analytic and nonvanishing in D^-. Thus

$$\kappa \equiv \frac{1}{2\pi i}[\log \det G]_C = \kappa_1 + \kappa_2 + \cdots + \kappa_n$$

The above nonconstructive approach for determining $X(z)$ can be turned into a constructive one provided that the individual indices κ_l, $l = 1, \ldots, N$ are known. For example, suppose that $\kappa_1 = \kappa_2 = \cdots = \kappa_N = 0$ (a necessary but not sufficient condition, for this is ind $\det G(t) = 0$). Then we show below that $X^-(t)$ solves the matrix Fredholm integral equation

$$X^-(t) - \frac{1}{2\pi i}\int_C \frac{[G^{-1}(t)G(\tau) - I_N]X^-(t)\,d\tau}{\tau - t} = I_N \qquad (7.5.5)$$

where I_N denotes the $N \times N$ identity matrix, and the integral is over the closed contour C. (In this and following sections, it will be clear from the context whether C is closed or open). Equation (7.5.5) is a matrix Fredholm integral equation of the second kind. The method of solution described in Section 7.3 for scalar Fredholm equations (see Eqs. (7.3.32)–(7.3.34) also applies to matrix Fredholm equations. We also note that if Eq. (7.5.5) has a **unique** solution $X^-(t)$, then all the individual indices are zero because it can be shown that the behavior of $X^-(z)$ at infinity is I_N.

Before deriving Eq. (7.5.5) we first show that a necessary and sufficient condition for the function $\Phi^-(t)$ to be the boundary value of a function analytic in D^- and tending to I_N at ∞ is

$$\frac{1}{2}\Phi^-(t) + \frac{1}{2\pi i}\oint_C \frac{\Phi^-(\tau)\,d\tau}{\tau - t} = I_N \qquad (7.5.6)$$

Let us call $\boldsymbol{\Phi}(z)$ the sectionally analytic function $\boldsymbol{\Phi}(z) = \{\boldsymbol{\Phi}^+(z)$ for $z \in D^+, \boldsymbol{\Phi}^-(z)$ for $z \in D^-\}$. First assume that $\boldsymbol{\Phi}^-$ is analytic in D^- and that it tends to I_N at ∞. Then using Cauchy's theorem, we have

$$\frac{1}{2\pi i} \int_{C-C_R} \frac{\boldsymbol{\Phi}^-(\tau)}{\tau - z} \, d\tau = -\boldsymbol{\Phi}^-(z),$$

where C_R denotes a large circular contour at infinity, and $z \in D^-$. This implies

$$-\boldsymbol{\Phi}^-(z) = \frac{1}{2\pi i} \int_C \frac{\boldsymbol{\Phi}^-(\tau)}{\tau - z} \, d\tau - I_N \qquad \text{for } z \text{ in } D^-$$

On the other hand, if $z \in D^+$, there is no singularity enclosed in $C - C_R$, and we have

$$0 = \frac{1}{2\pi i} \int_C \frac{\boldsymbol{\Phi}^-(\tau)}{\tau - z} \, d\tau - I_N \qquad \text{for } z \text{ in } D^+$$

We next take the limit to the boundary (i.e., the Plemelj formulae),

$$\frac{1}{2\pi i} \int_C \frac{\boldsymbol{\Phi}^-(\tau)}{\tau - z} \, d\tau \to \pm\frac{1}{2}\boldsymbol{\Phi}^-(z) + \frac{1}{2\pi i}\!\!\!\fint_C \frac{\boldsymbol{\Phi}^-(\tau)}{\tau - z} \, d\tau$$
$$\text{as } z \to C \quad \text{from } D^+ \text{or } D^-$$

Using this limit, we find that both of these equations yield Eq. (7.5.6). Conversely, assume that $\boldsymbol{\Phi}^-(t)$ satisfies Eq. (7.5.6) and that it satisfies a Hölder condition. Define a function $\boldsymbol{\Phi}(z)$ in D^- by

$$\boldsymbol{\Phi}(z) = -\frac{1}{2\pi i} \int_C \frac{\boldsymbol{\Phi}^-(\tau)}{\tau - z} \, d\tau + I_N, \qquad z \text{ in } D^- \qquad (7.5.7)$$

Clearly, $\boldsymbol{\Phi}(z) \to I_N$ as $z \to \infty$, and by its definition $\boldsymbol{\Phi}(z)$ is analytic in D^-. Its limit on the boundary is given by

$$\lim_{\substack{z \to t \\ z \in D^-}} \boldsymbol{\Phi}(z) = \frac{1}{2}\boldsymbol{\Phi}^-(t) - \frac{1}{2\pi i}\!\!\!\fint_C \frac{\boldsymbol{\Phi}^-(\tau)}{\tau - t} \, d\tau + I_N$$

which, using Eq. (7.5.6), equals $\boldsymbol{\Phi}^-(t)$. Thus $\boldsymbol{\Phi}^-(t)$ is the boundary value of the analytic function defined by Eq. (7.5.7).

Similar ideas show that $\Phi^+(z) = \frac{1}{2\pi i} \int_C \frac{\Phi^+(\tau)}{\tau - z} \, d\tau$ is analytic for $z \in D^+$ and that $\Phi^+(t)$ is the boundary value of a function analytic in D^+ if and only if

$$-\frac{1}{2}\Phi^+(t) + \frac{1}{2\pi i}\oint_C \frac{\Phi^+(\tau)}{\tau - t} \, d\tau = 0 \tag{7.5.8}$$

The derivation of Eq. (7.5.5) is a consequence of Eqs. (7.5.6) and (7.5.8). Using the RH problem, $\Phi^+(t) = G(t)\Phi^-(t)$, Eq. (7.5.8) becomes

$$-\frac{1}{2}G(t)\Phi^-(t) + \frac{1}{2\pi i}\oint_C \frac{G(\tau)\Phi^-(\tau)}{\tau - t} \, d\tau = 0$$

Multiplying this equation by $G(t)^{-1}$ and subtracting it from Eq. (7.5.6) we find Eq. (7.5.5), where we have denoted $\Phi^-(t)$ to be $X^-(t)$ as a fundamental solution.

To investigate the case when not all individual indices are zero, one needs to analyze several RH problems related to Eq. (7.5.1) (the so-called adjoint and accompanying RH problems). This leads to the study of integral equations of the type (7.5.5) where I_N is replaced by suitable forcing functions, (cf. Vekua, 1967). One then needs to use the results of Fredholm theory and, in particular, the Fredholm alternative theorem.

As mentioned earlier, Eq. (7.5.5) is a Fredholm integral equation of the second kind; because G satisfies a Hölder condition, its kernel satisfies

$$\frac{G^{-1}(t)G(\tau) - I_N}{\tau - t} = \frac{A(t, \tau)}{|\tau - t|^\alpha}, \qquad 0 \le \alpha < 1, \quad A \text{ Hölder} \tag{7.5.9}$$

where A satisfies a Hölder condition. The theory of such Fredholm integral equations is outside the scope of this book. (For $0 < \alpha < 1$, Eq. (7.5.9) is a weakly singular Fredholm equation.) Here we concentrate on those cases when the direct investigation of Fredholm equations can be bypassed. In particular, (a) if G is rational, then the RH problem can be solved in closed form; (b) if G is a triangular matrix, then the RH problem can also be solved in closed form; (c) if G satisfies certain symmetry conditions, then it can be shown directly that all the individual indices are zero (this also means that the homogeneous version of Eq. (7.5.5), that is, Eq. (7.5.5) with I_N replaced by 0, has only the zero solution).

7.5.1 The Riemann–Hilbert Problem for Rational Matrices

We assume that all the elements of the matrix $G(t)$ are rational functions, that is, $\{G(t)\}_{ij} = \frac{p_{ij}(t)}{q_{ij}(t)}$, where $p_{ij}(t)$, $q_{ij}(t)$ are polynomials. This $G(t)$ can be written as $G(t) = \frac{Q(t)}{r(t)}$, where $r(t)$ is a scalar polynomial and $Q(t)$ is a matrix whose elements are polynomials. Insight into the general case can be obtained by studying two particular cases.

Example 7.5.1 Solve Eq. (7.5.1) with $G(t) = \frac{Q(t)}{r(t)}$ where $\det Q(t)$ has no zeroes in D^+.

We write $r(t)$ as $r(t) = r_+(t)r_-(t)$, where $r_+(t)$ and $r_-(t)$ are polynomials that have no zeroes in D^+ and D^-, respectively. Then Eq. (7.5.1) becomes

$$r_+(t)Q^{-1}(t)\Phi^+(t) = \frac{\Phi^-(t)}{r_-(t)}$$

Because $\frac{1}{r_-(t)}$ is analytic in D^- and $Q^{-1}(t)r_+(t)$ is analytic in D^+, it follows that the above equation defines an analytic function in the entire complex z plane. Taking this function to be I_N, we obtain the fundamental solution $X(z)$:

$$X^-(z) = r_-(z)I_N, \qquad X^+(z) = \frac{Q(z)}{r_+(z)} \qquad (7.5.10)$$

We can verify that the solutions (7.5.10) satisfy Eqs. (7.5.3).

As a concrete illustration, take $r(t) = t(t-2)$ and let C be the unit circle. Then $r_+(t) = t - 2$, $r_-(t) = t$ and $X^-(z) = zI_N$, $X^+(z) = Q(z)/(z-2)$. We note that $\det G = \frac{\det Q}{t^N(t-2)^N}$, thus the total index $\kappa = \kappa_1 + \kappa_2 + \cdots + \kappa_N$ is $\kappa = -N$, by using the argument theorem (there are no zeroes and N poles inside C). On the other hand, $X^-(z) = zI_N$ implies that $\kappa_j = -1$, $1 \le j \le N$, so Eq. (7.5.2) is verified as well.

Example 7.5.2 Solve Eq. (7.5.1) with $G(t) = \frac{Q_+(t)Q_-(t)}{r(t)}$, where $\det Q_+(t)$ and $\det Q_-(t)$ are polynomials that have no zeroes in D^+ and D^-, respectively.

As in Example 7.5.1, we split $r(t) = r_+(t)r_-(t)$ where $r_\pm(t)$ have no zeroes in $D^\pm$. Equation (7.5.1) now becomes

$$r_+(t)Q_+^{-1}(t)\Phi^+(t) = \frac{Q_-(t)\Phi^-(t)}{r_-(t)}$$

and similar arguments as in Example 7.5.1 yield

$$X^-(z) = r_-(z)Q_-^{-1}(z), \qquad X^+(z) = \frac{Q_+(z)}{r_+(z)} \qquad (7.5.11)$$

As a concrete illustration, consider C to be the unit circle and let

$$Q(t) = \begin{pmatrix} t^2 - 1 & t - 2 \\ \frac{1}{2} & 2 \end{pmatrix}, \qquad r(t) = t^2 - 3t$$

We note that

$$Q(t) = \begin{pmatrix} t & -2 \\ 1 & 1 \end{pmatrix} \begin{pmatrix} t & 1 \\ \frac{1}{2} & 1 \end{pmatrix} = Q_+(t)Q_-(t), \quad r(t) = t(t-3) = r_-(t)r_+(t)$$

Thus

$$X^-(z) = \frac{z}{z - \frac{1}{2}} \begin{pmatrix} 1 & -1 \\ -\frac{1}{2} & z \end{pmatrix}, \qquad X^+(z) = \frac{1}{z - 3} \begin{pmatrix} z & -2 \\ 1 & 1 \end{pmatrix} \qquad (7.5.12)$$

From the behavior at infinity, Eq. (7.5.12) implies that $\kappa_1 = 0$, $\kappa_2 = -1$, and $\kappa = \kappa_1 + \kappa_2 = -1$. On the other hand $\det \frac{Q(t)}{r(t)} = \frac{(t+2)(t-\frac{1}{2})}{t^2(t-3)^2}$; therefore from the argument theorem we see (there are two poles and one zero inside C) that $\kappa = -1$, and Eq. (7.5.2) is again verified in two ways.

Using the ideas of the above example, one can solve the general case where, as before, $G(t) = Q(t)/r(t)$. We note that $Q(t)$ can always be written as $Q_+(t)D(t)Q_-(t)$, where $D(t)$ is a diagonal polynomial matrix and $\det Q^+$, $\det Q^-$ are polynomials that have no zeroes in D^+ and D^-, respectively (Gohberg and Krein (1958)). Letting $\Psi^+(t) = r_+(t)Q_+^{-1}(t)\Phi^+(t)$ and $\Psi^-(t) = \frac{Q_-(t)}{r_-(t)} \Phi^-(t)$, where as before $r(t) = r_+(t)r_-(t)$ for polynomials $r_\pm(t)$ with no zeroes in $D^\pm$, Eq. (7.5.1) becomes the following diagonal RH problem

$$\Psi^+(t) = D(t)\Psi^-(t), \qquad D = \operatorname{diag}(D_1, \ldots, D_N) \qquad (7.5.13)$$

The matrix RH problem (7.5.13) can be solved as follows. Let $D_j(t) = D_{j_+}(t)D_{j_-}(t)$, where $D_{j_+}(t)$ and $D_{j_-}(t)$ are polynomials that have no zeroes in D^+ and D^-, respectively. Then

$$(\Psi^+(z))_{ij} = \begin{cases} 0 & i \neq j \\ D_{j_+}(z) & i = j \end{cases} \quad (\Psi^-(z))_{ij} = \begin{cases} 0 & i \neq j \\ \frac{1}{D_{j_-}(z)} & i = j \end{cases} \qquad (7.5.14)$$

Having obtained Ψ; the solution for Φ follows from

$$\Phi^+(t) = \frac{1}{r_+(t)} Q_+(t)\Psi^+(t), \qquad \text{and} \qquad \Phi^-(t) = r_-(t)Q_-^{-1}(t)\Psi^-(t)$$

As an illustration of how to solve the diagonal RH problem (7.5.13), consider $D(t) = \operatorname{diag}(t, t-2)$ with contour C as the unit circle. Then $D_+(z) = (1, z-2)$,

$D_-(z) = (z, 1)$, and Eqs. (7.5.14) yield

$$\Psi^+(z) = \begin{pmatrix} 1 & 0 \\ 0 & z-2 \end{pmatrix}, \qquad \Psi^-(z) = \begin{pmatrix} \frac{1}{z} & 0 \\ 0 & 1 \end{pmatrix}$$

We note that $\kappa_1 = 1$, $\kappa_2 = 0$, and $\kappa = \text{ind}[t(t-2)] = 1$ for the RH problem defining Ψ.

7.5.2 Inhomogeneous Riemann–Hilbert Problems and Singular Equations

The method used in Section 7.3 for solving scalar inhomogeneous RH problems is also applicable to matrix RH problems. The solution of

$$\Phi^+(t) = G(t)\Phi^-(t) + f(t), \qquad t \text{ on } C$$

where G is a $N \times N$ matrix that satisfies a Hölder condition and is nonsingular on C and $f(t)$ is an N-dimensional vector that satisfies a Hölder condition on C, is derived in an analogous manner once the fundamental homogeneous solution $X(z)$ is obtained. Namely, $X^+(t)(X^-(t))^{-1} = G(t)$, which used in the above equation yields

$$(X^+(t))^{-1}\Phi^+(t) - (X^-(t))^{-1}\Phi^-(t) = (X^+(t))^{-1}f(t)$$

and from the Plemelj formulae we obtain the solution

$$\Phi(z) = X(z)\left[\frac{1}{2i\pi}\int_C \frac{[X^+(\tau)]^{-1}f(\tau)\,d\tau}{\tau - z} + P(z)\right] \qquad (7.5.15)$$

In Eq. (7.5.15), $P(z)$ is an N-dimensional vector with arbitrary polynomial components.

The equivalence established in Section 7.3.3 between RH problems and certain singular integral equations can also be extended to the matrix case. The singular integral equation

$$A(t)\varphi(t) + \frac{B(t)}{\pi i}\oint_C \frac{\varphi(\tau)}{\tau - t}\,d\tau = f(t), \qquad t \text{ on } C \qquad (7.5.16)$$

where $A(t)$ and $B(t)$ are $N \times N$ Hölder matrices such that $A \pm B$ are nonsingular on C and $f(t)$ is an N-dimensional Hölder vector, is equivalent to the matrix RH problem

$$A(t)(\Phi^+(t) - \Phi^-(t)) + B(t)(\Phi^+(t) + \Phi^-(t)) = f(t), \qquad t \text{ on } C \quad (7.5.17a)$$

or

$$\Phi^+(t) = (A(t) + B(t))^{-1} (A(t) - B(t)) \, \Phi^-(t)$$

$$+ (A(t) + B(t))^{-1} f(t), \qquad t \text{ on } C \qquad (7.5.17b)$$

$$\Phi(z) \to 0 \quad \text{as} \quad z \to \infty$$

where we have used the Plemelj formulae

$$\Phi^\pm(t) = \pm \frac{1}{2} \varphi(t) + \frac{1}{2\pi i} \oint_C \frac{\varphi(\tau)}{\tau - t} \, d\tau$$

in Eq. (7.5.16) to obtain Eq. (7.5.17a).

7.5.3 The Riemann–Hilbert Problem for Triangular Matrices

The RH problem the (7.5.1) where $G(t)$ is either an upper or lower triangular matrix can be solved in closed form. For this discussion it is more convenient to multiply G from the right. So we consider the matrix RH problem $\Phi^+(t) = \Phi^-(t)G(t)$, where $\Phi^\pm(t)$ are $N \times N$ matrices, and $G(t)$ is upper triangular; that is, we consider

$$\Phi^+(t) = \Phi^-(t) \begin{pmatrix} G_{11} & G_{12} & \cdots & G_{1N} \\ 0 & G_{22} & \cdots & G_{2N} \\ & & \vdots & \vdots \\ & 0 & & G_{NN} \end{pmatrix} \qquad (7.5.18)$$

Writing the matrix $\Phi(t)$ in terms of the vectors Φ_j, $1 \le j \le N$, that is, $\Phi(t) = (\Phi_1, \Phi_2, \cdots, \Phi_N)$, and decomposing G into the sum of its diagonal and "upper diagonal" parts, we can verify that Eq. (7.5.18) reduces to

$$\Phi^+(t) = \Phi^-(t)D + F$$

where D is a diagonal matrix and F is the matrix $F = (0, F_2, \ldots, F_N)$, with

$$D = \text{diag}(G_{11}, G_{22}, \ldots, G_{NN}), \qquad F_j \equiv \sum_{l=1}^{j-1} G_{lj} \Phi_l^-(t), \quad 2 \le j \le N$$

The above RH problem can be solved step by step: The equation

$$\Phi_1^+ = G_{11} \Phi_1^- \qquad (7.5.19a)$$

yields Φ_1, then the equation

$$\Phi_2^+ = G_{22}\Phi_2^- + G_{12}\Phi_1^- \tag{7.5.19b}$$

yields Φ_2

$$\Phi_3^+ = G_{33}\Phi_3^- + G_{13}\Phi_1^- + G_{23}\Phi_2^- \tag{7.5.19c}$$

yields Φ_3, etc.

We now give relevant formulae for $N = 2$; the generalization to arbitrary N follows analogously. Let us call $a_1(z)$ the solution of the scalar RH problem $a_1^+(t) = G_{11}(t)a_1^-(t)$; then

$$\Phi_1(z) = \begin{pmatrix} a_1(z) \\ 0 \end{pmatrix}$$

is a solution of Eq. (7.5.19a). Using this expression for Φ_1 in Eq. (7.5.19b) we find

$$(\Phi_2^+)_1 = G_{22}(\Phi_2^-)_1 + G_{12}a_1^-$$
$$(\Phi_2^+)_2 = G_{22}(\Phi_2^-)_2$$

where $(\Phi_2^+)_i$ is the ith component of the vector Φ_2^+.

Let $a_2(z)$ be a solution of the scalar RH problem $a_2^+(t) = G_{22}(t)a_2^-(t)$; then we have

$$\frac{\Phi_2^+(t)}{a_2^+(t)} - \frac{\Phi_2^-(t)}{a_2^-(t)} = \begin{pmatrix} \frac{G_{12}(t)a_1^-(t)}{a_2^+(t)} \\ 0 \end{pmatrix}$$

Thus

$$\frac{\Phi_2(z)}{a_2(z)} = \frac{1}{2\pi i} \int_C \frac{G_{12}(\tau)a_1^-(\tau)\,d\tau}{a_2^+(\tau)(\tau - z)} \begin{pmatrix} 1 \\ 0 \end{pmatrix} + \begin{pmatrix} 0 \\ 1 \end{pmatrix}$$

Therefore a solution of the RH problem for $\Phi_2(z)$ is given by

$$\Phi_2(z) = \begin{pmatrix} 0 \\ a_2(z) \end{pmatrix} + \frac{a_2(z)}{2\pi i} \int_C \frac{G_{12}(\tau)a_1^-(\tau)}{a_2^+(\tau)} \frac{d\tau}{\tau - z} \begin{pmatrix} 1 \\ 0 \end{pmatrix}$$

Thus a general solution to Eqs. (7.5.19a–c) in the case $N = 2$ is

$$\Phi(z) = P(z)\begin{pmatrix} a_1(z) & 0 \\ 0 & a_2(z) \end{pmatrix} + \begin{pmatrix} 0 & \frac{a_2(z)}{2\pi i}\int_C \frac{G_{12}(\tau)a_1^-(\tau)}{a_2^+(\tau)}\frac{d\tau}{\tau - z} \\ 0 & 0 \end{pmatrix} \tag{7.5.20}$$

where $P(z)$ is a scalar polynomial.

Example 7.5.3 Find the fundamental solution of the triangular RH problem (7.5.18) where $N = 2$ and $G_{11}(t) = t$, $G_{22}(t) = \frac{1}{(t - \frac{1}{2})}$, $G_{12}(t) = g(t)$ ($g(t)$ satisfies a Hölder condition), and C is the unit circle.

Using the above results and solving (by inspection) $a_1^+(t) = t a_1^-(t)$, we find $a_1^+(t) = 1$ and $a_1^-(t) = 1/t$. Similarly, $a_2^+(t) = a_2^-(t)/(t - \frac{1}{2})$ yields $a_2^+(t) = 1$ and $a_2^-(t) = t - \frac{1}{2}$. Substituting these expressions in Eq. (7.5.20) and letting

$$P(z) = I_2 = \begin{pmatrix} 1 & 0 \\ 0 & 1 \end{pmatrix}$$

we find

$$\Phi^-(z) = I_2 \begin{pmatrix} \frac{1}{z} & 0 \\ 0 & z - \frac{1}{2} \end{pmatrix} + \frac{1}{2\pi i} \int_C \begin{pmatrix} 0 & 1 \\ 0 & 0 \end{pmatrix} \frac{g(\tau)(z - \frac{1}{2})}{\tau - z} \, d\tau$$

From the behavior of $X^-(z)$ as $z \to \infty$ we see that $\kappa_1 = 1$ and $\kappa_2 = -1$, which is consistent with the fact that $\kappa = \kappa_1 + \kappa_2 = \operatorname{ind} \det G(t) = \operatorname{ind}\left(\frac{t}{t - 1/2}\right) = 0$.

7.5.4 Some Results on Zero Indices

In their investigation of certain matrix integral equations (of the form (7.4.31), see also Section 7.7), Gohberg and Krein(1958) were led to study the factorization problem

$$G(k) = G^+(k)G^-(k), \qquad k \text{ on the real axis;} \quad G^\pm(\infty) = I_N \quad (7.5.21)$$

where G is an $N \times N$ nonsingular matrix whose components belong to the set Λ_1 defined in Eq. (7.4.29), and G^+ and G^- are analytic in the upper and lower complex k plane (more precisely the components of $G^\pm$ belong to $\Lambda_1^\pm$, see the discussion in Section 7.4.1). The factorization problem (7.5.21) is clearly closely related with the homogeneous RH problem (7.5.1). The following remarkable result was proven in Gohberg and Krein (1958).

If the real or imaginary part of $G(k)$ is definite, where $\{G(k)\}_{ij} \in \Lambda_1$ and $\det G(k) \neq 0$, then all its individual indices are zero. The real and imaginary parts of a matrix G are defined by

$$G_R \equiv \frac{1}{2}(G + G^*), \qquad G_I \equiv \frac{1}{2i}(G - G^*), \qquad G^* = \overline{G^T} \quad (7.5.22)$$

where the superscript T denotes transpose and the over bar denotes complex conjugate. We recall that a matrix G is positive (negative) definite if $\xi^* G \xi$ is real and positive (negative) for all $N \times 1$ vectors $\xi \neq 0$.

The above result, in one or another form, has appeared in the literature many times and it provides a useful approach to establish the uniqueness and existence of solutions of certain RH problems. It has been considered for functions in a space other than Λ_1 and it has also been extended to more complicated contours. Such extensions are discussed below.

We consider the RH problem

$$\Phi^+(k) = \Phi^-(k)G(k), \qquad -\infty < k < \infty \tag{7.5.23a}$$

where $\Phi^\pm(k)$ are $N \times N$ matrices with the boundary condition

$$\Phi^\pm(\infty) = I_N \tag{7.5.23b}$$

We assume that G is nonsingular, that is, $\det G \neq 0$, and that $g(k) = G(k) - I_N$ satisfies $g(k), g'(k) \in L_2 \cap L_\infty$. (Recall that a function $f \in L_2$ satisfies $\int_{-\infty}^{\infty} |f(x)|^2 \, dx < \infty$, and $f \in L_\infty$ if $|f(x)| < M < \infty$ for all x; g and g' must satisfy both conditions.)

Letting $G = I_N + g$, Eq. (7.5.23a) reduces to

$$\Phi^+(k) - \Phi^-(k) = \Phi^-(k)g(k) \tag{7.5.24a}$$

or

$$\left(\Phi^+(k) - I_N\right) - \left(\Phi^-(k) - I_N\right) = \Phi^-(k)g(k), \qquad -\infty < k < \infty \tag{7.5.24b}$$

Using the Plemelj formulae and the boundary condition (7.5.23b), this equation yields

$$\Phi^-(k) = I_N - \frac{1}{2}\Phi^-(k)g(k) + \frac{1}{2\pi i}\fint_{-\infty}^{\infty} \frac{\Phi^-(\xi)g(\xi)}{\xi - k} \, d\xi$$

$$= I_N + \frac{1}{2\pi i}\int_{-\infty}^{\infty} \frac{\Phi^-(\xi)g(\xi)}{\xi - (k - i\epsilon)} \, d\xi \tag{7.5.25}$$

or

$$\Phi^-(k) = I_N + \int_{-\infty}^{\infty} K(k, \xi)\Phi^-(\xi) d\xi, \qquad K(k, \xi) = \frac{1}{2\pi i}\frac{g(\xi)}{\xi - (k - i\epsilon)} \tag{7.5.26}$$

It has recently been shown in Zhou (1989) that in the space where $g(k), g'(k) \in L_2 \cap L_\infty$ (i.e., $g(k)$ and $g'(k)$ are both square integrable and bounded), Eq. (7.5.26) is actually a Fredholm integral equation of the second kind. For such Fredholm equations it is known that a solution exists so long as the only solution to the homogeneous problem (i.e., replace I_N by 0 in the integral equation above) is the zero solution. This is equivalent to establishing uniqueness because the difference of any two solutions satisfies the homogeneous problem. Therefore the question of solvability reduces to the question of existence of nontrivial homogeneous solutions. We shall prove that if $g \pm g^*$ is definite, then there exists no nontrivial homogeneous solution.

Lemma 7.5.1 (**Vanishing Lemma**) Let $G(x) = I + g(x)$ where I denotes the identity matrix. Assume that g and its derivative belong to $L_2 \cap L_\infty$, and that either $\frac{G+G^*}{2}$ or $\frac{G-G^*}{2i}$ is definite, where G^* denotes the adjoint of G: $G^* = \overline{G^T}$. Then the RH problem

$$\Phi^+(x) = \Phi^-(x)G(x), \qquad -\infty < x < \infty, \qquad \Phi^\pm(\infty) = 0 \qquad (7.5.27)$$

has only the zero solution: $\Phi^\pm(x) = 0$.

Proof Recall from the symmetry principle (Section 5) that if $\Phi(z)$ is analytic in the lower z plane, then $\overline{\Phi(\bar{z})}$ is analytic in the upper half complex plane. Let $\Phi(z)$ be the solution of Eq. (7.5.27). Then $H \equiv \Phi^+(z)(\Phi^-(\bar{z}))^*$ is analytic in the upper half complex plane, and $H \to 0$ as $z \to \infty$. Multiplying Eq. (7.5.27) by $(\Phi^-(x))^*$ and integrating we find

$$\int_{-\infty}^{\infty} \Phi^+(x)(\Phi^-(x))^* \, dx = \int_{-\infty}^{\infty} \Phi^-(x)G(x)(\Phi^-(x))^* \, dx = 0 \qquad (7.5.28)$$

where we have used Cauchy's Theorem to evaluate the first integral by closing the contour in the upper half complex plane where H is analytic. Adding or subtracting the adjoint of Eq. (7.5.28), that is,

$$\int_{-\infty}^{\infty} \Phi^-(x)\,(G(x))^*\,(\Phi^-(x))^* dx = 0$$

to Eq. (7.5.28), we find that

$$\int_{-\infty}^{\infty} \Phi^-(x)(G(x) \pm G(x)^*)(\Phi^-(x))^* dx = 0. \qquad (7.5.29)$$

Thus if either $\frac{G+G^*}{2}$ or $\frac{G-G^*}{2i}$ is definite, then $\Phi^-(x) = 0$, as otherwise the left hand side of Eq. (7.5.29) would be either positive or negative. ∎

The theory presented above can be extended to more complicated contours. In particular, Lemma 7.5.3 can be generalized to cover any set of contours that remain invariant under the map $z \to \bar{z}$. Rather than quoting the general result we give a concrete example.

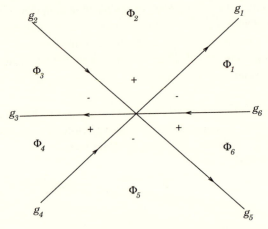

Fig. 7.5.1. RH problem defined in six sectors

Example 7.5.4 Consider the matrix RH problem defined on the rays $\frac{j\pi}{3}$, $0 \leq j \leq 5$ (see Figure 7.5.1).

$$\Phi_{j+1} = \Phi_j g_j, \qquad 1 \leq j \leq 5; \qquad \Phi_1 = \Phi_6 g_6 \qquad (7.5.30)$$

$$\Phi_j \to 0 \quad \text{as} \quad z \to \infty, \qquad 1 \leq j \leq 6 \qquad (7.5.31)$$

We assume that $g_j \in L_2 \cap L_\infty$, and that

$$\text{(i)} \quad g_2(z) = g_4^*(\bar{z}), \qquad g_1(z) = g_5^*(\bar{z}) \qquad (7.5.32)$$

$$\text{(ii)} \quad \frac{g_3 + g_3^*}{2} \left(\text{or } \frac{g_3 - g_3^*}{2i} \right) \quad \text{and} \quad \frac{g_6 + g_6^*}{2} \left(\text{or } \frac{g_6 - g_6^*}{2i} \right) \quad (7.5.33)$$

are both positive or both negative definite.

$$\text{(iii)} \quad \Pi_{j=1}^6 g_j = I_N \qquad (7.5.34)$$

Show that the only solution of this RH problem is zero.

We first note that Eq. (7.5.34) is a consistency condition. Indeed

$$\Phi_1 = \Phi_6 g_6 = \Phi_5 g_5 g_6 = \cdots = \Phi_1 \prod_{j=1}^{6} g_j$$

and hence Eq. (7.5.34) follows.

We define $H = X(z) X^*(\bar{z})$, and choose the gs and $X(z)$, $X^*(z)$ in such a way that H is analytic in the upper half plane. This is achieved by ensuring that H has no jumps across the contours: On the contour $\arg z = \frac{2\pi}{3}$ (note the notations "+" and "−" in Figure 7.5.1):

$$H^+(z) = \Phi_2(z)\Phi_5^*(\bar{z}) = \Phi_2(z)g_4^*(\bar{z})\Phi_4^*(\bar{z})$$

$$H^-(z) = \Phi_3(z)\Phi_4^*(\bar{z}) = \Phi_2(z)g_2(z)\Phi_4^*(\bar{z})$$

From Eq. (7.5.32) we see that $H^+(z) = H^-(z)$ on $\arg z = 2\pi/3$. Similarly, continuity across the contour $\arg z = \frac{\pi}{3}$, and using $g_1(z) = g_5^*(\bar{z})$, implies that

$$H^+(z) = \Phi_2(z)\Phi_5^*(\bar{z}) = \Phi_1(z)g_1(z)\Phi_5^*(\bar{z})$$

$$H^-(z) = \Phi_1(z)\Phi_6^*(\bar{z}) = \Phi_1(z)g_5^*(\bar{z})\Phi_5^*(\bar{z})$$

Because we have constructed a function H that is analytic in the upper half plane and $H \to 0$ as $z \to \infty$, Cauchy's Theorem implies that $\int_{-\infty}^{\infty} H(x)\,dx = 0$ or

$$\int_{-\infty}^{0} \Phi_3(x)\Phi_4^*(x)\,dx + \int_{0}^{\infty} \Phi_1(x)\Phi_6^*(x)\,dx = 0$$

or

$$\int_{-\infty}^{0} \Phi_3(x)g_3^*(x)\Phi_3^*(x)\,dx + \int_{0}^{\infty} \Phi_6(x)g_6(x)\Phi_6^*(x)\,dx = 0$$

Adding or subtracting to this equation its adjoint, we deduce the equation

$$\int_{-\infty}^{0} \Phi_3(x)(g_3^*(x) \pm g_3(x))\Phi_3^*(x)\,dx$$

$$+ \int_{0}^{\infty} \Phi_6(x)(g_6(x) \pm g_6^*(x))\Phi_6(x)\,dx = 0 \qquad (7.5.35)$$

From Eq. (7.5.33) we deduce that $\Phi_3(x) = \Phi_6(x) = 0$, and therefore $\Phi_i(x) = 0$ for all i.

Problems for Section 7.5

1. Consider the following matrix RH problem:

$$\Phi^-(k) = \Phi^+(k)V(k)$$

with

$$V(k) = \begin{pmatrix} 1 & b(k) \\ -\lambda\overline{b(k)} & 1 - \lambda|b(k)|^2 \end{pmatrix}, \qquad k \text{ real}$$

$$\Phi(k) \to I \qquad \text{as } k \to \infty$$

and where λ is either 1 or -1 and $b(k)$ and $b'(k)$ are square integrable. Assume that if $\lambda = 1$ then $|b|^2 < 1$.

(a) Show that

$$G(k) \equiv V(k) + \overline{V}(k)^T = \begin{pmatrix} 2 & (1-\lambda)b \\ (1-\lambda)\overline{b} & 2 - 2\lambda|b|^2 \end{pmatrix}$$

(b) Establish that both the eigenvalues of G are positive, and hence deduce that the above RH problem is always solvable.
Hint: If $\lambda = -1$, then $\det G = 4$; if $\lambda = 1$, then $\det G = 4(1 - |b|^2) > 0$. Also note that one of the diagonal elements of G is 2.

2. A natural generalization of a matrix RH problem is to replace the jump matrix by an operator. This yields a so-called *nonlocal RH problem*. A scalar nonlocal RH problem on the infinite line is defined as follows:

$$\varphi^+(k) - \varphi^-(k) = \int_{-\infty}^{\infty} f(k,l)\varphi^-(l)\,dl, \qquad k \text{ real}$$

$$\varphi(k) \to 1 \qquad \text{as } k \to \infty$$

where $f(k,l)$ is a given function that has the property that f and its derivatives are square integrable (i.e., f belongs to L_2). Show that

$$\varphi^-(k) = 1 + \frac{1}{2\pi i}\int_{-\infty}^{\infty}\int_{-\infty}^{\infty}\frac{f(k',l)\varphi^-(l)}{k'-(k-i0)}\,dl\,dk'$$

3. A particular case of a nonlocal RH problem is the so-called RH problem with a shift. This problem arises when $f(k,l) = \delta(l - \alpha(k))g(k)$, where $\delta(x)$ is a Dirac delta function and $f(k,l)$ is defined in Problem 7.5.2 above. We now study such a RH problem: Let C be a finite, closed contour, let $\alpha(t)$ be differentiable, and let $h(t)$, $\alpha'(t)$ satisfy a Hölder condition, continuous on C. Assume that $\alpha(t)$ maps C onto itself and that $\alpha'(t)$ does not vanish

anywhere on C. Consider the RH problem

$$\Phi^+[\alpha(t)] - \Phi^-(t) = h(t) \qquad \text{on } C \tag{1}$$

$$\Phi^-(z) \to 1 \qquad \text{as } z \to \infty$$

(a) Using a similar argument to that which led to Eq. 7.5.6 of the text show that

$$-\frac{1}{2}\Phi^+(t) + \frac{1}{2\pi i}\oint_C \frac{\Phi^+(\tau)}{\tau - t}\,d\tau = 0$$
$$\frac{1}{2}\Phi^-(t) + \frac{1}{2\pi i}\oint_C \frac{\Phi^-(\tau)}{\tau - t}\,d\tau = 1 \tag{2}$$

(b) Replace t with $\alpha(t)$ in equation (2), subtract (2) from (3), and use (1) to establish that

$$\Phi^-(t) - \frac{1}{2\pi i}\int_C \left[\frac{\alpha'(\tau)}{\alpha(\tau) - \alpha(t)} - \frac{1}{\tau - t}\right]\Phi^-(\tau)\,d\tau$$

$$= 1 - \frac{1}{2}h(t) + \frac{1}{2\pi i}\oint_C \frac{h(\tau)\alpha'(\tau)}{\alpha(\tau) - \alpha(t)}\,d\tau$$

It can be shown that the kernal of this equation has at the point $\tau = t$ a singularity of order lower than one (see Gakhov, 1966). Furthermore, it can be shown that this equation is solvable for any Hölder continuous $h(t)$.

4. Consider the matrix RH problem with the jumps indicated in Figure 7.5.2.

$$\Phi_{j+1} = \Phi_j \begin{pmatrix} e^{-iz^3} & 0 \\ 0 & e^{iz^3} \end{pmatrix} G_j \begin{pmatrix} e^{iz^3} & 0 \\ 0 & e^{-iz^3} \end{pmatrix} = \Phi_j DG_j D^{-1}$$

where

$$D = \begin{pmatrix} e^{-iz^3} & 0 \\ 0 & e^{iz^3} \end{pmatrix}, \qquad 1 \le j \le 6, \quad \Phi_7 \equiv \Phi_1$$

$$G_1 = \begin{pmatrix} 1 & 0 \\ a & 1 \end{pmatrix}, \qquad G_2 = \begin{pmatrix} 1 & b \\ 0 & 1 \end{pmatrix}, \qquad G_3 = \begin{pmatrix} 1 & 0 \\ c & 1 \end{pmatrix}$$

$$G_4 = \begin{pmatrix} 1 & a \\ 0 & 1 \end{pmatrix}, \qquad G_5 = \begin{pmatrix} 1 & 0 \\ b & 1 \end{pmatrix}, \qquad G_6 = \begin{pmatrix} 1 & c \\ 0 & 1 \end{pmatrix}$$

and a, b, and c are constants.

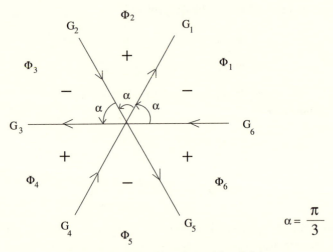

Fig. 7.5.2. RH problem for problem 7.5.4

(a) Show that by analytic continuation, a consistency condition in the neighborhood of $z = 0$ implies

$$\Phi_1 = \Phi_7 = \Phi_6 D^{-1} G_6 D = \Phi_5 D^{-1} G_5 D D^{-1} G_6 D = \cdots$$
$$= \Phi_1 D^{-1} G_1 G_2 \cdots G_6 D$$

Hence $G_1 G_2 \cdots G_6 = I$, and therefore,

$$a + b + c + abc = 0$$

(b) Show that the above RH problem is equivalent to the one depicted in Figure 7.5.3.
Hint: Note that e^{2iz^3} is analytic and decreasing in regions I, III, and V and e^{-2iz^3} is analytic and decreasing in regions II, IV, and VI where region I is the sector $0 \le \arg z < \pi/3$, etc. Because Φ_1 and e^{2iz^3} are analytic in region I, the equation $\Phi_2 = \Phi_1 D G_1 D^{-1}$ provides the analytic continuation of Φ_2 into region I. From $\Phi_1 = \Phi_6 D G_6 D^{-1}$ we have $\Phi_2 = \Phi_6 D G_6 D^{-1}$ etc.

(c) Show that the conditions for the vanishing lemma are satisfied provided that $G_2 = G_5^*$ and, $G_3 G_4 + (G_3 G_4)^*$, and $G_6 G_1 + (G_6 G_1)^*$ are positive definite.

(d) Establish that these conditions are fulfilled if

$$b = \bar{b}, \quad |a - \bar{c}| < 2$$

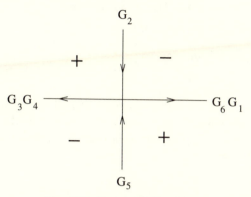

Fig. 7.5.3. Reduced RH problem

Note: It turns out that this result can be used to establish the solvability of the second Painlevé equation (PII):

$$y_{tt} = 2y^3 + ty$$

for y purely imaginary.

5. Consider the matrix RH problem

$$\Phi^- = \Phi^+ \begin{pmatrix} 1 & -be^{-\theta} \\ \bar{b}e^{\theta} & 1 - |b|^2 \end{pmatrix}, \qquad e^{\theta} = e^{i(kx+k^2t)}, \qquad k \text{ real}$$

$$\Phi \to I \qquad \text{as } k \to \infty \tag{1}$$

where $b(k)$ is a given complex valued function of k, which decays rapidly for large k.

Note that because

$$\begin{pmatrix} 1 & -be^{-\theta} \\ \bar{b}e^{\theta} & 1 - |b|^2 \end{pmatrix} = \begin{pmatrix} 1 & 0 \\ \bar{b}e^{\theta} & 1 \end{pmatrix} \begin{pmatrix} 1 & -be^{-\theta} \\ 0 & 1 \end{pmatrix}$$

it follows that

$$\Phi^- \begin{pmatrix} 1 & be^{-\theta} \\ 0 & 1 \end{pmatrix} = \Phi^+ \begin{pmatrix} 1 & 0 \\ \bar{b}e^{\theta} & 1 \end{pmatrix}$$

Let I, II, III, IV denote the first, second, third, and fourth quadrant of the k-complex plane. Show that if x and t are real and positive, and b is analytic for $k \in IV$, then the left-hand side of the above equation is analytic for $k \in$ IV, while the right-hand side is analytic for $k \in I$.

7.6 The DBAR Problem

We now consider the scalar equation

$$\frac{\partial \Phi(z, \bar{z})}{\partial \bar{z}} = f(z, \bar{z}), \qquad z \in D \tag{7.6.1}$$

(see also Section 2.6) where D is some simply connected domain of the complex z plane. We shall refer to this equation as a scalar DBAR (or $\bar{\partial}$) problem. Equation (7.6.1) is the complex form of the nonhomogeneous Cauchy-Riemann equations. Indeed, letting

$$\Phi = u + iv, \qquad f = \frac{g + ih}{2}, \qquad z = x + iy$$

where u, v, g, and h are real functions of the real variables x and y, and using

$$\frac{\partial}{\partial \bar{z}} = \frac{1}{2}\left(\frac{\partial}{\partial x} + i\frac{\partial}{\partial y}\right) \tag{7.6.2}$$

Eq. (7.6.1) yields

$$\frac{\partial u}{\partial x} - \frac{\partial v}{\partial y} = g(x, y), \qquad \frac{\partial u}{\partial y} + \frac{\partial v}{\partial x} = h(x, y) \tag{7.6.3}$$

If Φ is analytic for $z \in D$, then the Cauchy–Riemann equations must be satisfied. In fact, we see that if $g(x, y) = h(x, y) = 0$, then $\frac{\partial \Phi}{\partial \bar{z}} = 0$, and one recovers the Cauchy–Riemann equations. It was pointed out in Section 7.1 that a $\bar{\partial}$ problem can be considered to be a generalization of an RH problem. We recall that the solution of a scalar RH problem was obtained in closed form using a Cauchy type integral. The solution of a $\bar{\partial}$ problem can also be obtained in closed form.

In the following we introduce the so-called **wedge product**

$$d\xi \wedge d\xi = d\eta \wedge d\eta = 0, \qquad -d\eta \wedge d\xi = d\xi \wedge d\eta = d\xi \, d\eta$$

Lemma 7.6.1 Assume that $\Phi(z, \bar{z})$ is continuous and has continuous partial derivatives in a finite region D and on a simple closed contour C enclosing D. Then the solution of the $\bar{\partial}$ problem (7.6.1) is given by

$$\Phi(z, \bar{z}) = \frac{1}{2\pi i} \int_C \frac{\Phi(\zeta, \bar{\zeta}) \, d\zeta}{\zeta - z} + \frac{1}{2\pi i} \iint_D \frac{f(\zeta, \bar{\zeta})}{\zeta - z} \, d\zeta \wedge d\bar{\zeta} \tag{7.6.4}$$

where $d\zeta \wedge d\bar{\zeta} = (d\xi + i \, d\eta) \wedge (d\xi - i \, d\eta) = -2i \, d\xi \, d\eta$.

complex ζ -plane

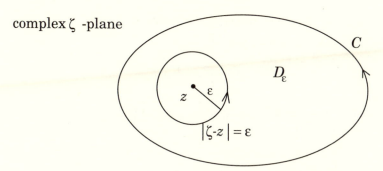

Fig. 7.6.1. Boundary C of finite region D in the complex ζ plane

Actually, we proved this theorem in Section 2.6 with Theorem 2.6.7. For the convenience of the reader we reproduce the proof using slightly different notation (see Figure 7.6.1).

Proof We first derive the complex form of Green's theorem (Theorem 2.5.1)

$$\int_C w(z, \bar{z}) \, dz = -\iint_D \frac{\partial w}{\partial \bar{z}} \, dz \wedge d\bar{z} \tag{7.6.5}$$

where $dz \wedge d\bar{z} = -2i \, dx \, dy$. Indeed, letting $w = u(x, y) + iv(x, y)$ and using Green's theorem, we find

$$\int_C w(z, \bar{z}) \, dz = \int_C (u + iv)(dx + i \, dy)$$

$$= \int_C (u \, dx - v \, dy) + i \int_C (v \, dx + u \, dy)$$

$$= -\iint_D \left(\frac{\partial v}{\partial x} + \frac{\partial u}{\partial y} \right) dx \, dy + i \iint_D \left(\frac{\partial u}{\partial x} - \frac{\partial v}{\partial y} \right) dx \, dy$$

$$= i \iint_D \left[\left(\frac{\partial u}{\partial x} - \frac{\partial v}{\partial y} \right) + i \left(\frac{\partial v}{\partial x} + \frac{\partial u}{\partial y} \right) \right] dx \, dy$$

$$= i \iint_D \left[\left(\frac{\partial u}{\partial x} + i \frac{\partial u}{\partial y} \right) + i \left(\frac{\partial v}{\partial x} + i \frac{\partial v}{\partial y} \right) \right] dx \, dy$$

$$= i \iint_D \left(\frac{\partial}{\partial x} + i \frac{\partial}{\partial y} \right) (u + iv) \, dx \, dy = 2i \iint_D \frac{\partial w}{\partial \bar{z}} \, dx \, dy$$

Let z be a fixed point in the complex ζ plane. Then the function $\frac{1}{(\zeta - z)}$ is an analytic function of ζ for $|\zeta - z| \geq \varepsilon$. Because the $\bar{\partial}$ derivative of an analytic function is zero, it follows that

$$\frac{\partial}{\partial\bar{\zeta}}\left(\frac{\Phi(\zeta,\bar{\zeta})}{\zeta-z}\right) = \left(\frac{1}{\zeta-z}\right)\frac{\partial\Phi(\zeta,\bar{\zeta})}{\partial\bar{\zeta}} \qquad \text{in } D_\varepsilon$$

where D_ε is the domain D punctured by a disc of radius ε at the point $\zeta = z$. Applying Eq. (7.6.5) in D_ε with $w = \frac{\Phi}{\zeta-z}$, we find

$$\int_C \frac{\Phi(\zeta,\bar{\zeta})\,d\zeta}{\zeta-z} - \int_{|\zeta-z|=\varepsilon} \frac{\Phi(\zeta,\bar{\zeta})\,d\zeta}{\zeta-z} = -\iint_{D_\varepsilon} \frac{\frac{\partial\Phi(\zeta,\bar{\zeta})}{\partial\bar{\zeta}}}{\zeta-z}\,d\zeta \wedge d\bar{\zeta}$$

But

$$\lim_{\varepsilon\to 0} \iint_{D_\varepsilon} \frac{\frac{\partial\Phi(\zeta,\bar{\zeta})}{\partial\bar{\zeta}}}{\zeta-z}\,d\zeta \wedge d\bar{\zeta} = \iint_D \frac{\frac{\partial\Phi(\zeta,\bar{\zeta})}{\partial\bar{\zeta}}}{\zeta-z}\,d\zeta \wedge d\bar{\zeta}$$

because $\iint_{D-D_\varepsilon} \to 0$ as $\varepsilon \to 0$ (see Eq. (2.6.25)).

Similarly, as $\varepsilon \to 0$, the second term in the above equation becomes $-2\pi i\,\Phi$ $(z,\bar{z})$, and Eq. (7.6.4) follows.   ∎

In many applications $\Phi \to 1$ as $z \to \infty$ and D is the entire complex plane. In this case, Eq. (7.6.4) reduces to

$$\Phi(z,\bar{z}) = 1 + \frac{1}{2\pi i} \iint_{\mathbb{R}^2} \frac{f(\zeta,\bar{\zeta})}{\zeta-z}\,d\zeta \wedge d\bar{\zeta} \qquad (7.6.6)$$

where $\mathbb{R}^2$ denotes integration over the entire complex plane. Actually Eq. (7.6.6) provides a solution of Eq. (7.6.1) provided that the integral in Eq. (7.6.6) makes sense. Assuming $f(\zeta,\bar{\zeta})$ is continuous, the only possible values of ζ for which this integral may have singularities are $\zeta = z$ and $\zeta = \infty$. It turns out that if $f \in L_{2+\varepsilon}$ (i.e., $\iint |f|^{2+\varepsilon}\,dx\,dy < \infty$), where ε is arbitrarily small, the integral is well behaved as $\zeta \to z$, and if $f \in L_{2-\varepsilon}$, then the integral is well behaved as $\zeta \to \infty$. In fact, a convenient class of functions for which Eq. (7.6.6) provides a solution to Eq. (7.6.1) is $f \in L_1 \cap L_\infty$. (The function $f \in L_1$ if $\iint |f(x,y)|\,dx\,dy < \infty$, and $f \in L_\infty$ if $|f(x,y)| \le M$; $f \in L_1 \cap L_\infty$ means that both $f \in L_1$ and $f \in L_\infty$.) In this case, $\Phi \to 1$ as $z \to \infty$ and Φ is continuous. Also Φ satisfies $\frac{\partial\Phi}{\partial\bar{z}} = f$, but in general only in a weak sense. In order for $\frac{\partial\Phi}{\partial\bar{z}}$ to exist in a strong sense, one needs some smoothness conditions on f (for example, $\partial f/\partial z, \partial f/\partial\bar{z} \in L_1 \cap L_\infty$).

7.6.1 Generalized Analytic Functions

An analytic function $\Phi(z)$ in a region $\mathcal{R}$ satisfies $\frac{\partial \Phi}{\partial \bar{z}} = 0$ in $\mathcal{R}$. A simple generalization of analytic functions, so-called **generalized analytic functions**, have been studied extensively by Vekua (1962). The equation $\frac{\partial \Phi}{\partial \bar{z}} = 0$ is now replaced by a special form of a $\bar{\partial}$ equation, namely

$$\frac{\partial \Phi}{\partial \bar{z}} = A(z, \bar{z})\Phi + B(z, \bar{z})\bar{\Phi} \qquad (7.6.7)$$

in a region $\mathcal{R}$, where $\bar{\Phi}$ is the complex conjugate of Φ and A, B are given functions of z and $\bar{z}$. Generalized analytic functions have had applications in differential geometry, in elasticity, and more recently in the solution of a certain class of multidimensional nonlinear PDEs, and in multidimensional inverse scattering.

If $B = 0$, Eq. (7.6.7) can be solved in closed form. A derivation of this solution can be obtained using the following formal argument. The first term in the right-hand side of Eq. (7.6.4) is analytic and therefore its $\bar{\partial}$ derivative is zero. Applying $\partial/\partial \bar{z}$ in Eq. (7.6.4), we find (note that $f = \partial \Phi / \partial \bar{z}$)

$$f(z, \bar{z}) = \frac{\partial}{\partial \bar{z}} \frac{1}{2\pi i} \iint_D \frac{f(\zeta, \bar{\zeta})}{\zeta - z} \, d\zeta \wedge d\bar{\zeta} \qquad (7.6.8)$$

Using Eq. (7.6.8), it follows that the general solution of $\frac{\partial \Phi}{\partial \bar{z}} = A(z, \bar{z})\Phi$ is

$$\Phi(z, \bar{z}) = w(z) e^{\frac{1}{2\pi i} \iint_D \frac{A(\zeta, \bar{\zeta})}{\zeta - z} d\zeta \wedge d\bar{\zeta}} \qquad (7.6.9)$$

where $w(z)$ is an arbitrary analytic function; (take $\partial/\partial \bar{z}$ of Eq. (7.6.9) and use Eq. (7.6.8)). Similarly, the same procedure shows that the general solution of Eq. (7.6.7) can be expressed (Vekua, 1962) as

$$\Phi(z, \bar{z}) = w(z) \exp \left(\frac{1}{2\pi i} \int \int_D \left[A(\zeta, \bar{\zeta}) + B(\zeta, \bar{\zeta}) \frac{\bar{\Phi}(\zeta, \bar{\zeta})}{\Phi(\zeta, \bar{\zeta})} \right] \frac{d\zeta \wedge d\bar{\zeta}}{\zeta - z} \right) \qquad (7.6.10)$$

Note that Eq. (7.6.7) is really a first-order PDE. Its general solution depends on an arbitrary function, which in this case is $w(z)$.

We now make some further comments about Eq. (7.6.7). We assume that D is the entire complex plane, $A, B \in L_1 \cap L_\infty$, and seek a solution Φ that tends to 1 as $z \to \infty$.

If $B = 0$, we know that $\int \int_{\mathbb{R}^2} \frac{A(\zeta, \bar{\zeta})}{\zeta - z} \, d\zeta \wedge d\bar{\zeta} \to 0$ as $z \to \infty$. Therefore, from Eq. (7.6.9), $w(z) \to 1$ as $z \to \infty$. Because $w(z)$ is analytic and bounded, by Liouville's Theorem, $w(z) = 1$. Thus we expect that

$$\Phi(z, \bar{z}) = \exp\left(\frac{1}{2\pi i} \iint_{R^2} \frac{A(\zeta, \bar{\zeta})}{\zeta - z} d\zeta \wedge d\bar{\zeta}\right) \qquad (7.6.11)$$

is the unique solution of equation $\frac{\partial \Phi}{\partial \bar{z}} = A\Phi$, $\Phi \to 1$ as $z \to \infty$.
In the general case, equation Eq. (7.6.6) implies

$$\Phi = 1 + \frac{1}{2\pi i} \iint_{\mathbb{R}^2} (A(\zeta, \bar{\zeta})\Phi(\zeta, \bar{\zeta}) + B(\zeta, \bar{\zeta})\bar{\Phi}(\zeta, \bar{\zeta})) \frac{d\zeta \wedge d\bar{\zeta}}{\zeta - z} \qquad (7.6.12)$$

It can be shown (Vekua, 1962) that this equation is a Fredholm integral equation, and therefore the question of its solvability reduces to the question of existence of nontrivial homogeneous solutions. We claim that Eq. (7.6.12) has no such solutions. Indeed, suppose that $\varphi(\zeta, \bar{\zeta})$ is a homogeneous solution of Eq. (7.6.12); that is, φ solves the same equation as Eq. (7.6.12) except 1 is replaced by 0. Then $\varphi \to 0$ as $z \to \infty$. But from Eq. (7.6.10), φ can be written as

$$\varphi(z, \bar{z}) = w(z) \exp\left(\frac{1}{2\pi i} \iint_{R^2} \left[A(\zeta, \bar{\zeta}) + B(\zeta, \bar{\zeta}) \frac{\bar{\varphi}(\zeta, \bar{\zeta})}{\varphi(\zeta, \bar{\zeta})}\right] \frac{d\zeta \wedge d\bar{\zeta}}{\zeta - z}\right)$$

and because $\varphi \to 0$ as $z \to \infty$, it follows that $w \to 0$ as $|z| \to \infty$. But because $w(z)$ is analytic and bounded, by Liouville's Theorem we find that $w = 0$ everywhere. (This is provided that the integral in the exponential is well behaved, which is indeed the case because for well-behaved functions A and B, $|\bar{\varphi}/\varphi| = 1$.)

Problems for Section 7.6

1. Note that the function $(z - \alpha)^{-1}$ is analytic everywhere except at $z = \alpha$. Thus its derivative with respect to $\bar{z}$ is zero everywhere except at $z = \alpha$. Indeed, it can be shown that the following identity is valid

$$\frac{\partial}{\partial \bar{z}}(z - \alpha)^{-1} = -2\pi i \delta(z - \alpha)$$

where δ denotes the dirac delta function. (See also Problem 2.6.11 in Section 2.6). This function has the property that if $f(z, \bar{z})$ is a sufficiently smooth function, and D contains the origin, then

$$\iint_D f(z, \bar{z})\delta(z) \, dz \wedge d\bar{z} = f(0, 0)$$

Use the above result to verify that the function

$$\Phi(z, \overline{z}) = w(z) \exp\left[\frac{1}{2\pi i} \iint_D \frac{A(\zeta, \overline{\zeta})}{\zeta - z} \, d\zeta \wedge d\overline{\zeta}\right]$$

where $w(z)$ is an arbitrary analytic function, satisfies

$$\frac{\partial \Phi}{\partial \overline{z}} = A(z, \overline{z})\Phi \quad \text{in } D$$

2. Consider the following PDE:

$$\Phi_x + i\Phi_y = \frac{2\Phi}{x^2 + y^2 + 1}, \qquad \Phi \to 1 \text{ as } x^2 + y^2 \to \infty$$

Show that

$$\Phi(z, \overline{z}) = \exp\left[\frac{1}{2\pi i} \iint_{R^2} \frac{d\zeta \wedge d\overline{\zeta}}{(|\zeta|^2 + 1)(\zeta - z)}\right]$$

Hint: Note that the above PDE can be written as

$$\frac{\partial \Phi}{\partial \overline{z}} = \frac{\Phi}{|z|^2 + 1}$$

3. Consider the following linear differential equation for the scalar function $\mu(x, y, k)$, with the scalar function $q(x, y)$ as a forcing function:

$$\mu_x + i\mu_y - 2k\mu = 2q, \qquad -\infty < x < \infty, \ -\infty < y < \infty \qquad (1)$$

where $q(x, y)$ and its derivatives decay sufficiently fast as $|x| + |y| \to \infty$. Show that the unique solution of this equation, which decays as $x^2 + y^2 \to \infty$ and which is bounded for all complex k, is given by

$$\mu(x, y, k_R, k_I) = \frac{1}{\pi} \iint_{R^2} \frac{e^{k(\overline{z} - \overline{\zeta}) - \overline{k}(z - \zeta)}}{z - \zeta} q(\xi, \eta) \, d\xi \, d\eta \qquad (2)$$

where $z = x + iy$, $\zeta = \xi + i\eta$, $k = k_R + ik_I$.
Hint: Note that (1) can be written as $\frac{\partial \mu}{\partial \overline{z}} - k\mu = q$. The Green's function of the operator $\frac{\partial}{\partial \overline{z}}$ is $\frac{1}{\pi z}$, thus the Green's function of the left-hand side of (1) is $\frac{c}{\pi z} e^{k\overline{z}}$ where c is independent of $\overline{z}$. Choose c so that this Green function is bounded for all complex k.

4.   In Section 7.4 the equation $\mu_x - ik\mu = q(x)$ was used to derive the direct
     and inverse Fourier transforms. Now we use equation (1) of Problem 7.6.3,
     that is

$$\frac{\partial \mu}{\partial \bar{z}} - k\mu = q(x, y), \qquad z = x + iy$$

to derive the direct and inverse two-dimensional Fourier transform. A
certain *nonlinearization* of these results is discussed in the exercises of
Section 7.7

(a) Let μ be defined by equation (2) of Problem 7.6.3. Show that

$$\lim_{z \to \infty} (z e^{\bar{k}z - k\bar{z}} \mu) = \hat{q}(k_I, k_R)$$

where

$$\hat{q}(k_I, k_R) \equiv \frac{1}{\pi} \iint_{R^2} e^{-2i(k_I x - k_R y)} q(x, y)\, dx\, dy \qquad (1')$$

(b) Show that equation (2) of Problem 7.6.3 implies

$$\frac{\partial \mu}{\partial \bar{k}} = e^{-k\bar{z} - \bar{k}z} \hat{q}, \qquad \mu = O\left(\frac{1}{k}\right) \text{ as } k \to \infty$$

(c) Establish that the unique solution of the $\bar{\partial}$ problem formulated in (b) is

$$\mu = \frac{-1}{\pi} \iint_{R^2} \frac{e^{2i(l_I x - l_R y)} \hat{q}(l_I, l_R)}{k - l}\, dl_R dl_I, \qquad l = l_R + il_I$$

(d) By comparing the large k assumptions of equation (1) of Problem 7.6.3
     and those of the above equation, show that

$$q = -\lim_{k \to \infty} (k\mu) = \frac{1}{\pi} \iint_{R^2} e^{2i(k_I x - k_R y)} \hat{q}(k_I, k_R)\, dk_R\, dk_I \qquad (2')$$

Equations $(1')$ and $(2')$ are the usual formulae for the two-dimensional
direct and inverse Fourier transforms.

*7.7 Applications of Matrix Riemann–Hilbert Problems and $\bar{\partial}$ Problems

It was shown in Section 7.4 (Example 7.4.6) that the inversion formulae for the
Fourier transform can be obtained by relating the linear ODE with a parameter k,

$\mu_x - ik\mu = q(x)$, to the solution of a certain scalar RH problem. Similarly, the inversion formulae associated with more general transforms can be obtained by connecting more general linear eigenvalue problems with the solution of certain vector RH and $\bar{\partial}$ problems. It is well known that the Fourier transform can be used to solve linear PDEs. It is interesting and significant that these more general transforms (called the inverse scattering transform: IST) can be used to solve certain nonlinear PDEs.

In Examples 7.7.2–7.7.5 below we study the RH and $\bar{\partial}$ problems that arise from the analysis of certain linear (scattering) equations associated with the so-called Korteweg–deVries, Kadomtsev–Petviashvili, and Painlevé IV equations. In these examples we concentrate on the relevant RH and $\bar{\partial}$ problems without addressing in detail (though we do give the interested reader some background concepts behind the derivations for Examples 7.7.2–7.7.3) the more difficult questions of how they are derived from the underlying linear equations.

Throughout this section we assume that the jump functions appearing in the RH and $\bar{\partial}$ problems we study are in the proper spaces so that the theory mentioned in Sections 7.5 and 7.6 applies, and then the associated linear integral equations are of the Fredholm type. We recall that, if the jump functions are in the spaces $L_2 \cap L_\infty$ and in $L_1 \cap L_\infty$ for RH and $\bar{\partial}$ problems, respectively, then it can be shown that the linear integral equations are indeed Fredholm integral equations. It is known that to establish the solvability of Fredholm integral equations of the second kind, it is sufficient to establish uniqueness, that is, to prove that the associated homogeneous equations have only the zero solution.

We begin this section by discussing an important extension of the example we considered in Section 7.4, namely, integral equations with displacement kernels. In this example we will consider systems of such equations. As mentioned earlier, Gohberg and Krein considered this problem in their seminal work (Gohberg and Krein, (1958)), and in this example we mention some of their main results.

Example 7.7.1 We consider the following system of N coupled integral equations

$$\varphi_\ell(x) - \int_0^\infty g_{\ell m}(x - t)\varphi_m(t)\, dt = f_\ell(x),$$
$$0 < x < \infty, \quad \ell, m = 1, 2, \ldots, N$$

where $f_\ell(x)$ and $g_{\ell m}(x)$ are continuous and belong to the space L_1 of absolutely integrable functions, $\int_0^\infty |f_\ell|\, dx < \infty$ and $\int_0^\infty |g_{\ell m}|\, dx < \infty$. It is more convenient to work with vector-matrix notation:

$$\varphi(x) - \int_0^\infty g(x - t)\varphi(t)\, dt = f(x),$$

where $\varphi(x)$ and $f(x)$ represent the column vectors $(\varphi_1, \ldots, \varphi_n)^T$ and $(f_1, \ldots, f_n)^T$, and $g(x)$ is the matrix whose coefficients are $\{g_{\ell m}(x)\}$. The vectors φ, f, and matrix g all belong to the space L_1 because their components do so. We assume that the Fourier transform of the matrix $g(x)$ (i.e., the transform of each coefficient), which we denote as $\hat{G}(k)$, satisfies $\det(I - \hat{G}) \neq 0$; that is, the matrix $I - \hat{G}(k)$ is not singular. The matrix $\hat{G}(k)$ is in the dual space of suitably bounded functions denoted by Λ_1.

In the same way as shown in Example 7.4.5 we can reduce the above integral equation to a RH problem – in this case a vector RH problem. Let us define $\tilde{\varphi}$ as $\tilde{\varphi}(x) = \varphi(x)$ for $x > 0$ and $\tilde{\varphi}(x) = 0$ for $x < 0$, and obtain the system

$$\tilde{\varphi}(x) - \int_{-\infty}^\infty g(x - t)\tilde{\varphi}(t)\, dt = f(x), \qquad x > 0$$

$$-\int_{-\infty}^\infty g(x - t)\tilde{\varphi}(t)\, dt = h(x), \qquad x < 0$$

where $h(x)$ is an unknown vector function to be found.

Calling

$$\tilde{\varphi}(x), \tilde{f}(x), \tilde{h}(x) = \begin{cases} \varphi(x), f(x), 0 & x > 0 \\ 0, 0, h(x) & x < 0 \end{cases}$$

the two vector-matrix equations above for $x > 0$ and $x < 0$ can be written as the single matrix equation

$$\tilde{\varphi}(x) - \int_{-\infty}^\infty g(x - t)\tilde{\varphi}(t)\, dt = \tilde{f}(x) + \tilde{h}(x)$$

for $-\infty < x < \infty$. Let us denote by $\hat{\Phi}(k)$, $\hat{G}(k)$, $\hat{F}(k)$, and $\hat{H}(k)$, the Fourier transforms of $\tilde{\varphi}, g, \to f$, and $\to h$, respectively, using the convention

$$\hat{\Phi}(k) = \int_{-\infty}^\infty \tilde{\varphi}(x)e^{-ikx}dx = \int_0^\infty \tilde{\varphi}(x)e^{-ikx}dx$$

Because all components of $\tilde{\varphi}$ and $\tilde{f}$ have support only for $x > 0$, and similarly $\tilde{h}$ has support only for $x < 0$, their Fourier transforms are analytically extendible to the lower half plane ($\hat{\Phi}^-(k)$, $\hat{F}^-(k)$) and the upper half plane ($\hat{H}^+(k)$), respectively. For convenience we call the unknown $\hat{H}^+(k)$ to be $\hat{\Phi}^+(k)$, hence the Fourier transform of the above integral equation for $\bar{\varphi}(x)$ yields

$$\hat{\Phi}^+(k) = (I - \hat{G}(k))\hat{\Phi}^-(k) - \hat{F}^-(k) \quad \text{on } -\infty < k < \infty, \quad G(\pm\infty) = 0$$

which is the vector RH analog of Eq. (7.4.33b).

If we assume that $\hat{G}(k)$ and $\hat{F}^-(k)$ satisfy a Hölder condition in addition to being in the space Λ_1, then we connect with the study of vector and matrix RH problems that was discussed in Section 7.5, suitably extended to the real line, $-\infty < k < \infty$ (see also Section 7.4.1). The critical issue involves solving the homogeneous matrix RH problem

$$\hat{X}^+(k) = (I - \hat{G}(k))\hat{X}^-(k), \qquad I - \hat{G}(\pm\infty) = I, \qquad \hat{X}^{\pm}(\pm\infty) = I,$$
$$-\infty < k < \infty$$

because with this in hand we can readily solve the inhomogeneous equation; that is

$$\hat{\Phi}^+(k) = \hat{X}^+(k)(\hat{X}^-(k))^{-1}\hat{\Phi}^-(k) - \hat{F}^-(k) \quad \text{on } -\infty < k < \infty$$

or

$$(\hat{X}^+(k))^{-1}\hat{\Phi}^+(k) - (\hat{X}^-(k))^{-1}\hat{\Phi}^-(k) = -(\hat{X}^+(k))^{-1}\hat{F}^-(k)$$

and therefore

$$\Phi(z) = \hat{X}(z)\left[-\frac{1}{2\pi i}\int_{-\infty}^{\infty} \frac{\hat{X}^+(k)\hat{F}^-(k)}{k - z} \, dk + D^{(1)}(z)P(z) \right]$$

where $P(z)$ is a vector of polynomials and $D^{(1)}(z)$ is an appropriate diagonal matrix that depends on the partial index associated with $I - \hat{G}(k)$ discussed below. Below is a brief summary of the results of Gohberg and Krein (1958); some readers may wish to skip the summary because the connection between systems of integral equations with displacement kernels and with RH problems has been established.

Call $M(k) = (I - \hat{G}(k))$ and write the homogeneous RH problem in the form

$$\hat{X}^+(k) = M(k)\hat{X}^-(k), \qquad -\infty < k < \infty$$

where $X^{\pm}(\pm\infty) = I$, $M(\pm\infty) = I$. Gohberg and Krein show that the matrix $M(k)$ can always be factorized into the following product:

$$M(k) = \boldsymbol{\eta}^+(k)D(k)\boldsymbol{\eta}^-(k)$$

(called the "left" standard form), where $\eta^{\pm}(k)$ are matrices that are analytically extendible into the upper (+) and lower (−) half planes where (a) $\eta^{\pm}(\pm\infty) = I$, (b) det $\eta^{\pm}(k)$ do not vanish anywhere in their analytically extended half planes, and (c) $D(k)$ is the diagonal matrix

$$D(k) = \text{diag}\left(\left(\frac{k-i}{k+i}\right)^{\kappa_1}, \left(\frac{k-i}{k+i}\right)^{\kappa_2}, \ldots, \left(\frac{k-i}{k+i}\right)^{\kappa_n}\right)$$

the integers $\kappa_1, \kappa_2, \ldots, \kappa_n$ are the partial indices associated with the RH problem, analogous to those discussed in Section 7.5. While the partial indices can be characterized, other than for special cases ($M(k)$ triangular, $M(k)$ polynomial, $M(k)$ satisfies $M \pm M^*$ definite, in which case $\kappa_1 = \kappa_2 = \cdots = \kappa_n = 0$), in general it is not known how to calculate the indices constructively. However, the total index can be calculated from $M(k) = \eta^+(k)D(k)\eta^-(k)$ by taking the logarithm of the determinant of this equation:

$$\text{ind}(M(k)) = \frac{1}{2\pi i}\left[\log\det M(k)\right]_{-\infty}^{\infty}$$
$$= \frac{1}{2\pi i}\left[\log(\det \eta^+(k)) + \log(\det D(k)) + \log(\det \eta^-(k))\right]_{-\infty}^{\infty}$$

Because $\log(\det \eta^{\pm}(k))$ never vanish in their respective half planes, their index is zero. Moreover, from Section 7.4.1 we know that the index of $\left(\frac{k-i}{k+i}\right)^{\kappa_j}$ is κ_j. Let us denote the index of $M(k)$ as κ. Thus the sum of the partial indices, referred to as the **total index**, is known a priori:

$$\kappa_1 + \kappa_2 + \ldots + \kappa_n = \text{ind}M(k) = \kappa$$

Calling $X^-(k) = (\eta^-(k)D(k))^{-1}$ and $X^+(k) = \eta^+(k)$, we see that $X^-(k)$ has a zero at $k = -i$ of order κ_j ($\kappa_j > 0$) in its jth column, and $\det(X^-(k))$ does not vanish anywhere else in the LHP. If $\kappa_j < 0$, then there is no solution to the homogeneous problem because $X^-(k)$ has a pole of order $|\kappa_j|$ at $k = -i$.

We also note that Gohberg and Krein(1958) show that corresponding to each partial index $\kappa_j > 0$ there corresponds a κ_j-dimensional family of homogeneous solutions to the original integral equation; that is, $\varphi(x) - \int_0^{\infty} g(x-t)\varphi(t)\,dt = 0$, and for each $\kappa_\ell < 0$ there corresponds a $|\kappa_\ell|$-dimensional family of solutions to the homogeneous adjoint equation

$$\Psi(x) - \int_0^{\infty} g^T(t-x)\Psi(t)\,dt = 0$$

On the other hand, if $M(k)$ satisfies $M \pm M^*$ definite, then it can be shown that $\kappa_1 = \kappa_2 = \cdots = \kappa_n = 0$, and then $X^{\pm}(k)$ satisfy Fredholm integral equations, similar in form to Eq. (7.5.5). (Via a similar derivation one finds that in Eq. (7.5.5) the contour C is replaced by an integral along the real axis $(-\infty, \infty)$, and the right-hand side of that equation must be modified to be $(M^{-1}(k)+I_N)I_N/2$.) In the forced solution of $\Phi(z)$ above, $P(z) = P_0$ a constant vector and $D^{(1)}(z) = I_N$. Other cases where the indices are positive or negative can be discussed, but the zero indices case is sufficiently representative of the essential ideas.

Example 7.7.2 (*The Time Independent Schrödinger Equation*) We now consider Problem 6 of the introduction to this section; given appropriate scattering data, reconstruct the potential $q(x)$ of the time independent Schrödinger equation

$$\psi_{xx} + (q(x) + k^2)\psi = 0, \qquad -\infty < x < \infty \qquad (7.7.1)$$

The results we quote assume that $q(x)$ is real, and $\int_{-\infty}^{\infty} dx(1+x^2)|q(x)| < \infty$.

This problem has a natural quantum mechanical interpretation. If a "potential" is bombarded by quantum particles, then its shape can be reconstructed from knowledge of how these particles scatter. To be more precise, if a wave e^{-ikx} impinges on the potential $q(x)$ from the right, this wave creates a transmitted wave to the left with amplitude $T(k)$ and a reflected wave to the right with amplitude $R(k)$, that is

$$\begin{aligned}
\psi(x, k) &\to T(k)e^{-ikx} & \text{as } x \to -\infty, \\
\psi(x, k) &\to e^{-ikx} + R(k)e^{ikx} & \text{as } x \to +\infty
\end{aligned} \qquad (7.7.2)$$

The potential $q(x)$ can also support "bound states," that is, there may exist a finite number of discrete eigenvalues, k_n, such that there exist square integrable solutions of Eq. (7.7.1):

$$\int_{-\infty}^{\infty} \psi_n^2(x)\, dx < \infty, \qquad n = 1, 2, \ldots, N \qquad (7.7.3)$$

It is significant that the reconstruction of $q(x)$ from scattering data can be turned into a matrix RH problem. In what follows we will only sketch the main ideas. The "inverse scattering" analysis requires us to study solutions of Eq. (7.7.1) with suitable boundary conditions imposed as $x \to \pm\infty$; that is

$$\begin{aligned}
\varphi_L(x, k) &\sim e^{-ikx} & \text{as } x \to -\infty \\
\hat{\psi}_R(x, k) &\sim e^{-ikx} & \text{as } x \to +\infty \\
\psi_R(x, k) &\sim e^{ikx} & \text{as } x \to +\infty
\end{aligned}$$

The solutions $\hat{\psi}_R$ and ψ_R are linearly independent because the Wronskian $\mathcal{W} = \mathcal{W}(\hat{\psi}_R, \psi_R) = \hat{\psi}_R \psi_{R_x} - \hat{\psi}_{R_x} \psi_R = 2ik \neq 0$ ($k \neq 0$). Thus φ_L is linearly dependent upon $\hat{\psi}_R$ and ψ_R:

$$\varphi_L(x, k) = a(k)\hat{\psi}_R(x, k) + b(k)\psi_R(x, k) \qquad (7.7.4)$$

The transmission and reflection coefficients, $T(k)$ and $R(k)$ in Eq. (7.7.2), are defined by $T(k) = 1/a(k)$, $R(k) = b(k)/a(k)$, and we see that the function $\varphi_L(x, k)/a(k) = T(k)\varphi_L(x, k)$ is identified with the function $\psi(x, k)$ in Eq. (7.7.2). Multiplying Eq. (7.7.4) by e^{ikx} and defining

$$\tilde{\Phi}(x, k) = \varphi_L(x, k)e^{ikx}$$

$$\hat{\Psi}(x, k) = \hat{\psi}_R(x, k)e^{ikx}$$

$$\Psi(x, k) = \psi_R(x, k)e^{-ikx}$$

we obtain the equation

$$\frac{\tilde{\Phi}(x, k)}{a(k)} = \Psi(x, -k) + R(k)e^{2ikx}\Psi(x, k) \qquad (7.7.5)$$

where we have used the fact that $\hat{\Psi}_R(x, -k) = \Psi_R(x, k)$ (note that $\hat{\Psi}_R(x, k)$ and $\Psi_R(x, k)$ satisfy the same ODE, and the boundary conditions for $\hat{\Psi}_R(x, -k)$ and $\Psi_R(x, k)$ are the same). When $q(x)$ is such that $\int_{-\infty}^{\infty}(1 + x^2)|q(x)|\, dx < \infty$, the following analytic properties can be established: $\tilde{\Phi}(x, k)$, $\Psi(x, k)$, $a(k)$ are extendible to analytic functions in the upper half k plane, and by the symmetry principle, $\Psi(x, -k)$ is extendible to a function analytic in the lower half k plane. The functions $\tilde{\Phi}(x, k)$, $\Psi(x, k)$, and $a(k)$ all tend to unity as $|k| \to \infty$.

The method usually employed to establish these analytic conditions is to study the integral equations (cf. Ablowitz and Clarkson, 1991) that govern $\tilde{\Phi}(x, k)$ and $\Psi(x, k)$; the function $a(k)$ is given by an integral over $q(x)$ and $\tilde{\Phi}(x, k)$. So for example, $\tilde{\Phi}(x, k)$ satisfies the ODE

$$\tilde{\Phi}_{xx} - 2ik\tilde{\Phi}_x = -q(x)\tilde{\Phi}$$

and has the following representation as an integral equation:

$$\tilde{\Phi}(x, k) = 1 + \int_{-\infty}^{\infty} G(x - \xi, k)q(\xi)\tilde{\Phi}(\xi, k)\, d\xi$$

where $G(x, k)$ is defined below.

The kernel $G(x, k)$ is often called the Green's function associated with $\tilde{\Phi}(x, k)$, and G satisfies the same ODE as $\tilde{\Phi}$ except the right-hand side of the ODE is replaced by a Dirac delta function, that is, $q(x)\tilde{\Phi}(x, k) \to \delta(x)$, so that $G_{xx} - 2ikG_x = -\delta(x)$. The kernel $G(x, k)$ is given by

$$G(x, k) = \frac{1}{2\pi} \int_{C_+} \frac{e^{ipx}}{p(p - 2k)} \, dp$$

$$= \frac{1}{2ik}(1 - e^{2ikx}) \, \theta(x)$$

where C_+ is the contour from $-\infty$ to ∞ indented below $p = 0$, $p = 2k$; $\theta(x) = \{1 \text{ if } x > 0, 0 \text{ if } x < 0\}$. The above integral equation for $\tilde{\Phi}(x, k)$ is a Volterra integral equation; note that the kernel (Green's function) is analytic in the upper half k plane (i.e., for $\operatorname{Im} k > 0$). When $q(x)$ satisfies the condition $\int_{-\infty}^{\infty}(1 + x^2)|q(x)| \, dx < \infty$, one can establish that the Volterra integral equation has a unique solution (i.e., its Neumann series converges; cf. Ablowitz and Clarkson (1991)).

In fact, with these analyticity results Eq. (7.7.5) is a generalized RH problem – the generalization being due to the fact that $\tilde{\Phi}(x, k)/a(k)$ is a *meromorphic* function in the upper half k plane, owing to the zeroes of $a(k)$. (The analyticity of $a(k)$ for $\operatorname{Im} k > 0$ follows from the analyticity of $\tilde{\Phi}(x, k)$ for $\operatorname{Im} k > 0$; in fact, it can be shown that $a(k) = 1 + \frac{1}{2ik}\int_{-\infty}^{\infty}q(x)\tilde{\Phi}(x, k) \, dx$, or alternatively from Eq. (7.7.4)

$$a(k) = \frac{1}{2ik}(\varphi_L \psi_{R_x} - \varphi_{L_x} \psi_R) = \frac{1}{2ik}W(\varphi_L, \psi_R)$$

where $\mathcal{W}(\phi_L, \psi_R)$ is the Wronskian.) Because $a(k)$ is analytic for $\operatorname{Im} k > 0$, $a(k) \to 1$ as $|k| \to \infty$, and is continuous for $\operatorname{Im} k \geq 0$, its zeroes must be isolated and finite in number. One can also show that all the zeroes of $a(k)$ are simple and lie on the imaginary axis: $a(k_n) = 0$, $k_n = ip_n$, $n = 1, 2, \ldots, N$. A more standard form for this generalized RH problem is obtained by subtracting the poles of $\tilde{\Phi}(x, k)/a(k)$ from both sides of Eq. (7.7.5). At a zero of $a(k)$, $a(k_n) = 0$, the bound states of $\varphi_L(x, k_n)$ and $\psi_L(x, k_n)$ are related: $\varphi_L(x, k_n) = c_n\psi_R(x, k_n)$, and the so-called normalization constants are given by $c_n = b_n/a'(k_n)$. Calling

$$\Phi^+(x, k) = \frac{\tilde{\Phi}(x, k)}{a(k)} - \sum_{n=1}^{N} \frac{c_n e^{2ik_n x} \Psi^+(x, k_n)}{k - k_n}$$

Eq. (7.7.5) is given by

$$\Phi^+(x, k) = \Psi^+(x, -k) + R(k)e^{2ikx}\Psi^+(x, k)$$
$$- \sum_{n=1}^{N} \frac{c_n e^{2ik_n x}\Psi^+(x, k_n)}{k - k_n}, \qquad -\infty < k < \infty \quad (7.7.6)$$

where $\Phi^+(x, k)$ and $\Psi^+(x, -k)$ tend to 1 as $k \to \infty$ (note that $G(x, k) \to 0$ as $k \to \infty$). We show below that Eq. (7.7.6) can be viewed as a vector RH problem.

It turns out that the above physical picture can be turned into a rigorous mathematical theory. Given a reflection coefficient $R(k)$, N discrete eigenvalues $\{k_n\}_1^N$ and N normalization constants $\{c_n\}_1^N$, there exists a unique way to reconstruct the potential $q(x)$ (as mentioned above, the discrete eigenvalues are pure imaginary: $k_n = ip_n$, $p_n > 0$):

$$q(x) = \frac{\partial}{\partial x}\left[-\frac{1}{\pi}\int_{-\infty}^{\infty} dk\, R(k)e^{2ikx}\Psi^+(x, k) + 2i \sum_{n=1}^{N} c_n e^{2ik_n x}\Psi^+(x, k_n) \right]$$
$$(7.7.7)$$

where Ψ^+ is a solution of the generalized RH problem (7.7.6). (Eq. (7.7.7) can be established by equating the $O(1/k)$ term in the large k expansion of Ψ^+ using the Greens function with the $O(1/k)$ term from the $+$ projection of Eq. (7.7.6).)

This inverse scattering analysis can be thought of as a generalized Fourier transform. On the "direct" side we give $q(x)$ and find $R(k)$, $T(k)$, $\{c_n, k_n\}_{n=1}^N$, $\Phi^+(x, k)$, $\Psi^+(x, -k)$, etc. On the "inverse" side we "recover" $q(x)$ via Eqs. (7.7.6)–(7.7.7) by giving $R(k)$, $\{c_n, k_n\}_{n=1}^N$. However, a significant complication is that even if $R(k)$, $\{c_n\}_{n=1}^N$ and $\{k_n\}_{n=1}^N$ are given, one first needs to solve the generalized RH problem (7.7.6) for $\Phi^+(x, k)$ and $\Psi^+(x, -k)$ before evaluating $q(x)$. In fact, Eq. (7.7.6) together with the equation obtained from it by letting $k \to -k$ defines a vector RH problem for the $\oplus$ and $\ominus$ functions $(\Phi^+(x, k), \Psi^+(x, k))^T$ and $(\Phi^+(x, -k), \Psi^+(x, -k))^T$, respectively. We show this concretely in the simpler case where there are no discrete eigenvalues. (We make this assumption in order to simplify the algebra.) Equation (7.7.6) for arguments k and $-k$ yields

$$\Phi^+(x, k) - \Psi^+(x, -k) = R(k)e^{2ikx}\Psi^+(x, k)$$
$$\Phi^+(x, -k) - \Psi^+(x, k) = R(-k)e^{-2ikx}\Psi^+(x, -k)$$

Substitution of $\Psi^+(x, k)$ in the first of these equations by using the second equation yields, in place of the first

$$\Phi^+(x, k) = (1 - R(k)R(-k))\,\Psi^+(x, -k) + R(k)e^{2ikx}\Phi^+(x, -k)$$

This equation and the second of the above pair yields

$$\begin{pmatrix} \Phi^+(x, k) \\ \Psi^+(x, k) \end{pmatrix} = \begin{pmatrix} 1 - R(k)R(-k) & R(k)e^{2ikx} \\ -R(-k)e^{-2ikx} & 1 \end{pmatrix} \begin{pmatrix} \Psi^+(x, -k) \\ \Phi^+(x, -k) \end{pmatrix}$$

$$= \hat{G}(k) \begin{pmatrix} \Psi^+(x, -k) \\ \Phi^+(x, -k) \end{pmatrix}$$

Actually, if $q(x)$ is real, then $R(-k) = \bar{R}(k)$, and it is possible to establish solvability for this RH problem provided that $|R(k)| < 1$. Indeed

$$G + G^* = \begin{pmatrix} 2\left(1 - |R(k)|^2\right) & 0 \\ 0 & 1 \end{pmatrix}$$

thus if $|R(k)| < 1$, $G + G^*$ is positive definite and Lemma 7.5.3 implies solvability. An alternative approach for solving this RH problem is to transform the RH problem to a certain Fredholm linear integral equation called the Gel'fand–Levitan–Marchenko equation (see, e.g., Ablowitz and Segur, 1981; Ablowitz and Clarkson, 1991) whose solution can be proven to exist.

Example 7.7.3 (The KdV equation.) In this example we consider the solution of the Korteweg-deVries (KdV) equation

$$q_t + q_{xxx} + 6qq_x = 0 \qquad -\infty < x < \infty, \quad t > 0 \qquad (7.7.8)$$

where we assume that $q(x, 0)$ is sufficiently differentiable and it satisfies $\int_{-\infty}^{\infty}(1 + x^2)|q(x)|\,dx < \infty$. This equation was first derived in 1895 as an approximation to the equations of water waves. The KdV equation has subsequently appeared in many physical applications. In the context of water waves, it describes the amplitude of long waves of small amplitude propagating in shallow water (it is assumed that viscosity is negligible and that waves propagate only in one direction).

It turns out that there is an intimate relationship between the KdV equation and the time independent Schrödinger equation, which was discussed in Example 7.7.2. Assume that the potential q appearing in Eq. (7.7.1) evolves in time according to the KdV equation. Then, the remarkable fact is that the evolution of the scattering data: $R(k)$, c_n, and k_n is very simple; namely (Ablowitz and Segur, 1981)

$$\frac{d}{dt}k_n = 0, \qquad \frac{d}{dt}R(k, t) = 8ik^3 R(k, t), \qquad \frac{d}{dt}c_n(t) = 8ik_n^3 c_n(t) \quad (7.7.9)$$

This means that Eqs. (7.7.6) and (7.7.7), with $R(k)$ and c_n replaced by $R(k, 0)$ e^{8ik^3t} and $c_n(0)e^{8ik_n^3t}$, respectively, can be used to integrate the KdV equation!

In particular, given initial data $q(x, 0)$, one first computes $R(k, 0), c_n(0)$, and k_n. Then Eq. (7.7.7), where Ψ^+ is found by solving the RH problem (7.7.6), describes the evolution of $q(x, t)$.

Before computing some explicit solutions we give an interpretation of the above construction. Suppose that the amplitude $q(x, t)$ of the water wave described by the KdV is frozen at a given instant of time. Knowing the scattering data, we can reconstruct $q(x, t)$ from knowledge of how particles would scatter off $q(x, t)$. In other words, the scattering data provides an alternative description of the wave at a fixed time. The time evolution of the water wave satisfies the KdV equation, which is a nonlinear equation. The above alternative description of the shape of the wave is useful because the evolution of the scattering data is linear, that is, Eq. (7.7.9). This highly nontrivial change of variables, from the physical to scattering space, provides a linearization of the KdV equation.

We now consider the special case when $R = 0$. Then the RH problem (7.7.6) reduces to

$$\Phi^+(x, t, k) - \Psi^+(x, t, -k) = -\sum_{n=1}^{N} \frac{c_n(0)e^{2ik_n x + 8ik_n^3 t}\Psi^+(x, t, k_n)}{k - k_n},$$

$$-\infty < k < \infty \qquad (7.7.10)$$

where we have integrated the equation for $c_n(t)$ in Eq. (7.7.9). The function $\Psi^+(x, t, k)$ is a $\oplus$ function (with respect to k), thus by the symmetry principle, $\Psi^+(x, t, -k)$ is a $\ominus$ function. The function on the right-hand side of Eq. (7.7.10) is also a $\ominus$ function because $k_n = ip_n$, $p_n > 0$, and the poles lie on the upper half plane. Thus the solution of Eq. (7.7.10) is immediately found to be

$$\Psi^+(x, t, -k) = 1 + \sum_{j=1}^{N} \frac{c_j(0)e^{2ik_j x + 8ik_j^3 t}\Psi^+(x, t, k_j)}{k - k_j}$$

and $\Phi^+(x, t, k) = 1$. (Recall the boundary conditions that $\Phi(x, k)$ and $\Psi^+(x, k)$ go to 1 as $|k| \to \infty$.) Evaluating the above equation at $k = -k_n$, we find the following linear system of algebraic equations for the vector $\Psi^+(k_n, x, t)$ for $n = 1, \ldots, N$:

$$\Psi^+(x, t, k_n) = 1 - \sum_{j=1}^{N} \frac{c_j(0)e^{2ik_j x + 8ik_j^3 t}\Psi^+(x, t, k_j)}{k_n + k_j} \qquad (7.7.11)$$

The $N \times N$ linear algebraic system (7.7.11) can be shown to be solvable because the relevant matrix is positive definite. Having obtained $\{\Psi^+(x, t, k_n)\}_{n=1}^N$, $q(x, t)$ follows from Eq. (7.7.7) with $R(k) = 0$:

$$q(x, t) = 2i \frac{\partial}{\partial x} \sum_{n=1}^N c_n(0) e^{8ik_n^3 t} \Psi^+(x, t, k_n) \qquad (7.7.12)$$

This solution is known as an N-soliton solution. As $t \to \infty$, it describes the interaction of N exponentially localized traveling waves. These localized waves have the remarkable property that they regain their initial amplitude and velocity upon interaction (i.e., they behave like particles).

The simplest situation occurs when $N = 1$, which is called the one-soliton solution. From Eq. (7.7.11) we have

$$\Psi^+(x, k_1) = \frac{1}{1 + \frac{c_1(0) e^{8ik_1^3 t}}{2k_1} e^{2ik_1 x}}$$

Calling $k_1 = ip_1$, p_1 real, the one-soliton solution is found from Eq. (7.7.12) to be

$$q(x, t) = 2p_1^2 \operatorname{sech}^2 \left(p_1 \left(x - 4p_1^2 t \right) - \alpha_0 \right) \qquad (7.7.13)$$

where α_0 is defined by $e^{2\alpha_0} = c_1(0)/(2k_1)$.

Example 7.7.4 (A $\bar{\partial}$ problem for a 2-spatial-dimensional generalization of the KdV equation – the so-called Kadomtsev–Petviashvili (KP) equation.)
We consider the KP equation

$$(q_t + 6qq_x + q_{xxx})_x + 3q_{yy} = 0, \qquad -\infty < x, y < \infty, \quad t > 0 \quad (7.7.14)$$

and we assume that q is real, $q(x, y, 0)$ is sufficiently smooth and bounded, and $q(x, y, 0) \in L_2(\mathbb{R}^2) \cap L_\infty(\mathbb{R}^2)$. This equation governs weakly dispersive nonlinear waves with slow transverse variations. It is a natural generalization of the one-dimensional KdV equation and arises in many physical applications such as water waves, stratified fluids, plasma physics, etc. We also note that the q_{yy} term can have a minus sign (formally, $y \mapsto iy$) but in that case the inverse scattering is quite different. Equation (7.7.14) is sometimes referred to as the KPII equation.

The KP equation is associated with the following linear "scattering" problem:

$$-\psi_y + \psi_{xx} + 2ik\psi_x + q(x, y)\psi = 0, \qquad -\infty < x < \infty, \quad -\infty < y < \infty$$
$$(7.7.15)$$

In what follows we sketch the essential ideas in the method of solution of the KP equation. (The main results are Eqs. (7.7.18) and (7.7.20) below.)

By considering the integral equation that governs a solution $\psi(x, y, k)$ ($\psi(x, y, k) \to 1$ as $|k| \to \infty$), one can derive a $\bar{\partial}$ equation that ψ satisfies. Namely, by taking

$$\frac{\partial}{\partial \bar{k}} = \frac{1}{2}\left(\frac{\partial}{\partial k_R} + i\frac{\partial}{\partial k_I}\right)$$

of the integral equation ($k = k_R + ik_I$)

$$\psi(x, y, k) = 1 + \int_{-\infty}^{\infty}\int_{-\infty}^{\infty} G(x - x', y - y', k)\psi(x', y', k)\, dx'\, dy' \quad (7.7.16)$$

where the Green's function is given by

$$G(x, y, k) = \frac{1}{(2\pi)^2}\int_{-\infty}^{\infty}\int_{-\infty}^{\infty} \frac{e^{i\xi x + i\eta y}}{\xi^2 + 2\xi k + i\eta}\, d\xi\, d\eta \quad (7.7.17)$$

one can establish (cf. Ablowitz and Clarkson, 1991) the following $\bar{\partial}$ equation for ψ:

$$\frac{\partial}{\partial \bar{k}}\psi(x, y, k) = \frac{\operatorname{sgn} k_R}{2\pi} e^{-2i(x - 2k_I y)k_R}\, T(k_R, k_I)\, \bar{\psi}(x, y, k) \quad (7.7.18)$$

where $T(k_R, k_I)$ plays the role of scattering data and where $\psi \to 1$ as $k \to \infty$. (Note that $\bar{\psi}(x, y, k) = \psi(x, y, -k_R, k_I)$.) Equation (7.7.18) is exactly in the form studied in Section 7.6. It is therefore a generalized analytic function and Eq. (7.7.18) is uniquely solvable given appropriate (decaying) scattering data $T(k_R, k_I)$.

A reconstruction formula for $q(x, y)$ is obtained by considering asymptotic formulae for $\psi(x, y, k)$ as $|k| \to \infty$ in an analogous way as done for one-dimensional problems. As $|k| \to \infty$, $\psi(x, y, k)$ has the expansion

$$\psi(x, y, k) \sim 1 + \frac{\psi^{(1)}(x, y)}{k} + \cdots$$

From Eq. (7.7.15) we find that $q(x, y) = -2i\psi_x^{(1)}$. On the other hand, Eq. (7.7.18) is equivalent to the integral equation (see, e.g., Eq. (7.6.6))

$$\psi(x, y, k) = 1 - \frac{1}{\pi}\int_{-\infty}^{\infty}\int_{-\infty}^{\infty} \frac{\frac{\partial \psi}{\partial \bar{z}}(x, y, z)}{z - k}\, dz_R\, dz_I \quad (7.7.19)$$

where $\partial\psi/\partial\bar{z}(x, y, z)$ satisfies Eq. (7.7.18). As $|k| \to \infty$ this integral equation simplifies:

$$\psi(x, y, k) \sim 1 + \frac{1}{\pi k} \int_{-\infty}^{\infty} \int_{-\infty}^{\infty} \frac{\partial\psi}{\partial\bar{z}}(x, y, z)\, dz_R\, dz_I$$

hence $q(x, y)$ is given by

$$q(x, y) = -\frac{2i}{\pi} \frac{\partial}{\partial x} \int_{-\infty}^{\infty} \int_{-\infty}^{\infty} dk_R\, dk_I\, \frac{\partial\psi}{\partial\bar{k}}(x, y, k_R, k_I) \tag{7.7.20}$$

As with the KdV equation, if q satisfies the KP equation, it can be shown that the evolution of the scattering data, $T(k_R, k_I)$ which we now call: $T(k_R, k_I, t)$, evolves simply:

$$\frac{dT(k_R, k_I, t)}{dt} = -4i(k^3 + \bar{k}^3)T(k_R, k_I, t) \tag{7.7.21}$$

Hence Eqs. (7.7.18) and (7.7.20), with T replaced by $T(k_R, k_I, 0)e^{-4i(k^3+\bar{k}^3)t}$, provide the solution of the KP equation.

Example 7.7.5 (A Special Solution for the Painlevé IV Equation: PIV) The mathematical and physical significance of the Painlevé equations was briefly discussed in Section 3.7. These equations were studied extensively by mathematicians at the turn of the century, and in recent years considerable interest in Painlevé equations has developed because of the deep connection with integrable equations and physical phenomena, for example, spin systems, relativity, field theory, and quantum gravity theory where special solutions of the PIV equation have appeared. In this example we discuss how these special solutions can be obtained via RH theory.

We consider the following special case of the PIV equation (see PIV in Section 3.7 where the parameters a and b are given by $a = -2\theta + 1$ and $b = -8\theta^2$)

$$q_{tt} = \frac{q_t^2}{2q} + \frac{3}{2}q^3 + 4tq^2 + 2(t^2 + 2\theta - 1)q - \frac{8\theta^2}{q}, \qquad 0 < \mathrm{Re}\,\theta < \frac{1}{2} \tag{7.7.22}$$

It turns out that the Painlevé equations can be solved via RH problems. The conceptual framework is similar to that discussed for the KdV equation; that is, one can relate the PIV equation to a linear problem that is intimately connected to a RH problem. (In this case, the linear problem is an ODE, in an auxiliary variable, say z, whose coefficients depend on the variable t in PIV. Such ODEs

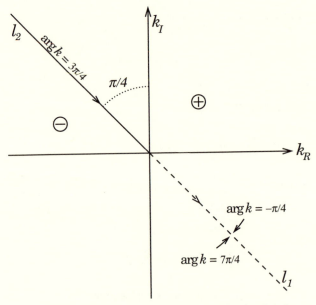

Fig. 7.7.1. RH problem defined on contour $l = l_1 + l_2$

are often called Monodromy problems, and the reconstruction procedure is referred to as the Inverse Monodromy Transform, IMT.) For a special choice of initial data, one can establish that a solution of PIV is expressible as

$$q(t) = -2t - \frac{\partial}{\partial t} \log\left[\lim_{k\to\infty} \left(\frac{1}{k}\right)^\theta (k\psi_{12}(t,k)) \right] \qquad (7.7.23)$$

where ψ_{12} is the 12 component of a matrix $\psi(t,k)$, which is found from a RH problem defined on the contour $\ell = \ell_1 + \ell_2$ where ℓ_2 is a line along the ray $\arg k = 3\pi/4$ and along line ℓ_1 we will have a branch cut with the rays $\arg k = -\pi/4, 7\pi/4$ on either side of the cut as depicted in Figure 7.7.1.

This RH problem is given by

$$\Psi^+(t,k) = \Psi^-(t,k) \begin{pmatrix} a(t,k) & b(t,k) \\ 0 & c(t,k) \end{pmatrix} \qquad (7.7.24)$$

where

$$\Psi(t,k) \to \begin{pmatrix} \left(\frac{1}{k}\right)^\theta & 0 \\ 0 & \left(\frac{1}{k}\right)^{-\theta} \end{pmatrix} \qquad \text{as } k \to \infty$$

and where

$$\{a, b, c\} = \begin{cases} \{e^{2i\pi\theta}, e^{2Q-2i\pi\theta}, e^{-2i\pi\theta}\} & \text{on } l_1, \\ \{1, e^{2Q}, 1\} & \text{on } l_2 \end{cases} \qquad Q = \frac{k^2}{2} + kt \quad (7.7.25)$$

The jump matrix of the above RH problem is upper triangular; hence we can reduce this matrix RH problem to a solvable system. Even though the coefficients a, b, c are discontinuous at $k = 0$ (and do not satisfy the Hölder condition), we can solve this problem in a manner analogous to the way we solved RH problems with open contours. In fact, discontinuous problems such as these can be thought of as the sum of two RH problems with open contours ℓ_1 and ℓ_2.

Letting $\Psi = (\Psi_1, \Psi_2)$, where Ψ_1 and Ψ_2 are each two-component vectors, Eq. (7.7.24) reduces to the vector equations

$$\Psi_1^+ = a\Psi_1^- \tag{7.7.26a}$$

$$\Psi_2^+ = c\Psi_2^- + b\Psi_1^- \tag{7.7.26b}$$

Let us consider the equation for Ψ_1 in Eq. (7.7.26a). We note that the boundary condition

$$\Psi_1 \sim \left(\frac{1}{k}\right)^\theta \begin{pmatrix} 1 \\ 0 \end{pmatrix}$$

in fact satisfies the homogeneous RH problem exactly. Namely, across ℓ_2 there is no jump; hence $\Psi_1^+ = \Psi_1^-$, and across ℓ_1,

$$\Psi_1^- = \left(\frac{1}{|k|}\right)^\theta e^{-7i\pi\theta/4}, \qquad \Psi_1^+ = \left(\frac{1}{|k|}\right)^\theta e^{-i\pi\theta/4}$$

hence

$$\Psi_1^+ = e^{2i\pi\theta}\Psi_1^-$$

thus

$$\Psi_1 = \left(\frac{1}{k}\right)^\theta \begin{pmatrix} 1 \\ 0 \end{pmatrix}$$

Next we consider the equation for Ψ_2 in Eq. (7.7.26b). The same arguments we just used for Ψ_1 hold for the homogeneous solution of Eq. (7.7.26b), that is, $\Psi_{2,H}^+ = c\Psi_{2,H}^-$, so that

$$\Psi_{2,H} = \left(\frac{1}{k}\right)^{-\theta} \begin{pmatrix} 0 \\ 1 \end{pmatrix}$$

Because Ψ_1 has a nontrivial contribution only in its first component, we have a scalar equation to solve:

$$\Psi_2^{+(1)} = c\Psi_2^{-(1)} + b\Psi_1^{-(1)} \qquad \text{along } \ell = \ell_1 + \ell_2$$

where the superscript $^{(1)}$ stands for the first component of $\Psi_2 = (\Psi_2^{(1)}, \Psi_2^{(2)})^+$. Using $c = \Psi_{2,H}^{+}{}^{(2)}/\Psi_{2,H}^{-}{}^{(2)}$, we obtain

$$\frac{\Psi_2^{+(1)}}{\Psi_{2,H}^{+}{}^{(2)}} - \frac{\Psi_2^{-(1)}}{\Psi_{2,H}^{-}{}^{(2)}} = b\frac{\Psi_1^{-(1)}}{\Psi_{2,H}^{+}{}^{(2)}} \qquad \text{along } \ell = \ell_1 + \ell_2$$

and by the Plemelj formulae (note that b depends explicitly on t)

$$\Psi_2^{(1)}(t, k) = \frac{\Psi_{2,H}^{(2)}(k)}{2\pi i}\int_{\ell=\ell_1+\ell_2}\frac{b\Psi_1^{-(1)}(k')}{\Psi_{2,H}^{+}{}^{(2)}(k')(k' - k)}\,dk'$$

Substituting $\Psi_{2,H}^{(2)}(k) = (\frac{1}{k})^{-\theta}$ and $\Psi^{(1)}(k) = (\frac{1}{k})^{\theta}$, where we understand that $k = |k|e^{7i\pi/4}$ along ℓ_1 in $\Psi_1^{-(1)}(k)$, we find

$$\Psi_2^{(2)}(t, k) = \frac{1}{2\pi i}\left(\frac{1}{k}\right)^{-\theta}\int_\ell \frac{e^{2Q(t,k')}\left(\frac{1}{k'}\right)^{2\theta}}{k' - k}\,dk' \qquad (7.27)$$

and therefore the matrix $\Psi(t, k) = (\Psi_1, \Psi_2)(t, k)$ is given by

$$\Psi(t, k) = \begin{pmatrix} \left(\frac{1}{k}\right)^{\theta} & \frac{1}{2\pi i}\left(\frac{1}{k}\right)^{-\theta}\int_\ell \frac{e^{2Q(t,k')}\left(\frac{1}{k'}\right)^{2\theta}}{k'-k}dk' \\ 0 & \left(\frac{1}{k}\right)^{-\theta} \end{pmatrix} \qquad (7.28)$$

Equation (7.7.23) therefore yields the special solution of PIV:

$$q(t) = -2t - \frac{1}{2\pi i}\frac{\partial}{\partial t}\log\int_\ell e^{2Q(t,k)}\left(\frac{1}{k}\right)^{2\theta}dk \qquad (7.29)$$

We note that the integral representation, $\int_\ell e^{2Q(t,k)}\left(\frac{1}{k}\right)^{2\theta}dk$, is expressible in terms of a classical special function, the Weber–Hermite function, and hence this special solution of PIV is also related to these functions.

Problems for Section 7.7

1. Consider the linear matrix eigenvalue equation

$$\mu_x + ik[\sigma_3, \mu] = Q\mu \qquad (1)$$

where $[A, B] \equiv AB - BA$,

$$\sigma_3 = \begin{pmatrix} 1 & 0 \\ 0 & -1 \end{pmatrix}, \qquad Q = \begin{pmatrix} 0 & q \\ \bar{q} & 0 \end{pmatrix}$$

and where the function $q(x)$ and all its derivatives are absolutely integrable. Let Φ and Ψ be defined by

$$\Phi = I + \int_{-\infty}^{x} e^{-ik(x-\xi)\sigma_3} Q \Phi e^{ik(x-\xi)\sigma_3} \, d\xi$$

$$\Psi = I - \int_{x}^{\infty} e^{-ik(x-\xi)\sigma_3} Q \Psi e^{ik(x-\xi)\sigma_3} \, d\xi$$

(a) Show that Φ and Ψ satisfy (1)

Hint: Note that the transformation $\mu = W e^{ikx\sigma_3}$ transforms equation (1) to

$$W_x + ik\sigma_3 W = QW \tag{2}$$

(b) Show that

$$\Phi = (\Phi^+, \Phi^-) = \begin{pmatrix} \Phi_1^+ & \Phi_1^- \\ \Phi_2^+ & \Phi_2^- \end{pmatrix}$$

$$\Psi = (\Psi^-, \Psi^+) = \begin{pmatrix} \Psi_1^- & \Psi_1^+ \\ \Psi_2^- & \Psi_2^+ \end{pmatrix},$$

where $+$ and $-$ denote analyticity in the upper and lower half k plane, respectively.

Hint: Note that $e^{\alpha \sigma_3} = \mathrm{diag}(e^{\alpha}, e^{-\alpha})$

(c) Show that

$$\det \Phi = \det \Psi = 1$$

and

$$\Phi_2^-(x, k) = \overline{\Phi_1^+(x, \bar{k})}, \qquad \Phi_1^-(x, k) = \overline{\Phi_2^+(x, \bar{k})}$$

$$\Psi_2^+(x, k) = \overline{\Psi_1^-(x, \bar{k})}, \qquad \Psi_1^+(x, k) = \overline{\Psi_2^-(x, \bar{k})}$$

2. (a) Show that the eigenfunctions Φ and Ψ introduced in Problem 7.7.1 satisfy

$$\Phi - \Psi = \Phi e^{-ikx\sigma_3} T e^{+ikx\sigma_3}, \qquad k \in \mathbb{R}$$

where

$$T(k) = \int_{-\infty}^{\infty} e^{ikx\sigma_3} Q(x) \Psi(x,t) e^{-ikx\sigma_3} \, dx, \qquad k \in \mathbb{R}$$

Hint: If W_1 and W_2 satisfy equation (2) of Problem 7.7.1, then $W_2 = W_1 C$, where C is an x-independent matrix.

(b) Establish that

$$\det(I - T(k)) = 1, \qquad k \in \mathbb{R}$$

and

$$T_{21}(k) = \overline{T_{12}}(k), \qquad k \in \mathbb{R}, \quad T_{22}(k) = \overline{T_{11}(\bar{k})}, \quad \text{Im } k > 0$$

where $T_{11}, T_{12}, T_{21}, T_{22}$ denote the $11, 12, 21, 22$ entries of the matrix T.

Hint: Use (c) of problem 7.7.1.

(c) Let $\alpha \equiv T_{12}, \beta \equiv T_{22}$. Use (b) to show that

$$(1 - \beta(k))(1 - \bar{\beta}(k)) = 1 + |\alpha(k)|^2, \qquad k \in \mathbb{R}$$

(d) This equation, together with $\beta(k) = O(\frac{1}{k})$ as $|k| \to \infty$, define a scalar RH problem for $\beta(k)$. Show that if $\beta(k)$ has no zeroes in the upper half plane, then

$$1 - \beta(k) = \exp\left[\frac{1}{2i\pi} \int_{-\infty}^{\infty} \frac{\log(1 + |\alpha(l)|^2)}{l - k} \, dl \right], \qquad \text{Im } k > 0$$

Hint: Note that $\beta(k)$ is analytic in the upper half plane, while $\overline{\beta(\bar{k})}$ is analytic in the lower half plane.

3. (a) Rearrange the equation obtained in part (a) of Problem 7.7.2 to show that

$$\left(\Psi^-, \frac{\Phi^-}{1 - \bar{\beta}} \right) = \left(\frac{\Phi^+}{1 - \beta}, \Psi^+ \right) \begin{pmatrix} 1 & \frac{\alpha}{1 - \bar{\beta}} e^{-2ikx} \\ -\frac{\bar{\alpha}}{1 - \beta} e^{2ikx} & 1 - \frac{|\alpha|^2}{|1 - \beta|^2} \end{pmatrix},$$

$$k \in \mathbb{R}$$

(b) This equation, together with $\Phi \to I + O(\frac{1}{k}), \Psi \to I + O(\frac{1}{k})$ as $|k| \to \infty$ defines a matrix RH problem. Show that this RH problem is equivalent to

$$\begin{pmatrix} \overline{\Psi_2^+(x, \bar{k})} \\ \overline{\Psi_1^+(x, \bar{k})} \end{pmatrix} = \begin{pmatrix} 1 \\ 0 \end{pmatrix} + \frac{1}{2\pi i} \int_{-\infty}^{\infty} \frac{e^{2ilx}}{l - k} \frac{\bar{\alpha}(l)}{1 - \beta(l)} \begin{pmatrix} \Psi_1^+(x, l) \\ \Psi_2^+(x, l) \end{pmatrix} dl,$$

$$\text{Im } k < 0$$

(c) Use the result of Problem 7.5.1 to show that the above RH problem is always solvable.

4. Combine the results of Problems 1–3 above to show that

$$q(x) = \frac{1}{\pi} \int_{-\infty}^{\infty} e^{-2ikx} \frac{\alpha(k)}{1 - \overline{\beta(k)}} \overline{\Psi_2^+}(x, k) \, dk$$

Hint: Equation (1) of Problem 7.7.1 implies $Q = i \lim_{k \to \infty} k[\sigma_3, \mu]$ or $q = 2i \lim_{k \to \infty} k\Psi_1^+$. Compute this limit by using (b) of Problem 7.7.3.

5. In Section 7.4 a scalar RH problem was used to derive the Fourier transform. Use the results of Problems 7.7.1–7.7.4 to derive the following *nonlinear Fourier transform.*

   Let $q(x) \in \mathcal{S}(\mathbb{R})$. (That is, $\mathcal{S}(\mathbb{R})$ denotes space where q and all its derivatives are absolutely integrable.) Given q, obtain $\Psi_1(x, k)$ and $\Psi_2(x, k)$ as the solution of

$$\Psi_1(x, k) = -\int_x^{\infty} e^{2ik(\xi - x)} q(\xi) \Psi_2(\xi, k) \, d\xi$$

$$\Psi_2(x, k) = 1 - \int_x^{\infty} \overline{q}(\xi) \Psi_1(\xi, k) \, d\xi, \qquad \text{Im } k > 0$$

A nonlinear Fourier transform of q denoted by $\alpha(k)$ is defined by

$$\alpha(k) \equiv \int_{-\infty}^{\infty} e^{2ikx} q(x) \Psi_2(x, k) \, dx \tag{1}$$

Conversely, given $\alpha(k) \in S(\mathbb{R})$ obtain $\beta(k)$, $\Psi_1(x, k)$ and $\Psi_2(x, k)$ by

$$1 - \beta(k) = \exp\left[\frac{1}{2\pi i} \int_{-\infty}^{\infty} dl \frac{\log(1 + |\alpha(l)|^2)}{l - k} \right], \qquad \text{Im } k > 0$$

and

$$\begin{pmatrix} \overline{\Psi_2(x, \overline{k})} \\ \overline{\Psi_1(x, \overline{k})} \end{pmatrix} = \begin{pmatrix} 1 \\ 0 \end{pmatrix} + \frac{1}{2\pi i} \int_{-\infty}^{\infty} \frac{e^{2ilx}}{l - k} \frac{\overline{\alpha}(l)}{1 - \beta(l)} \begin{pmatrix} \Psi_1(x, l) \\ \Psi_2(x, l) \end{pmatrix} dl,$$

$$\text{Im } k < 0$$

A nonlinear inverse Fourier transform of $\alpha(k)$ is defined by

$$q(x) \equiv \frac{1}{\pi} \int_{-\infty}^{\infty} e^{-2ikx} \frac{\alpha(k)}{1 - \overline{\beta(k)}} \overline{\Psi_2}(x, k) \, dk \tag{2}$$

6. Show that the *defocusing* nonlinear Schrödinger equation (NLS)

$$iq_t + q_{xx} - 2|q|^2 q = 0, \qquad -\infty < x < \infty,$$
$$0 < t < \infty, \quad q(x,0) = q_0(x)$$

is the compatibility condition (1) of Problem 7.7.1 and of

$$\mu_t + 2ik^2[\sigma_3, \mu] = (i\lambda|q|^2\sigma_3 - 2kQ + iQ_x\sigma_3)\mu \qquad (1)$$

Hint: use

$$\frac{\partial^2 \mu}{\partial x \partial t} = \frac{\partial^2 \mu}{\partial t \partial x}$$

7. (a) Show that if μ satisfies equation (1) of Problem 7.7.1 and equation (1) of Problem 7.7.6 and if $T(k), \alpha(k)$, and $\beta(k)$ are defined as in Problem 7.7.2, then

$$T_t + 2ik^2[\sigma_3, T] = 0$$

that is

$$\beta_t = 0, \ \alpha_t + 4ik^2\alpha = 0$$

Hint: Use $I - T = \lim_{x \to -\infty} e^{ikx\sigma_3} \Psi(x, k) e^{-ikx\sigma_3}$.

   (b) Deduce that the nonlinear Fourier transform defined in Problem 7.7.5 can be used to solve the initial-value problem of the NLS equation as follows: Let $\alpha(k, 0)$ be defined by equation (1) of Problem 7.7.5 where $q(x)$ is replaced by $q_0(x)$. Define $q(x, t)$ equation (2) of Problem 7.7.5 where $\alpha(k)$ is replaced by $\alpha(k, 0)e^{-4ik^2t}$. Then $q(x, t)$ solves the NLS equation with $q(x, 0) = q_0(x)$.

8. Consider the following linear eigenvalue system of equations:

$$\frac{\partial \mu_1}{\partial \bar{z}} = \tfrac{1}{2}q\mu_2, \qquad \frac{\partial(\overline{\mu_2 e^{kz}})}{\partial \bar{z}} = \tfrac{1}{2}q(\overline{\mu_1 e^{kz}}),$$
$$z = x + iy, \quad -\infty < x, y < \infty$$

where $q(x, y)$ and all its derivatives are absolutely integrable. Show that the unique solution of these equations that is bounded for all complex k and that satisfies the boundary condition

$$(\mu_1, \mu_2)^T \to (1, 0)^T \quad \text{as} \quad |x| + |y| \to \infty$$

is given by

$$\mu_1(x, y; k_R, k_I)$$

$$= 1 + \frac{1}{2\pi} \iint_{R^2} \frac{q(\xi, \eta)\mu_2(\xi, \eta, k_R, k_I)}{z - \zeta} \, d\xi \, d\eta, \qquad \zeta = \xi + i\eta$$

$$\overline{\mu_2}(x, y; k_R, k_I)$$

$$= \frac{1}{2\pi} \iint_{R^2} \frac{e^{\bar{k}(\bar{\zeta} - \bar{z}) - k(\zeta - z)} q(\xi, \eta)\overline{\mu_1}(\xi, \eta, k_R, k_I)}{z - \zeta} \, d\xi \, d\eta$$

Hint: see Problem 7.6.4.

9. (a) Show that the eigenvectors $(\mu_1, \mu_2)^T$ defined in Problem 7.7.8 satisfy the $\bar{\partial}$ equation

$$\frac{\partial}{\partial \bar{k}}(\mu_1 \pm \mu_2) = e^{\bar{k}\bar{z} - kz} \overline{\alpha} \overline{(\mu_2 \pm \mu_1)}$$

where

$$\alpha(k_I, k_R) \equiv \frac{1}{2\pi} \iint_{R^2} e^{-2i(k_I x + k_R y)} q(x, y)\overline{\mu_1}(x, y, k_R, k_I) \, dx \, dy$$

(b) The above equations, together with the limit $(\mu_1, \mu_2)^T \to (1, 0)$ as $k \to \infty$, define a $\bar{\partial}$ problem. Show that the unique solution of this $\bar{\partial}$ problem is

$$(\mu_1 \pm \mu_2)(x, y; k_R, k_I)$$

$$= 1 + \frac{1}{\pi} \iint_{R^2} \frac{e^{\bar{l}\bar{z} - lz}}{k - l} \overline{\alpha}(l_I, l_R) \overline{(\mu_2 \pm \mu_1)}(x, y, l_R, l_I) \, dl_R \, dl_I$$

where $l = l_R + i l_I$.

Hint: use the fact that $\mu_1 \pm \bar{\mu}_2$ is a generalized analytic function.

10. Combine the results of Problems 7.7.8 and 7.7.9 to show that

$$q(x, y) = \frac{2}{\pi} \iint_{R^2} e^{2i(k_I x + k_R y)} \alpha(k_I, k_R)\mu_1(x, y, k_R, k_I) \, dk_R \, dk_I$$

Hint: Use $\bar{q} = 2 \lim_{k \to \infty} k\bar{\mu}_2$.

Note: The expressions for $\alpha(k_I, k_R)$ in part (a) of Problem 7.7.9 and $q(x, y)$ above define a nonlinear two-dimensional Fourier transform. These equations can be used to solve a certain two-dimensional generalization of the nonlinear Schrödinger equation.

11. Let $\mu^+(k)$ and $\mu^-(k)$ be analytic in the upper and lower half planes, respectively, and let $\mu^\pm = 1 + O\left(\frac{1}{k}\right)$ as $|k| \to \infty$. Suppose that μ^+ and μ^- are related to a function $\mu(k)$ by

$$\mu^+(k) = \mu(k) + \int_{-\infty}^{\infty} \alpha(k, l)\mu(l)\, dl$$

$$\mu^-(k) = \mu(k) + \int_{-\infty}^{\infty} \beta(k, l)\mu(l)\, dl$$

where α and β are given functions with suitable decay properties. Show that μ satisfies the following linear integral equation:

$$\mu(k) = 1 + \iint_{R^2} \left(\frac{\alpha(k', l)}{k - (k' - i0)} - \frac{\beta(k', l)}{k - (k' + i0)} \right)\mu(l)\, dk\, dl$$

Hint: Note that because $\mu^+ - 1$ is a $\oplus$ function, $P^-(\mu^+ - 1) = 0$, where

$$(P^- f)(k) = \frac{1}{2i\pi} \int_{-\infty}^{\infty} \frac{f(l)}{l - (k - i0)}\, dl$$

similarly, $P^+(\mu^- - 1) = 0$.

Note: this result is useful for the linearization of the so-called Kadomtsev–Petviashvili I equation.

Appendix A
Answers to Odd-Numbered Exercises

Chapter 1

Section 1.1, Page 3

1. (a) $e^{2\pi i k}$, $\quad k = 0, \pm 1, \pm 2, \ldots$

   (b) $e^{\frac{3\pi i}{2} + 2\pi i k}$, $\quad k = 0, \pm 1, \pm 2, \ldots$

   (c) $\sqrt{2}\, e^{i\left(\frac{\pi}{4}\right) + 2k\pi i}$, $\quad k = 0, \pm 1, \pm 2, \ldots$

   (d) $e^{i\left(\frac{\pi}{3}\right) + 2k\pi i}$, $\quad k = 0, \pm 1, \pm 2, \ldots$

   (e) $e^{i\left(\frac{-\pi}{3}\right) + 2k\pi}$, $\quad k = 0, \pm 1, \pm 2, \ldots$

3. (a) $z_\kappa = 4^{1/3} e^{2\pi i \kappa/3}$, $\quad \kappa = 0, 1, 2$

   (b) $z_\kappa = e^{i(\pi + 2\kappa\pi)/4}$, $\quad \kappa = 0, 1, 2, 3$

   (c) $z_\kappa = \dfrac{c^{1/3} e^{2\pi i \kappa/3} - b}{a}$, $\quad \kappa = 0, 1, 2$

   (d) $z_1 = 2^{1/4} e^{i3\pi/8}$, $\quad z_2 = 2^{1/4} e^{i11\pi/8}$, $\quad z_3 = 2^{1/4} e^{5i\pi/8}$, $\quad z_4 = 2^{1/4} e^{13i\pi/8}$

Section 1.2, Page 8

1. (a) compact;

   (b) open, bounded;

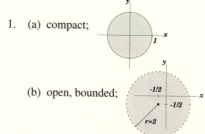

627

(c) closed, unbounded;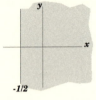

(d) closed, unbounded;

(e) bounded, not open or closed;

5. (a) $\displaystyle\sum_{j=0}^{\infty} \frac{(-1)^j z^{2j}}{(2j+1)!}$
    (b) $\displaystyle\sum_{j=0}^{\infty} \frac{z^{2j}}{(2j+2)!}$
    (c) $\displaystyle\sum_{j=0}^{\infty} \frac{z^{j+1}}{(j+2)!}$

11. circle passing through the north pole

Section 1.3, Page 20

1. (a) 0
    (b) $\left(\dfrac{1}{z_o}\right)^m$
    (c) $i\sin(1)$
    (d) 1

    (e) doesn't exist
    (f) $\dfrac{1}{9}$
    (g) 0

Chapter 2

Section 2.1, Page 32

1. (a) C-R not satisfied
    (b) C-R satisfied; $f(z) = i(z^3 + 2)$
    (c) C-R not satisfied

3. (a) analytic except at singular points $z = (\pi/2 + n\pi), n \in \mathbf{Z}$

    (b) entire; singularity at ∞

    (c) analytic except at singular point $z = 1$

(d) analytic nowhere

(e) analytic except at $z = \exp(i(\pi/4 + n\pi/2))$, $n = 0, 1, 2, 3$

(f) entire; singularity at ∞

Section 2.2, Page 46

1. (a) Branch points at $z = 1, \infty$. Possible branch cuts: (i)$\{z : \mathrm{Re}\, z \geq 1, \mathrm{Im}\, z = 0\}$ obtained by letting $z - 1 = re^{i\theta}$, $0 \leq \theta < 2\pi$; (ii)$\{z : \mathrm{Re}\, z \leq 1, \mathrm{Im}\, z = 0\}$ obtained by letting $z - 1 = re^{i\theta}$, $-\pi \leq \theta < \pi$.

   (b) Branch points at $z = -1+2i, \infty$. Possible branch cut: $\{z : \mathrm{Re}\, z \geq -1, \mathrm{Im}\, z=2\}$ obtained by letting $z + 1 - 2i = re^{i\theta}$, $0 \leq \theta < 2\pi$.

   (c) Branch points at $z = 0, \infty$. Possible branch cut: $\{z : \mathrm{Re}\, z \geq 0, \mathrm{Im}\, z = 0\}$ obtained by letting $z = re^{i\theta}$, $0 \leq \theta < 2\pi$.

   (d) Branch points at $z = 0, \infty$. Possible branch cut: $\{z : \mathrm{Re}\, z \geq 0, \mathrm{Im}\, z = 0\}$ obtained by letting $z = re^{i\theta}$, $0 \leq \theta < 2\pi$.

3. (a) $z = e^{2\pi i n/5}$, $n = 0, 1, 2, 3, 4$ (b) $z = \frac{1}{e} + i$ (c) $z = \pi/4$

7. $\begin{aligned} \phi &= \kappa \log \rho \\ \psi &= \kappa\theta \end{aligned}$ where $z - z_o = \rho e^{i\theta}$

Section 2.3, Page 61

1. (a) Branch points at $z = \pm i$. Branch cut: $\{z : \mathrm{Re}\, z = 0, -1 \leq \mathrm{Im}\, z \leq 1\}$ obtained by letting $z - i = r_1 e^{i\theta_1}$ and $z + i = r_2 e^{i\theta_2}$ for $\frac{-3\pi}{2} \leq \theta_1, \theta_2 < \frac{\pi}{2}$.

   (b) Branch points at $z = -1, 2, \infty$. Branch cut: $\{z : \mathrm{Re}\, z \geq -1, \mathrm{Im}\, z = 0\}$ obtained by letting $z + 1 = r_1 e^{i\theta_1}$ and $z - 2 = r_2 e^{i\theta_2}$ for $0 \leq \theta_1, \theta_2 < 2\pi$.

3. Branch points at $z = \pm i, \infty$. Branch cut: $\{z : \mathrm{Re}\, z = 0, \mathrm{Im}\, z \leq 1\}$ obtained by letting $z + i = r_1 e^{i\theta_1}$ and $z - i = r_2 e^{i\theta_2}$ for $\frac{-\pi}{2} \leq \theta_1, \theta_2 < \frac{3\pi}{2}$.

Section 2.4, Page 70

1. (a) 0 (b) 0 (c) $2\pi i$

3. (a) 0 (b) πi (c) $8i$ (d) $4i$

11. (b) $\Gamma = 0$, $F = 2\pi\kappa$

Section 2.5, Page 81

1. (a) 0 (b) 0 (c) $2\pi i$ (d) 0 (e) 0 (f) 0

3. (a) 0 (b) $-2i$ (c) $-4i$ (d) $-2i$

Section 2.6, Page 91

1. (a) 0   (b) 0   (c) 0   (d) $2\pi i$ (e) $-2\pi i$

3. 0

Chapter 3

Section 3.1, Page 109

"UC" means uniformly convergent

1. (a) $\lim_{n\to\infty} \frac{1}{nz^2} = 0$; UC   (b) $\lim_{n\to\infty} \frac{1}{z^n} = 0$ for $1 < \alpha \le |z| \le \beta$; UC

   (c) $\lim_{n\to\infty} \sin\left(\frac{z}{n}\right) = 0$; UC   (d) $\lim_{n\to\infty} \frac{1}{1+(nz)^2} = 0$; UC

3. $\lim_{n\to\infty} \int_0^1 nz^{n-1} dz = 1$; $\int_0^1 \lim_{n\to\infty} nz^{n-1} dz = 0$; not a counterexample since convergence is not uniform.

Section 3.2, Page 114

1. (a) $R = 1$ (b) $R = \infty$ (c) $R = 0$ (d) $R = \infty$ (e) $R = e$

3. (a) $R = \pi/2$

   (b) $E_0 = 1$, $E_1 = 0$, $E_2 = -1$, $E_3 = 0$, $E_4 = 5$, $E_5 = 0$, $E_6 = -61$

7. $\log(1 + z) = \sum_{k=0}^{\infty} \frac{(-1)^k z^{k+1}}{k+1}$

9. $\sum_{k=m-1}^{\infty} \frac{k(k-1)\cdots(k-(m-2))z^{k-(m-1)}}{(m-1)!}$

Section 3.3, Page 127

1. (a) $\sum_{n=0}^{\infty} (-1)^n z^{2n}$ (b) $\sum_{n=0}^{\infty} \frac{(-1)^n}{z^{2n+2}}$

3. (a) $\left(\frac{2}{5} - \frac{i}{5}\right) \sum_{n=0}^{\infty} \frac{(2i)^n - 1}{2^n} z^n$

   (b) $\frac{-2+i}{5} \left(\sum_{n=0}^{\infty} \left(\frac{z}{2}\right)^n - (-1)^n \left(\frac{i}{z}\right)^{n+1} \right)$

   (c) $\frac{2-i}{5} \sum_{n=0}^{\infty} \left(\left(\frac{2}{z}\right)^{n+1} + (-1)^n \left(\frac{i}{z}\right)^{n+1} \right)$

Section 3.4, Page 137

1. (a) yes (b) yes (c) no (d) Cauchy sequence for Re $z > 0$.

Section 3.5, Page 144

1. (a) $z = 0$ is a removable singularity; $z = \infty$ is an essential singularity.

   (b) $z = 0$ is a simple pole; $z = \infty$ is an essential singularity.

   (c) $z = \frac{\pi}{2} + n\pi$ are essential singularities for $n \in \mathbf{Z}$; $z = \infty$ is a nonisolated singularity; it is a cluster point.

   (d) Simple poles at $z = -\frac{1}{2} \pm i\sqrt{\frac{3}{2}}, \infty$.

   (e) Branch points at $z = 0, \infty$. Depending on which branch of $z^{\frac{1}{3}}$ is chosen, may have simple pole at $z = 1$. Letting $z = re^{i\theta}$ for $0 \le \theta < 2\pi$, $z = 1$ is a removable singularity; letting $z = re^{i\theta}$ for $2\pi \le \theta < 4\pi$, $z = 1$ is a simple pole; letting $z = re^{i\theta}$ for $4\pi \le \theta < 6\pi$, $z = 1$ is a simple pole.

   (f) $z = 0, \infty$ are branch points. If we let $z = re^{i\theta}$ for $2\pi \le \theta < 4\pi$, then $z = 1$ is an additional branch point.

   (g) $|z| = 1$ is a boundary jump discontinuity.

   (h) $|z| = 1$ is a natural boundary.

   (i) Simple poles at $z = i(\pi/2 + n\pi)$ for $n \in \mathbf{Z}$, and $z = \infty$ is a cluster point.

   (j) Simple poles at $z = i/(n\pi)$ for $n \in \mathbf{Z}$, and $z = 0$ is a cluster point.

3. (a) Simple poles at $z = 2^{\frac{1}{4}} e^{i\pi(1+2n)/4}$, $n = 0, 1, 2, 3$, with strength $\frac{-1}{4\sqrt{2}}, \frac{1}{4\sqrt{2}}, \frac{-1}{4\sqrt{2}}, \frac{-1}{4\sqrt{2}}$, respectively.

   (b) Simple poles at $z = \pi/2 + n\pi$, $C_{-1} = -1$.

   (c) Poles of order two at $z = n\pi$; $C_{-2} = n\pi$.

   (d) Pole of order two at $z = 0$; $C_{-2} = \frac{1}{2}$.

   (e) Simple pole at $w = \sqrt{2}$, $C_{-1} = 1/2$; simple pole at $w = -\sqrt{2}$, $C_{-1} = 1/2$.

Section 3.6, Page 157

1. (a) Converges for $|z| < 1$. (b) Converges for all z.

   (c) Diverges. (d) Converges for all z.

Section 3.7, Page 173

1. (a) $z = 0$ fixed singularity, no movable singularities; $w(z) = -z + cz^2$.

(b) $z = 0$ fixed singularity, $z = e^c$ a movable singularity, where $w(z) = (1)/(c - \log z)$.

(c) No fixed singular points; movable singular point at z : $c - 2\int_{z_0}^{z} a(z)\,dz = 0$, $c = 1/\omega_0^2$ where $\omega(z_0) = \omega_0$, and

$$w(z) = \cfrac{1}{\left[\frac{1}{w_0^2} - 2\int_{z_0}^{z} a(z')\,dz'\right]^{\frac{1}{2}}}$$

(d) $z = 0$ fixed singularity, no movable singularities; $w(z) = [c_1 \sin(\log z) + c_2 \cos(\log z)]$.

Chapter 4

Section 4.1, Page 205

1. (a) $-\frac{3}{10}$ (b) 1   (c) 0   (d) $\frac{1}{2}\log\left(\frac{3}{2}\right)$ (e) $\frac{1}{2}$

3. (a) Pole of order m (b) branch point   (c) simple pole

    (d) branch point   (e) branch point   (f) essential singularity

    (g) simple pole   (h) analytic   (i) branch point

    (j) branch point

Section 4.2, Page 216

1. (a) $\dfrac{\pi}{2a}$ (b) $\dfrac{\pi}{4a^3}$ (c) $\dfrac{\pi}{2ab(b+a)}$ (d) $\dfrac{\pi}{3}$

Section 4.4, Page 258

1. (a) $I = n$ (b) $I = 0$

    (c) The theorem can't be applied; singularity on contour.   (d) $I = N - M$

    (e) The theorem can't be applied; essential singularity.

3. (a) 3 times   (b) 3 times   (c) $2\pi N$

Section 4.5, Page 266

1. (a) $\hat{F}(k) = \dfrac{2}{1 + k^2}$

    (b) $\hat{F}(k) = \dfrac{\pi}{a}e^{-|k|a}$

(c) $\hat{F}(k) = \dfrac{\pi e^{-|k|a}}{2a^2}\left(\dfrac{1}{a} + |k|\right)$

(d) $\hat{F}(k) = \begin{cases} \dfrac{\pi}{2ic}\left(e^{(-ib-c)(-k+a)} - e^{(-ib-c)(-k-a)}\right) & \text{for } k < -a \\[2mm] \dfrac{\pi}{2ic}\left(e^{(-ib-c)(-k+a)} - e^{(-ib+c)(-k-a)}\right) & \text{for } -a < k < a \\[2mm] \dfrac{\pi}{2ic}\left(e^{(-ib+c)(-k+a)} - e^{(-ib+c)(-k-a)}\right) & \text{for } a < k \end{cases}$

5. (a) $\hat{F}(k) = \begin{cases} \pi, & |k| < w \\ 0, & |k| > w \end{cases}$

7. (a) $\hat{F}_s(k) = \dfrac{k}{w^2 + k^2}\sqrt{2}$

(b) $\hat{F}_s(k) = \dfrac{\pi}{\sqrt{2}}e^{-k}$

(c) $\hat{F}_s(k) = \begin{cases} \dfrac{1}{\sqrt{2}}\pi e^{-w}\sinh k & k < w \\[2mm] \dfrac{1}{\sqrt{2}}\pi e^{-k}\sinh w & w < k \end{cases}$

11. (a) $f(x) = \cos wx$

(b) $f(x) = xe^{-wx}$

(c) $f(x) = \dfrac{x^{n-1}}{(n-1)!}e^{-wx}$

(d) $f(x) = \dfrac{1}{(n-1)!}[(n-1)x^{n-2}e^{-wx} - wx^{n-1}e^{-wx}]$

(e) $f(x) = \dfrac{e^{-w_1 x}}{w_2 - w_1} + \dfrac{e^{-w_2 x}}{w_1 - w_2}$

(f) $f(x) = \dfrac{x}{w^2} - \dfrac{\sin wx}{w^3}$

(g) $f(x) = \dfrac{e^{-w_1 x}(\sin w_2 x)}{w_2}$

(h) $f(x) = \dfrac{e^{wx}}{4w^3}(xw - 1) + \dfrac{e^{-wx}}{4w^3}(xw + 1)$

Section 4.6, Page 284

5. $u(x, t) = \dfrac{e^{-i\pi/4}}{2\sqrt{\pi t}}\displaystyle\int_{-\infty}^{\infty} e^{i(x-\xi)^2/4t} f(\xi)d\xi$

9. $\Phi(x, y, t) = \displaystyle\int_{-\infty}^{\infty}\int_{-\infty}^{\infty} f(x', y')\dfrac{1}{4\pi t}\exp\left(\dfrac{-\left[(x - x')^2 + (y - y')^2\right]}{4t}\right)dx'dy'$

Chapter 5

Section 5.2, Page 307

1. The function $w = u + iv = (1)/(x + iy)$ gives the relation

$$u = \frac{x}{x^2 + y^2}, \qquad v = \frac{-y}{x^2 + y^2}$$

and noticing that $u^2 + v^2 = (1)/(x^2 + y^2)$, we obtain

$$x = \frac{u}{u^2 + v^2}, \qquad y = \frac{-v}{u^2 + v^2}$$

Thus when $x = c_1$ and $y = c_2$, we have

$$\frac{u}{u^2 + v^2} = c_1, \qquad \frac{-v}{u^2 + v^2} = c_2$$

Upon completing the square in the above equation, we obtain

$$\left(u - \frac{1}{2c_1}\right)^2 + v^2 = \left(\frac{1}{2c_1}\right)^2, \qquad u^2 + \left(v + \frac{1}{2c_2}\right)^2 = \left(\frac{1}{2c_2}\right)^2$$

which are the desired image circles of the lines $x = c_1 \neq 0$ and $y = c_2 \neq 0$

3. $w = 2z - 2 + \frac{3i}{2}$

Section 5.3, Page 312

1. $x^2 - y^2 = c_1, xy = c_2$
3. (a) $w = (1 + i)(z^2 + \bar{z}^2) + (2 - 2i)(z\bar{z} + 8iz)$ (b) $w = z^3$

    Only (b) can define a conformal mapping.

Section 5.6, Page 339

1. (a) $w = z^{5/4}$ (b) $w = \log z$

5. $q^2 = \dfrac{u_0^2}{1 + \left(\frac{2\pi}{d}\right)^2 e^{\frac{4\pi x}{d}} + \frac{4\pi}{d} e^{\frac{2\pi x}{d}} \cos \frac{2\pi y}{d}}$

    $w = z + e^{\frac{2\pi z}{d}}, z = x + iy, w = u + iv, q$ is the speed of the flow (an explicit formula for z in terms of w, i.e., $x = x(u, v)$ and $y = y(u, v)$, does not exist).

7. $F(w) = \dfrac{hq}{\pi} f^{-1}(w) = \dfrac{hq}{\pi} z$
    where

$$w = f(z) = \frac{h}{\pi}\left((z^2 - 1)^{1/2} + \cosh^{-1} z\right).$$

9. $w = \left(\dfrac{2 \sin z + 1}{2 \sin z - 1} \right)^{1/2}$

Section 5.7, Page 360

1. $w = \dfrac{(8 - 3i)z + i}{(3 - i)z + 1}$

3.   Radius of the inner circle

$$\delta = \frac{\rho}{c + \sqrt{c^2 - \rho^2}}$$

5. $w = \dfrac{e^{\pi/z} - i}{e^{\pi/z} + i}$

7. $\phi(x, y) = \dfrac{2}{\pi} \left(\dfrac{\pi}{4} - \tan^{-1} \dfrac{x - y}{x(x - 1) + y(y - 1)} \right)$

Chapter 6

Section 6.1, Page 407

1.   (a) $f(\epsilon) \sim \displaystyle\sum_{n=0}^{\infty} \dfrac{\epsilon^n}{n!}$

     (b) $f(\epsilon) \sim 0, \quad \epsilon > 0$ does not exist if $\epsilon < 0$.

3.   (a) $k - \dfrac{1}{18}k^3 + \dfrac{1}{600}k^5 - \cdots + (-1)^{n-1} \dfrac{k^{2n-1}}{(2n-1)(2n-1)!} + \cdots$

     (b) $\dfrac{4}{3}k^{\frac{3}{4}} - \dfrac{4}{7}k^{\frac{7}{4}} + \dfrac{2}{11}k^{\frac{11}{4}} - \dfrac{2}{45}k^{\frac{15}{4}} + \cdots$

     (c) $\dfrac{1}{4}\Gamma\left(\dfrac{1}{4}\right) - k + \dfrac{k^5}{5} - \dfrac{k^9}{9} + \cdots$

5. $I(z) \sim - \displaystyle\sum_{0}^{\infty} \dfrac{\int_a^b u(x)x^n dx}{z^{n+1}}$

Section 6.2, Page 418

1.   (a) $e^{-k} \left[\dfrac{1}{k} \sin 1 + \dfrac{1}{k^2} \cos 1 + \cdots \right]$

     (b) $e^{-5k} \left[\dfrac{1}{5k} - \dfrac{1}{25k^2} + \cdots \right]$

3. $I(k) \sim \sum_{n=0}^{\infty} \frac{(-1)^n \Gamma\left(\frac{2}{3} + 2n\right)}{(2n)! \, k^{\frac{2}{3}+2n}}$

5. $I(k) \sim \dfrac{\Gamma\left(\frac{1}{4}\right)}{2k^{\frac{1}{4}}}$

Section 6.3, Page 434

1. (a) $-\dfrac{i}{k}(\sin 2 + 2)e^{2ik} + \dfrac{1}{k^2}(\cos 2 + 1)e^{2ik} - \dfrac{2}{k^2}$ (b) $\dfrac{i}{k}$

3. (a) $\dfrac{\sqrt{\pi}\, e^{i\pi/4}}{4k^{1/2}}$ (b) $\sqrt{\dfrac{2\pi}{k}}\, e^{-\frac{2ik}{3}} e^{i\pi/4}$

Section 6.4, Page 444

1. (a) $\dfrac{i\sqrt{\pi}}{2\sqrt{k}} e^{-\frac{2k}{3}}$ (b) $2\sqrt{\dfrac{\pi}{k}}\, e^{-\frac{4k}{5\sqrt{2}}} \cos\left(\dfrac{4k}{5\sqrt{2}} - \dfrac{3\pi}{8}\right)$

3. $\dfrac{e^{\frac{i\pi}{6}} \Gamma\left(\frac{1}{3}\right)}{3^{\frac{2}{3}} k^{\frac{1}{3}}}$

Bibliography

Ablowitz, M.J., and Clarkson, P.A., *Solitons, Nonlinear Evolution Equation, and Inverse Scattering*, p. 335, London Math. Society Lecture Notes #149, Cambridge University Press, New York, 1991.

Ablowitz, M.J., and Segur, H., *Solitons and the Inverse Scattering Transform*, SIAM Studies in Applied Mathematics, 1981.

Ablowitz, M.J., Bar Yaacov, D., and Fokas, A.S., *Stud. in Appl. Math.* **69**, 1983, p. 135.

Abramowitz, M., and Stegun, I., *Handbook of Mathematical Functions*, Dover Publications, New York, 1965.

Berry, M.V., *Proc. Roy. Soc. of London A* **422**, 1989, p. 7.

Bender, C.M., and Orszag, S.A., *Advanced Mathematical Methods for Scientists and Engineers*, McGraw-Hill, New York, 1978

Bleistein, N., and Handelsman, R.A., *Asymptotic Expansions of Integrals*, Dover Publications, New York, 1986.

Buck, R.C., *Advanced Calculus*, McGraw-Hill, New York, 1956.

Carleman, T., *Math Z* **15**, 1922a, p. 161.

Carleman, T., *Arkiv Main. Astron. Fysik* **16**, 1922b, p. 141.

Carrier, G.F., Krook, M. and Pearson, C.E., *Functions of a Complex Variable: Theory and Technique*, McGraw-Hill, New York, 1966.

Chazy, J., *Acta Math.* **34**, 1911, p. 317.

Darboux, G., *Ann. Sci. Éc. Normal. Supér.* **7**, 1878, p. 101.

Debye, P., *Collected Papers*, Interscience Publishers, New York, 1954.

Deift, P., and Zhou, X., *Ann. of Math.* **137**, 1993, p. 295.

Dingle, R.B., *Asymtotic Expansions: Their Derivation and Interpretation*, Academic Press, New York, 1973.

Erdelyi, A., *Asymptotic Expansions*, Dover Publications, New York, 1956.

Fokas, A.S., and Novikov, R.G., *C.R. Acad. of Science Paris* **313**, 1991, p. 75.

Fokas, A.S., and Zakharov, V.E., editors, *Important Developments in Soliton Theory*, Springer-Verlag, Berlin, Heidelburg, 1993.

Fornberg, B., *SIAM J. Sci. Stat. Computing* **1**, 1980, p. 386.

Freund, L.B., *Dynamic Fracture Mechanics*, Cambridge University Press, New York, 1990.

Gaier, D., *Computational Aspects of Complex Analysis*, edited by Werner, H., et al., Reidel Publishers, 1983, p. 51.

Gakhov, F.D., *Boundary Value Problems*, Pergamon Press, New York, 1966.

Garabedian, P.R., *Amer. Math. Monthly* **98**, 1991, p. 824.

Gohberg, I.T., and Krein, M., *Upsekhi Mat. Nauk.* **13**(2), 1958, p. 3.

637

Halphen, G.H., *C.R. Acad. of Science Paris* **92**, 1881, p. 1001.
Henrici, P., *Applied and Computational Complex Analysis*, Volumes I, II, III, Wiley-Interscience, New York, 1977.
Hille, E., *Ordinary Differential Equations in the Complex Plane*, Wiley-Interscience, New York, 1976.
Ince, E.L., *Ordinary Differential Equations*, Dover Publications, New York, 1956.
Jeffreys, H., and Jeffreys, B., *Methods of Mathematical Physics*, Cambridge University Press, New York, 1962.
Korteweg, D.J., and DeVries, G., *Phil. Mag.* **39**, 1895, p. 422.
Levinson, N., and Redheffer, R.M., *Complex Variables*, Holden-Day, San Francisco, 1970.
Lighthill, M.J., *An Introduction to Fourier Analysis and Generalized Functions*, Cambridge University Press, New York, 1959.
Muskhelishvili, N.I., *Singular Integral Equations*, Noordhoff International Publishing, Leyden, The Netherlands, 1977.
Nehari, Z. *Conformal Mapping*, McGraw-Hill, New York, 1952.
Olver, F.W.J., *Introduction to Asymptotics and Special Functions*, Academic Press, New York, 1974.
Rudin, W., *Theory of Real and Complex Functions*, McGraw-Hill, New York, 1966.
Stokes, G.G., *Trans. Comb. Phil. Soc.* **X**, 1864, p. 106.
Stokes, G.G., *Mathematical and Physical Papers*, Vol. 4, Cambridge University Press, New York, 1904.
Titchmarsh, E.C., *Introduction to the Theory of Fourier Integrals*, 2nd ed., Oxford University Press, New York, 1948.
Trefethen, L.N., editor, *Numerical Conformal Mapping*, North-Holland, Amsterdam, 1986.
Vekua, I.N., *Generalized Analytic Functions*, International series of monographs on pure and applied mathematics **25**, Pergamon Press, Oxford, New York, 1962.
Vekua, N.P., *Systems of Singular Integral Equations*, F. Noordhoff, Groningen, The Netherlands, 1967.
Wegmann, R., *J. Comput. Appl. Math.* **23**, 1988, p. 323.
Whitham, G.B., *Linear and Nonlinear Waves*, Wiley-Interscience, New York, 1974.
Whittaker, E.T., and Watson, G.N., *A Course of Modern Analysis*, 4th edition, Cambridge University Press, New York, 1927.
Zhou, X., *SIAM J. Math. Anal.* **20**, 1989, p. 966.

The following is a short list of **supplementary** books which have basic and advanced material, applications, and numerous exercises.

Ahlfors, L.V., *Complex Analysis: an Introduction to the Theory of Analytic Functions of One Complex Variable*, 3rd edition, McGraw-Hill, New York, 1979.
Batchelor, G.K., *An Introduction to Fluid Dynamics*, Cambridge University Press, 1967.
Boas, R.P., *Invitation to Complex Analysis*, 1st edition, Random House, New York, 1987.
Bowman, F., *Introduction to Elliptic Functions with Applications*, English University Press, London, 1953.
Churchill, R.V., and Brown, J.W., *Complex Variables and Applications*, 5th edition, McGraw-Hill, New York, 1990.
Copson, E.T., *Asymptotic Expansions*, Cambridge University Press, England, 1965.

Copson, E.T., *An Introduction to the Theory of Functions of a Complex Variable*, Clarendon Press, Oxford, 1955.

Hille, E., *Analytic Function Theory*, 2nd edition, Chelsea Publishing Co., New York, 1973.

Markushevich, A.I., *Theory of Functions of a Complex Variable*, 2nd edition, Chelsea Publishing Co., New York, 1977.

Milne-Thomson, L.M., *Theoretical Hydrodynamics*, 5th edition, Macmillan, London, 1968.

Saff, E.B., and Snider, A.D., *Fundamentals of Complex Analysis for Mathematics, Science, and Engineering*, Prentice-Hall, Englewood Cliffs, N.J., 1976.

Silverman, R.A., *Complex Analysis with Applications*, Prentice-Hall, Englewood Cliffs, N.J., 1974.

Spiegel, M.R., *Schaum's Outline of Theory and Problems of Complex Variables*, McGraw-Hill, New York, 1964.

Thron, W.J., *Introduction to the Theory of Functions of a Complex Variable*, Wiley, New York, 1953.

Index